Advanced Protocols in Oxidative Stress I

METHODS IN MOLECULAR BIOLOGY

Advanced Protocols in Oxidative Stress I

Edited by

Donald Armstrong

Humana Press

Editors
Donald Armstrong
University at Buffalo,
Buffalo, NY, USA; Showa
University School of
Medicine, Tokyo, Japan;
University of Florida,
Gainesville, FL, USA

Series Editor
John M. Walker
University of Hertfordshire
Hatfield, Herts, UK

ISBN: 978-1-61737-881-2
ISSN: 1064-3745
e-ISBN: 978-1-60327-517-0
e-ISSN: 1940-6029

Cover illustration: Figure 1: Comparison of transient *gusA* gene expression in oil palm embryogenic calli after bombardment with plasmids carrying different promoters. A. none (bombarded without plasmid DNA) B. pEmuGN (Emu), C. pAHC25 (Ubi1), D. pAct1-F4 (Act1), E. pGH24 (CaMV 35S) and F. pBARGUS (Adh1).

Printed on acid-free paper

9 8 7 6 5 4 3 2 1

springer.com

METHODS IN MOLECULAR BIOLOGY

Advanced Protocols in Oxidative Stress I

Edited by

Donald Armstrong

Humana Press

Editors
Donald Armstrong
University at Buffalo,
Buffalo, NY, USA; Showa
University School of
Medicine, Tokyo, Japan;
University of Florida,
Gainesville, FL, USA

Series Editor
John M. Walker
University of Hertfordshire
Hatfield, Herts, UK

ISBN: 978-1-61737-881-2
ISSN: 1064-3745
e-ISBN: 978-1-60327-517-0
e-ISSN: 1940-6029

Softcover reprint of the hardcover 1st edition 2008

Cover illustration: Figure 1: Comparison of transient *gusA* gene expression in oil palm embryogenic calli after bombardment with plasmids carrying different promoters. A. none (bombarded without plasmid DNA) B. pEmuGN (Emu), C. pAHC25 (Ubi1), D. pAct1-F4 (Act1), E. pGH24 (CaMV 35S) and F. pBARGUS (Adh1).

Printed on acid-free paper

9 8 7 6 5 4 3 2 1

springer.com

Preface

The present volume is an attempt to cover the field of oxidative stress with state-of-the-art technology on oxidative stress reactions in biology and medicine. It provides the reader with an assemblage of up-to-date techniques to utilize in their research. Contributors for this volume are internationally recognized for developing new procedures and methods, and their inclusion signifies the commitment of the Advanced Protocols volumes to invite prominent scientists to participate in future compilations.

In terms of background information on the number of citations covering free radical reactions, antioxidant supplementation and biosystems technology are reflected in previous volumes of the *Methods in Molecular Biology* series (108, 186 and 196). Those volumes covered this through the year 2000 and totaled 12,000 reports. A review of papers on oxidative stress complied from a 2007 PubMed search combining oxidative stress and each of the headings listed below indicates a 30-fold increase over the 2000 figure and provides a breakdown of specialized areas where free radical and antioxidant reactions have been published from both in vitro and in vivo studies:

Oxygen free radicals		48790
Nitrogen free radicals		6244
Carbon-centered free radicals		5507
Lipid oxidation		36995
Protein oxidation		71383
Nucleic acid oxidation		9685
Carbohydrate oxidation		29396
Mitochondrial oxidation		14192
Oxidative stress and gene regulation		5591
Oxidative stress and transcription		4533
Oxidative stress and metabolomics		39200
Lipid soluble antioxidants		2271
Water soluble antioxidants		2068
Antioxidant enzymes		96946
Use of multiple antioxidants		5549
Genetic engineering of antioxidants		1433
Oxidative stress and cell ultrastructure		1341
HPLC/MS applications		2625
Imaging following oxidative stress		395
Oxidative stress and -; cardiovascular disease		7638
; cancer	4852	
; aging	3938	
; diabetes	3848	
; hepatic disease	1805	
; renal disease	1788	
; pulmonary disease	1478	
; ophthalmic disease	806	

The overall distribution from this list shows that 65% covers free radical biomarkers and 35% covers antioxidants. One can see from these lists that papers related to antioxidant enzymes rank first, followed by protein oxidation, oxygen and nitrogen radicals, metabolomics, lipid peroxides, and oxidized carbohydrate. Also of note is that the number of papers on lipid- and water-soluble antioxidants are roughly equal. It is interesting that the number of papers on genetic engineering now totals over 1400, indicating this rapidly increasing area is the wave-of-the-future in oxidative stress research. Finally, there are over 18,500 papers involving medical issues, with cardiovascular disease ranking the highest, followed by cancer, aging, and diabetes. Under aging, the main studies are on osteoporosis, menopause, and macular degeneration. We can expect clinical applications to grow due to the popularity of exploring the use of antioxidants to control free radicals in treatment of disease.

This present volume (I) is the first book in a new sub-series that presents high-tech methodology in the above-mentioned areas, and gives a perspective on the diversity of applications in the ever-emerging field of free radical and antioxidant reactions. New topics presented in Parts III, IV, and V of this volume represent 22% of the book and include methods for analyzing gene expression, the exciting new area of oxidative stress and stem cell differentiation, and the specific biostatistical evaluation of biomarkers. We anticipate these latter areas will continue to expand and represent a larger percentage of subsequent volumes.

Since oxidative stress has become such a dynamic topic, Humana Press is preparing additional books on Advanced Protocols in Oxidative Stress to form a multi-volume collection, designated volumes I, II, III, IV, V, etc. Up-to-date research techniques for the basic and clinical scientist will be covered in each volume, divided into four parts, which will be a standard format throughout the series. Our expectation is to create a library of biotechnology-relevant procedures helpful to the scientist wishing to expand his or her research. The book should also be useful in graduate education and as a desk reference for methodologies used in oxidative stress.

I am especially thankful to David P. Armstrong for his assistance with computer support, programming expertise and ideas for improving this book. I also acknowledge the many friends, colleagues, family, and Patrick Marton who encouraged me in this effort.

Donald Armstrong

Contents

Contributors

IAN N. ACWORTH • *ESA Biosciences, Inc, Chelmsford, MA, USA*

MOHAMMAD AFZAL • *Department of Biological Sciences, Faculty of Science, Kuwait University, Safat, Kuwait*

SAMEERA AL-AWADI • *Department of Biological Sciences, Faculty of Science, Kuwait University, Safat, Kuwait*

DRAGAN AMIĆ • *Faculty of Agriculture, The Josip Juraj Strossmayer University, Osijek, Croatia*

ANGELA AMORESANO • *Department of Clinical and Organic Biochemistry, Federico II University of Naples, Naples, Italy*

REŞAT APAK • *Department of Chemistry, Faculty of Engineering, Istanbul University, Turkey*

DONALD ARMSTRONG • *Department of Biotechnology and Clinical Laboratory Science, University at Buffalo, Buffalo, NY, USA; Department of Pharmacology, Showa University School of Medicine, Tokyo, Japan; College of Medicine, University of Florida, Gainesville, FL, USA*

MICHAEL AVIRAM • *Lipid Research Laboratory, Technion Faculty of Medicine and Department of Laboratory Medicine, Rambam Medical Center, Haifa, Israel*

ANIL BAMEZAI • *Department of Biology, Villanova University, Villanova, PA, USA*

BURCU BEKTAŞOĞLU • *Department of Chemistry, Faculty of Engineering, Istanbul University, Istanbul, Turkey*

MUSTAFA BENER • *Department of Chemistry, Faculty of Engineering, Istanbul University, Istanbul, Turkey*

TIMOTHY S. BLACKWELL • *Division of Allergy, Pulmonary and Critical Care Medicine and Department of Cancer Cell Biology and Cell and Developmental Biology, Vanderbilt University, Nashville, TN, USA*

YEONG-RENN CHEN • *Davis Heart and Lung Research Institute, Department of Molecular and Cellular Biochemistry, College of Medicine, Ohio State University, Columbus, Ohio*

GIOVANNI CHIAPPETTA • *Department of Clinical and Organic Biochemistry, Federico II University, Naples, Italy*

SRIDAR V. CHITTUR • *Center for Functional Genomics, Department of Biomedical Sciences, School of Public Health, University at Albany, Rensselaer, NY, USA*

GRAEME L. CLOSE • *Division of Metabolic and Cellular Medicine, School of Clinical Sciences, University of Liverpool, Liverpool, UK*

L. ASHLEY COWART • *Department of Biochemistry and Molecular Biology, Medical University of South Carolina, and Ralph H. Johnson Veteran's Affairs Medical Center, Charleston, SC, USA*

ROCHELLE D'SOUZA • *Department of Physiology, Justus Liebig University, Giessen, Germany*

NATHAN A. ELLIOTT • *NanoString Technologies, Seattle, WA, USA*

E. ANN ELLIS • *Microscopy and Imaging Center, Texas A&M University, College Station, Texas*

PAUL H. GAMACHE • *ESA Biosciences, Inc, Chelmsford, MA, USA*

KUBILAY GÜÇLÜ • *Department of Chemistry, Faculty of Engineering, Istanbul University, Turkey*

H. DAVID GUTHRIE • *Animal Bioscience and Biotechnology Laboratory, Agricultural Research Service, US Department of Agriculture, Beltsville, Maryland*
HIDEO IWAHASHI • *Department of Chemistry, Wakayama Medical University, Wakayama, Japan*
SHINICHI IWAI • *Department of Pharmacology, School of Medicine, Showa University, Tokyo, Japan*
MALCOLM J. JACKSON • *Division of Metabolic and Cellular Medicine, School of Clincal Sciences, University of Liverpool, Liverpool, UK*
KELLY L. JORDAN-SCIUTTO • *Department of Pathology, University of Pennsylvania School of Dental Medicine, Philadelphia, PA, USA*
COLLEEN KENNEDY • *Department of Biology, Villanova University, Villanova, PA, USA*
KATHRYN A. LINDL • *Robert Schattner Center, School of Dental Medicine, Department of Pathology, University of Pennsylvania, Philadelphia, PA, USA*
XIYONG LIU • *Department of Clinical & Molecular Pharmacology, City of Hope National Medical Center, Duarte, CA, USA*
YANXIA LIU • *Department of Medicinal Chemistry, School of Pharmacy, University of Florida, Gainesville, FL, USA*
BONO LUČIĆ • *The Rugjer Bošković Institute, Zagreb, Croatia*
HENDRIK LUESCH • *Department of Medicinal Chemistry, School of Pharmacy, University of Florida, Gainesville, FL, USA*
GENNARO MARINO • *Department of Organic Chemistry and Biochemistry, Federico II University of Naples, Naples, Italy*
RAFAEL MEDINA-NAVARRO • *Centro de Investigacion Biomedica de Michoacan, Michoacan, Mexico*
JUAN A. MELENDEZ • *Center for Immunology & Microbial Disease, Albany Medical College, Albany, NY, USA*
NAOKI MIYATA • *Department of Organic and Medicinal Chemistry, Graduate School of Pharmaceutical Sciences, Nagoya City University, Japan*
JÖRG P. MÜLLER • *Institute of Molecular Biology, Medical Faculty, Friedrich Schiller University, Jena, Germany*
JOHN A. MULNIX • *Mulnix Animal Clinic, Fort Collins, CO, USA; and Animal Health Options, Golden, CO, USA*
HIDEHIKO NAKAGAWA • *Department of Organic and Medicinal Chemistry, Graduate School of Pharmaceutical Sciences, Nagoya City University, Japan*
TAKAKO NAKANISHI-UEDA • *Department of Pharmacology, Showa University School of Medicine, Tokyo, Japan*
KATSUJI OGUCHI • *Department of Pharmacology, School of Medicine, Showa University, Tokyo, Japan*
SOSAMMA OOMMEN • *Department of Biological Sciences, Faculty of Science, Kuwait University, Safat, Kuwait*
MUSTAFA ÖZYÜREK • *Department of Chemistry, Faculty of Engineering, Istanbul University, Turkey*
VICTOR PAROMOV • *Department of Pediatrics, James H. Quillen College of Medicine, East Tennessee State University, Johnson City, TN, USA*
GHULAM KADIR AHMAD PARVEEZ • *Advanced Biotechnology and Breeding Centre, Biological Research Division, Malaysian Palm Oil Board, Kuala Lumpur, Malaysia*
PIERO PUCCI • *Department of Organic Chemistry and Biochemistry, Federico II University of Naples, Naples, Italy*

MIRA ROSENBLAT • *The Lipid Research Laboratory, Rambam Medical Center, Haifa, Israel*
LESLIE ROSENTHAL • *Division of Epidemiology, Statistics and Prevention Research, National Institute of Child Health and Human Development , Rockville, MD, USA*
CAROLA RUHE • *Department of Internal Medicine, Cardiology Division, Friedrich Schiller University, Jena, Germany*
RUXANA T. SADIKOT • *Department of Pulmonary, Critical Care and Sleep Medicine, Department of Veterans Affairs and University of Illinois, Chicago, IL, USA*
GHASSAN M. SAED • *Departments of Obstetrics and Gynecology & Cell Biology and Anatomy, Wayne State University School of Medicine, Detroit, MI, USA*
HEINRICH SAUER • *Department of Physiology, Justus-Liebig-University, Giessen, Germany*
ENRIQUE SCHISTERMAN • *Division of Epidemiology, Statistics and Prevention Research, National Institute of Child Health and Human Development, Rockville, MD, USA*
MAIKE SCHMELTER • *Department of Physiology, Justus Liebig University, Giessen, Germany*
MILTON SMITH • *Department of Pediatrics, James H. Quillen College of Medicine, East Tennessee State University, Johnson City, TN, USA*
ROLAND STOCKER • *Centre for Vascular Research, Bosch Institute and Discipline of Pathology, The University of Sydney, Sydney, Australia*
BROOK E. STODDARD • *Animal Health Options, Golden, CO, USA*
WILLIAM L. STONE • *Department of Pediatrics, James H. Quillen College of Medicine, East Tennessee State University, Johnson City, TN, USA*
SITA SUBBARAM • *Center for Immunology & Microbial Disease, Albany College of Medicine, Albany, NY, USA*
NENAD TRINAJSTIĆ • *The Rugjer Bošković Institute, Zagreb, Croatia*
PAUL A. ULLUCCI • *ESA Biosciences, Inc, Chelmsford, MA, USA*
CRISTINA VASSALLE • *Instituto di Fisiologia Clinica, CNR, Pisa, Italy*
JACOB VAYA • *Laboratory of Natural Medicinal Compounds, Migal-Galilee Technology Center, Kiryat Shmona, Israel*
MICHAEL R. VOLKERT • *Department of Molecular Genetics and Microbiology, University of Massachusetts Medical School, Worcester, MA, USA*
JEN-YEU WANG • *Institute of Molecular Biology, Academica Sinica, Taipei, Taiwan, R.O.C*
XIAO SUO WANG • *Centre for Vascular Research, Bosch Institute and Discipline of Pathology, The University of Sydney, Sydney, Australia*
MARIA WARTENBERG • *Department of Internal Medicine, Cardiology Division, Fredrich Sciller University, Jena, Germany*
GLENN R. WELCH • *Animal Bioscience and Biotechnology Laboratory, Agriculture Research Service, U.S. Department of Agriculture, Beltsville, Maryland*
ALAN J. WILDER • *Department of Biochemistry and Molecular Biology, Medical University of South Carolina, Charleston, SC, USA*
LIJUN XUE • *Department of Clinical & Molecular Pharmacology, City of Hope National Medical Center, Duarte, CA, USA*
HONGSONG YANG • *Department of Pediatrics, James H. Quillen College of Medicine, East Tennessee State University, Johnson City, TN, USA*
YUN YEN • *Department of Clinical & Molecular Pharmacology, City of Hope National Medical Center, Duarte, CA, USA*

Part I
Reactive Oxygen and Nitrogen Techniques

Chapter 1

Novel Designed Probes for the Characterization of Oxidative Stress in Biological Fluids, Cells, and Tissues

Jacob Vaya

Abstract

Oxidative stress (OS) is linked to the development of human diseases. Early identification of OS-associated diseases is essential in the control of their progression and treatment. Efforts have been undertaken to identify reliable endogenous markers, which correlate with the progression of a disease in an organ undergoing OS. An ideal biomarker must be validated, utilize noninvasive sampling, and have a simple, specific and highly sensitive detection method. Among the currently used markers assessing OS, are those that are nonspecific (peroxide value [PV], conjugated dienes [CD], thiobarbitoric acid reactive substances [TBARS]), and others that measure end-products of oxidized degradation biomolecules (isoprostanes, oxysterols, keto-proteins, 8-oxodeoxyguanosine), whose accumulation is not necessarily correlated with augmented OS. The search for a more reliable marker necessitates new approaches to fulfill such requirements and overcome many of the obstacles associated with the current markers. We suggest a new strategy of using designed exogenous novel reporters, constructed from endogenous subunits, that are sensitive to reactive oxygen and nitrogen species (ROS/RNS) and commonly known to react with them, forming specific oxidized products. These subunits are tyrosine (representing proteins), bonded covalently to linoleic acid (representing polyunsaturated fatty acids) forming an amide bond, which can be further connected through an ester bond to a third unit, either to cholesterol (representing sterols) or to 2′-deoxyguanosine (representing DNA). Oxidation of the designed probe can outline, in real time, the formation of oxidation products and distinguish them from intrinsic biomolecules, provide information about the relative subunit susceptibilities to a specific oxidant challenge, and allow for the assessment of the utility of intervention, such as antioxidant supplementation. By utilizing such markers, it may be possible to correlate between the damaged fingerprints of the marker and the specific pathological conditions. The above markers were tested to characterize OS in *in vitro* and in *in vivo* experiments, such as in those carried out in human fluids (blood, serum, saliva), tissues (brain or muscle homogenates), and cells (macrophages, astrocytes, neurons), pertaining to OS-associated diseases, such as atherosclerosis, diabetes, and Alzheimer's disease.

Key words: Biomarkers, exogenous markers, free radical, oxidative stress, ROS.

From: *Methods in Molecular Biology, vol. 477: Advanced Protocols in Oxidative Stress I*,
Edited by: D. Armstrong, DOI: 10.1007/978-1-60327-517-0_1, © Humana Press, New York, NY

1. Introduction

Oxidative stress (OS) is proposed to be involved in several human diseases, such as cardiovascular and neurological diseases, cancer, inflammation-related diseases, and aging. Reactive oxygen and nitrogen species (ROS/RNS) are all products of normal metabolic pathways in humans, and are formed during the destruction of invading organisms. The identification of reliable biomarkers, e.g., modified endogenous compounds formed in cells and organs as a result of OS, is essential for the characterization of OS and for the prediction of the early development of pathological conditions. Many such modified endogenous compounds, and methods for their detection, have been proposed during the years. Examples are peroxide values, thiobarbituric acid reactive substances, isoprostanes and oxysterols for polyunsaturated fatty acids (PUFA), keto-proteins, chloro- or nitro- tyrosine for proteins, and 8-oxodeoxyguanosine (8-oxodG) for DNA. However, most of these biomarkers, and the proposed methods for their detection, bear several limitations. Some are not specific, measuring uncharacterized products of OS. The in vivo application of others is questionable since their levels are determined by a dynamic process involving formation, accumulation, and removal, and may not be due to the oxidative process within the system, but rather to alternative processes. Furthermore, in many instances, measuring biomarker levels can provide only a global picture of events occurring in the body, without showing where the phenomenon took place *(1)*.

In the present study, a new methodology for the characterization of OS in biological systems is proposed. It includes the design and synthesis of sensitive molecules, which are constructed from various endogenous subunits, connected covalently together to form a novel probe (not present as such in organs). The designed markers consist of the amino acid tyrosine (T) connected to linoleic acid (L), forming an amide bond (N-linoleoyl tyrosine, LT) *(2)*. Additional endogenous units, which can be added to the LT, forming an ester with the tyrosine carboxylic group may include either cholesterol, creating linoleoyl tyrosine cholesteryl ester (LTC), or 2′-deoxyguanosine, producing linoleoyl tyrosine 2′-deoxyguanosyl ester (LTG) (**Fig. 1.1**) *(3)*. These molecules, LT, LTC, and LTG, contain the major groups of which the body is composed: linoleic acid represents the PUFA, the tyrosine residue denotes proteins, the cholesterol represents the sterols, and the hydrophilic subunit 2′-deoxyguanosine (2′-deoxyG) represents the DNA. Each component of the synthetic marker is well known to be easily oxidized and to form specific products, depending on the type of ROS/RNS present. Such reporters can simultaneously provide information about (i) oxidative modifications that may occur to PUFA, amino acids, sterols, and nucleic acids, (ii) the type of products formed, (iii) the relative oxidative sensitivity of the elements constructing

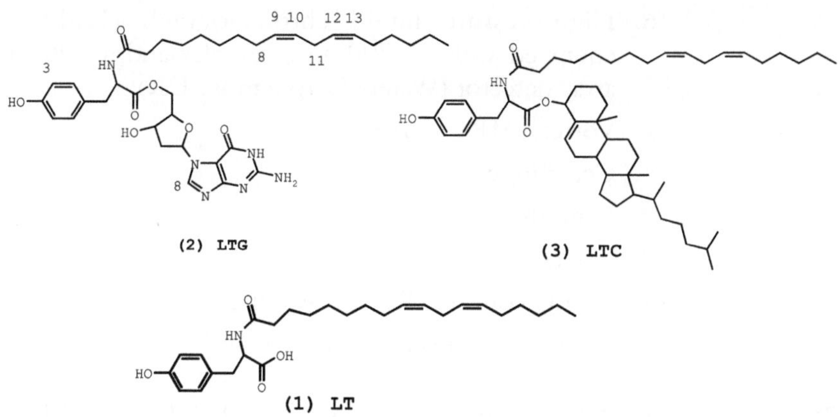

Fig. 1.1. The structures of the synthetic probes containing two [(1) – LT, linoleoyl tyrosine)] or three [(2) – LTG and (3) – LTC, linoleoyl tyrosine 2′-deoxyguanosyl ester and linoleoyl tyrosine cholesteryl ester, respectively)] endogenous subunits, known to be easily oxidized in the presence of ROS/RNS and to generate selective oxidative fingerprint characteristics in response to the stress employed *(2, 3, 4, 5, 6)*.

the probe under defined OS conditions, and (iv) the effectiveness of specific intervention (such as supplementation of antioxidants), which is aimed to decrease OS and its damaging consequences *(4)*. Furthermore, such exogenous markers can reveal data related to the kinetics of the specific oxidative changes undergone by the probe, and may be designed to be target-oriented, by modifying their chemical stability and physical properties (hydrophobic/hydrophilic properties). By utilizing such markers, it may be possible to correlate between the damage fingerprints of the marker and specific pathological conditions *(5)*.

Exogenous markers of these types were tested in biological fluids (urine, serum, whole blood, saliva), cell systems (macrophages, endothelial, Astrocytes, neurons), *(2, 3, 4, 5, 6)* and *in vivo*, by injecting them (intramuscularly or intravenously) to specific animals, which are known to develop OS-associated diseases (unpublished data).

2. Materials

2.1. Equipment

1. Rotary evaporator unit (Buchi, Switzerland)
2. Tank of nitrogen gas
3. Capillary gas chromatography column (DB-5MS, CA, USA)
4. Gas chromatograph/mass spectrometer (GC/MS) (HP 5890, Palo Alto, CA, USA)
5. Liquid chromatograph/mass spectrometer (LC/MS) (Micromass Quattro LC, Manchester, UK)

6. High-pressure liquid chromatograph (HPLC) (Waters, equipped with a 2790 auto sampler and a 996 photodiode array detector (Waters Corporation, UK)
7. Vortex (MRC, Israel)
8. Centrifuge
9. Sonicator
10. Filters
11. HPLC column (RP-18, Waters, MA, USA)
12. HP ChemStation data system (HP, CA, USA)

2.2. Reagents

1. Synthetic markers (probes) – The synthesis of LT and LTG are described in references 2 and 3, respectively, whereas the synthesis of LTC is a modified form of LTG and will be described elsewhere.
2. Dimethyl sulfoxide (DMSO), spectroscopic grade, extra dry (Bio Lab Ltd., Israel)
3. Heptadecanoic acid (internal standard) (Sigma-Aldrich, Israel)
4. Oxysterols – All oxysterols used as standards in this study are from Steroids Inc., USA.
5. Hexane, AR
6. 2-Propanol, HPLC grade
7. Ethanol, AR
8. Diethyl ether, AR
9. Acetonitrile, HPLC grade
10. Acetic acid, HPLC grade
11. Citric acid
12. Butylated hydroxyl toluene (BHT) (Sigma-Aldrich)
13. N,O-bis(trimethylsilyl)acetamide (BSA) (Sigma-Aldrich)
14. 4 Å molecular sieves (Sigma-Aldrich)
15. Aluminum oxide (Merck KGaA, Germany)
16. Anhydrous sodium sulfate (Sigma-Aldrich)

3. Method of Assay

3.1. Incubation of the Exogenous Probes with Biological Fluids, Tissues, or Cells

Fluids (whole blood, serum, saliva), tissues (atherosclerotic lesions, brain homogenates, muscle homogenates), or cultured cells (macrophages, endothelial cells, astrocytes, neurons) (*see* **Note 1**) are incubated with the synthetic redox reporter molecules, [8 μL of each, from a stock solution of 20 mM, dissolved in DMSO (for LT, LTC), or in ethanol (for LTG)], with each probe separately (LT, LTC, LTG) or together in a mixture (*see* **Note 2**), and with internal standards (5 μg/mL of heptadecanoic acid and 19-hydroxycholesterol [19-OH] from a stock solution of 5 mg/mL in ethanol) at 37°C for 1 h to overnight (*see* **Note 3**). Control flasks are supplemented with standards and a DMSO (or ethanol) vehicle. After incubation, a mixture of organic solvents is added to the biological sample (*see* **Note 4**) for probe extraction (*see* **Section 3.2**).

3.2. Extraction of the Probes and Their Oxidized Products

3.2.1. Fluid

Whole blood, serum, saliva, or any other fluid incubated with the probes, LT, LTC, LTG (separately or all together), as well as with the internal standards, are extracted by the addition of three volumes of the extracting solvents, hexane:2-propanol (3:2, v/v) and 10 μL butylated hydroxytoluene (BHT, from a stock solution of 5 mg/mL in ethanol), by vigorous vortexing for 2 min. The resulting two phases are separated, if necessary by centrifugation (1900 g × 4 min), as in the case of whole blood, and the upper phase is collected. The extraction procedure is repeated once more with three additional volumes of hexane:2-propanol. The collected organic solvents are combined, dried (with 0.3 g anhydrous sodium sulfate), filtered, and evaporated to dryness with purging nitrogen. Dried samples are stored at −20°C until analyzed.

3.2.2. Whole-Cell

Cell pellets from each experimental group are sonicated three times in 1 mL PBS containing 10 μL BHT (from a stock solution of 5 mg/mL in ethanol) for 20 s (80 W). Samples (15 μL) are removed for the determination of total protein concentrations by a protein assay based on the Lowry protocol (Bio-Rad Laboratories, Hercules, CA, USA). Afterwards, 3 mL of the extracting solvents, hexane:2-propanol (3:2 v/v), is added, and the mixture is vortexed for 1 min. The resulting two phases are separated, if necessary by centrifugation (1900 g × 3 min), and the upper phase collected. The extraction procedure is repeated with an additional 3 mL hexane:2-propanol, and the collected organic solvents are combined, dried (with 0.3 g anhydrous sodium sulfate), filtered, and evaporated to dryness with purging nitrogen. Cell extracts are stored under nitrogen at −20°C until analysis.

3.3. The Analysis of the Probes (LT and LTG) and Their Oxidized Products (Ox-LT and Ox-LTG) by the LC/MS Method

3.3.1. Instrument Conditions

The detection of LT and Ox-LT/total LT, and LTG and Ox-LTG/total LTG is performed by an LC/MS, equipped with an HPLC (Waters 2790) and a Waters photodiode array detector (model 996) connected to an MS (Micromass Quattro Ultima MS, UK). The HPLC column is 3.5 μm C18 ODS XTerra (Waters, MA, USA), and the eluents are comprised of a gradient of solution A (0.1% acetic acid in acetonitrile) and solution B (0.1% acetic acid in DDW) as follows: first 40% A, which is increased to 60% A for 2 min, followed by an increase to 80% A for 10 min and in the end the column is washed with a solution of 98% A (*see* **Note 5**). MS/MS analysis of the oxidized products is performed in the scan mode, using electrospray negative ions (ES^-). The MS source temperature is set at 150°C, with a cone gas flow of 22 L/h and a desolvation gas flow of 600 L/h. Peak spectra are monitored between m/z of 30 to 900. Collision-induced dissociation MS is performed with a collision energy of 25–30 eV and 3–3.5 kV capillary voltages, and multiple-reaction-monitoring is performed under the same conditions. Calibration curves of LT and LTG are run with each set of analyses.

3.3.2. Probe Analysis

Each dried sample of probes, after their incubation (*see* **Section 3.1**) and extraction (*see* **Section 3.2**), is dissolved in a methanol (100 μL): acetonitrile (400 μL) mixture, injected into the LC in the above gradient of eluents (*see* **Section 3.3.1**), and monitored by the MS detector. Each compound of the LC chromatogram is further fragmented in the mass spectrum to reveal its specific molecular ion and its characteristic fragmentations (**Fig. 1.2** for the LT and its Ox-LT and **Fig. 1.3** for LTG and LT-8-oxodG).

3.4. Hydrolysis Condition of LTC Probe Containing Linoleic Acid, Tyrosine, and Cholesterol

3.4.1. Analysis of LT/Ox-LT and Cholesterol/Oxysterols from Linoleoyl Tyrosine Cholesteryl Ester (LTC) Marker, by LC/MS and GC/MS, Respectively

The markers LTC and Ox-LTC are hydrolyzed to LT and Ox-LT, as well as to cholesterol and oxysterols, during LTC oxidation, prior to their analysis *(7)* (*see* **Note 6**).

3.4.2. The Saponification Procedure

The dry residue of the extracted sample (extracted from biological fluids, tissues, and cells – **Section 3.2**) is dissolved in 0.5 mL diethyl ether, and 0.5 mL 20% KOH in 70% methanol (v/v). The remaining head space of the vial is filled with nitrogen, and the reaction mixture is left in the dark at 37°C for 3 h, following which the mixture is neutralized by adding 0.5 mL 25% citric acid in water, and the upper organic phase is collected. The remaining aqueous layer is again washed twice with 1.5 mL diethyl ether, and the organic layers from all three extractions are combined, dried (with 1 g anhydrous sodium sulfate), filtered, divided into two equal portions, and evaporated to dryness under nitrogen (*see* **Note 7**).

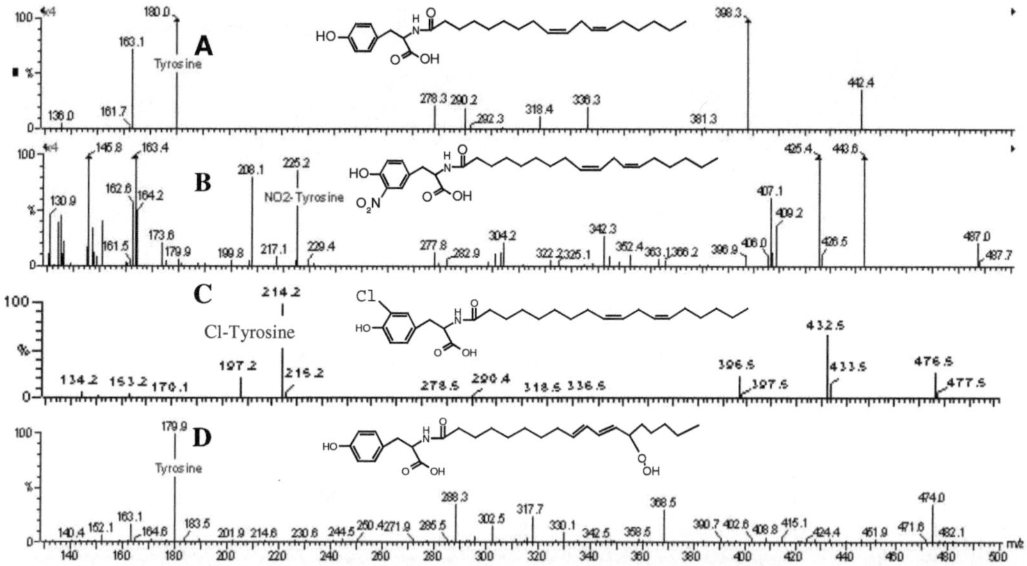

Fig. 1.2. *LC MS/MS analysis of linoleoyl tyrosine (LT) oxidized products.* The oxidized mixture is run in an LC (HP-1100), equipped with a diode array detector, on an RP-8 column, with acetonitrile:water as eluents, and afterwards is analyzed by MS/MS analysis. Compounds A–D are identified in the scan mode, and the molecular ions are further fragmented by collision-induced dissociation MS. (**A**) linoleoyl tyrosine (M^{-1}, *m/z* 442) and its fragmentations ions at *m/z* 398 (M^{-1}-CO_2) and *m/z* 180 (tyrosine [M^{-1}-linoleic acid]); (**B**) 3-nitro-linoleoyl tyrosine (M^{-1}, *m/z* 487) and its fragmentation ions at *m/z* 425 (M^{-1}-CO_2) and *m/z* 225 (nitrotyrosine). (**C**) 3-chloro-linoleoyl tyrosine (M^{-1}, *m/z* 476.5) and its fragmentation ions at *m/z* 432.5 (M^{-1}-CO_2) and at *m/z* 214 (chloro-tyrosine). (**D**) linoleoyl tyrosine hydropyroxide (M^{-1}, *m/z* 474) and its fragmentation ions at *m/z* 415 (M^{-1}-CO_2) and at *m/z* 180 (tyrosine).

3.4.3. Analysis of LT and Ox-LT by LC/MS

LT and its oxidized products are analyzed without derivatization by LC/MS, similarly to the procedure in **Section 3.3**.

3.4.4. Silylation of Cholesterol and Oxysterols

The free cholesterol and oxysterols are silylated and analyzed by GC/MS. Dried sample extracts are subjected to a silylating reagent, N,O-bis(trimethylsilyl) acetamide (BSA) (200 μL) (*see* **Note 8**), and to 200 μL 1,4-dioxane as a solvent (the dioxane is first dried on 4 Å molecular sieves and then passed through aluminum oxide to remove any possible peroxides), after which they are heated to 80°C for 60 min, cooled to room temperature, and injected into the GC/MS.

3.5. GC/MS Method for the Detection of Oxidized Cholesterol (Oxysterols)

Cholesterol and oxysterols are analyzed by GC-MS, using an HP Model 5890 Series II gas chromatograph (Waldbronn, Germany) fitted with an HP-DB-5MS trace analysis capillary column (0.32 mm I.D., 0.25 μm film thickness, 5% phenyl methyl silicone, CA, USA) and a model 5972 mass selective detector (Waldbronn, Germany), linked to an HP ChemStation data system (CA, USA). The GC is operated in a splitless mode for 0.8 min, and afterwards at a split ratio of 1:1. Helium is used

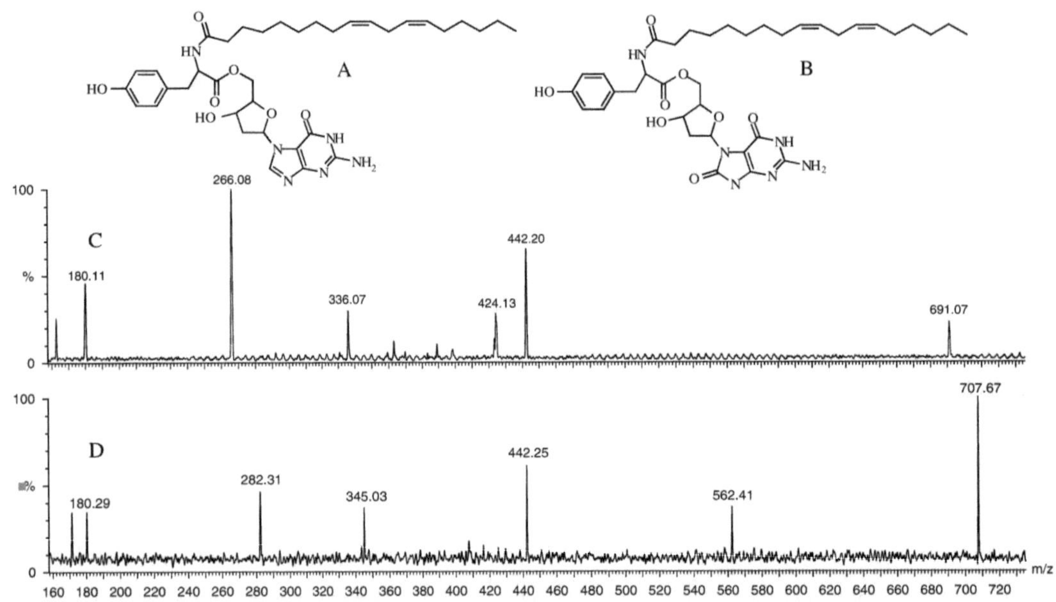

Fig. 1.3. *LC/MS analysis of linoleoyl tyrosine 2'-deoxyguanosine (LTG) and its oxidized product (LT-8-oxodG).* The structures of LTG (**A**) and LT-8-oxodG (**B**) are shown, as are the LC/MS spectrum of LTG (M^{-1}, $m/z = 691$) **(C)** and its fragmentation ions at m/z 442 (LT, M^{-1}-G), 266 (G, M^{-1}-LT), and 180 (T, (M^{-1}-LG). The LC/MS spectrum of LT-8-oxodG (M^{-1}, $m/z = 707.7$) and its fragmentation ions at m/z 442 (LT, M^{-1}-8-oxodG), 282 (8-oxoG, M^{-1}-LT), and 180 (T, [M^{-1}-L-8-oxodG]) are also presented.

as the carrier gas at a flow rate of 0.656 mL/min, with 10.4 psi pressure, and at a linear velocity of 31 cm/s. The MS transfer line is maintained at 280°C, the injector is set at 300°C, and the detector at 330°C. The column is gradient-heated, starting at 200°C, increasing to 250°C at 10°C/min, followed by 5°C/min, up to 300°C, and then maintained for an additional 15 min at 300°C. Samples are detected in the GC-MS in a total ion monitor (TIM) from which two to four of the most representative ions are selected for re-injection in the single ion monitoring mode (SIM), which enhances the limit of detection. The mean quantity of each oxysterol is calculated from calibration curves of its standard. Under the above conditions, the limit of detection for each oxysterol is determined from the analysis of its standard. Corresponding areas, equal to 10 times the area measured in the blanks, are set as the limit of detection, and 19-OH is used as an internal standard for oxysterols. In the above procedure, the following oxysterols may be detected: 7-ketocholesterol (7-keto), 7α and 7β-hydroxycholesterol (7α-OH, 7β-OH), 5α,6α and 5β,6β-epoxycholesterol (α-epoxy, β-epoxy), 24-hydroxycholesterol (24-OH), 4β-hydroxycholesterol (4β-OH), 25-hydroxycholesterol (25-OH), and 27-hydroxycholesterol 27-OH) (**Fig. 1.4**).

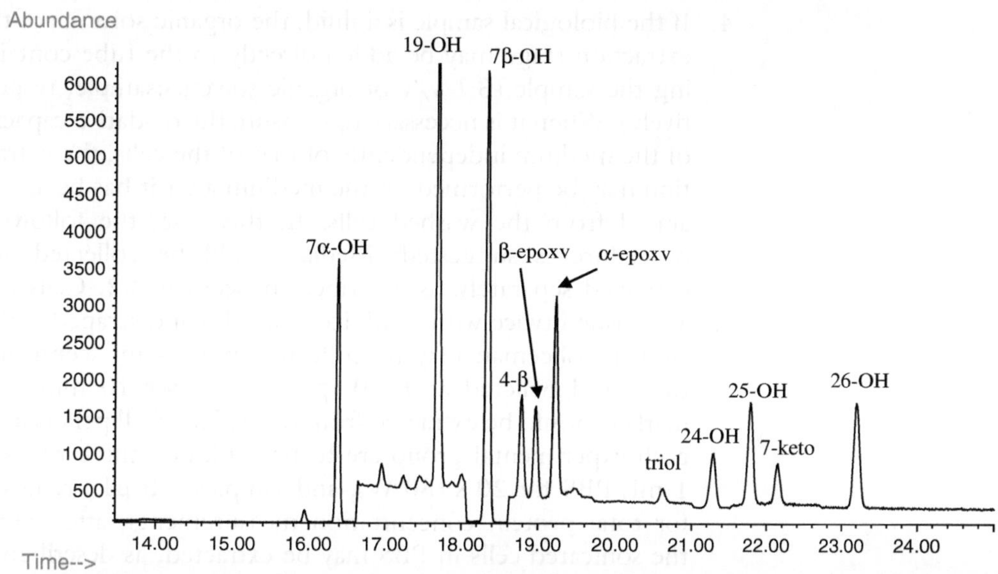

Fig. 1.4. GC/MS chromatogram of the silylated oxysterol standards, each at a concentration of 5 μg/mL. The sample is injected in a total ion monitoring mode (TIM), from which the most representative ions of each oxysterol are deduced, and re-injected in SIM to enhance the limit of detection. For example, the molecular ion of the silylated 7α-OH and 7β-OH are both at *m/z* 546, (they differ in their retention times) with major fragments at *m/z* 456. Two ions are selected, 546 and 456, for their analysis in SIM, whereas in that of 7-keto cholesterol, the molecular ion is at *m/z* 472 (mono-silylated) and three ions are selected for SIM injection e.g., *m/z* 472, 382, and 367.

4. Notes

1. The markers of **Fig. 1.1** have been tested as sensors of oxidative stress in blood *(4)*, serum *(2)*, saliva (unpublished results), cells, mouse brains *(5, 6)*, and animals (rats, mice, unpublished data) that have been injected with them, as well as in many *in vitro* experiments (regarding LDL oxidation, human lesions, cigarette smoke, etc).
2. The different markers of **Fig. 1.1** are synthesized from endogenous components, tyrosine, linoleic acid, cholesterol or 2′-deoxyguanosine, as described *(2, 3)*. They differ in their structure, physical properties (solubility), and oxidizing sensitivity. It has been shown that when the components, constructing these markers are exposed to different ROS/RNS, alone, or in a mixture (as they may be present in biological systems and in the markers themselves), they generate different oxidative products *(6)*.
3. The time of incubation depends on the type of the biological sample under investigation, e.g., for whole blood, serum and saliva, 1 h of incubation is sufficient, while for cells or tissues, 3 h of incubation up to overnight at 37°C is recommended.

4. If the biological sample is a fluid, the organic solvents of the extraction stage may be added directly to the tube containing the sample (3:1 v/v of organic solvents:sample, respectively). When it is necessary to measure the oxidative capacity of the medium independently of that of the cells, the extraction may be performed on the medium after it has been separated from the washed cells. In this case, the following procedure is suggested: Media should be collected and extracted separately, as described in **Section 3.2.** Cells that are washed twice with 5 mL ice-cold PBS and scraped with a rubber policeman may be collected in a 15 mL centrifuge tube, and pelleted at 1300 rpm at 4°C for 10 min. The marker should be extracted from the pellets. Cell pellets from each experimental group are to be sonicated three times in 1 mL PBS for 20 s (80 W), and samples (15 μL) removed for total protein concentration measurements, after which the sonicated cells in PBS may be extracted, as described in **Section 3.2.** When it is not necessary to differentiate between the OS level of the medium and that of the cells, the extraction may be carried out on both together.
5. When LTG and Ox-LTG are analyzed, the acetic acid from solvents A and B is to be omitted.
6. When the analyzed sample contains an endogenous source of cholesterol and oxysterols (blood, serum, cells, etc.), the use of LTC should be handled differently.

In these instances the levels of cholesterol and oxysterols as analyzed in the GC/MS are the sum of cholesterol and oxysterols present originally in the analyzed sample, and the extra cholesterol and oxysterols contributed from the LTC and oxidized LTC after their hydrolysis. In order to separate between the two sources, an analysis of control sample which is not incubated with LTC should also be added. The analytical methods, developed for the detection of LT/Ox-LT and LTG/Ox-LTG, are performed on the intact molecules, LT and LTG, without the need to hydrolyze the ester bond (in the LTG).

7. In a separate experiment, a standard of cholesteryl linoleate was subjected to identical saponification and extraction conditions, and cholesterol was recovered with a 98% (± 5%) yield. The stability of the oxysterols under the experimental conditions was also verified (extraction, hydrolysis, re-extraction, and silylation). Thus, β-epoxy, 7-keto and cholestane-3,5,6-triol (triol) were subjected to saponification under identical conditions, followed by extraction and GC-MS analysis using 19-OH as an internal standard. The recovery was almost quantitative (98% [RSD] ± 6.8%).
8. Silylation with BSA in 1,4-dioxane is milder and more superior, in terms of reaction conversion, reproducibility, and suitability

to GC-MS, than other reagents that have been examined, such as trimethylsilylchloride in pyridine and BSA in other solvents (diethyl ether, dimethyl formamide, chloroform) *(7)* (**Fig. 1.4**).

Acknowledgement

This work was supported by a grant from the UJIA-United Jewish Israel Appeal.

References

1. Azzi, A. (2007) Oxidative stress: a dead end or a laboratory hypothesis? *Biochem. Biophys. Res. Commun.* **362**, 230–2.
2. Szuchman, A., Aviram, M., Soliman, K., Tamir, S., Vaya, J. (2006) Exogenous N-linoleoyl tyrosine marker as a tool for the characterization of cellular oxidative stress in macrophages. *Free Radic. Res.* **40**, 41–52.
3. Khatib, S., Musa, R., Vaya, J. (2007) Linoleoyl tyrosine 2′-deoxyguanosyl ester as a novel designed synthetic probe for the characterization of oxidative stress. *Bioorg. Med. Chem.* **15**, 3661–6.
4. Szuchman, A., Aviram, M., Soliman, S., Vaya, J. (2007) Characterization of oxidative stress in blood from diabetic versus hypercholesterolemic patients, using a novel synthesized marker. *Biomarkers* August **30**, 1–13.
5. Vaya, J., Song, W., Khatib, S., Geng, G., Schipper, H. (2007) Effects of heme oxygenase-1 expression on sterol homeostasis in rat astroglia. *Free Radic. Biol. Med.* **42**, 864–71.
6. Szuchman, A., Aviram, M., Tamir, S., Vaya, J. (2003) Cholesterol, linoleic acid or/and tyrosine yield different spectra of products when oxidized alone or in a mixture: studies in various oxidative systems. *Free Radic. Res.* **37**, 1277–88.
7. Vaya, J., Aviram, M., Mahmood, S., Hayek, T., Grenadir, E., Hoffman, A., Milo, S. (2001) Selective distribution of oxysterols in atherosclerotic lesions and human plasma lipoproteins. *Free Radic. Res.* **34**, 485–97.

Chapter 2

A Rapid and Selective Mass Spectrometric Method for the Identification of Nitrated Proteins

Angela Amoresano, Giovanni Chiappetta, Piero Pucci, and Gennaro Marino

Abstract

The nitration of protein tyrosine residues represents an important posttranslational modification during development, oxidative stress, and biological aging. The major challenge in the proteomic analysis of nitroproteins is the need to discriminate modified proteins, usually occurring at substoichiometric levels, from the large amount of nonmodified proteins. Moreover, precise localization of the nitration site is often required to fully describe the biological process. Identification of the specific targets of protein oxidation was previously accomplished using immunoprecipitation techniques followed by immunochemical detection. Here, we report a totally new approach involving dansyl chloride labeling of the nitration sites which relies on the enormous potential of MS^n analysis. The tryptic digest from the entire protein mixture is directly analyzed by MS on a linear ion trap mass spectrometer. Discrimination between nitro- and unmodified peptide is based on two selectivity criteria obtained by combining a precursor ion scan and a MS3 analysis. The novel labeling procedure was successfully applied to the identification of 3-nitrotyrosine residues in complex protein mixtures.

Key words: Protein nitration, proteomics, dansyl chloride, MS3, mass spectrometry, 3-nitrotyrosine, precursor ion scan.

1. Introduction

Oxidative stress is now recognized as accountable for redox regulation involving reactive oxygen species (ROS) and reactive nitrogen species (RNS). Its role is pivotal for the modulation of critical cellular functions, notably for neurons astrocytes and microglia, such as apoptosis program activation, ion transport, and calcium mobilization involved in excitotoxicity *(1)*.

From: *Methods in Molecular Biology, vol. 477: Advanced Protocols in Oxidative Stress I*,
Edited by: D. Armstrong, DOI: 10.1007/978-1-60327-517-0_2, © Humana Press, New York, NY

Oxidative stress is caused by an imbalance in the pro-oxidant and antioxidant systems. Classically, oxidative stress is described as an imbalance between generation and elimination of ROS and RNS. Oxidative stress may cause reversible and/or irreversible modifications on sensitive proteins leading to structural, functional, and stability modulations *(2, 3)*. Protein modifications such as carbonylation, nitration, and protein–protein cross-linking are generally associated with loss of function and may lead to either the unfolding and degradation of the damaged proteins, or aggregation leading to accumulation as cytoplasmic inclusions, as observed in age-related neurodegenerative disorders *(4)*.

Proteomics has the potential to identify novel targets of tyrosine nitration in cells and tissues and depict the nitro- and phosphoproteome, or to identify proteins undergoing S-nitrosylation in vivo. The nitration of a tyrosine residue to form 3-nitrotyrosine (the addition of a nitro ($-NO_2$) group to position 3 of the phenolic ring of a tyrosine residue *(5)*) is an important post-translational protein modification that is associated with different biological processes and pathologies *(6, 7, 8)* including cancer, Parkinson's disease, Alzheimer's disease, Huntington's disease, lung infection, retinal ischemia, aging, and oxidative stress.

Methods for separation, detection, and quantitation of *o*-nitrotyrosine in biological samples include immunochemical techniques using anti-*o*-nitrotyrosine antibodies, HPLC in combination with various detection systems, and GC/MS *(9)*. A combination of MS techniques had been used to identify the specific tyrosine residues nitrated *in vitro* in model proteins *(10)*. More recently, a higher throughput characterization of protein targets for tyrosine nitration in cells and several tissues, including aged tissues, has been attempted using proteomic methodologies *(11, 12)*. These approaches essentially consist of protein fractionation by two dimensional polyacrylamide gel electrophoresis (2D PAGE), partial transfer onto poly(vinylidene difluoride) membranes, and western blot analysis using anti-nitrotyrosine antibody to identify the modified proteins. However, such methods are both time-consuming and laborious, and as noted require a good guess to the identity of the protein at the beginning. Moreover, post-translational modification can sometimes change the structure of proteins, which could then prevent the formation of the appropriate antigen–antibody complex. With advances in technology, proteomics coupled to mass spectrometry has been a major methodological development that allows the identification of large number of proteins at once *(13)*.

Alignment of the western blots with the 2D PAGE gels enables identification of immunopositive protein spots. These are then excised, trypsin digested, and identified by either peptide mass fingerprinting procedures using MALDI mass spectrometry or capillary LC-MS/MS analyses *(14)*. Recently, a novel nitroproteomics analytical system was developed for the current study. A highly specific nitrotyrosine affinity column (NTAC) was used to preferentially enrich and isolate endogenous nitroproteins and nitroprotein–protein complexes from a human pituitary tumor proteome *(15)*.

Here, a novel approach to selectively label *o*-nitrotyrosine residues in proteins using dansyl modification coupled with tandem mass spectrometry experiments in precursor ion /MS3 scan mode *(16)* is presented. The methodology, developed on the basis of a general derivatization method – the reporter ion generation tag (RIGhT) strategy *(17, 18)*, was first tested on *in vitro* nitrated BSA as a model protein and then applied to more complex matrices. Because of its operational simplicity avoiding long-lasting and time-consuming fractionation procedures, this new strategy seems to be well suited for large-scale proteomic profiling of nitration sites.

2. Materials

2.1. Equipment

1. Voyager DE-STR MALDI-TOF mass spectrometer with 337 nm nitrogen laser and reflector analyzer, Applied Biosystems, Foster City, California
2. Hybrid 4000 Q-Trap instrument with linear ion trap, Applied Biosystems, Foster City, California
3. Model 1100 nano HPLC system, Agilent Technologies, Santa Clara, California

2.2. Chemicals

Tri(hydroxymethyl)aminomethane (Tris), 5-N,Ndimethylaminophtalene-1-sulfonyl chloride (dansyl chloride, DNSCl), ammonium hydrogen carbonate (AMBIC), and iodoacetamide were purchased from Fluka (St. Louis, MO), tetranitromethane (TNM), bovine serum albumin (BSA), ethylenediaminetetraacetic acid (EDTA), sodium dithionite ($Na_2S_2O_4$), urea, sodium acetate, trypsin, and dithiothreitol (DTT) were from Sigma (St. Louis, MO) and acetonitrile (ACN) was purchased from Baker (Phillipsburg, NJ). Trifluoroacetic acid HPLC grade was from Carlo Erba. All solvents were of the highest purity available from Baker. All other reagents and proteins were of the highest purity available from Sigma.

2.3. In Vitro Nitration of BSA

1. Bovine serum albumin (BSA) solution (10 mg/mL) from Sigma was prepared in 200 mM Tris buffer (pH 8.0); stored at −20°C.
2. 350 mM Tetranitromethane (TNM) solution was prepared in acetonitrile. Made fresh as required.
3. Sephadex G-25M size exclusion chromatography column (Amersham).
4. 50 mM AMBIC buffer (pH 8.0).

2.4. Tryptic Digestion

1. Denaturant buffer contained 6 M urea, 10 mM Tris, 125 mM EDTA, pH 8.0. Stored at room temperature.
2. 50 mM AMBIC buffer (pH 8.0). Made fresh as required.
3. Reversed-phase HPLC C4 column (100 × 4.6 mm, 5 μm; Phenomenex).
4. 0.1% TFA in water (solvent A)
5. 0.07% TFA in 95% acetonitrile (solvent B).
6. Trypsin solution was prepared dissolving the enzyme at 1mg/ml in 50 mM AMBIC buffer (pH 8.0). Made fresh as required.

2.5. Reduction of Nitrotyrosine to Aminotyrosine

1. 200 mM Tris buffer (pH 8.0). Made fresh as required.
2. 200 mM $Na_2S_2O_4$ as described by McIntyre *(19)*. Freshly prepared.
3. Samples cleaning was performed using reversed-phase Zip-Tips C18 from Millipore (Billerica, MA) using the procedure recommended by the manufacturer.

2.6. Synthesis of o-Aminodansyltyrosine

1. Reversed-phase HPLC C18 column (100 × 4.6 mm, 5 μm; Phenomenex).
2. 0.1% TFA in water (solvent A)
3. 0.07% TFA in 95% acetonitrile (solvent B).
4. 100 mM sodium acetate pH 5.0 buffer, 0.9% sodium chloride. Stored at 4°C.
5. Labeling solution was prepared by dissolving DNS-Cl at 18.5 ng/μL in ACN. Made fresh as required.

2.7. Labeling of Bovine Milk Protein Extract

1. 100 mM TNM solution was prepared and stored at room temperature.
2. Precipitation with the Amersham Clean Up kit as recommended.

3. Denaturant buffer (20 mM DTT, 40 mM iodoacetamide, 6 M urea, pH 8.0)
4. 200 mM solution $Na_2S_2O_4$ in denaturant buffer.
5. 50 mM AMBIC buffer, pH 8.0. Made fresh as required.
6. Labeling solution was prepared by dissolving DNS-Cl at 18.5 ng/μL in ACN. Made fresh as required.

2.8. MALDI Mass Spectrometry

1. 10 mg/mL solution of R-cyano-4-hydroxycinnamic acid in acetonitrile/50 mM citrate buffer (2/3, v/v). Made fresh as required and stored at 4°C for 1 day.

2.9. NanoLC-MS2 and MS3 Mass Spectrometry

1. Agilent reversed-phase precolumn cartridge (Zorbax 300 SB-C18, 5 × 0.3 mm, 5 μm) working at 10 μL/min
2. Agilent reversed-phase column (Zorbax 300 SBC18, 150 mm × 75 μm, 3.5 μm), at a flow rate of 0.2 μL/min
3. Solvent A was 0.1% formic, ACN in water acid solution
4. Solvent B was 98% ACN, 2% 0.1% formic acid solution
5. Uncoated silica tip (o.d. 150 μm, i.d. 20 μm, tip diameter 10 μm) from New Objectives (Ringoes, NJ).

3. Methods

A challenging task in the proteomic analysis of nitroproteins is the need to discriminate nitrated proteins, usually in very low concentrations, from the large amount of nonmodified proteins. Moreover, precise localization of the nitration site is often required to fully describe the biological process.

The strategy described here is based on dansyl chloride labeling and relies on the enormous potential of MSn analysis. Peptide analysis is carried out by LC-MS/MS, and the ions of interest are discriminated by two selectivity criteria based on two subsequent tandem MS experiments, a precursor ion scan followed by an MS3 scan mode, a sort of "instrumentally driven" bidimensional selection. This instrumental selectivity approach needs an appropriate derivatizing agent, like dansyl chloride, producing stable and diagnostic fragment ions (**Scheme 2.1**) to be used in the MS/MS scan modes and capable of improving ionization of modified peptides.

3.1. In Vitro Nitration of BSA

1. A BSA solution (10 mg/mL) in 200 mM Tris buffer (pH 8.0) was nitrated by addition of 350 mM TNM in acetonitrile using a Tyr/TNM ratio of 1/1 (mol/mol). The reaction mixture was stirred at room temperature for 30 min.

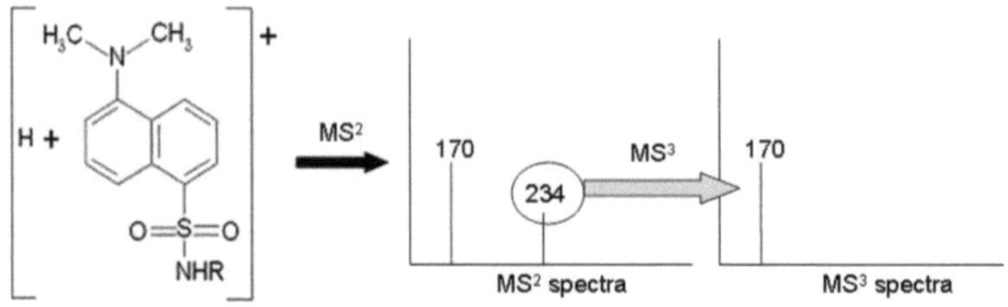

Scheme 2.1. Peculiar electrospray dansyl derivative fragmentation. Distinctive m/z 170 and 234 product ions are observed in MS2 mode and the diagnostic m/z 234→170 fragmentation is observed in the MS3 mode.

2. The excess reagent was rapidly removed by size exclusion chromatography as on a Sephadex G-25M column equilibrated and eluted with 50 mM AMBIC buffer. Protein elution was monitored at 280 and 350 nm.
3. The protein-containing fractions were manually collected, lyophilized, and stored at −20°C.

3.2. Trypsin Digestion

1. Aliquots of the BSA and N-BSA mixture were dissolved in denaturant buffer, reduced with DTT (10-fold molar excess on the cysteine residues) for 2 h at 37°C and then alkylated with iodoacetammide (5-fold molar excess on the thiol residues) for 30 min at room temperature in the dark.
2. Protein samples were desalted by reversed-phase (RP)-HPLC by means of a linear gradient from 5% to 95% of solvent B in 10 min. Fractions were lyophilized.
3. Reduced and alkylated proteins were dissolved in 50 μl of 50 mM AMBIC buffer
4. Trypsin digestion was carried out using an enzyme/substrate ratio of 1/50 (w/w) at 37°C for 18 h.
5. The resulting peptide mixtures were directly analyzed by MALDI mass spectrometry on a Voyager DE-STR MALDI-TOF mass spectrometer (Applied Biosystems, Framingham, MA) equipped with a nitrogen laser (337 nm). Typically, 1 μL of the total mixture was mixed (1/1, v/v) with matrix solution to determine TNM-induced modifications. Spectra were acquired by monitoring positive ions using accelerating voltage of 70 kV, grid 63%, and grid wire 0.01% in respect of accelerating voltage; delay time was 100 ns. Mass range was 400–4000 m/z. Mass calibration was performed using external peptide standards by Applied Biosystems. Raw data were analyzed using the computer software provided by the manufacturer and reported as monoisotopic masses. As an example, a mass signal recorded at m/z 972.5

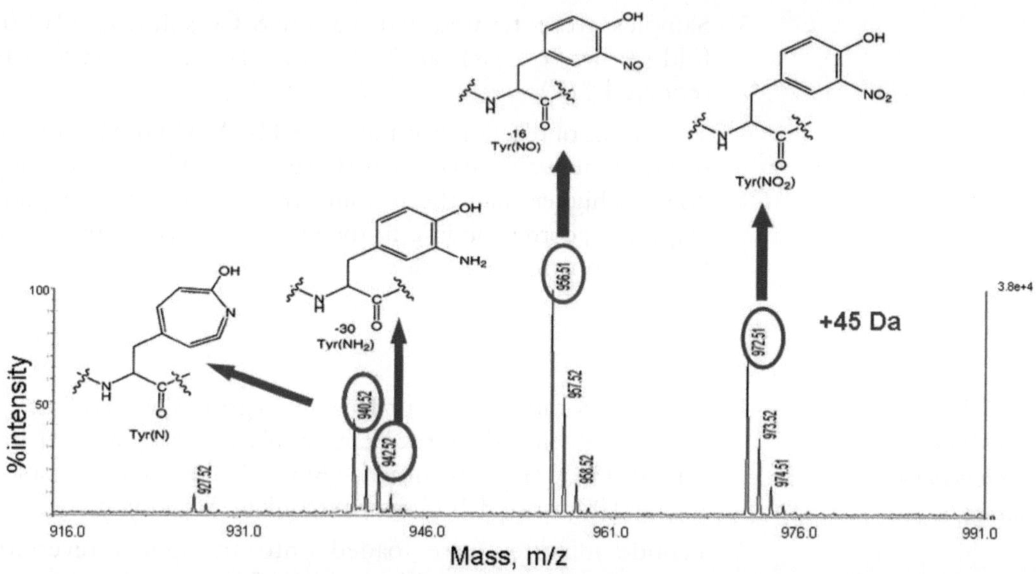

Fig. 2.1. Partial MALDI-MS spectrum of peptide mixture from *in vitro* nitrated BSA. Satellite peaks at −16 and −30 and −32 Da, revealing the typical MALDI photodecomposition pattern of nitrated peptides, are present and the corresponding structures are reported.

did not correspond to any peptide within the BSA sequence. This value occurred 45 Da higher than the signal corresponding to the peptide 137–143 suggesting the nitrated derivative of the fragment (**Fig. 2.1**).

3.3. Reduction of Nitrotyrosine to Aminotyrosine

1. Conversion of nitropeptides into their *o*-dansylamino derivatives was accomplished by $Na_2S_2O_4$ treatment. The sample was dissolved in 200 mM $Na_2S_2O_4$ solution for 1 min at room temperature.
2. Before MALDI-TOF analysis, cleaning of the samples was performed using reversed-phase Zip-Tips C18 following the procedure suggested by the manufacturer.
3. The extent of reduction was monitored by MALDI-MS performed as previously described. MALDIMS spectrum showed the disappearance of the classical *o*-nitrotyrosine photodecomposition pattern, and its replacement by a new signal at m/z 942.4 attributed to peptides m/z 137–143 containing an amino-Tyr residue.

3.4. Synthesis of o-Aminodansyltyrosine

1. The mixture of reduced BSA tryptic peptides was desalted by RP-HPLC by means of a linear gradient from 5% to 65% of solvent B in 5 min.
2. Peptide fractions were lyophilized and then dissolved in 100 mM sodium acetate buffer.

3. Samples were treated with a DNS-Cl solution (1000-fold molar excess) at 65°C for 16 h as previously reported *(17)*.
4. The extent of reaction was monitored by MALDI-MS analysis showing the presence of a new signal at m/z 1175.6 occurring 233 Da higher than the *o*-aminotyrosine-containing peptide (**Fig. 2.2**), corresponding to the expected dansyl derivative of the *o*-aminotyrosine peptide.

3.5. LC-MS/MS Analysis of o-Aminodansyl-Tyr Peptides

1. The exact location of the original nitro groups was assessed by LCMS/MS analysis of the peptide mixture using a linear ion trap 4000 Q-Trap instrument (Applied Biosystems) coupled to an 1100 nano HPLC system (Agilent Technologies).
2. Peptide mixtures were loaded onto an Agilent reversed-phase pre-column cartridge at 10 μL/min with solvent A using a loading time of 7 min.
3. Peptides were then separated on a Agilent reversed-phase column, at a flow rate of 0.2 μL/min.
4. The elution was accomplished by a 5–65% linear gradient of solvent B in 60 min.
5. A micro-ionspray source was used at 2.5 kV with liquid coupling, with a declustering potential of 50 V using an uncoated silica tip.

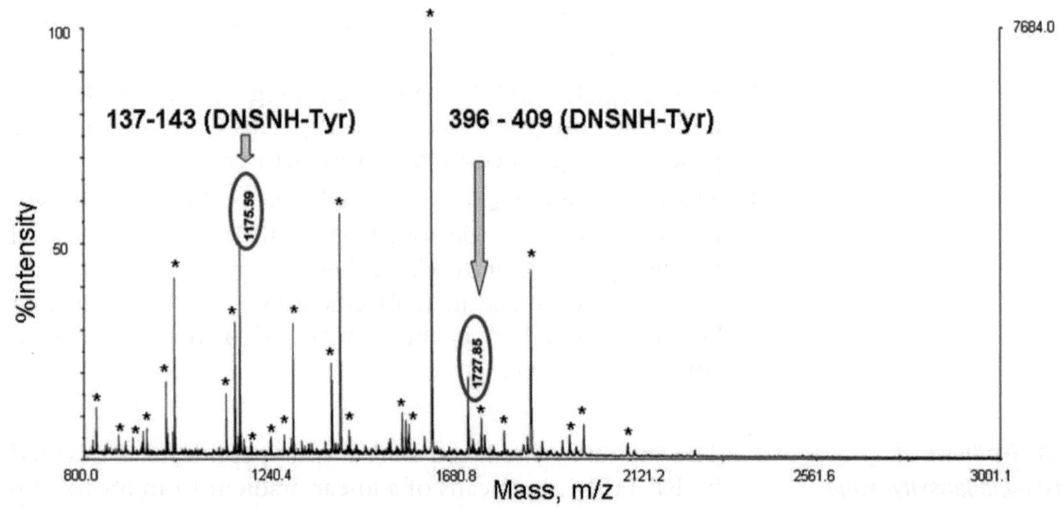

Fig. 2.2. Partial MALDI-MS spectrum of dansylated *o*-aminotyrosine-containing peptides. The modified peptides are indicated by arrows. Other peaks present in the spectrum and labeled with asterisks correspond to theoretical peptide fragments within BSA sequence.

6. Spectra acquisition was based on a survey precursor ion scan for m/z 170. The Q1 quadrupole was scanned from m/z 500 to 1000 in 2 s with resolution "low", and the precursor ions were fragmented in q2 using a linear gradient of collision potential from 30 to 70 V. Finally, Q3 was set to transmit only ions at m/z 170 with resolution "unit".
7. The survey precursor ion scan was followed by an enhanced resolution experiment for the ions of interest and then by MS3 and MS2 acquisitions of the two most abundant ions. The entire cycle duration was 5.3 s.
8. MS2 spectra were acquired using the best collision energy calculated on the basis of m/z values and charge state (rolling collision energy).
9. MS3 spectra were performed on the fragment ion at m/z 234 and acquired using Q0 trapping, with a trapping time of 150 ms and an activation time of 100 ms, scanning from m/z 160 to 210. The MS3 TIC (**Fig. 2.3A**) showed the presence of only three peaks related to doubly charged ions at m/z 588.2, 792.9, and 986.8. The corresponding MS2 fragmentation spectra led to the determination of the entire sequence, identifying the fragments m/z 137–143,

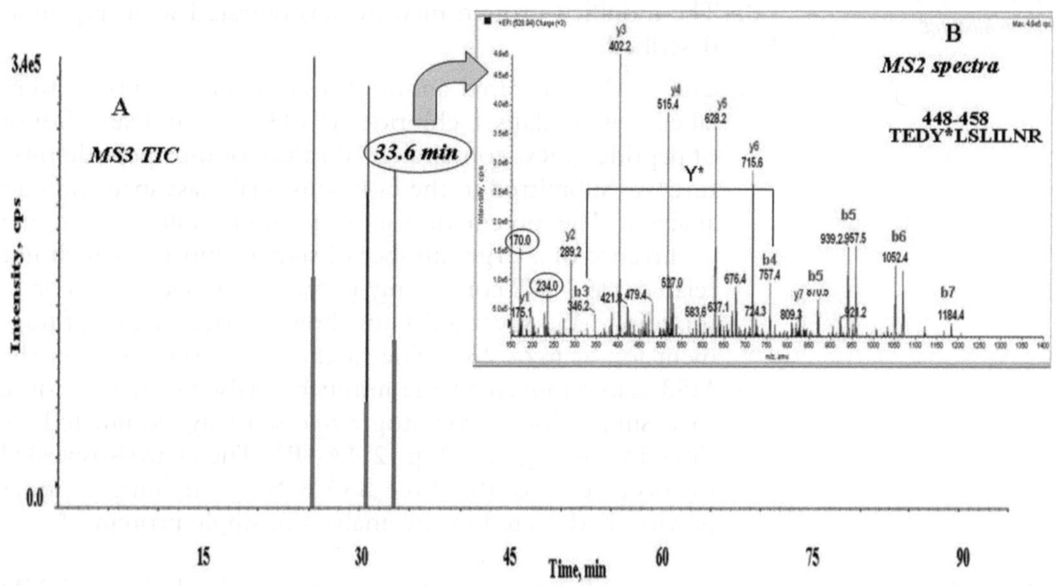

Fig. 2.3. MS3 TIC profile of tryptic nitrated BSA digest. As indicated, the MS3 analysis showed the presence of only three peaks related to doubly charged ions at m/z 588.2, 792.9, and 986.8 (Panel **A**). The corresponding MS2 fragmentation spectra led to the determination of the entire sequence of these species. The insert (**B**) showed the MSMS spectrum, properly annotated, leading to the reconstruction of the sequence 445–448. This sequence is related to the fragment 448–458, modified at level of Tyr 451 allowing for the exact identification of the original nitration site.

396–409, and 445–458, respectively, containing *o*-aminodansyl-Tyr residues.

10. Identification of nitrated tyrosine residues was carried out by taking advantage of the flexibility of the MASCOT software. In fact, a variable modification of 411.0 Da corresponding to *o*-aminodansyltyrosine residues was introduced into the Modification File within the MASCOT software.
11. The peak list used for database search only consisted in the MS2 spectra of the peptide species that had generated a signal both in the precursor ion TIC and in the MS3 ion scan.
12. Modified Tyr residues could easily be detected, by the interpretation of the MS2 spectra, allowing for the exact identification of the original nitration sites. As an example, **Fig. 2.3B** shows the MS/MS spectrum of the peptide m/z 445–458 carrying a dansyl moiety. The modified ion is stable during collision-induced dissociation. Both the y and b fragment ions still retained the modifying group linked to the amino Tyr residue, thus localizing the original nitration site.

3.6. Applications

3.6.1. Location of BSA o-Aminotyrosine-containing Peptides in Complex Protein Mixtures

1. The feasibility of the developed strategy was probed by mixing 100 μg of the mixture of BSA and nitrated BSA with 10 mg of the entire cellular extract from Escherichia coli.
2. Nitrotyrosines were reduced with dithionite, as described before.
3. The modified protein mixture was digested with trypsin as described.
4. The newly generated amino Tyr-containing peptides were labeled with dansyl chloride at pH 5.0, and an aliquot of peptide digest containing 100 fmol of nitropeptide mixture was submitted to the bidimensional mass spectrometry analysis. The precursor ion scan mode still showed the occurrence of a large number of signals, most of which not related to 3-NT-containing peptides. In fact, the corresponding MS2 spectra do not show the occurrence of fragment ion at m/z 234. The further selection based on the MS3 scan removed a large number of false positives leading to a simple ion chromatogram essentially dominated by three intense signals (**Fig. 2.4A, B**). The analysis revealed the occurrence of the three BSA 3-NT-containing peptides previously detected in the analysis of single protein.

3.6.2. Labeling and Analysis of a Nitrated Protein Extract of Bovine Milk

1. Bovine milk was reacted with a 100 mM solution of TNM for 30 min at room temperature.
2. Modified milk proteins were purified by precipitation with the Amersham Clean Up kit and dissolved in denaturant buffer.

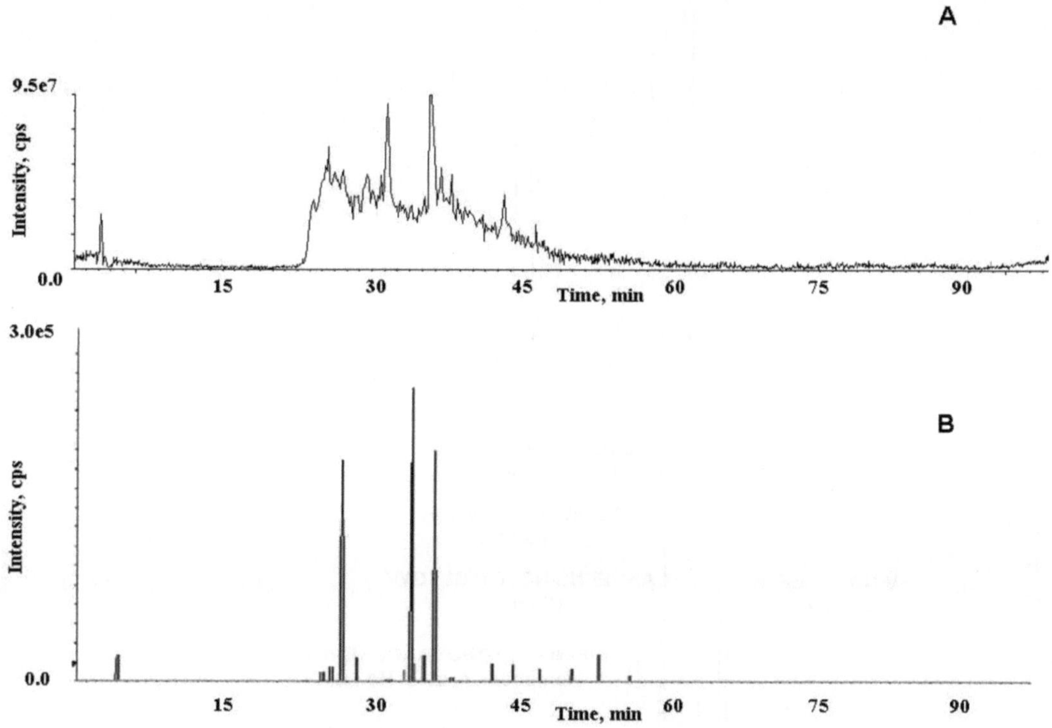

Fig. 2.4. LC-MS/MS (Panel **A**) and MS3 (Panel **B**) traces of bidimensional mass spectrometry experiment performed on the entire *E. coli.* cellular extract spiked with a mixture of BSA and nitrated BSA.

3. Reduction of both the SH and nitro groups of the protein mixture was carried out in "one pot" using 20 mM DTT, 40 mM iodoacetamide, and a 200 mM solution of $Na_2S_2O_4$ in the dark for 30 min.
4. The protein mixture was purified by size exclusion chromatography on a Sephadex G-25 M column equilibrated and eluted with 50 mM AMBIC.
5. Protein fractions were concentrated and then digested with trypsin as already described.
6. The resulting peptide mixture was selectively labeled with dansyl chloride solution.
7. 150 fmol of the labeled peptide mixture was directly submitted to LC-MS/MS analyze using the double selectivity criteria. As an example, **Fig. 2.5** (panels A and B) reports the MS, and MS3 profiles for the milk mixture.
8. Identification of nitrated milk proteins was carried out using the modified MASCOT software as previously described (*see* **Fig. 2.5B**).

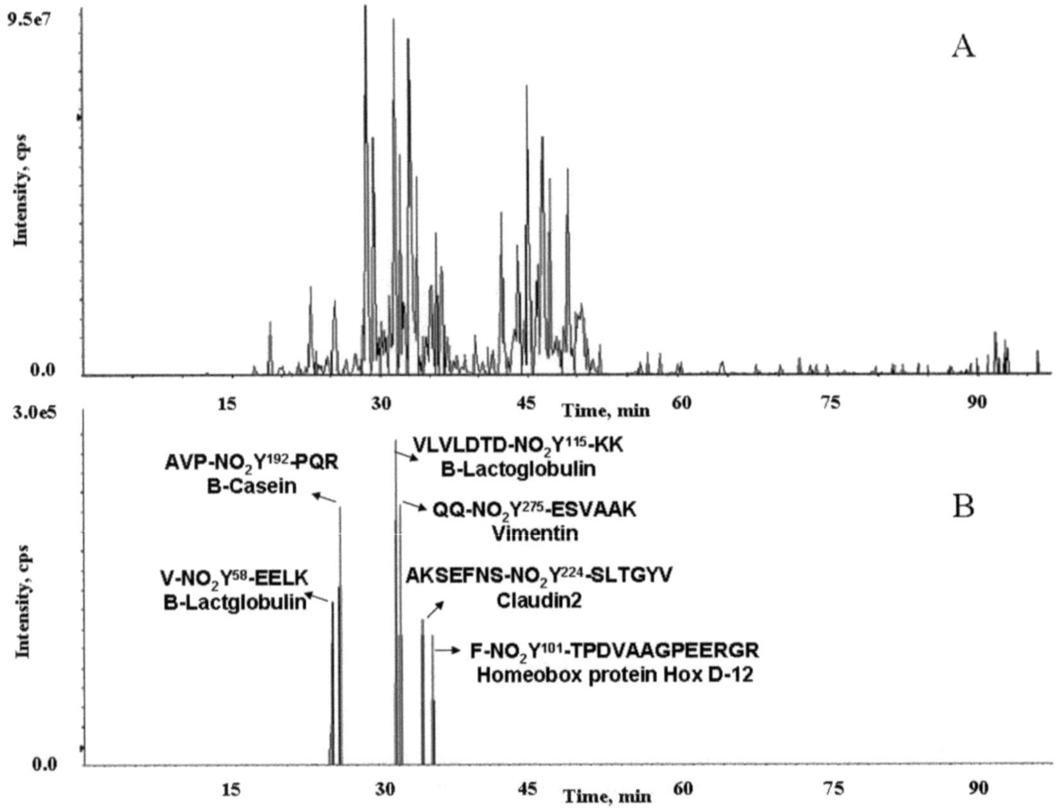

Fig. 2.5. Reconstructed ion chromatogram for the precursor ion scan (Panel **A**) and the selective dansyl transition in MS3 mode (Panel **B**) of bidimensional mass spectrometry experiment performed on *in vitro* nitrated bovine milk proteins. Nitrated proteins identified and localization of the nitration sites are reported.

4. Notes

1. Pay attention when handling TNM. We generally use two couples of gloves.
2. All solutions were prepared by using bi-distilled water by using Milli-Q system (Waters).
3. Aliquots of standard protein solution contain about 1 mg of intact protein.
4. Aliquots of 10 pmol and 150 fmol of tryptic digest were used for MALDIMS and LCMSMS analyses, respectively.
5. A sample of commercially available bovine milk was used for the labeling experiments.
6. The analysis is carried out at the level of peptides following tryptic digest of the whole protein mixture rather then at intact protein level.

7. Discrimination between nitropeptides and unmodified peptides is achieved by taking advantage of the instrumental features of a hybrid (triple quadrupole/linear ion trap) mass spectrometer.
8. The presence of a possible nitropeptide in the MALDIMS analysis is suggested by satellite peaks at −16 and −32 Da, revealing the typical photodecomposition pattern of nitrated peptides previously described to occur during the MALDI analysis of nitropeptides.
9. Check pH during Na_2SO_4 treatment. A slightly acid value could give rise to dithionite dismutation.
10. Aminotyrosine residues were chemoselectively labeled with DNS-Cl at pH 5.0 exploiting the different p*Ka* values of the *o*-aminotyrosine (4.7), 15 which is partly deprotonated and is therefore amenable to reaction with DNS-Cl at variance with aliphatic (10.4–11.1) and N-terminal amino groups (6.8–8.0), which are largely protonated.
11. In the MALDIMS analysis, the occurrence of signals corresponding to the unmodified *o*-aminotyrosine peptides indicated that the extent of dansyl reaction was about 60% as estimated by the relative intensity of the corresponding peaks.
12. No modification at the N-terminus or Lys residues should be detected when operating in this condition.
13. The LC-MS/MS analysis revealed the occurrence of a further nitration site within the peptide m/z 448–458 at level of Tyr451, which escaped previous MALDI analysis, probably due to the different chemical-physical properties of the peptide under different ionization conditions.
14. Analysis of chemically nitrated BSA following tryptic digestion confirmed the ability of the bidimensional selection to simplify the peptide ion chromatogram, leading to selective identification of *o*-nitrotyrosine-containing peptides
15. The set up strategy underlined the usefulness of the two selectivity criteria in that the precursor ion scan still showed the presence of nonnitrated peptide ions that were completely ruled out in the MS3 scan mode. In fact, the precursor ion scan mode still showed the occurrence of a large number of signals, most of which not related to 3-NT-containing peptides. (the corresponding MS2 spectra do not show the occurrence of fragment ion at m/z 234). The further selection based on the MS3 scan removed a large number of false positives leading to a simple ion chromatogram essentially dominated by three intense signals.

Acknowledgments

This work was supported by grants from the Ministero dell'Universita' e della Ricerca Scientifica (Progetti di Rilevante Interesse Nazionale 2002, 2003, 2005, 2006; FIRB 2001). Support from the National Center of Excellence in Molecular Medicine (MIUR – Rome) and from the Regional Center of Competence (CRdC ATIBB, Regione Campania – Naples) is gratefully acknowledged.

References

1. Emerit, J., Edeasc, M. and Bricairea, F. (2004) Neurodegenerative diseases and oxidative stress. *Biomed. Pharmacother.* **58**, 39–46.
2. Nakamura, K., Hori, T., Sato, N., Sugie, K., Kawakami, T. and Yodoi, J. (1993) Redox regulation of a src family protein tyrosine kinase p56lck in T cells. *Oncogene* **8**, 3133–3139.
3. Staal, F.J., Anderson, M.T., Staal, G.E., Herzenberg, L.A. and Gitler, C. (1994) Redox regulation of signal transduction: tyrosine phosphorylation and calcium influx. *Proc. Natl. Acad. Sci. U. S. A.* **91**, 3619–3622.
4. Dalle-Donne, I., Scaloni, A., Giustarini, D., Cavarra, E., Tell, G., Lungarella, G., Colombo, R., Rossi, R. and Milzani, A. (2005) Proteins as biomarkers of oxidative/nitrosative stress in diseases: the contribution of redox proteomics. *Mass Spectrom. Rev.* **24**, 55–99.
5. Zhan, X. and Desiderio, D.M. (2004) The human pituitary nitroproteome: detection of nitrotyrosyl-proteins with two-dimensional Western blotting, and amino acid sequence determination with mass spectrometry, *Biochem. Biophys. Res. Commun.* **325**, 1180–1186.
6. Haddad, I.Y., Pataki, G., Hu, P., Galliani, C., Beckman, J.S. and Matalon, S. (1994) Quantitation of nitrotyrosine levels in lung sections of patients and animals with acute lung injury, *J. Clin. Invest.* **94**, 2407–2413.
7. Tsukahara, H. (2007) Biomarkers for oxidative stress: clinical application in pediatric medicine. *Curr. Med. Chem.* **14**, 339–351.
8. Bian, K., Ke, Y., Kamisaki, Y. and Murad, F. (2006) Proteomic modification by nitric oxide. *J. Pharmacol. Sci.* **101**, 271–279.
9. Soderling, A.S., Ryberg, H., Gabrielsson, A., Larstad, M., Toren, K., Niari, S. and Caidahl, K. (2003) A derivatization assay using gaschromatography/negative chemical ionization tandem mass spectrometry to quantify 3-nitrotyrosine in human plasma. *J. Mass Spectrom.* **38**, 1187–1196.
10. Lee, S.J., Lee, J.R., Kim, Y.H., Park, Y.S., Park, S.I., Park, H.S. and Kim, K.P. (2007) Investigation of tyrosine nitration and nitrosylation of angiotensin II and bovine serum albumin with electrospray ionization mass spectrometry. *Rapid Commun. Mass Spectrom.* **21**, 2797–2804.
11. Kanski, J., Alterman, M.A. and Schoöneich, C. (2003) Proteomic identification of age-dependent protein nitration in rat skeletal muscle. *Free Radic. Biol. Med.* **35**, 1229–1239.
12. Kanski, J. and Schoneich, C. (2005) Protein nitration in biological aging: proteomic and tandem mass spectrometric characterization of nitrated sites. *Meth. Enzymol.* **396**, 160–171.
13. Butterfield, D.A., Perluigi, M. and Sultana R. (2006) Oxidative stress in Alzheimer's disease brain: new insights from redox proteomics. *Eur. J. Pharmacol.* **545**, 39–50.
14. Aulak, K.S., Koeck, T., Crabb, J.W. and Stuehr, D.J. (2004) Proteomic method for identification of tyrosine-nitrated proteins. *Methods Mol. Biol.* **279**, 151–165.
15. Zhan, X. and Desiderio, D.M. (2006) Nitroproteins from a human pituitary adenoma tissue discovered with a nitrotyrosine affinity column and tandem mass spectrometry. *Anal. Biochem.* **354**, 279–289.
16. Amoresano, A., Monti, G., Cirulli, C. and Marino, G. (2006) Selective detection and identification of phosphopeptides by dansyl MS/MS/MS fragmentation. *Rapid Commun. Mass Spectrom.* **20**, 1400–1404.

17. Cirulli, C., Marino, G. and Amoresano, A. (2007) Membrane proteome in Escherichia coli probed by MS3 mass spectrometry: a preliminary report. *Rapid Commun. Mass Spectrom.* **21**, 2389–2397.
18. Amoresano, A., Chiappetta, G., Pucci, P., D'Ischia, M. and Marino, G. (2007) Bidimensional tandem mass spectrometry for selective identification of nitration sites in proteins. *Anal. Chem.* **79**, 2109–2117.
19. McIntyre, J.C., Schroeder, F. and Behnke, W.D. (1990) Synthesis and characterization of the dansyltyrosine derivatives of porcine pancreatic colipase. *Biochemistry* **29**, 2092–2101.

Chapter 3

An Easy and Reliable Automated Method to Estimate Oxidative Stress in the Clinical Setting

Cristina Vassalle

Abstract

During the last few years, reliable and simple tests have been proposed to estimate oxidative stress in vivo. Many of them can be easily adapted to automated analyzers, permitting the simultaneous processing of a large number of samples in a greatly reduced time, avoiding manual sample and reagent handling, and reducing variability sources. In this chapter, description of protocols for the estimation of reactive oxygen metabolites and the antioxidant capacity (respectively the d-ROMs and OXY Adsorbent Test, Diacron, Grosseto, Italy) by using the clinical chemistry analyzer SYNCHRON, CX 9 PRO (Beckman Coulter, Brea, CA, USA) is reported as an example of such an automated procedure that can be applied in the clinical setting. Furthermore, a calculation to compute a global oxidative stress index (Oxidative-INDEX), reflecting both oxidative and antioxidant counterparts, and, therefore, a potentially more powerful parameter, is also described.

Key words: Oxidative stress score, hydroperoxides, total antioxidant capacity, automated methods.

1. Introduction

Reactive oxygen species (ROS) are physiologically produced in the cell consequently to energy production through molecular oxygen reduction to water during aerobic metabolism *(1)*. However, excess production or impaired elimination of free radicals lead to increased oxidative stress. In fact, these species may target a wide range of molecules, including proteins, nucleic acids, lipids, and carbohydrates and cause marked alterations to a variety of cellular pathways *(1)*. Consequently, the role of oxidative stress in critical illness, including cardiovascular disease, is supported by a considerable body of basic science literature *(2, 3)*.

From: *Methods in Molecular Biology, vol. 477: Advanced Protocols in Oxidative Stress I,*
Edited by: D. Armstrong, DOI: 10.1007/978-1-60327-517-0_3, © Humana Press, New York, NY

To date, techniques available for the oxidative stress estimation often required the use of special equipment and/or a skilled operator, and are expensive and time-consuming, which make these procedures reserved for specialized research laboratory and not for large-scale applications *(4, 5)*. However, during the last few years, some assays have been developed, with reliable results, and are easy to perform, quick, and inexpensive, thus promising for application to routine clinical diagnosis and monitoring *(5)*. Many of them could be adjusted to automated analyzers without great problems, permitting the simultaneous processing of a large number of samples in a greatly reduced time, avoiding manual sample and reagent handling, and reducing variability sources. In addition, this last step would permit optimal standardization and validation through multicenter large-scale clinical trials.

In this chapter, a detailed description of protocols for the spectrophotometric estimation of reactive oxygen metabolites and the total antioxidant status (d-ROMs and OXY Adsorbent Test, respectively, from Diacron, Grosseto, Italy) by using the clinical chemistry analyzer SYNCHRON, CX 9 PRO (Beckman Coulter, Brea, CA, USA) is reported as an example of such an automated procedure that can be easily applied to the clinical target.

Furthermore, a calculation to compute a global oxidative stress index (Oxidative-INDEX), able to reflect both oxidative and antioxidant components, and therefore potentially more powerful, is also described. Results obtained by using this comprehensive parameter were tested in a healthy population in relation to age, gender, and smoking habit.

2. Materials

2.1. Equipment

1. An automated clinical chemistry analyzer, SYNCHRON CX 9 PRO, Beckman Coulter, Brea, CA, USA.
2. Refrigerate centrifuge.
3. Multi-channel pipettes and tips.

2.2. Reagents Included in the d-ROMs Kit, Diacron, Grosseto, Italy

1. R1 reagent = chromogenic mixture: aromatic alkyl-amine with concentration <10%.
2. R2 reagent = acetate buffer, pH 4.8, preservatives and stabilizers (do not contain sodium azide) (*see* **Note 1**).

2.3. Reagents Included in the OXY Adsorbent Kit, Diacron, Grosseto, Italy

1. R1 reagent = oxidant solution (HClO) with concentration <0.1‰.
2. R2 reagent = chromogen (N,N-diethylparaphenilendiamide) stabilizers with concentration <10% (*see* **Note 1**).

2.4. Lyophilized Human Serum Control for d-ROMs and OXY Calibration, Diacron, Grosseto, Italy (see Note 2)

3. Methods

3.1. User-defined Chemistries on the SYNCHRON, CX 9 PRO Platform

The routine automated clinical chemistry analyzer, SYNCHRON CX 9 PRO permits the creation of up to 100 user-defined chemistries by defining a set of parameters that fully characterize the analyte of interest such as test name, reaction type, wavelength, mathematical model, calibrator number, volume, and time at which reagent and sample will be pipetted into the reaction cuvette, i.e., reaction read time. Once determined, parameters are stored in memory according to the test name designed in the setup.

The reagents may be placed in a cartridge with separated compartments, which can be loaded on the system manually.

3.2. Automated Colorimetric Determination of Reactive Oxygen Metabolites Utilizing the SYNCHRON, CX 9 PRO Platform: The d-ROMs Test

This assay is based on the capacity of transition metals to catalyze peroxides in the sample and form alkoxy and peroxy radicals, which then react with an amine, leading to the formation of colored species that can spectrophotometrically detected. The results are expressed as arbitrary units (AU). We previously evaluated the analytical and clinical performance of the automated version of this test in a population of healthy subjects and coronary artery disease patient cohorts *(6, 7, 8)*.

The automated procedure developed using the routine automated clinical chemistry analyzer SYNCHRON, CX 9 PRO uses a reaction mixture obtained by adding 3 μL of N,N-diethyl-para-phenylendiamine (DEPPD) to 300 μL of acetate buffer solution (pH 4.8) for each sample, both provided by the manufacturer *(6)* (*see* **Notes 3** and **4**). Then, 3 μL of serum was added to the reaction mixture and samples are subjected to repeated spectrophotometric readings at 520 nm taken over a total of 18 successive analytical spin cycles (*see* **Notes 5** and **6**). The final concentration is automatically calculated from the mean slope (the rate of change in absorbance) (*see* **Note 7**).

Each laboratory should determine its own reference value; however, expected values are generally lower than 320 AU, while higher values should indicate progressively elevated oxidative stress levels.

As we previously reported, the limit of quantification for the d-ROM test, defined as the concentration corresponding to the mean value of 10 determinations of the zero calibrator +2 SD was determined to be 40 AU *(6)*.

The standard calibration curves for the d-ROMs test resulted linear up to 475 AU, with correlation coefficients greater than 0.99. Samples with different concentrations were evaluated to estimate within- and between-run coefficients of variation (CVs). Within-run imprecision was assessed by evaluating results of 10 samples analyzed on the same day for three different concentration levels. Between-run imprecision was evaluated over 10 days by analysis of d-ROM on the same serum sample for three different concentration levels. The between-run imprecision (CV) of the methods ranged between 2.9% and 4.7%, and the within-run imprecision resulted, 1.4–4% *(6)*.

Recovery obtained after addition of different amounts of known concentration to six different aliquots of two serum samples ranged between 97% and 105% *(6)*.

To evaluate the effect of storage, measurements were performed on 10 fresh serum samples and repeated after 24 h on aliquots of the same samples stored at 4°C without observing any significant difference. Moreover, samples showed a very good degree of stability over a 3-month storage period at −80°C *(6, 9, 10)*.

3.3. Automated Colorimetric Determination of Antioxidant Barrier to Hypochlorite-induced Oxidation Utilizing the SYNCHRON, CX 9 PRO Platform: The OXY Adsorbent Test

This assay is based on the ability of endogenous antioxidant capacity to oppose the oxidant action of added hypochlorous acid *(11)*.

All standards and samples should be diluted 1:100 with distilled water before the analysis (*see* **Notes 5** and **6**). During the automated procedure, sample (3 µl) to be tested undergoes the oxidant action of a know-title highly concentrated HClO solution (Rl reagent included in the kit, 300 µl). After 10-min incubation, residual HClO undergoes the reaction with an alkyl-substituted aromatic amine (A-NH2, solubilized in a chromogenic mixture, the R2 reagent included in the kit, 6 µl), previously diluted with distilled water (1:2) leading to the formation of pink-colored species that can be spectrophotometrically detected (520 nm) (*see* **Notes 3** and **4**).

The concentration of the colored complex is directly proportional to the concentration of HClO and indirectly proportional to the antioxidant capacity (*see* **Note** 7). The calculation of data must be done off-line, using the absorbance produced by photometric reading immediately after the addition of R2, according to the following formula:

Sample concentration equal to:

$$\frac{(\text{Blank abs} - \text{Sample abs})}{(\text{Blank abs} - \text{Standard abs})} \times \text{Standard concentration}$$

where abs = absorbance

The results are expressed as μmol of HClO consumed by 1 ml of sample (μmol HClO/ml).

Each laboratory should determine its own reference value; however, expected values are generally higher than 350 μmol HClO/ml of sample, while lower values should indicate a reduced antioxidant capacity.

The standard calibration curves for the OXY test were linear up to 440 μmol HClO/ml, with correlation coefficients greater than 0.99.

Samples with different concentrations were evaluated to estimate within- and between-run coefficients of variation (CVs) (**Table 3.1**). Within-run imprecision was assessed by evaluating results of 10 samples analyzed on the same day for three different concentration levels. Between-run imprecision was evaluated over 10 days by analysis of OXY on the same serum sample for three different concentration levels. The between-run imprecision (CV) of the methods ranged between 5.4% and 9.3%,

Table 3.1
Imprecision for the OXY test in samples of different concentration

	Sample 1	Sample 2	Sample 3
Within-run imprecision			
Mean, μmol HClO/ml	418	397	358
SD, μmol HClO/ml	10	8	9
Minimum, μmol HClO/ml	407	380	347
Maximum, μmol HClO/ml	443	408	371
CV, %	2.5	2.1	2.6
Between-run imprecision			
Mean, μmol HClO/ml	442	287	224
SD, μmol HClO/ml	24	23	20
Minimum, μmol HClO/ml	398	249	200
Maximum, μmol HClO/ml	482	316	256
CV, %	5.4	8.2	9.3

and the within-run imprecision was 2.1–2.6%. Mean, standard deviation, minimum and maximum values, together with CVs, are calculated and reported in **Table 3.1**.

To evaluate the effect of storage, measurements were performed on 10 fresh serum samples and repeated after 24 h on aliquots of the same samples stored at 4°C without observing significant statistical difference. However, sample concentration progressively decreases over time, with a significant reduction for long storage period, and a significant mean loss of 15–20% after 2 years at −80°C.

3.4. Calculation of the Oxidative-INDEX

The imbalance between the oxidant and antioxidant systems can be estimated using the Oxidative-INDEX, which is calculated as follows: in order to consider parameters with different measurement units and variability, the standardized value of the oxidant and antioxidant indices are calculated according to the usual formula ($sv_{var} = (v_{var} - m_{var})/ds_{var}$) where sv_{var} represents the standard value of a certain parameter, v_{var} its original value, and m_{var} and ds_{var}, the mean and standard deviation of the parameter. The Oxidative-INDEX is then calculated subtracting the OXY standardized variable from the ROMs standardized variable (*see* **Note 8**).

The Oxidative-INDEX was calculated in 101 healthy volunteers (32 men and 69 women, mean age: 52 ± 16 and 51 ± 14 years, 2 and 6 current smokers, respectively), all free from acute or chronic inflammatory disease, immunological disease, and history or evidence of malignancy. None of the subjects was receiving vitamin and/or antioxidant therapies.

A significant positive correlation was observed between age and the Oxidative-INDEX in the whole population ($r = 0.25$, $p < 0.05$; **Fig. 3.1**). A significant difference was observed

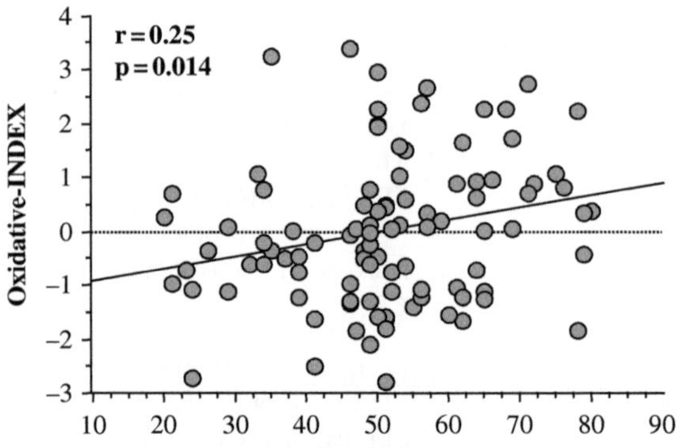

Fig. 3.1. Relationship between age and levels of oxidative stress in the whole population.

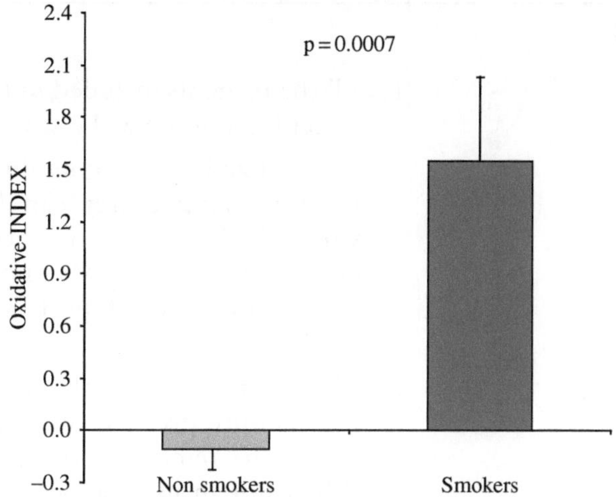

Fig. 3.2. Levels of oxidative stress according to smoking habit.

between oxidative stress levels in women and men (0.22 ± 0.16 *versus* -0.40 ± 0.22, respectively, $p < 0.05$). Current smokers showed significantly higher levels of oxidative stress with respect to nonsmokers (1.55 ± 0.5 vs -0.11 ± 0.13, $p < 0.001$) (**Fig. 3.2**).

Subjects were then grouped into two subsets according to 75th percentile of the Oxidative-INDEX parameter, which corresponds to a score of 0.8. Logistic analysis performed using the statistical package Statview, version 5.0.1 (SAS Institute, Abacus Concept, Inc., Berkeley, CA, USA), identified aging and smoking habit as independent factors for elevated oxidative stress in our population (**Table 3.2**).

Table 3.2
Multivariate model predicting Oxidative-INDEX

	OR	95% CI	*p*
Age > 62 years (75th percentile)	4.9	1.8–13.5	0.0021
Smoking habit	5.6	1.1–28	0.03
Sex	1.6	0.8–4.3	Ns

Odds ratio derived from logistic regression analysis including all the variables listed in the table.

4. Notes

1. All the reagents included in the kits must be kept away from direct light and may be stored at 4°C.
2. Lyophilized serum containing a concentration provided by the manufacturer can be reconstituted and diluted with distilled H_2O to construct standard calibration curves both for ROMs and OXY. Once the serum solution is prepared, aliquots should be made and stored at −80°C.
3. All reagents and samples should be allowed to reach room temperature prior to use. However, do not allow reagent bottles to remain at room temperature for a long time. When not in use, or once the necessary reagent amounts have been put into the cartridge compartments, replace the bottles at 4°C.
4. Do not leave the reagent bottles open. Replace the caps as soon the desired volume is removed.
5. To avoid any possible influence due to anticoagulant addition, serum was considered preferable for the measurements.
6. Proper procedures adopted during blood collection and sample storage are essential for consistent and accurate results. For optimum conditions, samples, unless immediately dosed, must be kept on ice soon after collection and rapidly separated by centrifugation at 4°C. Then hemolysis-free serum samples should be frozen and preferably maintained at −80°C.
7. Each sample and calibrator should be assayed at least in duplicate.
8. The estimation of the oxidative stress load, due to an increased production of oxygen-derived species and/or an exhaustion of the stores of antioxidants, can be better done by quantifying the pro- and the anti-oxidant counterparts in most cases. However, this assumption is not true in all situations. As an example, total antioxidant capacity appears increased in patients with chronic renal failure, principally related to high serum urate *(12)*. This finding may be not indicative of the real oxidative stress status, because they also present high malondialdehyde values together with a marked fall in ascorbate levels *(12)*. Consequently, the sensibility of the researcher must drive the choice of appropriate tests and parameters to be used in each specific experimental or clinical setting.

Acknowledgments

The author wants to thank Mr Claudio Boni for his invaluable assistance, cooperation and support in the laboratory analyses.

References

1. Griendling KK, FitzGerald GA (2003) Oxidative stress and cardiovascular injury: Part I: basic mechanisms and in vivo monitoring of ROS. *Circulation* **108**: 1912–1916.
2. Griendling KK, FitzGerald GA (2003) Oxidative stress and cardiovascular injury: Part II: animal and human studies. *Circulation* **108**:2034–2040.
3. Madamanchi NR, Vendrov A, Runge MS (2005) Oxidative stress and vascular disease. *Arterioscler Thromb Vasc Biol* **25**:29–38.
4. Ridker PM, Brown NJ, Vaughan DE, Harrison DG, Mehta JL (2004) Established and emerging plasma biomarkers in the prediction of first atherothrombotic events. *Circulation* **109**:IV6–IV19.
5. Vassalle C, Andreassi MG (2004) 8-iso-prostaglandin F2alpha as a risk marker in patients with coronary heart disease. *Circulation* **110**:49–50.
6. Vassalle C, Boni C, Di Cecco P, Ndreu R, Zucchelli GC (2006) Automation and validation of a fast method for the assessment of in vivo oxidative stress levels. *Clin Chem Lab Med* **44**:1372–1375.
7. Vassalle C, Landi P, Boni C, Zucchelli G (2007) Oxidative stress evaluated using an automated method for hydroperoxide estimation in patients with coronary artery disease. *Clin Chem Lab Med* **45**:367–371.
8. Vassalle C, Maffei S, Boni C, Zucchelli GC (2008). Gender-related differences in oxidative stress levels among elderly patients with coronary artery disease. *Fertil Steril* **89**: 608–613.
9. Iamele L, Fiocchi R, Vernocchi A (2002) Evaluation of an automated spectrophotometric assay for reactive oxygen metabolites in serum. *Clin Chem Lab Med* **40**:673–676.
10. Cavalleri A, Colombo C, Venturelli E, Miceli R, Mariani L, Cornelli U, Pala V, Berrino F, Secreto G (2004) Evaluation of reactive oxygen metabolites in frozen serum samples. Effect of storage and repeated thawing. *Int J Biol Markers* **19**:250–253.
11. Vassalle C, Masini S, Carpeggiani C, L'Abbate A, Boni C, Zucchelli GC (2004) In vivo total antioxidant capacity: comparison of two different analytical methods. *Clin Chem Lab Med* **42**:84–89.
12. Jackson P, Loughrey CM, Lightbody JH, McNamee PT, Young IS (1995) Effect of hemodialysis on total antioxidant capacity and serum antioxidants in patients with chronic renal failure. *Clin Chem* **41**: 1135–1138.

Chapter 4

Correlative Transmission Microscopy: Cytochemical Localization and Immunocytochemical Localization in Studies of Oxidative and Nitrosative Stress

E. Ann Ellis

Abstract

Microscopy studies of oxidative stress have often consisted of only immunocytochemical localization or only cytochemical localization studies. However, correlative studies on the same sections provide more useful data in interpreting oxidative and nitrosative stress. Cellular sites of superoxide and hydrogen peroxide production can be identified using cerium-based cytochemical localization of NADH oxidase enzymatic activity, while areas of nitrosative stress can be identified by immunocytochemical localization of nitrotyrosine in the same sections of tissue.

Key words: cerium hydroperoxide, correlative imaging, cytochemistry, immunocytochemistry, NADH oxidase, nitrotyrosine.

1. Introduction

In diabetic hyperglycemia, there is increased flux of glucose via the sorbitol pathway resulting in increased cytosolic NADH/ NAD^+ and increased superoxide ($O_2^{\bullet-}$) and hydrogen peroxide (H_2O_2) described as hyperglycemia induced pseudohypoxia *(1)*. Nitric oxide ($^\bullet$NO) is a highly labile molecule and its effects depend on the microenvironment. One mechanism for the role of $^\bullet$NO in complications of diabetes is reaction with $O_2^{\bullet-}$ to form the toxic species, peroxynitrite ($ONOO^-$). Peroxynitrite oxidizes sulfhydryls and modifies tyrosines to form 3-nitrotyrosine, a useful product for localizing $^\bullet$NO-mediated toxicity *(2)*.

From: *Methods in Molecular Biology, vol. 477: Advanced Protocols in Oxidative Stress I,*
Edited by: D. Armstrong, DOI: 10.1007/978-1-60327-517-0_4, © Humana Press, New York, NY

Localization of sites of production of reactive oxygen and nitrogen species plays an important role in understanding vascular complications in diabetes and cardiovascular diseases. Many studies have used western blots or immunohistochemical localization procedures of the end products of the oxidative or nitrosative stress. Correlative studies consist of examining the same specimen with different techniques and relating the different results in the final interpretation. Correlative studies of oxidative and nitrosative stress have provided better understanding of the source of vascular complications in diabetic retinopathy *(3, 4)*. NADH oxidase is the major source of $O_2^{\bullet -}$ and H_2O_2 production in vascular tissue (see reaction below) *(6, 7)* while colloidal gold-based immunocytochemical localization of nitrotyrosine identifies sites of peroxynitrite production *(2)*.

$$\mathrm{NADH} + 2\mathrm{O}_2 \xrightarrow{\text{NADH OXIDASE}} 2\mathrm{O}_2^{\bullet -} + \mathrm{NAD}^+ + \mathrm{H}^+$$

$$2\mathrm{O}_2^{\bullet -} + 2\mathrm{H}^+ \xrightarrow{\text{SUPEROXIDE DISMUTASE}} \mathrm{H}_2\mathrm{O}_2 + \mathrm{O}_2$$

$$\mathrm{H}_2\mathrm{O}_2 + \mathrm{CeCl}_3 \rightarrow \mathrm{Ce(OH)}_2\mathrm{OOH}$$

2. Materials

2.1. Equipment

1. Fume hood (minimum flow rate of 100 ft/min).
2. Shaking water bath at 37°C.
3. Transmission electron microscope (JEOL 1200EX).
4. Ultramicrotome (Leica Reichert Ultracut E).

2.2. Reagents

Unless otherwise noted, reagents can be purchased from Sigma Chemical Co. (St. Louis, MO, USA) and/or Electron Microscopy Sciences (Hatfield, PA, USA).

1. Acrolein (fixative) (Electron Microscopy Sciences cat. # 10102).
2. 3-Amino-1, 2, 4-triazole (inhibitor of catalase) (Sigma cat. # A8056).
3. Cerium chloride (chromagen) (Sigma cat. # C8016).
4. Cold water fish gelatin (blocker component) (Sigma cat. # G7765)
5. DER 736 (Electron Microscopy Sciences cat. # 13000).
6. Dimethyl sulfoxide (DMSO) (Sigma cat. # D8779).

7. Donkey anti-rabbit IgG secondary antibody labeled with 12 nm colloidal gold (Jackson ImmunoResearch, West Grove, PA cat. # 711–205–152).
8. ERL 4221 epoxy resin (Electron Microscopy Sciences cat. #15004).
9. Glycine (Sigma cat. #G7126).
10. β-nicotinamide adenine nucleotide, reduced form (β-NADH) (substrate for NADH oxidase) (Sigma cat. # N6005).
11. 200-mesh nickel grids (Electron Microscopy Sciences cat. # HD200-Ni).
13. Nonenyl Succinic Anhydride (NSA) (Electron Microscopy Sciences cat. # 19050).
14. Osmium tetroxide (Electron Microscopy Sciences cat. # 19150).
13. Periodic acid (oxidizing agent for immunolabel) (Sigma cat. # P5463)
15. Phosphate buffered saline (PBS), pH 7.4; recipe given in text.
16. Polyclonal rabbit anti-nitrotyrosine antibodies (primary antibody) (Molecular Probes, Eugene, OR cat. # A-21285)
17. Potassium metabisulfite (neutralizing agent for acrolein) (Sigma cat. # P2522)
18. Quetol 651 epoxy resin (Electron Microscopy Sciences cat. #20440).
19. Sodium cacodylate buffer, pH 7.4 (Electron Microscopy Sciences, cat. # 11652).
20. Tris buffered saline (TBS), pH 8.2, recipe given in text.
21. Tris base (buffer component) (Sigma cat. # T6066).
22. Tris Maleate (buffer component) (Sigma cat. # M0375).
23. Triton X-100 (Sigma cat. # T9284).

3. Methods

3.1. Tissue Procurement and Fixation

1. Immediately after the animal is sacrificed, dissect out the desired tissue and fix the specimen in cold 5% (v/v) acrolein in 0.1 M sodium cacodylate buffer for 1 h (*see* **Note 1**). Phosphate buffers should not be used in any of the stages, including fixation and buffer washes, of cerium-based localization procedures (*see* **Note 2**).
2. Wash the specimen 4 × 15 min in cold 0.1 M sodium cacodylate buffer plus 1% (v/v) DMSO and 5% sucrose

(w/v). The initial buffer wash contains sucrose and DMSO (1% v/v), which aids in rapid removal of the fixative and protects enzyme and antigenic activity. Specimens can be kept in cold buffer wash (refrigerator temperature, 4°C) overnight or up to several weeks.

3. Add 0.1 M glycine to the last two buffer washes as the specimen is brought to room temperature just prior to the NADH oxidase localization procedure. Glycine in the final buffer washes helps to remove any unbound aldehydes from tissue.

3.2. Localization of NADH Oxidase

1. Preincubation steps (done at 37°C in a shaking water bath for 30 min) are critical to successful localizations. Buffers for the reaction media can be made the day before; however, all incubation mixtures should be prepared fresh and filtered immediately before use through a 0.45-μm filter. Buffers for all incubation steps should be kept at room temperature. Preincubations with the chromagen ($CeCl_3$) and appropriate inhibitors are essential to insure adequate penetration of these reagents into subcellular sites of the enzyme. Cerium has slow penetration into cells and tissues and penetration can be enhanced by addition of 0.0001–0.0002% Triton X-100 to the reaction medium *(8)*.
2. The inhibitor of catalase and glutathione peroxidase, 3-amino-1, 2, 4-triazole (10 mM), inhibits the removal of H_2O_2 generated by the NADH oxidase reaction and **must** be included in the preincubation medium as well as in the complete reaction medium [*see* **Note 3**]. Controls for specificity of the reaction are initiated in separate vials during the preincubation step. Two controls for specificity of the reaction include samples in which:
   a. substrate is omitted;
   b. a specific inhibitor of NADH oxidase, diphenyleneiodonium (DPI), is included in one control sample in both the preincubation and complete reaction steps.
3. The second stage in the localization procedure involves inclusion of the substrate, β-NADH, in a new batch of incubation medium that contains all the components that were used in the preincubation step. Incubation is done at 37°C in a shaking water bath for 30 min to 1 h. For optimal results, the complete reaction medium is changed after 30 min and the incubation is continued for an additional 30 min.
4. Reactions are stopped by placing the vials in an ice bath and washing immediately with **ice cold** buffer (the same Tris-maleate buffer that was used for the preincubation and incubation media). Specimens are then washed with a quick rinse in **ice cold** 0.1 M sodium cacodylate buffer, pH 7.4. Tissues can then be postfixed overnight in the cold in 1%

OsO_4 (v/v) followed by dehydration, infiltration, and embedding in epoxy resin for TEM.

3.3.1. Procedure

1. Incubate tissue for 30 min in a shaking water bath at 37°C in preincubation medium consisting of:

    0.1 M Tris-maleate buffer, pH 7.5 10 ml

    2.0 mM cerium chloride 7.45 mg

    10 mM aminotriazole 8.41 mg

    0.0002% Triton X-100 1–2 drops of 1% aqueous solution of Triton X-100 (v/v)

Filter the reaction medium through a 0.45-μm filter before use.

2. Incubate tissue in complete reaction medium in shaking water bath at 37°C for 30 min. Complete reaction medium consists of pre-incubation medium plus 5.69 mg NADH per 10 ml of medium. After 30 min, replace the medium with freshly prepared medium and continue incubation for 30 min.
3. Stop the reaction by plunging the vials into an ice bath. Rinse once with ice cold Tris-maleate buffer. Rinse once with ice cold 0.1 M sodium cacodylate buffer, pH 7.4.
4. Postfix in 1% OsO_4 (v/v) in 0.1 M sodium cacodylate buffer plus 5% sucrose (w/v) overnight in the refrigerator [*see* **Note 4**].
5. Dehydrate in an ethanol series in 10% (v/v) steps for 10 min/step starting at 10% and ending with two changes of 100% ethanol.
6. Replace the ethanol with propylene oxide and infiltrate in 25–30% steps (v/v) with the following epoxy resin *(9)* over a period of 2–3 days. Better infiltration is achieved if samples in epoxy resin are gently agitated on a shaker or blood rocker.

    Quetol 651 – 1.40 g

    ERL 4221 – 2.22 g

    DER 736 – 1.43 g

    NSA – 6.38 g

    BDMA – 0.20 ml

7. Embed in freshly prepared resin mixture and polymerize overnight at 55–60°C.

3.3. Preparation of Grids for Immunolocalization

1. Cut gold (80–100 nm) sections using either a diamond knife or a glass knife and pick up sections on the dull side of 200-mesh nickel grids cleaned with 1N HCl followed by several rinses in ethanol. Do **not** use copper grids; copper grids will be damaged by the oxidation step in the immunolocalization procedure.
2. Allow the sections to dry on the grids on a 55–60°C slide warmer or hot plate.

3.4. Immunolocalization of Nitrotyrosine

1. Prepare phosphate buffered saline (PBS) solution as follows:
   a. Dissolve 2.175 g of NaCl, 0.068 g of KH_2PO_4, and 0.284 g of Na_2HPO_4 in 200 ml of deionized water. Check the pH and adjust to 7.4 using either 1N HCl or 1N NaOH. Bring the final volume to 250 ml with deionized water.
2. Prepare Tris buffered saline (TBS) as follows:
   a. Dissolve 2.175 g of NaCl and 2.518 g of Tris base in 200 ml of deionized water. Adjust the pH to 8.2 with concentrated HCl and bring to a final volume of 250 ml with deionized water.
3. Oxidize sections on 100 μl drops of freshly prepared 1% aqueous periodic acid (w/v) with section side down on the drops for 30 min at room temperature in a spot plate (*see* **Note 5**).
4. Rinse grids at room temperature, section side down, 2 × 3 min by floating on deionized water in spot plate. Set the plate on a magnetic stirrer and turn on to rotate nickel grids slowly.
5. Rinse grids 3 × 3 min on PBS on magnetic stir plate as described above.
6. Place grids on 100 μl drops of PBS blocker [PBS plus 4% cold water fish gelatin (w/v)] for 30 min at room temperature.
7. Incubate grids in spot plate on 20 μl drops of primary antibody diluted 1:100 with PBS blocker. Place the spot plate in a damp chamber (sealed Rubbermaid food container with a small, open beaker of deionized water) and place the damp chamber in the refrigerator overnight at 5°C.
8. Rinse grids at room temperature, 2 × 3 min on PBS followed by 3 × 3 min on TBS.
9. Place grids on 100 μl drops of TBS blocker (TBS plus 4% cold water fish gelatin) for 30 min at room temperature.
10. Incubate grids on 20 μl drops of donkey anti-rabbit IgG labeled with 12 nm colloidal gold diluted 1:30 with TBS blocker at room temperature for 2 h.
11. Rinxe 3 × 3 min with TBS at room temperature.
12. Rinse 3 × 3 min. with deionized water.
13. Dry grids and examine and photograph in TEM.

3.5. Results

The electron-dense cerium perhydroxide [$Ce(OH)_2OOH$] precipitates of NADH oxidase production of H_2O_2 and the colloidal gold particles used in nitrotyrosine immunolocalization can be identified simultaneously in the same section (**Fig. 4.1**); this makes correlative studies very useful in studying oxidative and nitrosative stresses in cardiovascular pathology.

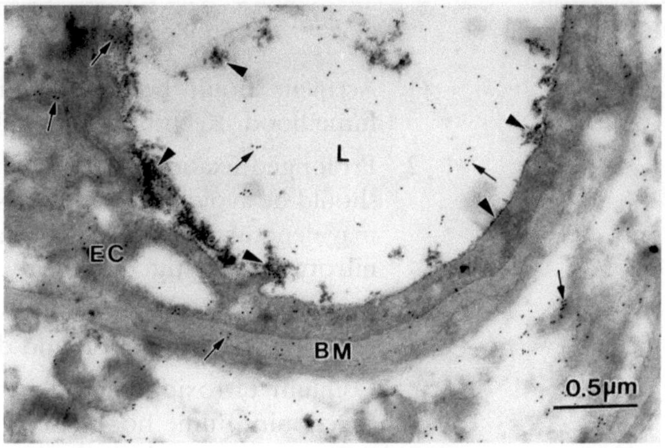

Fig. 4.1. Cytochemical localization of NADH oxidase production of $O_2^{\bullet-}$ and H_2O_2 (arrowheads) and immunocytochemical localization of nitrotyrosine (arrows) in a retinal vessel from a diabetic BBZ/Wor rat. Nitrotyrosine localizes in areas where there is abundant production of $O_2^{\bullet-}$ and there is reduced availability of •NO due to formation of $ONOO^-$. BM, basement membrane; EC, endothelial cell; L, vessel lumen. 30,000×.

3.6. Quantitation

3.6.1. Semiquantitation of NADH Oxidase

Cerium perhydroxide precipitates show actual sites of $O_2^{\bullet-}$ and H_2O_2 localization as a result of NADH oxidase enzyme activity. The reaction product lends itself to a number of quantitative methods. Briggs et al. *(10)* used a semiquantitative method to determine amounts of cerium perhydroxide in chronic granulomatous PMNs vs normal control cells. Cells that contained cerium perhydroxide were scored positive (+) and the results were expressed as a percentage of positive cells divided by the total number of cells examined. The most practical method for vascular tissue *(11)* is to examine blood vessels at a magnification of 10,000× and score either positive (+) (cerium perhydroxide is present) in any of the following sites: vessel lumen, endothelial cell cytoplasm, plasmalemma, basement membrane] or negative (−) (cerium perhydroxide is absent). Express results as the percentage of positive vessels relative to the total number of vessels. Ten or more vessels or cells should be examined to generate data for statistical analysis.

3.6.2. Morphometric Analysis of Nitrotyrosine

Semiquantitation of nitrotyrosine immunoreactivity in and around blood vessels or in individual cells is done by examining negatives at a magnification of 10,000×. Count the number of colloidal gold particles and express the counts per total area of the structure or counts per μm^2. A minimum of ten negatives or micrographs is usually done to generate data for statistical analysis.

4. Notes

1. Acrolein should be handled only in a properly functioning fume hood. Keep bisulfite available to neutralize acrolein.
2. Prolonged fixation is not recommended and glutarladehyde should be avoided since it cross-links tissue components and may denature the NADH oxidase enzymatic activity and the nitrotyrosine antigenic activity.
3. Aminotriazole is toxic to thyroid function. Use gloves when handling this compound and do not inhale the powder.
4. Osmium tetroxide should be handled only in a properly functioning fume hood. Keep corn oil available to neutralize osmium tetroxide.
5. Do not allow grids to dry at any point during the immunolabeling procedure; this results in nonspecific binding of the primary and/or secondary antibodies.

References

1. Williamson, J. R., Chang, K., Frangos, M., Hasan, K. S., Ido, I., Kawamura, T., Nyengaard, J. R., Van den Enden, M., Kilo, C., and Tilton, R. G. (1993) Hyperglycemic pseudohypoxia and diabetic complications. *Diabetes* **42**, 801–813.
2. Beckman, J. S., Ye, Y. Z., Anderson, P., Chen, J., Accavetti, M. A., Tarpey, M. M., and White, C. R. (1994) Extensive nitration of protein tyrosines in human atherosclerosis detected by immunohistochemistry. *Biol. Chem. Hoppe-Seyler* **375**, 81–88.
3. Ellis, E. A., Sengupta, N., Caballero, S., Guthrie, S. M., Mames, R. N., and Grant, M. B. (2005) Nitric oxide synthases modulate progenitor and resident endothelial cell behavior in galactosemia. *Antioxid. Redox Signal.* 7, 1413–1422.
4. Ellis, E. A., Guberski, D. L., Hutson, B., and Grant, M. B. (2002) Time course of NADH oxidase, inducible nitric oxide synthase and peroxynitrite in diabetic retinopathy in the BBZ/Wor rat. *Nitric Oxide* **6**, 295–304.
5. Cave, A. C., Brewer, A. C., Narayanapanicker, A., Rasy, R., Grieve, D. J., Walker, S., and Shah, A. M. (2006) NADPH oxidases in cardiovascular health and disease. *Antioxid. Redox Signal.* **8**, 691–728.
6. Briggs, R. T., Karnovsky, M. L., and Karnovsky, M. J. (1975) Localization of NADH oxidase on the surface of human polymorphonuclear leukocytes by a new cytochemical method. *J. Cell Biol.* **67**, 566–586.
7. Ellis, E. A., and Grant, M. B. (2002) Cytochemical localization of H_2O_2 in biological tissues, in *Oxidants and Antioxidants: Ultrastructure and Molecular Biology Protocols* (Armstrong, D., ed.) Humana Press, Totowa, New Jersey, pp. 3–12.
8. Robinson, J. M. (1985) Improved localization of intracellular sites of phosphatases using cerium and cell permeabilization. *J. Histochem. Cytochem.* **33**, 749–754.
9. Ellis, E. A. (2006) Solutions to the problem of substitution of ERL 4221 for vinyl cyclohexene dioxide in Spurr low viscosity embedding formulations. *Microscopy Today* **14**, 32–33.
10. Briggs, R. T., Karnovsky, M. L., and Karnovsky, M. J. (1977) Hydrogen peroxide in chronic granulomatous disease: a cytochemical study of reduced pyridine nucleotide oxidases. *J. Clin. Invest.* **59**, 1088–1098.
11. Ellis, E. A., Grant, M. B., Murray, F. T., Wachowski, M. B., Guberski, D. L., Kubalis, P. S., and Lutty, G. A. (1998) Increased NADH oxidase activity in the retina of the BBZ/Wor diabetic rat. *Free Rad. Biol. Med.* **24**, 111–120.

Chapter 5

Detection of Specifically Oxidized Apolipoproteins in Oxidized HDL

Xiao Suo Wang and Roland Stocker

Abstract

Atherosclerosis is associated with dysfunctional HDL, and oxidation of HDL is thought to give rise to HDL becoming dysfunctional. Lipoprotein oxidation represents a complex series of processes that can be assessed by various methods. In general, oxidation mediated by 1-electron or radical oxidants gives rise to lipid hydroperoxides (LOOHs) as the primary product. These LOOHs may then undergo further reactions giving rise to secondary lipid oxidation products and/or oxidation of lipoprotein-associated proteins. Thus, LOOHs specifically oxidize Met residues of apolipoprotein (apo) A-I and A-II (the major proteins of HDL) to MetO. Here we describe an HPLC-based method to detect oxidized HDL containing specifically oxidized forms of apoA-I and apoA-II. This method may be useful to assess the early stages of HDL oxidation in biological samples.

Key words: Oxidized apolipoproteins, oxidized HDL, lipid peroxidation, LOOH, MetO, antioxidants, ultracentrifugation, HPLC.

1. Introduction

Atherosclerosis is characterized by increased oxidative damage to lipids and proteins in the affected arterial wall, and this damage appears to be comparable to all classes of lipoproteins *(1)*. While oxidized low-density lipoprotein (LDL) has received much attention *(2)*, there is increasing evidence that atherosclerosis is also associated with dysfunctional high-density lipoprotein (HDL), and that oxidation of HDL-associated proteins (apolipoproteins) contributes to the lipoprotein becoming dysfunctional.

From: *Methods in Molecular Biology, vol. 477: Advanced Protocols in Oxidative Stress I,*
Edited by: D. Armstrong, DOI: 10.1007/978-1-60327-517-0_5, © Humana Press, New York, NY

In vitro experiments show that lipids in HDL become oxidized before those in LDL when isolated plasma is exposed to peroxyl radicals, and HDL is the major carrier of endogenous lipid hydroperoxides in human plasma *(3)*. It has also been reported that lipids in HDL and LDL, isolated from human atherosclerotic lesions, are oxidized to a comparable extent *(4)*. Furthermore, apolipoprotein A-I (apoA-I), the major protein associated with HDL appears to be a selective target in the arterial wall for oxidation mediated by myeloperoxidase and its principal oxidant hypochlorous acid *(5, 6)*. Such oxidation is associated with loss of the ability of apoA-I to stimulate efflux of cellular cholesterol mediated by ABCA1 *(5, 6, 7, 8, 9)*, although other forms of oxidation have been reported to have no effect on *(10)*, or even enhance, the ability of apoA-I to promote efflux of cellular cholesterol *in vitro (11)*. Oxidized HDL formed in the arterial wall may be able to re-enter the circulation, as HDL is small compared to LDL. Indeed, there is evidence that HDL-associated, oxidized apoA-I is present in circulation *(5, 12, 13)*.

As implied above, oxidation of lipoproteins is a complex process, the outcome of which depends on several factors, including the oxidant involved and the extent to which the oxidation process proceeds. In general, oxidation mediated by 1-electron (or radical) oxidants initially results in lipid peroxidation, defined as the formation of lipid hydroperoxides (LOOHs), whereas 2-electron oxidants (such as hypochlorous acid) initially react with apolipoproteins associated with lipoproteins *(1)*. LOOHs may degrade and/or become further oxidized to secondary lipid oxidation products, such as F_2-isoprostanes and fragmented lipid containing reactive groups, and these secondary processes commonly lead to protein oxidation. In addition, LOOHs may be reduced by apolipoprotein-associated amino acid residues, particularly Met (see below). Furthermore, primary protein oxidation products, such as those generated during lipoprotein oxidation mediated by hypochlorous acid, can also degrade giving rise to radical species that themselves then may initiate lipoprotein lipid peroxidation *(14)*. Given this complexity, it is not surprising that there is considerable diversity in the methods used to evaluate lipoprotein oxidation.

The present protocol provides an HPLC-based method to detect specifically oxidized forms of apolipoproteins generated during the initial stages of HDL oxidation, *i.e.*, when the lipoprotein still contains α-tocopherol. The oxidative modification detected is limited to methionine sulfoxide (MetO) contained in apoA-I and apoA-II. In the absence of cysteine residues, methionines in apoA-I are the major reducing amino acids that react with LOOH, giving rise to the corresponding MetO and lipid hydroxides, respectively *(15, 16, 17, 18)*. ApoA-I contains three methionine residues at positions 86, 112, and 148, whereas apoA-II contains a single methionine residue at position 26. The HPLC-based method described here is specific for HDL containing apoA-I_{+16} (MetO^{86} or MetO^{112}), apoA-I_{+32} (MetO^{86} plus MetO^{112}), and apoA-II_{+16} (apoA-II dimmer containing a single MetO^{26}). This method may be useful to document mildly, radical-oxidized HDL in biological fluids and tissues.

2. Materials

2.1. Instruments

1. Centrifuge (Allegra™ 6R, Beckman, USA).
2. Ultracentrifuge (Beckman Optima TLX, USA) equipped with TLA 100.4 and/or TLA 100.2 rotors (Beckman, USA).
3. Shaking and temperature-controlled water bath (SB-24, EYELA, Tokyo, Japan).
4. Microplate reader (Titertek multiskan MCC/340, Labsystems, Finland).
5. Incubator (Axyos Technologies, Australia).
6. HPLC-UV (Waters alliance 2690 separation module equipped with Waters 2487 dual absorbance detector).
7. Electrospray ionization mass spectrometry (ESI-MS) equipped with API QStar Pulsar *i* hybrid tandem mass spectrometer (TOF) (Applied Biosystems, Foster City, CA, USA).
8. Speedvac (AES1010, Thermo Savant, USA).

2.2. Reagents

1. Phosphate-buffered saline, pH 7.4, 10 mM (PBS) prepared from tablets (Oxoid, England) of density (ρ) = 1.006 g/mL, containing 1 mM EDTA (*see* **Note 1**).
2. Potassium bromide (KBr, Sigma).
3. 2,2′-Azobis(2-amidinopropane) dihydrochloride (AAPH) (Wako Pure Chemical Ind., Osaka, Japan).
4. BCA protein reagent A (Pierce).
5. Bovine serum albumin (BSA) (Sigma).
6. 5% $CuSO_4$ (Sigma).
7. Milli-Q Ultrapure water (*see* **Note 2**).
8. HPLC-grade water and trifluoroacetic acid (TFA) (British Drug House, England) (*see* **Note 3**).
9. HPLC-grade formic acid (Merck, Germany)
10. HPLC-grade solvents including methanol and acetonitrile (Merck, Germany).

2.2.1. Isolation of Native HDL From Human Plasma

1. Heparin-coated vacutainer (Becton Dickinson, USA).
2. Centrifuge cooled to 4°C prior to use.
3. Quick seal tubes (11 × 32 mm or 13 × 51 mm, Beckman).
4. Heat sealer (Beckman).
5. Quick seal tubes rack (Beckman).
6. 1 and 3 mL Syringes (Terumo, USA).
7. Needles (18G 1½ without bevel and 21G 1½) (B-D, USA).
8. NAP™-10 column (GE Health care, Sweden).
9. Human plasma (1 mL–50 mL), prepared from freshly obtained heparinized blood collected from overnight fasted healthy subjects (*see* **Note 4**).

2.2.2. Selective Oxidation of Isolated HDL by AAPH

1. Water bath set at 37°C.
2. Eppendorf tubes.
3. AAPH, 200 mM prepared in Milli-Q water and stored at −20°C.
4. HDL in PBS isolated from fresh human plasma.

2.2.3. Measurement of Protein in HDL

1. 96-Well microtiter plates (Nunc, Denmark or SARSTEDT, USA).
2. Incubator set at 37°C.
3. BCA protein reagent A (Pierce) for HDL protein determination.
4. BSA standard (2 mg/mL, Sigma, P0834), or 1 mg/mL BSA standard prepared by dissolving BSA (Sigma, A6003) into Milli-Q water.
5. 10 mM PBS, pH 7.4 used to dilute HDL samples.

2.2.4. HPLC Detection of Native and Oxidized Apolipoproteins in HDL

1. HPLC-UV system (Assumed Waters).
2. RP C18 column (250 × 4.6 mm, 5 μm, Vydac).
3. Guard column (C18, 4.6 ID, 5 μm, Vydac).
4. HPLC vials (Waters).
5. Mobile phase A, containing 0.1% TFA in HPLC-grade water.
6. Mobile phase B, containing 0.1% TFA in acetonitrile (HPLC grade, Merck) (*see* **Note 5**).
7. 10% Methanol (HPLC-grade) in HPLC-grade or Milli-Q water.
8. 40% Acetonitrile (HPLC-grade) in HPLC-grade water.

2.2.5. Mass Spectrometry Confirmation on Specifically Oxidized Apolipoproteins

1. HPLC-UV system (Assumed Waters).
2. Semi-preparative HPLC column (10 × 250 mm, 5 μm, Vydac) connected with a guard column (C18, 5 μm, Vydac).
3. HPLC vials (Waters).
4. API QStar Pulsar *i* hybrid tandem mass spectrometer.
5. Speedvac set at vapornet cryopumping mode prior to use.
6. 0.1% Formic acid in water/acetonitrile (1:1, vol:vol).

2.2.6. Oxidized Apolipoproteins in Human Plasma

1. Ultracentrifuge (Optima TLX) with TLA100.2 rotor and heat sealer (Beckman).
2. Quick seal tubes (11 × 32 mm, Beckman).
3. HPLC-UV(Waters) equipped with a Guard column (C18, 5 μm, Vydac) followed with a RP-C18 column (4.6 × 250 nm, 5 μm, Vydac).
4. Mobile phase A and mobile phase B as described in **Section 2.2.4**.
5. NAP-10 column (GE Healthcare, Sweden).
6. Human plasma samples (for details see "Methods" section).
7. Cord blood samples collected post-partum after normal full-term pregnancy and uncomplicated delivery from subjects recruited consecutively within the study period.

3. Methods

Lipid peroxidation is initiated conveniently and in a controlled manner by thermo-labile azo-initiators (R–N = N–R) *(19)*, such as AAPH. AAPH is a water-soluble initiator that temperature-dependently generates aqueous alkylperoxyl radicals ($ROO^\bullet$) at known and constant rates (Reaction 1). Under aerobic conditions, and in the presence of lipoproteins that contain fatty acids with bisallylic hydrogen atoms (LH), $ROO^\bullet$ initiates lipid peroxidation by generating a carbon-centered lipid radical ($L^\bullet$) (Reaction 2), to which molecular oxygen adds, resulting lipid peroxyl radical ($LOO^\bullet$) (Reaction 3). Lipid peroxidation is then propagated by $LOO^\bullet$ abstracting a bisallylic hydrogen atom from another lipid molecule, resulting in the formation of LOOH (Reaction 4).

$$R{-}N{=}N{-}R \xrightarrow[-N_2]{\text{Heat}} 2R^\bullet \xrightarrow{2O_2} 2ROO^\bullet \qquad \text{(Reaction 1)}$$

$$ROO^\bullet + LH \longrightarrow ROOH + L^\bullet \qquad \text{(Reaction 2)}$$

$$L^\bullet + O_2 + \longrightarrow LOO^\bullet \qquad \text{(Reaction 3)}$$

$$LOO^\bullet + LH \longrightarrow LOOH + L^\bullet \qquad \text{(Reaction 4)}$$

In the case of HDL, most LOOHs are in the form of hydroperoxides of phospholipids and cholesterylesters. These LOOHs can then become reduced to the corresponding LOH via direct two-electron transfer from the sulfide of thioethers of oxidation-prone methionine residues of apolipoproteins (Reaction 5), resulting in specifically oxidized apoA-I and apoA-II[18]. Formation of these specifically oxidized apoA-I/apoA-II is assessed by immediate analysis of the oxidized HDL by HPLC (see below). Alternatively, oxidized HDL may be stored at −80°C for prolonged periods of time prior to HPLC analysis.

$$\text{LOOH} + \text{Met} \longrightarrow \text{LOH} + \text{MetO} \qquad \text{(Reaction 5)}$$

3.1. Isolation of Native HDL From Human Plasma

1. Centrifuge freshly obtained heparin blood at 2,500 rpm for 20 min at 4°C.
2. Carefully remove plasma to new collection tube and place on ice (*see* **Note 6**).
3. Turn on ultracentrifuge ~30 min prior to use, to equilibrate temperature.
4. Use 0.4 mL (for 11 × 32 mm or 1.9 mL quick seal tube) or 1.9 mL plasma (13 × 51 mm, 5.1 mL tube); if insufficient plasma is available, make up required volume with 10 mM PBS (ρ = 1.006 g/mL).
5. Using an analytical balance, weigh out and add 381.6 mg KBr per mL plasma, *i.e.*, 152.6 or 725.1 mg for 0.4 or 1.9 mL plasma, respectively.
6. Dissolve KBr by gently and slowly rolling the tubes placed on ice and avoiding formation of bubbles (that tend to denature lipoproteins).
7. Place 1.3 mL (for 1.9 mL tube) or 3 mL (5.1 mL tube) of ice-cold 10 mM PBS (ρ = 1.006 g/mL) into each Quick Seal centrifuge tube using a 18G1½ gauge blunt needle fitted to an appropriate syringe.
8. Carefully underlay the density-adjusted plasma (0.4 or 1.9 mL) into the appropriate tube, using an 18G1½ gauge blunt needle with a 1 mL syringe and avoiding formation of air bubbles (*see* **Note 7**).
9. Top up tube with 10 mM ice-cold PBS to neck of tube (~0.2 mL).
10. Ensure absence of bubbles in tube, then cap and heat-seal. Ensure tubes are not leaking. Throughout, avoid disturbing density layers and keep tubes on ice as much as possible.
11. Remove TLA 100.2 (1.9 mL tubes) or TLA 100.4 (5.1 mL tubes) rotor from fridge and load tubes carefully. Insert spacers as needed; finger tighten rotor lid and secure rotor by pressing rotor security button.

12. Centrifuge at 100,000 rpm at 15°C for 3 h. Wait until full speed is reached (*see* **Note 8**).
13. Following centrifugation, remove rotor and tubes immediately and carefully, avoiding disturbing the gradient.
14. Identify and mark HDL band (located above yellow colored layer of plasma proteins and below orange-yellow colored LDL band); place tubes carefully on ice.
15. Prior to HDL removal, release pressure from each tube by piercing a hole at the top using 21 1½ Gauge needle. Remove HDL (0.3–0.5 mL for small tube, 0.8–2.0 mL for large tube) by piercing (using a fresh needle) below HDL band and by carefully drawing material into syringe (avoiding uncontrolled suction) (*see* **Note 9**).
16. Remove needles with utmost care and discard immediately in sharp bin (without recapping) placed on lab bench.
17. Transfer HDL solution into plastic tubes and place on ice (*see* **Note 10**).
18. Equilibrate NAP-10 column with at least three column volumes of 10 mM PBS; allow PBS to completely enter gel bed, and then load HDL (up to 1.0 mL). If sample volume is <1.0 mL, wait until HDL has entered gel completely, then add PBS to make up 1.0 mL. Allow PBS to enter gel completely.
19. Place new collection tube (or HPLC vial) underneath column, and elute HDL with 1.5 mL PBS.
20. Analyze HDL collected immediately by HPLC; if this is not possible, overlay HDL sample collected with argon and store at 4°C (if analyzed within 24 h) or freeze at −80°C.

3.2. Selective Oxidation of HDL Induced by AAPH In Vitro

1. Determine protein concentration of isolated HDL (BCA assay) and adjust to 1.2–1.5 mg protein per mL PBS.
2. Oxidize this HDL (typically 1 mL) by adding AAPH (1 mM final concentration, or 5 μL of 200 mM AAPH stock added to 1 mL HDL) and incubating the reaction mixture under air at 37°C for 3 h; the extent of oxidation may be adjusted by changing the AAPH concentration and/or incubation time.
3. Terminate oxidation by gel filtration of the reaction mixture using a pre-conditioned NAP-10 column (*see* **Section 3.1**), eluting the HDL with 1.5 mL PBS.
4. Determine the protein concentration of oxidized HDL (BCA assay).

3.3. Protein Determination of HDL and Oxidized HDL

1. Prepare fresh set of protein standards by diluting 2.0 mg/mL BSA stock standard (stock) in 10 mM PBS to working stock ranging from 0–1000 μg/mL in triplicates.
2. Prepare BCA working solution by mixing BCA protein assay reagent A with 5% $CuSO_4$ at ratio of 50:1.
3. Pipette 25 μL of each standard or (oxidized) HDL (diluted 5–10 times in 10 mM PBS) into microwell plate wells. Use 25 μL of diluent (10 mM PBS) for blank wells.
4. Add 200 μL working solution to each well and mix by placing plate on a shaker for 30 s.
5. Cover plate, incubate for 30 min at 37°C, and then cool plate to room temperature.
6. Read absorbance at 562 nm using a plate reader (*see* **Note 11**).
7. Subtract average $A_{562\ nm}$ of blanks from $A_{562\ nm}$ reading for each standard or sample.
8. Prepare standard curve by plotting average blank-corrected $A_{562\ nm}$ reading for each BSA standard vs. its concentration in μg/mL. Using standard curve obtained, determine protein concentration for each unknown sample.

3.4. HPLC Determination of Apolipoproteins in Native and Oxidized HDL

1. Use appropriate HPLC-UV system (Waters or equivalent) equipped with reverse phase C18 column (Vydac, 250 × 4.6 mm).
2. Gradient HPLC is performed on a Waters alliance 2690 separation module equipped with a Waters 2487 dual absorbance detector.
3. Turn on the instrument for 1–2 h prior to running samples. Set degasser to continuous mode while washing the needles (10% methanol) and purging system (mobile phases) (*see* **Note 12**).
4. Set degasser to normal mode, sample chamber to 4°C and column to 50°C after purging instrument. Equilibrate the system by running the mobile phase at 25% Solvent A for ~20 min (*see* **Note 13**).
5. Condition the column further by a mock run using the solvent gradient but without injecting a sample.
6. Transfer 1.5 mL native or oxidized HDL into 2 mL HPLC vials, load vials onto sample carousal set at 4°C, and then subject samples to guard column (5 μm, 4.6 ID, Vydac) connected to RP C18 column (250 × 4.6, 5 μm, Vydac) eluted at 0.5 mL/min and 50°C and monitored at 214 nm.
7. Analyze samples using the following gradient *(12)* and with in- line degasser turned on:
   a. Increase Solvent A linearly from 25% to 45% over 5 min
   b. Increase Solvent A linearly to 55% over 32 min
   c. Increase Solvent A linearly to 95% over 10 min

d. Increase Solvent A to 100% in 1 min
e. Decrease Solvent A to 25% for 10 min for column re-equilibration and before injection of next sample

8. For native HDL, apoA-I and apoA-II are the major proteins detected (**Fig. 5.1A**). For AAPH-oxidized HDL, oxidized forms of apoA-I and apoA-II elute prior to their respective nonoxidized forms (**Fig. 5.1B**). Oxidation products are formed time-dependently (**Fig. 5.2**).
9. Upon completion of HPLC analyses, equilibrate system for 5–10 min, then store column in 40% aqueous acetonitrile.

3.5. Mass Spectrometry Confirmation of Specifically Oxidized Apolipoproteins

To confirm specificity of the reaction, oxidized apoA-I and apoA-II are purified by HPLC and subjected to mass spectrometry. Following mass confirmation, respective retention times of individual oxidized apolipoproteins are used for identification of apolipoproteins and their oxidized forms in further HDL samples. MALDI-TOF, single or quadruple MS with ESI, is suitable to characterize apolipoproteins. In this protocol, ESI-MS-TOF (API QStar Pulsar) is used.

1. Subject freshly isolated, nonoxidized HDL to HPLC (*see* **Section 3.4**) and collected fractions corresponding to native apoA-I and apoA-II. It is recommended to use a semi-preparative HPLC column (RP C18, 10 mm × 250 mm) eluted at 2 mL/min while retaining all other settings described in **Section 3.4**.

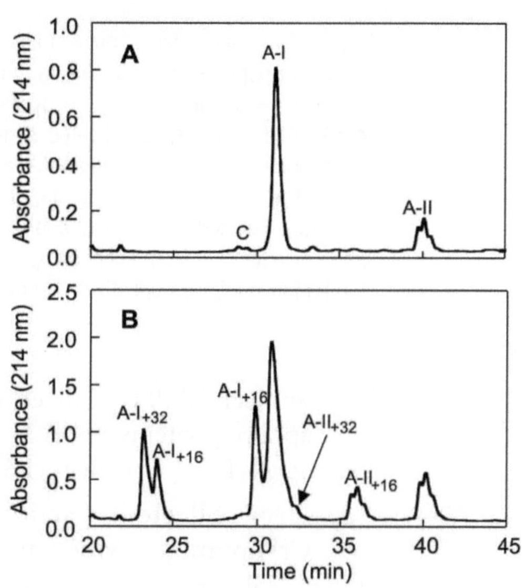

Fig. 5.1. Representative HPLC chromatograms of native HDL and peroxyl radical-oxidized HDL. (**A**) Gel-filtered native HDL (60 μg protein) isolated by density ultracentrifugation and subjected to HPLC-UV as described in **Section 3.4** - contains apoA-I (A-I) and apoA-II (A-II) as major apolipoproteins. (**B**) Oxidized HDL (1 mg protein) prepared by incubation with 1 mM AAPH for 3 h at 37°C - contains several specifically oxidized form of apoA-I and apoA-II designated as apoA-I_{+32}, apoA-$I_{+16,}$ and apoA-II_{+16}.

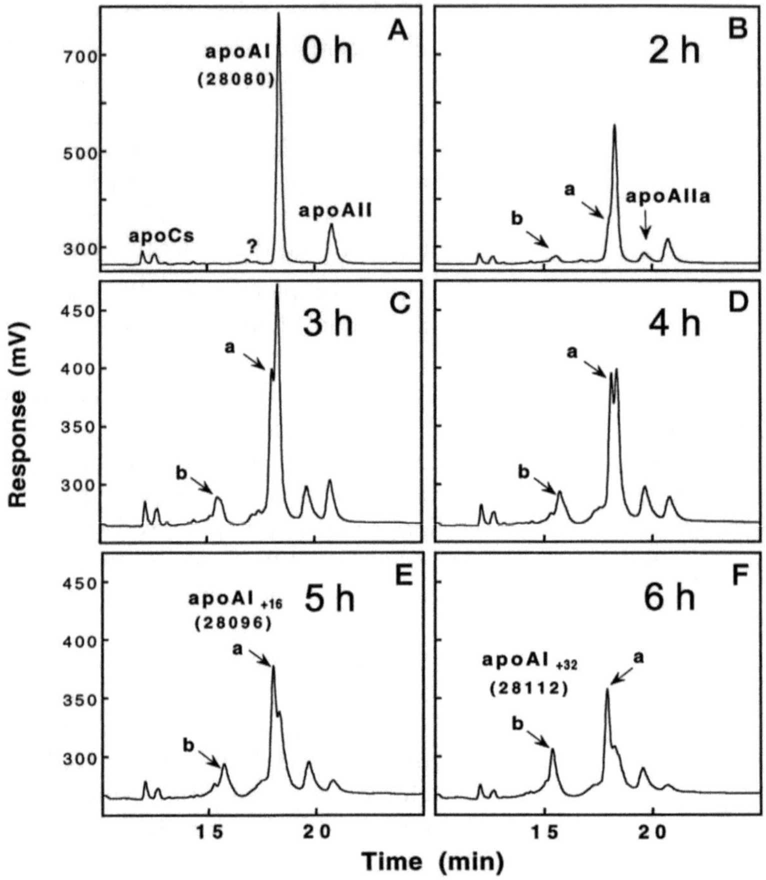

Fig. 5.2. Time-dependent changes to the content of nonoxidized and specifically oxidized major apolipoproteins in HDL during *in vitro* oxidation of the lipoprotein by peroxyl radicals. HDL was oxidized at 37°C by exposure to the peroxyl radical generator AAPH (4 mM). At the time points indicated, samples (50 μL) were taken and analyzed by HPLC. The major forms of apoA-I and apoA-II are labeled. The molecular masses of the various forms of apoA-I, determined by MS, are given in parentheses; the arrows indicate the appearance of oxidized forms of apoA-I, *i.e.* apoA-I_{+16} (*see arrow a*) and apoA-I_{+32} (*see arrow b*). A through F represent samples taken at 0 (*i.e.* directly after AAPH addition) and 2–6 h of incubation, respectively (From reference 17 with permission).

2. Subject AAPH-oxidized HDL to HPLC and collect oxidized apolipoproteins apoA-I_{+16} (Met^{86}, Met^{112}), apoA-I_{+32} and apoA-II_{+16} (*see* **Note 15**).
3. Dry collected fractions using speedvac in Vapornet Cryopumping Mode and with drying rate set to low.
4. Re-dissolve dried samples in water/acetonitrile (1:1, vol/vol) containing 0.1% formic acid.
5. Load aliquot (1 μL) of each sample into nanospray needles (Proxeon, Odense, Denmark) with the tip positioned ~10 mm from the orifice.

6. Use nitrogen as the curtain gas and apply a potential of 900 V to needle.
7. Acquire TOF-MS scans (m/z 550–2000, 1 s) and accumulate for ~1 min into single file.
8. De-convolute spectra using the Bayesian reconstruct method contained in the Analyst QS software (Applied Biosystems).
9. Retention times of native apoA-I and apoA-II and their specifically oxidized forms are listed (**Table 5.1**) for the instrument conditions and column specified in this protocol. Masses of these apolipoproteins and their oxidized forms are determined by ESI-MS, as shown in **Table 5.1**.

Table 5.1
ESI-MS Analysis of apoA-I and apoA-II and their oxidized forms using API Q-Star

Apolipoproteins	Retention time (min)	Predicted mass	ESI-MS
ApoA-I	~31	28,078.6	28,079
ApoA-II			
Species 1	~39.4	17,123	17,123.2
Species 2	~39.7	17,252	17,252.6
Species 3	~40.2	17,380	17,379.6
ApoA-I_{+32}	~22	28,111	28,112
ApoA-I_{+16} ($MetO^{112}$)	~23	28,095	28,095
ApoA-I_{+16} ($MetO^{86}$)	~30	28,095	28,097
ApoA-II_{+16} ($MetO^{26}$)			
Species 1	~35.3	17,139	17,139.8
Species 2	~35.6	17,268	17,267.5
Species 3	~36.1	17,396	17,395.8

Native and AAPH-oxidized HDL samples were subjected to HPLC, the major apolipoproteins isolated and their masses determined using ESI-MS. Spectra were acquired using an API QStar Pulsar I hybrid tandem mass spectrometer (TOF) (Applied Biosystems). ApoA-I contains three Met residues of which Met^{86} and Met^{112} become oxidized readily when HDL is exposed to AAPH. ApoA-II exists as a disulfide-linked homodimer, and elutes as three distinct species, each with different molecular mass. ApoA-II contains one Met residue (Met^{26}), and oxidation of HDL gives rise to the three corresponding apoA-II species 16 Da heavier than the corresponding native forms. Listed are the retention times and mass of each form of the apolipoproteins.

3.6. Oxidized HDL in Human Plasma

To apply the above-described HPLC method to the detection of oxidized HDL in biological materials, two examples of blood plasma are given in the following. First, samples obtained from the Fletcher challenge study, a prospective observational study examining the relationship of socio-demographic, psychological, and several blood factors with the risk of coronary heart disease in a New Zealand population *(20)*. Second, cord blood samples collected *post-partum* from subjects with abnormal (eNOSa/b) and normal endothelial nitric oxide synthase genotype (eNOSb/b), or cord blood samples without and with exposure to cigarette smoking during pregnancy.

1. For the Fletcher Challenge samples, blood was collected into EDTA coated tubes and centrifuged immediately at 4°C. The resulting plasma was stored at −80°C until used.
2. Immediately after thawing, HDL was isolated by ultracentrifugation and subjected to HPLC-UV system as described above (*see* **Sections 3.1 and 3.3**). **Figure 5.3** shows representative HPLC traces from samples representing controls and cases.
3. Cord blood was centrifuged within 10 min of blood collection and serum samples stored at −70°C within a further 10 min.
4. HDL was isolated immediately after thawing the sample using the method described above.
5. Aliquots (typically 100 μL) of freshly isolated HDL were applied to the HPLC-UV system. **Figure 5.4** indicates the relationship between apoA-I_{+32} and eNOS genotypes and exposure to cigarette smoking.

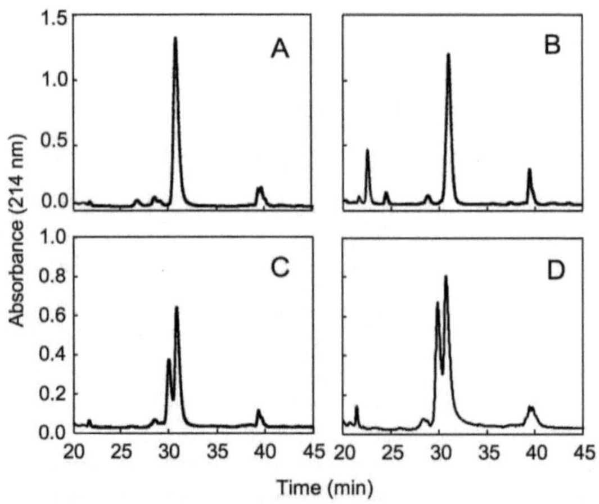

Fig. 5.3. Representative HPLC chromatograms of HDL isolated from human plasma samples of the Fletcher Challenge study. Isolated HDL (1 mg protein) was immediately subjected to HPLC. (**A**) ApoA-I and apoA-II appeared as native forms in most samples; (**B**) A few samples showed the presence of apoA-I_{+32} only. (**C**) and (**D**) ApoA-I_{+16} (Met[86]) was observed in 22% of the samples although the proportion of apoA-I present as apoA-I_{+16} varied.

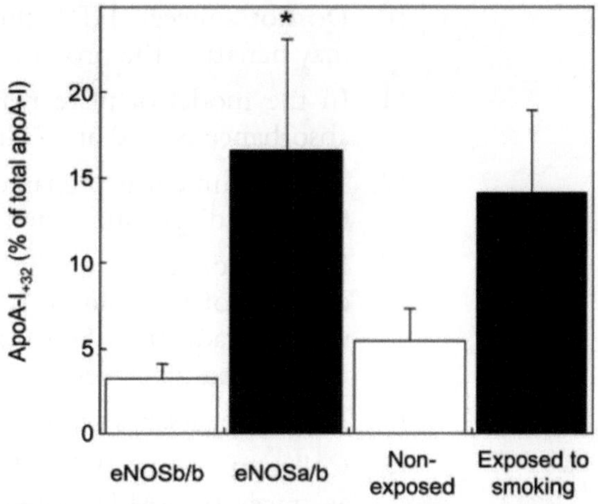

Fig. 5.4. Relationship between apoA-I_{+32} and eNOS genotype and exposure to cigarette smoking. The concentration of apoA-I_{+32} measured by HPLC, was determined in HDL isolated from serum of cord blood in subjects of eNOSa/b ($n = 4$) and eNOSb/b genotype ($n = 8$), and in cord blood samples without ($n = 12$) and with exposure to cigarette smoking during pregnancy ($n = 5$). Both eNOSa allele carriers and cigarette smokers are associated with increased risk for vascular diseases. Results are expressed as percentage of total apoA-I concentration and represent mean ± SD. Obtained from reference 12 with permission.

4. Notes

1. PBS may be substituted with 0.9% NaCl.
2. Water is purified by a Milli-Q Ultrapure water system (Australia); all aqueous solutions should be prepared using Milli-Q water.
3. HPLC grade water and solvents should be used throughout.
4. Freshly obtained human plasma should be analyzed immediately or stored at −80°C until analyzed. Repeated freeze/thawing cycles should be minimized to avoid auto-oxidation of samples.
5. Trifluoroacetic acid is corrosive and should be handled in a fume hood with care.
6. Unless stated otherwise, all procedures should be carried out on ice.
7. Use blunt, not sharp needle for this step.
8. In case of emergency, switch off centrifuge by pressing the power button.
9. The volume removed may depend on the desired concentration of HDL collected.

10. Do not squeeze HDL through the tip of needles, as this may denature the protein.
11. In the model of plate reader referred to in this protocol, absorbance is read at 570 nm.
12. The instrument is prepared by choosing appropriate menus from the diagnostic panel.
13. If retention times of the proteins of interest vary largely, cleaning of the column is required. For this, wash column with a gradient of Solvents A and B, increasing Solvent B from 5% to 95% over 30 min.
14. In-line degasser is on during the run.
15. Collect native apoA-I and apoA-II from nonoxidized HDL, as traces of AAPH potentially remaining in the collected apolipoprotein fractions subsequently promote oxidation.

Acknowledgments

We thank our past and present collaborators of the work referred to in the present manuscript. This work was supported by University Postgraduate Award (the University of Sydney) and the National Heart Foundation of Australia. RS is supported by the National Health & Medical Research Council (Senior Principal Research Fellowship) and the University of Sydney (Professorial Fellowship and Medical Foundation Fellowship).

References

1. Stocker R, Keaney JF, Jr (2004) Role of oxidative modifications in atherosclerosis. *Physiol Rev* **84**, 1381–1478.
2. Steinberg D, Parthasarathy S, Carew TE, Khoo JC, Witztum JL (1989) Beyond cholesterol: Modifications of low-density lipoprotein that increase its atherogenicity. *N Engl J Med* **320**, 915–924.
3. Bowry VW, Stanley KK, Stocker R (1992) High density lipoprotein is the major carrier of lipid hydroperoxides in fasted human plasma. *Proc Natl Acad Sci USA* **89**, 10316–10320.
4. Niu X, Zammit V, Upston JM, Dean RT, Stocker R (1999) Co-existence of oxidized lipids and α-tocopherol in all lipoprotein fractions isolated from advanced human atherosclerotic plaques. *Arterioscler Thromb Vasc Biol* **19**, 1708–1718.
5. Zheng L, Nukuna B, Brennan ML, et al. (2004) Apolipoprotein A-I is a selective target for myeloperoxidase-catalyzed oxidation and functional impairment in subjects with cardiovascular disease. *J Clin Invest* **114**, 529–541.
6. Bergt C, Pennathur S, Fu X, et al. (2004) The myeloperoxidase product hypochlorous acid oxidizes HDL in the human artery wall and impairs ABCA1-dependent cholesterol transport. *Proc Natl Acad Sci USA* **101**, 13032–13037.
7. Peng DQ, Wu Z, Brubaker G, et al. (2005) Tyrosine modification is not required for myeloperoxidase-induced loss of apolipoprotein A-I functional activities. *J Biol Chem* **280**, 33775–33784.
8. Oram JF, Heinecke JW (2005) ATP-binding cassette transporter A1: a cell

cholesterol exporter that protects against cardiovascular disease. *Physiol Rev* **85**, 1343–1372.

9. Shao B, Oda MN, Bergt C, et al. (2006) Myeloperoxidase impairs ABCA1-dependent cholesterol efflux through methionine oxidation and site-specific tyrosine chlorination of apolipoprotein A-I. *J Biol Chem* **14**, 9001–9004.
10. Panzenböck U, Kritharides L, Raftery M, Rye KA, Stocker R (2000) Oxidation of methionine residues to methionine sulfoxides does not decrease potential anti-atherogenic properties of apolipoprotein A-I. *J Biol Chem* **275**, 19536–19544.
11. Francis GA, Oram JF, Heinecke JW, Bierman EL (1996) Oxidative tyrosylation of HDL enhances the depletion of cellular cholesteryl esters by a mechanism independent of passive sterol desorption. *Biochemistry* **35**, 15188–15197.
12. Pankhurst G, Wang XL, Wilcken DE, et al. (2003) Characterization of specifically oxidized apolipoproteins in mildly oxidized high density lipoprotein. *J Lipid Res* **44**, 349–355.
13. Pennathur S, Bergt C, Shao B, et al. (2004) Human atherosclerotic intima and blood of patients with established coronary artery disease contain high density lipoprotein damaged by reactive nitrogen species. *J Biol Chem* **279**, 42977–42983.
14. Hazell LJ, Davies MJ, Stocker R (1999) Secondary radicals derived from chloramines of apolipoprotein B-100 contribute to HOCl-induced lipid peroxidation of low-density lipoproteins. *Biochem J* **339**, 489–495.
15. Sattler W, Christison JK, Stocker R (1995) Cholesterylester hydroperoxide reducing activity associated with isolated high- and low-density lipoproteins. *Free Radic Biol Med* **18**, 421–429.
16. Mashima R, Yamamoto Y, Yoshimura S (1998) Reduction of phosphatidylcholine hydroperoxide by apolipoprotein A-I: purification of the hydroperoxide-reducing proteins from human blood plasma. *J Lipid Res* **39**, 1133–1140.
17. Garner B, Witting PK, Waldeck AR, Christison JK, Raftery M, Stocker R (1998) Oxidation of high density lipoproteins. I. Formation of methionine sulfoxide in apolipoproteins AI and AII is an early event that correlates with lipid peroxidation and can be enhanced by α-tocopherol. *J Biol Chem* **273**, 6080–6087.
18. Garner B, Waldeck AR, Witting PK, Rye K-A, Stocker R (1998) Oxidation of high density lipoproteins. II. Evidence for direct reduction of HDL lipid hydroperoxides by methionine residues of apolipoproteins AI and AII. *J Biol Chem* **273**, 6088–6095.
19. Barclay LRC, Locke SJ, MacNeil JM, VanKessel J (1984) Autoxidation of micelles and model membranes. Quantitative kinetic measurements can be made by using either water-soluble or lipid-soluble initiators with water-soluble or lipid-soluble chain-breaking antioxidants. *J Am Chem Soc* **106**, 2479– 2481.
20. MacMahon S, Norton R, Jackson R, et al. (1995) Fletcher Challenge-University of Auckland Heart & Health Study: design and baseline findings. *N Z Med J* **108**, 499–502.

Chapter 6

High Performance Liquid Chromatography/Electron Spin Resonance/Mass Spectrometry Analyses of Lipid-Derived Radicals

Hideo Iwahashi

Abstract

High performance liquid chromatography/electron spin resonance/mass spectrometry (HPLC/EPR/MS) analyses of radicals is performed for the reaction mixture of 13-hydroperoxy-(9Z,11E)-octadeca-9,11-dienoic acid (13ZE-OOH) with ferrous ions under an aerobic condition, or an anaerobic condition. Radicals are identified from 13ZE-OOH by using high performance liquid chromatography/electron spin resonance spectrometry (HPLC/EPR) and HPLC/EPR/MS. The pentyl radical and isomers of epoxylinoleic acid radicals from 13ZE-OOH are identified under an anaerobic condition and the 7-carboxyheptyl radical and pentyl radical from 13ZE-OOH under an aerobic condition. These results suggest that the formation of the respective radical species depends to a great extent on oxygen concentration in the reaction mixtures.

Key Words: Lipid peroxidation, HPLC/EPR/MS, HPLC/EPR, Free radicals, Linoleic acid.

1. Introduction

Using spin trapping, the structures of radicals have been determined based on electron paramagnetic resonance (EPR) hyperfine coupling constants *(1)*. Similar hyperfine coupling constants, however, are often observed for different radical adducts, which makes determination of radical adduct structure difficult. Thus, in order to obtain comprehensive knowledge of the molecular structures of the radical adducts, information other than EPR hyperfine coupling constants is necessary. High performance liquid

From: *Methods in Molecular Biology, vol. 477: Advanced Protocols in Oxidative Stress I,*
Edited by: D. Armstrong, DOI: 10.1007/978-1-60327-517-0_6, © Humana Press, New York, NY

chromatography/ electron spin resonance spectrometry (HPLC/EPR) *(2, 3, 4)*, and high performance liquid chromatography/electron spin resonance/mass spectrometry (HPLC/EPR/MS) *(5)* have been employed for detection and identification of the radical adducts *(6, 7, 8, 9)*. The retention times and molecular weights of radical adducts are useful information for identification.

2. Materials

2.1. Equipment

1. Electron paramagnetic resonance spectrometer, a model JES-FR30 Free Radical Monitor (JEOL Ltd., Tokyo, Japan) equipped with a model CLASS-LC10 LC work station (Shimadzu Co., Kyoto, Japan).
2. High performance liquid chromatography consisted of a model 7125 injector (Reodyne, Cotati, CA, USA) with a 5 ml sample loop and a model 655A-11 pump with a model L-5000 LC controller (Hitachi Ltd., Ibaragi, Japan).
3. Column (30 mm × 10 mm I.D.) packed with TSKgel ODS-120T, 5 μm particle size and 120 A pore size (Tosoh Co., Tokyo, Japan).
4. Liquid chromatography/mass spectrometer (LC-MS), model M-1200AP LC-MS equipped with electrospray ionization (ESI), (Hitachi Ltd., Ibaragi, Japan).
5. 3VP-C rotary pump (Hitachi, Ltd., Ibaragi, Japan).

2.2. Reagents

1. 50 mM acetic acid (mobile phase A)
2. 50 mM acetic acid/acetonitrile (20:80, v/v). (mobile phase B)
3. 10 mM $FeSO_4(NH_4)_2SO_4$.
4. 1.0 M α-(4-pyridyl-1-oxide)-*N-tert*-butylnitrone (4-POBN) for the EPR spin trapping.
5. 1.4 mM 13-hydroperoxy-(9Z,11E)-octadeca-9,11-dienoic acid (13ZE-OOH)]
6. 10 mM EDTA
7. 50 mM phosphate buffer (pH 7.4).

2.3. Supplies

1. Thunberg tubes (18 mm × 160 mm, SANSYO Ltd., Tokyo, Japan).
2. Millex-GS filters (Millipore Products Div., Bedford, MA, USA).

3. Methods

3.1. Preparation of the Reaction Mixture of 13ZE-OOH for HPLC/EPR/MS

1. The complete reaction mixture contains 0.14 mM 13ZE-OOH, 0.33 mM $FeSO_4(NH_4)_2SO_4$, 0.33 mM EDTA, 0.1 M 4-POBN, and 38 mM phosphate buffer, pH 7.4.
2. The reactions are started by adding $FeSO_4(NH_4)_2SO_4$, and then incubating for 15 min at 20°C under anaerobic, or aerobic conditions.
3. Anaerobic conditions are obtained with a 3VP-C rotary pump (Hitachi) in Thunberg tubes.
4. Two milliliters of the reaction mixtures is filtered through a Millipore filter (Millex-GS, Millipore).
5. The filtrate is directly applied to the HPLC/EPR or the HPLC/EPR/MS (**Fig. 6.1**).

3.2. HPLC/EPR/MS Analysis

1. The EPR spectrometer is connected to the HPLC with a Teflon tube, which passes through the center of the EPR cavity. The mass spectra are obtained by introducing the eluent from the EPR detector into the LC-MS system just before the respective peaks are eluted (*see* **Note 1**). The flow rate is kept at 50 μl/min while the eluent is being introduced into the LC-MS system.
2. The HPLC used in the HPLC/EPR/MS (**Fig. 6.1**) consists of a model 7125 injector (Reodyne) with a 5 ml sample loop and a model 655A-11 pump with a model L-5000 LC controller (Hitachi). A column (300 mm × 10 mm I.D.) is packed with TSKgel ODS-120T (Tosoh Co). The TSKgel ODS-120T is a nonsilanol type disposable with an end cap to a remaining silanol. For the HPLC/EPR analyses, two solvents are used: solvent A, consisting of 50 mM ammonium acetate and solvent B, consisting of 50 mM ammonium acetate/acetonitrile (36:64, v/v) (*see* **Note 2**). A combination of an isocratic and linear gradient is used: 0–40 min, 100% A to 0% A (linear gradient) at a flow rate of 2.0 ml/min and 40–60 min, 0% A (isocratic) at the same flow rate of 2.0 ml/min.

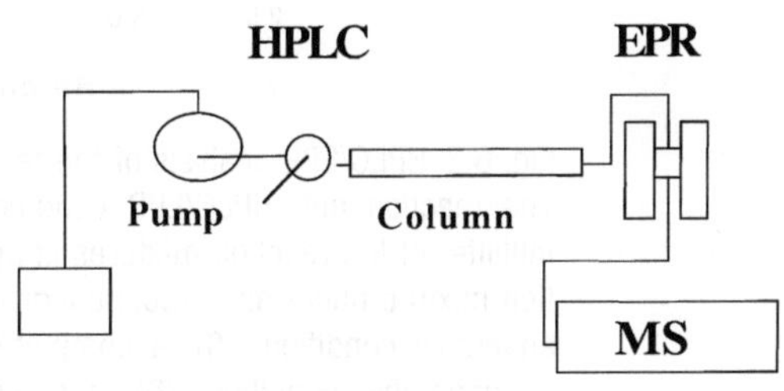

Fig. 6.1. Schematic of HPLC/EPR/MS system.

3. The operating conditions of the EPR spectrometer used in the HPLC/EPR /MS are: power of 4 mW, modulation width of 0.2 mT, and time constant of 1 s. The magnetic field is fixed at the third EPR peak in the characteristic six lines spectrum of the 4-POBN radical adduct throughout the experiments.
4. The operating conditions of the mass spectrometer are: nebulizer, 180°C; aperture 1, 120°C; N_2 controller pressure, 2.0 kgf/cm^2; drift voltage, 70 V; multiplier voltage, 1800 V; needle voltage, 3000 V; polarity, positive; resolution, 48.

4. Results

4.1. HPLC/EPR/MS Analysis of the Reaction Mixture of 13ZE-LOOH with Ferrous Ions

The complete reaction is performed under aerobic conditions using HPLC/EPR (**Fig. 6.2A**). Two peaks (P-1 and P-5) are detected at the retention times of 29.8 min and 46.2 min on the HPLC/EPR elution profile of the complete reaction mixture (**Fig. 6.2A**). The reactions are also performed for the complete reaction mixture, the complete reaction mixture without 13ZE-OOH, and the complete reaction mixture without ferrous ions

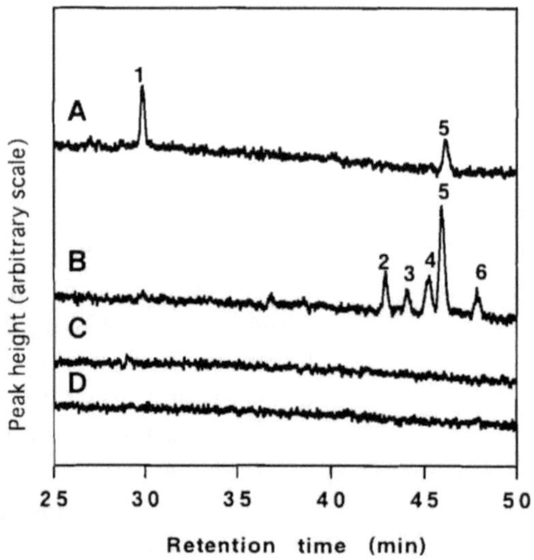

Fig. 6.2. HPLC/EPR analysis of the reaction mixture of 13ZE-OOH with ferrous ions. The reaction and HPLC/EPR conditions are as described under **Methods**. Two milliliter of the reaction mixtures is applied to the HPLC/EPR. (**A**) a complete reaction mixture under an aerobic condition; (**B**) a complete reaction mixture under an anaerobic condition; (**C**), a complete reaction mixture without 13ZE-OOH under an anaerobic condition; (**D**), a complete reaction mixture without ferrous ions under an anaerobic condition.

under an anaerobic condition (**Fig. 6.2B–D**). HPLC/EPR elution profiles of these reaction mixtures are shown in **Fig. 6.2B–D**. The HPLC/EPR analysis of the complete reaction mixture of 13ZE-OOH under an anaerobic condition gives five peaks (P-2, P-3, P-4, P-5, and P-6) at the retention times of 43.0 min, 44.1 min, 45.2 min, 45.9 min, and 47.9 min (**Fig. 6.2B**). The HPLC/EPR gave no peak for the complete reaction mixture without 13ZE-OOH and the complete reaction mixture without ferrous ions (**Fig. 6.2C, D**).

4.2. HPLC/EPR/MS Analyses of the Complete Reaction Mixtures of 13ZE-OOH with Ferrous Ions Under Anaerobic, or Aerobic Conditions

The HPLC/EPR/MS analyses of the complete reaction mixtures of 13ZE-OOH are performed under anaerobic, or aerobic conditions. Mass and EPR spectra of the respective HPLC/EPR peaks are as obtained in **Fig. 6.3**. The HPLC/EPR/MS analysis of the P-1 gives ions at m/z 251 and m/z 338 (**Fig. 6.3**, P-1). The ions at m/z 338 correspond to the protonated molecules of 4-POBN/7-carboxyheptyl radical adducts, $[M + H]^+$. The fragment ions at m/z 251 correspond to the loss of $(CH_3)_3C(O)N$ from the protonated molecules (**Fig. 6.3**, P-1) (*see* **Note 3**). The HPLC/EPR/MS analysis of the P-2 gives ions at m/z 490 (**Fig. 6.3**, P-2). A possible radical adduct is as follows:4-POBN/1-(7-carboxyheptyl)-4,5-epoxy-2-decenyl radical adduct or 1-(1,2-epoxyheptyl)-10-carboxy-2-decenyl radical adduct or 1-pentyl-12-carboxy-2, 4-dodecadienyloxyl radical adduct $[M + H]^+$ (**Fig. 6.4**). The HPLC/EPR/MS analysis of the peaks 3, 4, and 6 also gives ions at m/z 490 (**Fig. 6.3**, P-3, P-4, and P-6), suggesting that these are also isomers of 4-POBN/1-pentyl-12-carboxy-2,4-dodecadienyloxyl radical adducts [or 1-(7-carboxyheptyl)-4,5-epoxy-2-decenyl radical or 1-(1,2-epoxyheptyl)-10- carboxy-2-decenyl radical adducts]. The HPLC/EPR/MS analysis of the P-5 gives ions at m/z 179, m/z 266, and m/z 531 (**Fig. 6.3**, P-5). The ions at m/z 266 correspond to the protonated molecules of 4-POBN/pentyl radical adducts, $[M + H]^+$. The fragment ions at m/z 179 correspond to the loss of $(CH_3)_3C(O)N$ from the protonated molecules (**Fig. 6.3**, P-5) (*see* **Note 3**). The ions at m/z 531 correspond to the protonated dimmer, $[2M + H]^+$ of 4-POBN/pentyl radical adducts.

HPLC/EPR/MS analyses of peaks 2, 3, 4, and 6 show ions at m/z 490, which correspond to isomers of epoxylinoleic acid radicals. These isomers are indistinguishable based on information obtained by HPLC/EPR/MS. MS/MS spectra observed by tandem mass spectrometry may give information to solve the problem *(10)*.

The hyperfine-coupling constants of the respective radical adducts are obtained by computer simulations of the EPR spectra (**Fig. 6.3**, **Table 6.1**). These six radical adducts show almost the same hyperfine coupling constants of α-nitrogen atom a^N_α (15.5G to 15.1G). On the other hand, the hyperfine coupling constants of

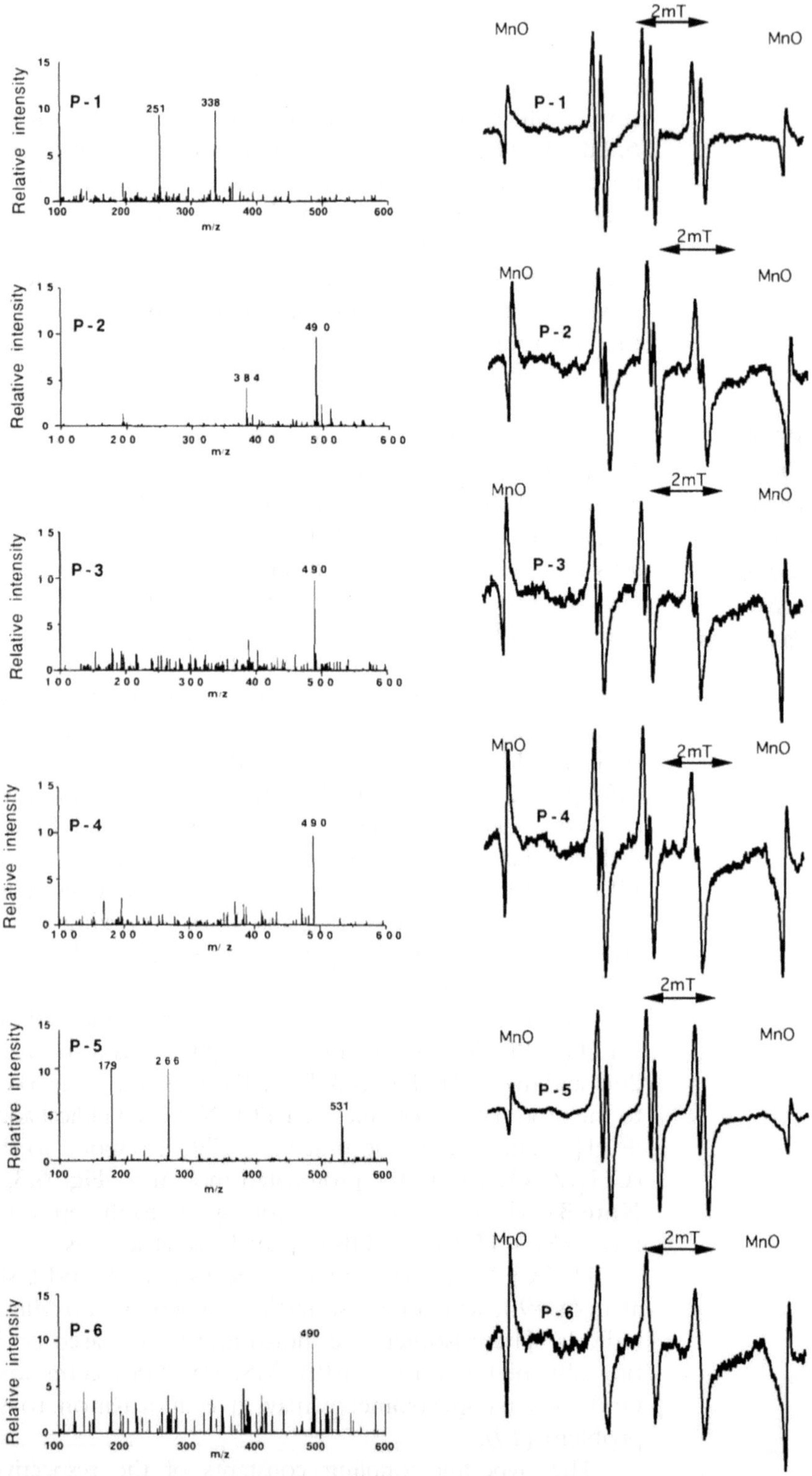

Fig. 6.3. HPLC/EPR/MS analysis of the reaction mixture of 13ZE-OOH with ferrous ions. The reaction, HPLC/EPR, and HPLC/EPR/MS conditions are as described under **Methods**. P-1, mass (left) and EPR (right) spectra of the peak P-1; P-2, mass (left) and EPR (right) spectra of the peak P-2; P-3, mass (left) and EPR (right) spectra of the peak P-3; P-4, mass (left) and EPR (right) spectra of the peak P-4; P-5, mass (left) and EPR (right) spectra of the peak P-5; P-6, mass (left) and EPR (right) spectra of the peak P-6.

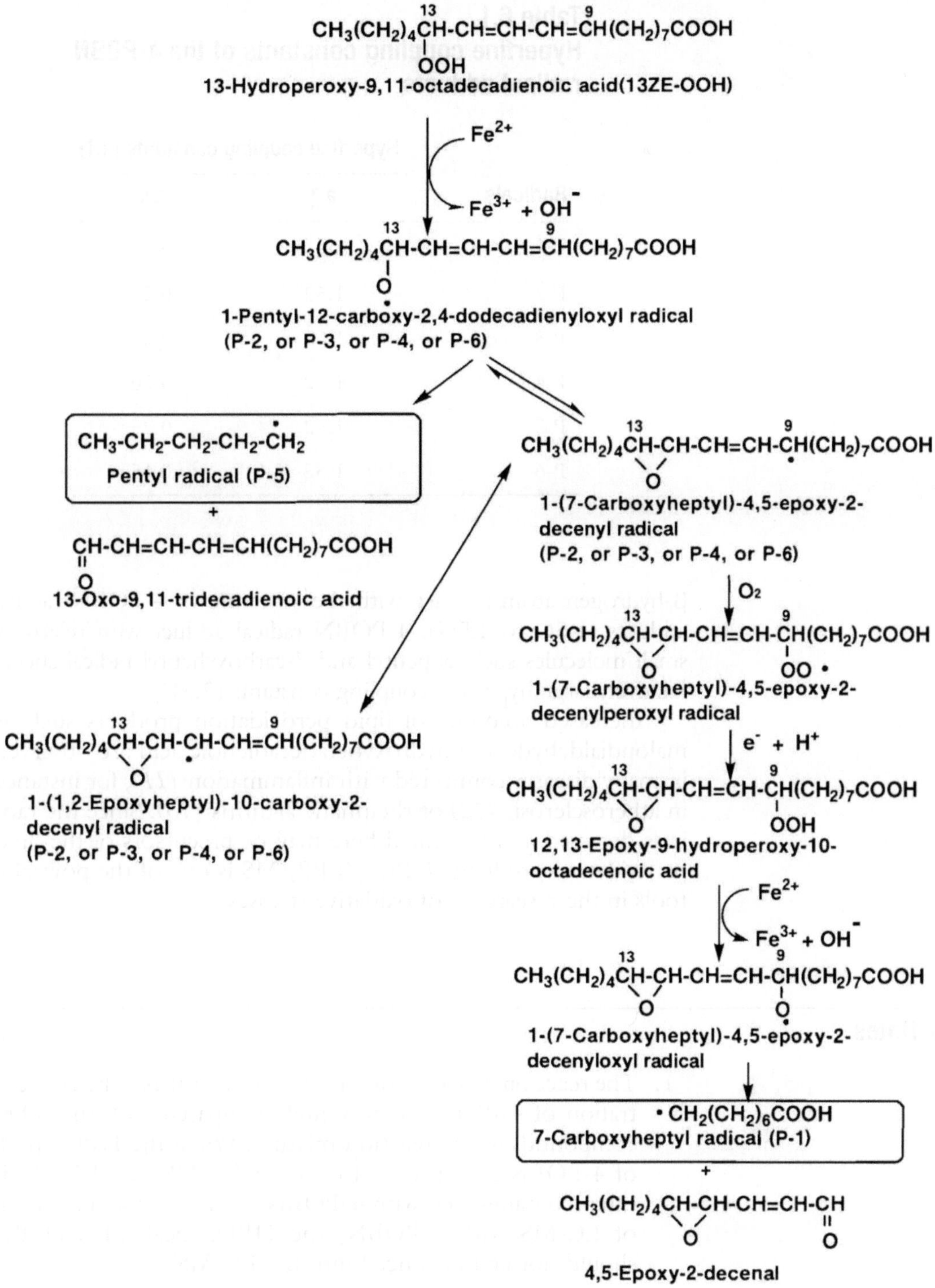

Fig. 6.4. Possible reaction paths for the formation of 7-carboxyheptyl radical (P-1), 1-(7-carboxyheptyl)-4,5-epoxy-2-decenyl radical (P-2, 3, 4, or 6), 1-(1,2-epoxyheptyl)-10-carboxy-2-decenyl radical (P-2, 3, 4, or 6), 1-pentyl-12-carboxy-2,4-dodecadienyloxyl radical (P-2, 3, 4, or 6), and pentyl radical (P-5).

Table 6.1
Hyperfine coupling constants of the 4-POBN radical adducts

	Hyperfine coupling constants (mT)	
Radicals	a^N_α	a^H_β
P-1	1.55	0.25
P-2	1.52	0.20
P-3	1.51	0.21
P-4	1.52	0.20
P-5	1.52	0.25
P-6	1.53	0.15

β-hydrogen atom a^H_β vary with the structure of 4-POBN radical adducts (1.5G to 2.5G). 4-POBN radical adduct with relatively small molecules such as pentyl and 7-carboxyheptyl radical shows relatively large hyperfine-coupling constants (2.5G).

Increased amounts of lipid peroxidation products such as malondialdehyde and hydroxyoctadecadienoic acid are observed in many diseases connected with inflammation *(11)*, for instance in atherosclerosis *(12)* or rheumatic arthritis *(13)*. Since the radicals detected and identified here may be precursors of the lipid peroxidation products, HPLC/EPR/MS is one of the powerful tools in the researches of oxidative stresses.

5. Notes

1. The reaction mixture contains 0.1 M 4-POBN. The concentration of 4-POBN is fairly high compared with the other compounds in the reaction mixture. When the HPLC peak of 4-POBN is introduced into the LC-MS, the LC-MS is often contaminated with 4-POBN. To avoid contamination of LC-MS with 4-POBN, the HPLC peak of 4-POBN should not be introduced into the LC-MS.

2. More than 30 min is needed for the HPLC/EPR/MS analysis. The radical adducts sometimes disappear during the HPLC/EPR/MS analysis. The life time of the radical adduct depends on pH to a great extent. Particular

attention should be given to the pH of mobile phase of the HPLC/EPR/MS.

3. A 4-POBN spin trapping reagent is used for the HPLC/EPR/MS analyses. ESI-MS spectra of the 4-POBN radical adducts show characteristic fragment ions. The fragment ions correspond to the loss of $(CH_3)_3C(O)N$ from the protonated molecules. The fragment ions can be employed as an indicator of 4-POBN radical adduct.

References

1. Li, A.S.W., Cummings, K.B., Roethling, H.P., Buettner, G.R., and Chignell, C.F. (1988) A spin-trapping database implemented on the IBM PC/AT. J. Magn. Reson., **79**, 140–142.
2. Rokushika, S., Taniguchi, H., and Hatano, H. (1975) Flow ESR detector for liquid chromatography of radicals. Anal. Lett., **8**, 205–213.
3. Makino, K., and Hatano, H. (1979) Separation and characterization of short-lived radicals in DL-methionine aqueous solution by high speed liquid chromatograph equipped with ESR spectrometer. Chem. Lett., 119–122.
4. Iwahashi, H., Negoro, Y., Ikeda, A., Morishita, H., and Kido, R. (1986) Inhibition by chlorogenic acid of haematin-catalysed retinoic acid 5,6-epoxidation. Biochem. J., **239**, 641–646.
5. Iwahashi, H., Parker, E.C., Mason, R.P., and Tomer, K.B. (1992) Combined liquid chromatography/electron paramagnetic resonance spectrometry/electrospray ionization mass spectrometry for radical identification. Anal. Chem., **64**, 2244–2252.
6. Iwahashi, H., Hirai, T., and Kumamoto, K. (2006) High performance liquid chromatography/electron spin resonance/mass spectrometry analyses of radicals formed in anaerobic reaction of 9- (or 13-) hydroperoxide octadecadienoic acids with ferrous ions. J. Chromotogr. A, **1132**, 67–75.
7. Minakata, K., Okuno, E., Nakamura, M., and Iwahashi, H. (2007) Identification of radicals formed in the reaction mixtures of rat liver microsomes with ADP, Fe^{3+} and NADPH Using HPLC-EPR and HPLC-EPR-MS. J. Biochem., **142**, 73–78.
8. Iimura, S., and Iwahashi, H. (2006) Enhancement by cigarette smoke extract of the radical formation in a reaction mixture of 13-hydroperoxide octadecadienoic acid and ferric ions. J. Biochem., **139**, 671–676.
9. Kumamoto, K., Hirai, T., Kishioka, S., and Iwahashi, H. (2007) Identification of a radical formed in the reaction mixture of rat brain homogenate with a ferrous ion/ascorbic acid system using HPLC-EPR and HPLC-EPR-MS. Free Radic. Res., **41**, 650–654.
10. Iwahashi, H., Deterding, L.J., Parker, C.E., Mason, R.P., and Tomer, K.B. (1996) Identification of radical adducts formed in the reactions of unsaturated fatty acids with soybean lipoxygenase uing continuous flow fast atom bombardment with tandem mass spectrometry. Free Radic. Res., **25**, 255–274.
11. Spiteller, G. (1996) Enzymic lipid peroxidation – consequence of cell injury? Free Radic. Biol. Med., **25**, 1003–1009.
12. Jira, W., Spiteller, G., Carson, W., and Schramm, A. (1998) Strong increase in hydroxy fatty acids derived from linoleic acid in human low density lipoproteins of atheroscleorotic patients. Chem. Phys. Lipids., **91**, 1–11.
13. Jira, W., Richter, A., and Spiteller, G. (1997) Increased levels of lipid oxidation products in low density lipoproteins patients suffering from rheumatoid arthritis. Chem. Phys. Lipids., **87**, 81–89.

Chapter 7

EPR Spin-Trapping and Nano LC MS/MS Techniques for DEPMPO/$^{\bullet}$OOH and Immunospin-Trapping with Anti-DMPO Antibody in Mitochondrial Electron Transfer System

Yeong-Renn Chen

Abstract

Mitochondrial superoxide ($O_2^{\bullet-}$) production is an important mediator of oxidative cellular injury and pathogenesis of many diseases such as myocardial ischemia/reperfusion. The $O_2^{\bullet-}$ generated in mitochondria acts as a redox signal triggering cellular events including apoptosis, proliferation, and senescence. The molecular mechanism of $O_2^{\bullet-}$ produced by electron transport chain components isolated from the inner membrane is investigated by the technique of EPR spin trapping with 5-diethoxylphosphoryl-5-methyl-1-pyrroline *N*-oxide (DEPMPO), indicating that FMN/FMN-binding domain (complex I), ubiquinone (complex I and III), FAD/FAD-binding domain (complex II), and cytochrome *b* (complex III) control the mediation of $O_2^{\bullet-}$ production in mitochondria. $O_2^{\bullet-}$ generation by ETC also induces oxidative damage with protein radical formation. Immunospin-trapping with anti-DMPO antibody and subsequent mass spectrometry are used to define the specific site of oxidative damage, indicating cysteine-206 and tyrosine-177 of complex I/51 kDa FMN-binding subunit and cysteine-655 of complex II/70 kDa FAD-binding subunit are involved in specific protein radical formation caused by $O_2^{\bullet-}$ attack.

Key words: Mitochondria, Electron transport chain, Superoxide, EPR Spin trapping, Immunospin-trapping, Protein radical.

1. Introduction

Mitochondria are the major cellular source of oxygen free radicals. The generation of reactive oxygen species (ROS) is particularly relevant under the physiological conditions of low oxygen tension such as state IV respiration or certain pathological conditions such as inflammation and ischemia-reperfusion injury. Elucidating the

From: *Methods in Molecular Biology, vol. 477: Advanced Protocols in Oxidative Stress I,*
Edited by: D. Armstrong, DOI: 10.1007/978-1-60327-517-0_7, © Humana Press, New York, NY

fundamental mechanisms of mitochondria-derived $O_2^{\bullet-}$ generation is the key step to understand disease processes associated with oxidative stress.

Electron leakage for $O_2^{\bullet-}$ production is mediated by the electron transport chain during mitochondrial respiration. Two segments of ETC have been implicated in $O_2^{\bullet-}$ generation. One, on complex I (or NQR, NADH ubiquinone reductase), operates via electron leakage of $FMNH_2$ in NADH dehydrogenase (NDH, or flavoprotein subcomplex) *(1)* and of ubisemiquinone in hydrophobic subcomplex. The other, on complex III, mediates $O_2^{\bullet-}$ generation through the Q cycle mechanism *(2)*, in which electron leakage results in autooxidation of unstable ubsemiquinone and reduced heme b_{566} of low potential cytochrome *b*. Production of $O_2^{\bullet-}$ by complex II (or SQR, succinate-ubiquinone reductase) through $FADH_2$ or ubisemiquinone has also been reported *(2, 3, 4, 5)*. Overproduction of $O_2^{\bullet-}$ with a defect in complex II plays a pathological role in a wide variety of disease.

It is widely recognized that EPR spin-trapping is a sensitive and reliable approach to measure $O_2^{\bullet-}$ production in biological systems *(1, 2, 6)*. To obtain direct evidence for $O_2^{\bullet-}$ production mediated by ETC enzyme complex, we employed this technique to measure $O_2^{\bullet-}$ generation unambiguously and directly. Of the available spin traps, DEPMPO is ideal for quantitating $O_2^{\bullet-}$ production by ETC enzyme complex based on the following advantages: (a) DEPMPO is 40-fold more sensitive than the cyt *c* assay for the detection of $O_2^{\bullet-}$ *(6, 7)*, (b) DEPMPO traps $O_2^{\bullet-}$ with an efficiency of 60–70% *(6, 7)*, and (c) the $O_2^{\bullet-}$ adduct of DEPMPO/$^{\bullet}$OOH is more stable than that of DMPO/$^{\bullet}$OOH *(6, 8)*.

Immunospin-trapping with a polyclonal antibody against DMPO (5, 5-dimethyl-1-pyrroline-*N*-oxide) nitrone adduct has developed as a powerful approach to detecting protein radicals resulting from the oxidative stress of $O_2^{\bullet-}$ and other reactive oxygen species *(1, 3, 9)*. The antibody exhibits the advantage of high specificity and high sensitivity in the detection of protein-derived radicals. In combination of mass spectrometry, this approach is exclusively unique to define the molecular mechanism of oxidative damage involved in the specific protein radical formation.

2. Materials and Reagents

2.1. Equipment

1. Bruker EMX spectrometer operating at 9.86 GHz (X-band) with 100 kHz modulation frequency. EMX spectrometer is equipped with HS (high sensitivity) cavity (Bruker Instrument, Billerica, MA).

2. Shimadzu UV-visible dual beam recording spectrometer, model 2401.
3. Capillary liquid chromatography tandem mass spectrometry is performed on a Thermo Finnigan LTQ mass spectrometer equipped with a nanospray source operated in positive ion mode. The LC system was an UltiMate™ Plus system from LC-Packings (Sunnyvale, CA) with a Famos autosampler and Switchos column switcher. A 5-cm 75 μm ID BioBasic C18 column packed directly in the nanospray tip is used for chromatographic separation.

2.2. Reagents and Supplies

5-diethoxylphosphoryl-5-methyl-1-pyrroline N-oxide (DEPMPO) and anti-DMPO polyclonal antibody were purchased from ALEXIS Biochemicals (San Diego, CA). Around 50-μl capillary was purchased from VWR International (West Chester, PA). CRISTOSEAL was purchased from OXFORD LABWARE (St. Louis, MO). 5, 5-dimethyl pyrroline *N*-oxide (DMPO), β-nicotinamide adenine dinucleotide (NADH), Ubiquinone-1 (Q_1), ubiquinone-2 (Q_2), sodium succinate, horse heart cytochrome *c* (prepared without using trichloroacetic acid), Zn/Cu superoxide dismutase (SOD-1), diphenyleneiodinium chloride (DPI), thenoyl trifluoroacetone (TTFA), antimycin A, *N*-ethyl maleimide (NEM), glycine, and glycerol were from Sigma-Aldrich (St. Louis, MO). Nitrocellulose Pure Transfer Membranes, Amersham Enhanced Chemiluminescent ECL™ Western Blotting Detection Reagents, Amersham Hyperfilm ECL™ (8 × 10 inches), and Amersham ECL™-horseradish peroxidase linked anti-rabbit IgG antibody were purchased from GE Healthcare. Tween-20, β-mercaptoethanol, nonfat dry milk, TRANS-BLOT® paper (15 × 20 cm), bromophenol blue, Coomassie Brilliant Blue R-250, pre-stained molecular weight markers: dual color precision plus protein pre-stained standards were purchased from BIO-RAD (Hercules, CA). Sequencing grade trypsin was from Promega (Madison, WI). Chymotrypsin was from Roche Diagnostics (Indianapolis, IN). Montage In-Gel Digestion Kit was purchased from Millipore, Bedford, MA).

2.2.1. EPR Section

1. The spin trap, DEPMPO (~4.25 M colorless oil), is diluted with double distilled water (dd water, HPLC grade water) to 1 M stock solution and stored under argon at −80°C until needed.
2. NADH is dissolved at 15 mM ($\varepsilon_{340\ nm}$ = 6.22 mM^{-1}cm^{-1}) in dd water and stored at −80°C. Q_1 ($\varepsilon_{275\ nm–290\ nm}$ = 12.25 mM^{-1}cm^{-1}) from Sigma-Aldrich is dissolved in 100% ethanol and stored at −20°C.

3. Succinic acid from Sigma is prepared in dd water in 500 mM, and adjusted to pH 8 by 1N NaOH, and stored at room temperature. Q_2 ($\varepsilon_{275\ nm-290\ nm}$ = 12.25 $mM^{-1}cm^{-1}$, $OD_{290\ nm}$ is measured by $NaBH_4$ reduction) is dissolved in 100% ethanol and stored at −20°C.
4. Horse heart cytochrome *c* is dissolved in 50 mM phosphate buffer, pH 7.0, at 1 mM ($\varepsilon_{550\ nm}$ = 18.5 $mM^{-1}cm^{-1}$) stock solution and stored in −80°C (*see* **Note 1**).
5. SOD-1 is dissolved in 50 mM phosphate buffer, pH 7.0, containing 20% glycerol at a final concentration of 30 units/μl and stored in −80°C (*see* **Note 2**)
6. Inhibitors: DPI is dissolved in DMSO at a concentration 5 mM. TTFA is dissolved in DMSO at a final concentration of 100 mM. They can be stored at −20°C.

2.2.2. Immunospin Section

1. The spin trap, DMPO (10 M colorless oil, *see* **Note 3**), is vacuum-distilled twice and stored under argon at −80°C until needed.
2. Thiol blocking (or alkylating) reagent, NEM, is dissolved in water at a concentration of 250 mM.

2.2.3. SDS–Polyacrylamide Gel Electrophoresis

1. Novex® 4–20% Tris-Glycine pre-cast gels 1.0 mm is purchased from Invitrogen (Carlsbad, CA).
2. Sample buffer (4×): 0.25 M Tris-HCl, pH 6.8, 40% (v/v) glycerol, 8% (w/v) SDS, 0.1% (w/v) bromophenol blue, and 0.4% (v/v) β-mercaptoethanol.
3. Running buffer (10×): 2.9% (w/v) Tris base, 14.4% glycine (w/v), and 1% (w/v) SDS in distilled water (do not adjust pH with base). The pH of 1× running buffer should be ~8.3.
4. Pre-stained molecular weight marker is used as purchased.

2.2.4. Western Blot Section

1. Transfer buffer: 25 mM Tris (do not adjust pH), 190 mM glycine, 20% (v/v) methanol, and stored at 4°C.
2. Tris-buffered saline with Tween-20 (TTBS): 25 mM Tris-HCl, pH 7.4, 137 mM NaCl, 0.1% Tween-20.
3. Blocking buffer: 5% (w/v) non-fat dry milk in TTBS.
4. Antibody dilution buffer: TTBS supplemented with 5% non-fat dry milk.

2.2.5. Mass Spectrometry Section

1. 0.25% Coomassie Brilliant Blue R-250 in solution containing 45.4% (v/v) methanol and 9.2% (v/v) acetic acid.

3. Methods

Cautious control experiments are required. To obtain the reliable quantitative measurement, it is important to carry out computer simulation due to the complexity of multi-line spectrum of DEPMPO/•OOH adduct. Occasionally, the decay of DEPMPO/•OOH to DEPMPO/•OH is a concern for certain enzymatic systems; however, computer simulation is extremely powerful to identify and quantitate each adduct from the obtained spectrum. For the additional purpose of calculating the basal activity of enzyme-mediated superoxide generation activity (SGA), spin quantitation can be performed using 4-hydroxyl-2,2,6,6-tetramethylpiperidinyloxy (TEMPOL) as a standard *(10)*. Furthermore, confirmation of $O_2^{\bullet-}$ production by the approaches other than DEPMPO spin-trapping is suggested, but this is optional. The available optional assay approaches include acetylated cytochrome *c* reduction assay *(11, 12)*, electrochemical detection by NO electrode *(2)*, and UV-VIS spectral analysis *(2)*. However, $O_2^{\bullet-}$ detection by these approaches requires a confirmation by the addition of SOD-1.

Protein radical can be induced by $O_2^{\bullet-}$ attack [detected in the flavoprotein of complex I and complex II *(1, 3)*] or H_2O_2 [detected in the hemoglobin *(13)* or cytochrome *c* *(14)*]. Therefore, it is important to confirm the detected signal by SOD-1 or catalase when the immunospin trapping approach was employed. The stability of DMPO-protein radical adduct is the major concern of immunospin trapping technique. Normally, we have observed that the intensity of Western signal is decreased with the prolonged incubation time *(14)*. This signal decrease is due to the loss or degradation of the spin trap with time, which potentially limits the detection of mass spectrometric analysis.

3.1. Preparations of Samples for Assay of Superoxide Generation by EPR Spin-Trapping with DEPMPO

3.1.1. Superoxide Generation Mediated by NADH Dehydrogenase (NDH) (reference 1)

1. NDH (Fp subcomplex of complex I) is prepared from bovine heart submitochondrial particles (SMP). Purified NDH (50 μg/ml, 0.62 μM) was incubated with DEPMPO (20 mM) in phosphate-buffered saline. The $O_2^{\bullet-}$ production is initiated by the addition of NADH (0.5 mM) (**Fig. 7.1A**).
2. The EPR spectrum is immediately recorded after signal averaging three scans. EPR measurement is performed on a Bruker EMX spectrometer operating at 9.86 GHz with 100 kHz modulation frequency at room temperature. The reaction mixture was transferred to a 50-μl capillary and sealed with CRISTOSEAL.

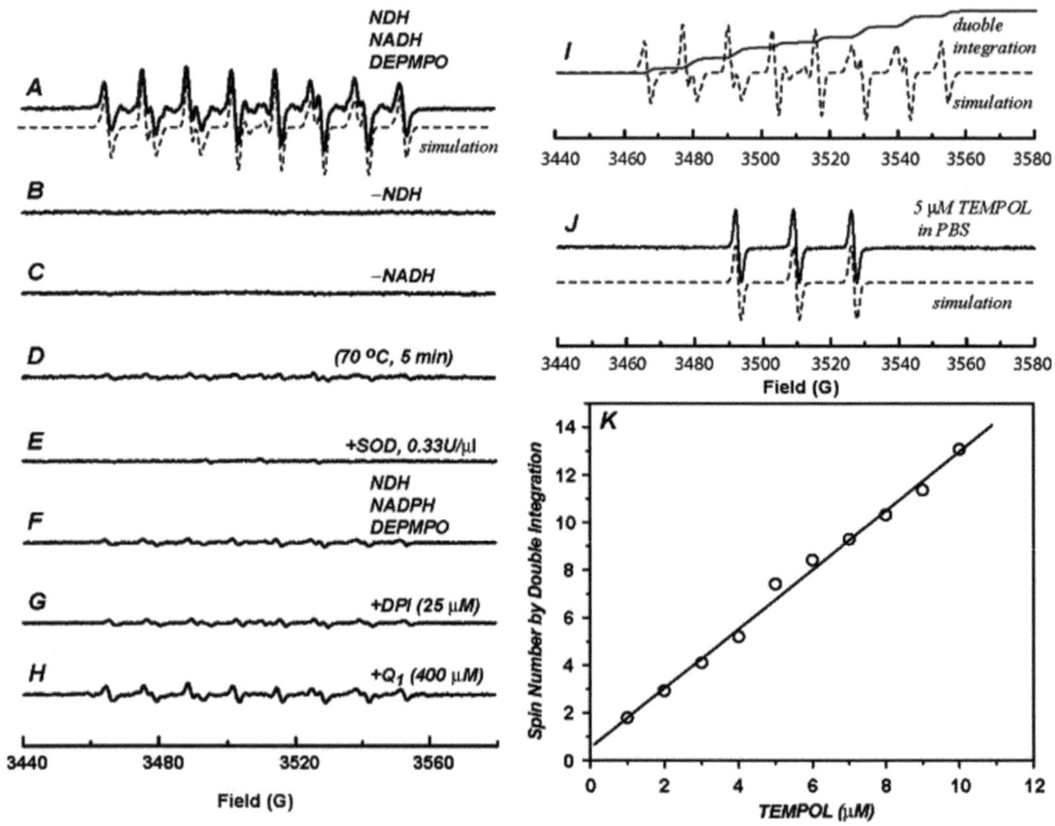

Fig. 7.1. EPR spin-trapping of superoxide generated from NDH in the presence of DEPMPO (**A–H**) and spin quantitation of DEPMPO/•OOH using TEMPOL as a standard (**I–J**). (**A**) The computer simulation (dashed line) superimposed on the experimental spectrum (solid line) of DEPMPO/•OOH obtained from NDH, DEPMPO, and NADH in PBS. (**B**) The same as **A** except that NDH is omitted from the system. (**C**) The same as **A**, except that the substrate NADH is omitted from the system. (**D**) The enzyme is heated at 70°C for 5 min prior to EPR measurement. (**E**) The same as **a**, except that SOD was added to mixture before the reaction is initiated by NADH. (**F**) The same as **A**, except that NADH is replaced with NADPH. (**G**) The same as **A**, except that NDH was pre-treated with DPI prior to NADH initiation. (**H**) The same as **A**, except Q_1 was included in the mixture prior to the addition of NDH. (**I**) Double integration (solid line) of simulated spectrum of DEPMPO/•OOH. (**J**) Computer simulation (dashed line) superimposed with experimental EPR spectrum (solid line) obtained from 5 µM TEMPOL in PBS. (**K**) Spin number obtained from double integration of simulated spectrum as a function of TEMPOL concentration.

3. The sample contained capillary is then positioned in the HS cavity (Bruker Instrument, Billerica, MA). The sample is scanned using the following instrumental setting: center field, 3510 G; sweep width, 140 G; power, 20 mW; receiver gain, 2×10^5; modulation amplitude, 1 G; time of conversion, 81.92 ms; time constant, 327.68 ms; signal average, three scans (*see* **Fig. 7.1**).
4. Control experiments should be immediately carried out by omitting the enzyme (NDH, *see* **Fig. 7.1B**) and omitting the NADH (*see* **Fig. 7.1C**) from the complete system.
5. Confirmation of DEPMPO/•OOH with SOD-1: Repeat the steps 1–3, but SOD-1 is added in the reaction mixture

at a final concentration of 0.33 U/μl prior to addition of NADH (0.5 mM) (*see* **Fig. 7.1E** and **Note 4**).

6. Effect of NADPH: NADH (0.5 mM) is replaced with NADPH (0.5 mM) in the complete system (**Fig. 7.1F**).
7. Confirmation of enzyme-mediated $O_2^{\bullet-}$ generation: NDH (50 μg/ml, 0.62 μM) in PBS is heated at 70°C for 5 min, which will inactivate enzymatic activity greater than 90%. Repeat steps 1–3 after the temperature of heat-denatured NDH returns to room temperature (*see* **Fig. 7.1D**).
8. Effect of inhibitors on $O_2^{\bullet-}$ production: NDH (50 μg/ml, 0.62 μM) in PBS is pre-incubated with DPI (25 μM) at room temperature for 3 min. DEPMPO (20 mM) and NADH (0.5 mM) are then added to DPI-treated NDH prior to EPR measurement (*see* **Fig. 7.1G** and **Note 4**).
9. Effect of electron acceptor, Q_1, on the $O_2^{\bullet-}$ production: NDH (50 μg/ml, 0.62 μM) in PBS is incubated with DEPMPO (20 mM) and Q_1 (400 μM), and the reaction is initiated with the addition of NADH (0.5 mM) prior to EPR measurement (**Fig. 7.1H**).

3.1.2. Superoxide Generation Mediated by Complex I (reference 10)

1. Complex I (0.2 mg/ml) is incubated with Q_1 (0.2 mM), DEPMPO (20 mM), and DTPA (1 mM) at room temperature. $O_2^{\bullet-}$ generation is initiated with NADH (0.5 mM) and subjected to EPR measurement as described in the steps 1–3 of **Section 3.1.1**.
2. The control experiment is carried out by excluding complex I from the reaction system described in step 1.

3.1.3. Superoxide Generation Mediated by Complex II (3).

1. Complex II (1–2 μM, based on heme *b*) is incubated with Q_2 (0.18 mM), DEPMPO (20 mM), and DTPA (1 mM) at room temperature. $O_2^{\bullet-}$ generation is initiated with succinate (0.18 mM) and subjected to EPR measurement as described in the steps 1–3 of **Section 3.1.1**.
2. The control experiment is carried out by excluding complex II from the reaction system described in step 1.
3. Effect of inhibitor on $O_2^{\bullet-}$ production by complex II: TTFA (100 mM stock in DMSO) is added to mixture containing complex II, Q_2, DEPMPO, and DTPA to a final concentration of 1 mM. DEPMPO/$^{\bullet}$OOH production is initiated by the addition of succinate and measured with EPR.

3.1.4. Simulation of the EPR Spectrum of DEPMPO/OOH for Spin Quantitation

1. The simulation of experimental spectrum is operated by Winsim2002 program, which is available at the NIEHS website: http://www.niehs.nih.gov/research/resources/software/tools/index.cfm *(15)*: The parameters of hyperfine coupling constants for DEPMPO/$^{\bullet}$OOH simulation

are: isomer 1: $a^N = 13.14$ G, $a^H_\beta = 11.04$ G, $a^H_\gamma = 0.96$ G, $a^P = 49.96$ G (80% relative concentration); isomer 2: $a^N = 13.18$ G, $a^H_\beta = 12.59$ G, $a^H_\gamma = 3.46$ G, $a^P = 48.2$ G (20% relative concentration) (**Fig. 7.1A**, dashed line) *(8)*. The parameter of DEPMPO/•OH simulation is $a^N = 14.03$ G, $a^H_\beta = 13.34$ G, $a^P = 47.19$ G *(8)*. However, the addition of DEPMPO/•OH to the parameters of simulation program is optional.

2. The simulated spectrum is optimized by the Winsim program. The correlation coefficient with more than 0.95 is considered a good simulation.
3. The simulated spectrum is exported as LMB binary file, and then operated with EPR application program (available at the NIEHS website) to obtain double integration for further spin quantitation (**Fig. 7.1I**, solid line).
4. Spin quantitation with TEMPOL as an internal standard *(10)*: Prepare a series of concentrations of TEMPOL dissolved in PBS from 1 to 10 μM. The three-line spectra of TEMPOL are measured with EPR and then subjected to computer simulation using the hyperfine coupling constant, $a^N = 16.99$ G (**Fig. 7.1J**). The simulated spectra are subjected to double integration and the standard curve for spin quantitation is plotted by spin number vs the concentration of TEMPOL (**Fig. 7.1K**)

3.2. Oxidative Stress-induced Protein Radicals of NDH and SQR (Complex II) Probed by Immunospin-trapping with DMPO(1, 3)

3.2.1. Spin-trapping with DMPO

1. NDH (0.5 m/ml)/or SQR (1 mM, DTT-treated) is mixed with a nitrone spin trap, DMPO (100 mM) in PBS containing DTPA (1 mM), and the reaction is initiated by the addition of NADH (1 mM)/or succinate (0.18 mM) at 37°C. The reaction mixture is allowed to incubate for 1-h prior to SDS–PAGE. The reaction is terminated by addition of sample buffer, and then heated at 70°C for 5 min.

3.2.2. SDS–PAGE

1. These instructions assume the use of XCell *SureLock*™ Mini-Cell gel electrophoresis system and Novex® Pre-Cast Gel of 4–20% Tris-Glycine ZOOM Gel 1.0 mm, IPG well. They are easy to operate and provide a highly reproducible experiment.
2. Prepare 1× running buffer by diluting 80 ml of 10× running buffer to ~800 ml with distilled water.
3. Complete the assembly of the pre-cast gel and XCell *Surelock*™ Mini-Cell unit, and connect to a power supply.
4. Add running buffer to the upper buffer chamber and lower buffer chamber.
5. Load the sample of NDH (10 μg)/or SQR (30 pmol) in a well, including one well for 5 μl of pre-stained marker.

6. Gel is run at room temperature at 100 V for 2 h [current 30–40 mA/gel (start); 12–13 mA/gel (end)].

3.2.3. Western Blotting for DMPO Adduct of Protein Radical(s)

1. The protein bands that have been separated by SDS–PAGE are electrophoretically transferred to nitrocellulose membrane. These directions assume the use of XCell II™ Blot Module transfer system (Invitrogen).
2. Prepare 1000 ml of transfer buffer containing 25 mM Tris, 192 mM glycine, and 20% methanol.
3. Pre-soak the blotting pads, pre-cut nitrocellulose membrane, and pre-cut filter paper with ~500 ml of transfer buffer for at least 1-h.
4. Remove the gel immediately upon completion of SDS–PAGE (**Section 3.2.2**).
5. Wet the surface of gel with transfer buffer and place pre-soaked transfer membrane on the gel, and then sandwiched with two pre-soaked filter papers on top of membrane and two pre-soaked blotting pads. The membrane/gel system is assembled in the cassette of blot module filled with transfer buffer and the electrophoresis tank filled with 500 ml transfer buffer.
6. Transfer is accomplished at 50 V for 90 min at 4°C.
7. The nitrocellulose membrane is then removed and subjected to blocking with 25 ml of TTBS containing 5% dry milk for 1-h at room temperature on a rocking platform.
8. The blots are then incubated with 1:2000 dilution of anti-DMPO antibody in blocking buffer overnight at 4°C (*see* **Note 5**).
9. The primary antibody is then removed. The membrane is then washed with TTBS three times at room temperature, and incubated with 1:3000 dilution of HRP-conjugated anti-rabbit IgG antibody in blocking buffer for 1-h at room temperature.
10. The secondary antibody is then removed, and the membrane is washed twice in TTBS and twice in TBS, and then visualized using ECL Western blotting detection agent.
11. With the above approaches (**Sections 3.2.1–3.2.3**), $O_2^{\bullet -}$-induced protein radical(s) detected as DMPO adducts of NDH (**Fig. 7.2A–C**), NQR (**Fig. 7.2D**), and SQR (**Fig. 7.2E**) are shown in **Fig. 7.2A–E**.

3.3. Mass Spectrometry to Identify the Specific DMPO Binding Site (1, 3, 13)

1. Remove the gel upon completion of SDS–PAGE (**Section 3.2.2**), and the gel is rinsed in the fixing solution (methanol: acetic acid: water = 40:10:50) for 30 min at room temperature.

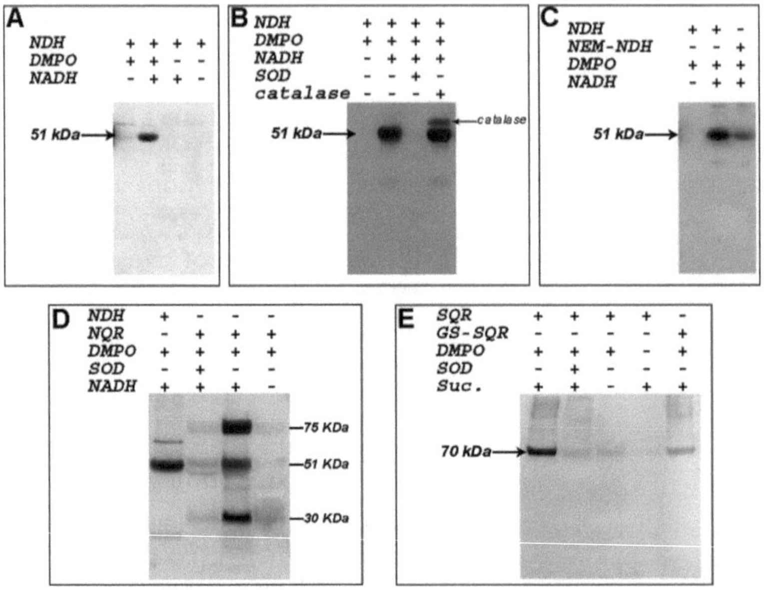

Fig. 7.2. Detection of the DMPO adducts of the NDH-derived protein radicals (**A–C**) Complex I (or NQR)-derived protein radicals (**D**) and Complex II (or SQR)-derived protein radicals (**E**) by Western blot using anti-DMPO nitrone adduct polyclonal antibody. NEM-NDH denotes that the cysteines on the surface of NDH is alkylated with NEM (*N*-ethyl maleimide). GS-SQR denotes S-glutathionylated SQR obtained by exposure of SQR to GSSG.

3.3.1. In-Gel Digestion

2. The fixing solution is discarded and the gel is incubated with 20 ml of Coomassie blue staining solution for 1-h at room temperature.
3. The staining solution is discarded and the gel is de-stained by incubation with solution containing 0.5% methanol and 0.75% acetic acid overnight at room temperature.
4. The gel is equilibrated with 50 ml of water with three changes of water.
5. The bands of 51 kDa FMN-binding subunit (for NDH) or 70 kDa FAD-binding subunit (for SQR) were trimmed as closely as possible to minimize background polyacrylamide material (*see* **Note 6**).
6. The trimmed gels are washed twice with 50% methanol/5% acetic acid for several hours, and the gels are dehydrated with acetionitrile.
7. The gels are reconstituted with dithiothretol to reduce cysteinyl residues, and iodoacetamide is then added to alkylate cysteine sulfhydryls.
8. Gels are washed with cycles of acetonitrile and ammonium bicarbonate buffer. Gels are then dried by speed vac.

9. 50 μl of sequencing grade trypsin (20 ng/μl) or chymotrypsin (25 ng/μl) are added to the dehydrated gel.
10. The gels are set on ice 10 min for rehydration, and 20 μl of 50 mM ammonium bicarbonate buffer are added. The mixture is incubated at room temperature for overnight.
11. The peptides in gels are extracted with 50% acetonitrile with 5% formic acid several times, and extracted peptides are pooled together and concentrated in a speed vac to ~25 μl for further analysis.

3.3.2. Nano-LC MS/MS

1. Capillary liquid chromatography tandem mass spectrometry is performed on a Thermo Finnigan LTQ mass spectrometer equipped with a nanospray source operated in positive ion mode. The LC system was a UltiMate™ Plus system from LC-Packings (Sunnyvale, CA) with a Famos autosampler and Switchos column switcher. A 5-cm 75 μm ID BioBasic C18 column packed directly in the nanospray tip is used for chromatographic separation.
2. Mobile phase preparation for capillary LC system is as follows: solvent A is water containing 50 mM acetic acid, solvent B is acetonitrile.
3. 2.5-μl aliquots of each sample are injected onto the column for the analysis. Peptides are eluted off the column into the LTQ system using a gradient of 2–80% solvent B over 48 min with a flow rate of 300 nl/min. A total run time is 65 min.
4. The scan sequence of the mass spectrometer was programmed for a full scan and MS/MS scans of the 10 most abundant peaks in the spectrum. Dynamic exclusion was used to exclude multiple MS/MS of the same peptide after detecting it three times.
5. Peak list (mgf file) from MS/MS data is generated using the Mascot 2.0 active perl script with standard data processing parameters. Data base searches are performed using the MASCOT 2.0 (Matrix Science, Boston, MA) and PEAKS (Bioinformatics Solutions Inc., Waterloo, ON, Canada) programs.
6. The mass accuracy of the precursor ions is set to 1.5 Da to accommodate accidental selection of the C13 ion and the fragment mass accuracy is set to 0.5 Da. Number of missed cleavages permitted in the search is 2 for both tryptic and chymotryptic digestions. Considered modifications (variable) are oxidation (met), carbamidomethylation (cys) and DMPO modification (tyr cys, and his).
7. The MS/MS spectra of DMPO-binding peptide containing C_{206} of NDH 51 kDa and DMPO-binding peptide containing C_{655} of SQR 70 KDa are shown in **Fig. 7.3**.

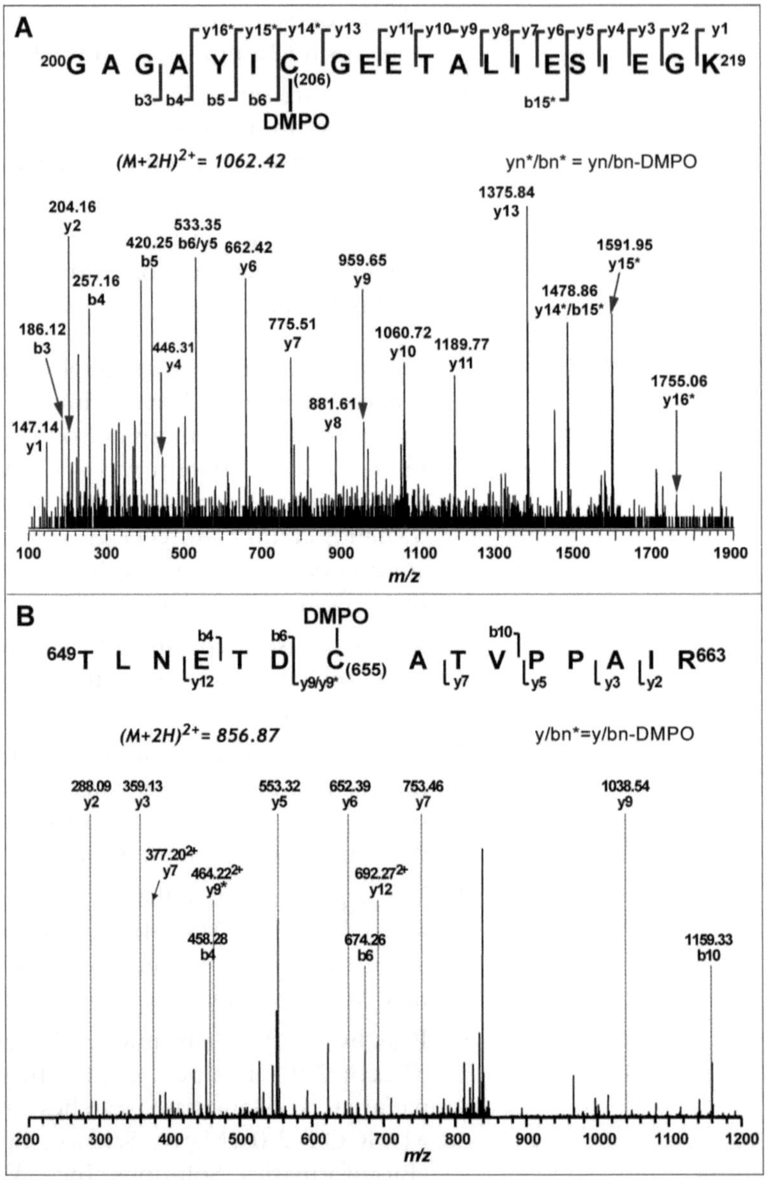

Fig. 7.3. Tandem mass spectra (MS/MS) of the doubly protonated molecular ions of the DMPO-binding peptides. (**A**) ^{200}GAGAYICGEETALIESIEGK^{219} from the trypsic digests of 51 kDa polypeptide of NDH. (**B**) ^{649}TLNETDCATVPPAIR^{663} from trypsic digests of 70 kDa polypeptide of SQR. Sequencing-specific ions are labeled as *y* and *b* ions. The amino acid residues involved in DMPO binding, C_{206} of NDH, and C_{655} of SQR are identified.

4. Notes

1. Horse heart cytochrome *c* should be highest grade available in commercial, and prepared without using trichloroacetic acid.

2. The use of Mn-SOD (SOD-2) is optional (from Sigma-Aldrich). It works as well as SOD-1.
3. Most DMPO available in commercial (Sigma-Aldrich, or Alexis Biochemicals) comes with yellowish color, and contains high amounts of impurity. Therefore, it requires vacuum distilled at 77°K (the temperature of liquid nitrogen). Alternatively, highly purified DMPO crystal can be purchased from Dojindo Molecular Technology, Inc. (http://www.dojindo.com, Gaithersburg, MD) and further purification is not necessary.
4. The optimal dosage of SOD-1 or inhibitor of DPI should be tested by titration.
5. The titer of anti-DMPO antibody may vary with different protein/DMPO adducts. It should be tested by ELISA to obtain the optimal titer.
6. In-gel digestion and nano-LC/MS/MS analysis should be conducted as soon as possible once gel de-staining is completed.

Acknowledgments

The author would like to thank Dr. Ronald P. Mason (NIEHS/NIH, Research Triangle Park, NC) for development of anti-DMPO polyclonal antibody and for his advice; Dr. Bradley E. Sturgeon (Monmouth College, Monmouth, IL) for his advice with regards to computer simulation of DEPMPO/•OOH. Drs Liwen Zhang and Kari B. Green-Church (MS and Proteomics Facility, CCIC, The Ohio State University) for their expertise in Nano-LC/MS/MS. This work was supported by RO1 HL83237.

References

1. Chen, Y. R., Chen, C. L., Zhang, L., Green-Church, K. B., and Zweier, J. L. (2005) Superoxide generation from mitochondrial NADH dehydrogenase induces self-inactivation with specific protein radical formation. J Biol Chem 280, 37339–37348.
2. Chen, Y. R., Chen, C. L., Yeh, A., Liu, X., and Zweier, J. L. (2006) Direct and indirect roles of cytochrome b in the mediation of superoxide generation and NO catabolism by mitochondrial succinate-cytochrome c reductase. J Biol Chem 281, 13159–13168.
3. Chen, Y. R., Chen, C. L., Pfeiffer, D. R., and Zweier, J. L. (2007) Mitochondrial complex II in post-ischemic heart: Oxidative injury and the role of protein S-glutathionylation. J Biol Chem 282, 32640–32654.
4. Guo, J., and Lemire, B. D. (2003) The ubiquinone-binding site of the Saccharomyces cerevisiae succinate-ubiquinone oxidoreductase is a source of superoxide. J Biol Chem 278, 47629–47635.
5. Messner, K. R., and Imlay, J. A. (2002) Mechanism of superoxide and hydrogen peroxide formation by fumarate reductase, succinate dehydrogenase, and aspartate oxidase. J Biol Chem 277, 42563–42571.

6. Vasquez-Vivar, J., Kalyanaraman, B., Martasek, P., Hogg, N., Masters, B. S., Karoui, H., Tordo, P., and Pritchard, K. A., Jr. (1998) Superoxide generation by endothelial nitric oxide synthase: the influence of cofactors. Proc Natl Acad Sci U S A 95, 9220–9225.
7. Roubaud, V., Sankarapandi, S., Kuppusamy, P., Tordo, P., and Zweier, J. L. (1997) Quantitative measurement of superoxide generation using the spin trap 5-(diethoxyphosphoryl)-5-methyl-1-pyrroline-N-oxide. Anal Biochem 247, 404–411.
8. Frejaville, C., Karoui, H., Tuccio, B., Le Moigne, F., Culcasi, M., Pietri, S., Lauricella, R., and Tordo, P. (1995) 5-(Diethoxyphosphoryl)-5-methyl-1-pyrroline N-oxide: a new efficient phosphorylated nitrone for the in vitro and in vivo spin trapping of oxygen-centered radicals. J Med Chem 38, 258–265.
9. Detweiler, C. D., Deterding, L. J., Tomer, K. B., Chignell, C. F., Germolec, D., and Mason, R. P. (2002) Immunological identification of the heart myoglobin radical formed by hydrogen peroxide. Free Radic Biol Med 33, 364–369.
10. Chen, C. L., Zhang, L., Yeh, A., Chen, C. A., Green-Church, K. B., Zweier, J. L., and Chen, Y. R. (2007) Site-specific S-glutathiolation of mitochondrial NADH ubiquinone reductase. Biochemistry 46, 5754–5765.
11. Boveris, A. (1984) Determination of the production of superoxide radicals and hydrogen peroxide in mitochondria. Methods Enzymol 105, 429–435.
12. Taylor, E. R., Hurrell, F., Shannon, R. J., Lin, T. K., Hirst, J., and Murphy, M. P. (2003) Reversible glutathionylation of complex I increases mitochondrial superoxide formation. J Biol Chem 278, 19603–19610.
13. Deterding, L. J., Ramirez, D. C., Dubin, J. R., Mason, R. P., and Tomer, K. B. (2004) Identification of free radicals on hemoglobin from its self-peroxidation using mass spectrometry and immuno-spin trapping: observation of a histidinyl radical. J Biol Chem 279, 11600–11607.
14. Chen, Y. R., Chen, C. L., Liu, X., Li, H., Zweier, J. L., and Mason, R. P. (2004) Involvement of protein radical, protein aggregation, and effects on NO metabolism in the hypochlorite-mediated oxidation of mitochondrial cytochrome c. Free Radic Biol Med 37, 1591–1603.
15. Duling, D. R. (1994) Simulation of multiple isotropic spin-trap EPR spectra. J Magn Reson B 104, 105–110.

Chapter 8

Determination of High Mitochondrial Membrane Potential in Spermatozoa Loaded with the Mitochondrial Probe 5,5′,6,6′-Tetrachloro-1,1′,3,3′-Tetraethylbenzimidazolyl-Carbocyanine Iodide (JC-1) by Using Fluorescence-Activated Flow Cytometry

H. David Guthrie and Glenn R. Welch

Abstract

A flow cytometric method was developed to identify viable, energized sperm cells with high mitochondrial inner transmembrane potential ($\Delta\Psi_m$), >80–100 mV using the mitochondrial probe 5,5′,6,6′-tetrachloro-1,1′,3,3′-tetraethylbenzimidazolylcarbocyanine iodide (JC-1) and the impermeant nuclear stain propidium iodine (PI). This flow cytometric method is described in detail here. When in contact with membranes possessing a high $\Delta\Psi_m$, JC-1 forms aggregates (J_{agg}) that are fluorescent at 590 nm in response to 488 nm excitation. We found that the reactive oxygen species generator, menadione reduced sperm motility and reduced $\Delta\Psi_m$ in a dose responsive fashion that was closely correlated with the loss of motility.

Key words: 5,5′,6,6′-Tetrachloro-1,1′,3,3′-tetraethylbenzimidazolylcarbocyanine iodide, flow cytometry, mitochondrial transmembrane potential, sperm motility.

1. Introduction

Mitochondria are the major sites of intracellular reactive oxygen species (ROS) formation and their formation results in a disruption of electron transport *(1)*. The coupling of electron transport to oxidative phosphorylation maintains a high mitochondrial transmembrane potential ($\Delta\Psi_m$) that is required for mitochondrial ATP production in somatic cells *(2)*. Treatment of somatic cells with menadione has served as an

From: *Methods in Molecular Biology, vol. 477: Advanced Protocols in Oxidative Stress I,*
Edited by: D. Armstrong, DOI: 10.1007/978-1-60327-517-0_8, © Humana Press, New York, NY

experimental model to generate intracellular superoxide as a consequence of the disruption of mitochondrial NADH-ubiquinone oxidoreductase *(3)*. We hypothesized that in sperm, ROS formation would uncouple electron transport and oxidative phosphorylation resulting in a decrease in the number of sperm that maintain a high $\Delta\Psi_m$ and sperm motility *(4)*.

The percentage of cells with a high $\Delta\Psi_m$ can be determined flow cytometrically using the mitochondrial probe 5,5′,6,6′-tetrachloro-1,1′,3,3′-tetraethylbenzimidazolylcarbocyanine iodide (JC-1). The probe JC-1 is freely permeable to cells and undergoes reversible transformation from a monomer to an aggregate form (J_{agg}) when it binds to membranes having a high $\Delta\Psi_m > 80$ to 100 mV *(5)*. The J_{agg} form is fluorescent at 590 nm in response to 488 nm excitation. Flow cytometric analysis of J_{agg} fluorescence has been used to identify sperm cells with high $\Delta\Psi_m$ in a number of experiments *(6, 7, 8)*. Our method is an improvement over these methods because we specifically enumerate the viable sperm with high $\Delta\Psi_m$ by electronically gating on viable sperm that contain low PI fluorescence *(4)*. The ability to distinguish between viable and nonviable sperm cells is particularly important for evaluation of sperm subjected to long-term hypothermic liquid storage or cryopreservation where viability is reduced compared to cells in freshly collected semen.

2. Materials

2.1. Equipment

1. SpermaCue micro-cuvette (cat. # 12300/1100), Minitube of America
2. Spermacue photometer (cat. # 12300/0500), Minitube of America
3. Epics XL-MCL analyzer, Beckman-Coulter
4. IVOS version-12 motion analysis system, Hamilton Thorne Biosciences
5. Twenty-micron four-chamber glass counting slide (SC 20-01 FA), Leja Products

2.2. Reagents

2.2.1. Boar Sperm Extender

1. Beltsville Thawing Solution (BTS) (Minitube of America, Verona, WI)

2.2.2 Percoll and Suspending Media

1. Percoll (P-1644, Sigma-Aldrich, Milwaukee, WI)
2. "10X saline-Hepes medium" consists of NaCl, 1.37 M; Hepes, 200 mM; glucose, 100 mM; and KOH, 25 mM.
3. "1X saline-Hepes medium" consists of one volume of 10X saline-Hepes medium diluted with 9 volumes of deionized water (approximately 290 mOsm/kg, pH 7.4)
4. Prepare "Percoll stock suspension" by mixing 56.3 mL of Percoll with 5 mL of 10X saline-Hepes medium (final osmolality 311 mOsm/kg).
5. The 70% Percoll suspension is prepared by adding 13.16 mL of 1X saline-Hepes medium to 30.65 mL (half) of the Percoll stock suspension (final osmolality typically 307 mOsm/kg).
6. The 35% Percoll suspension is prepared by adding 30.65 mL of Percoll stock solution to 56.92 mL of 1X saline-Hepes medium (final osmolality typically 300 mOsm/kg). Kanamycin is added to each Percoll suspension (50 μg/mL).
7. Four milliliters of 35% Percoll and 2 mL of 70% Percoll were transferred into 15 mL conical and 12 × 75 mm tubes, respectively, and stored at 4°C until use.

2.2.3. Modified Tyrode's Sperm Medium (TYR)

Noncapacitating basal medium excluded bicarbonate and consists of 116 mM NaCl, 3.1 mM KCl, 0.4 mM $MgSO_4$, 0.3 mM NaH_2PO_4, 5 mM glucose, 21.7 mM sodium lactate, 1 mM sodium pyruvate, 20 mM Hepes, 0.1% polyvinyl chloride, and 50 μg kanamycin/mL. Osmolarity is adjusted to 300 mOsm/kg with NaCl. Pyruvate is added on the day of use. Medium pH was adjusted to 7.3–7.4 at 37°C.

2.2.4. Working Solution for Menadione

Menadione (M5625, Sigma Chemical Co., St. Louis, MO) is dissolved in dimethyl sulfoxide, 200 mM.

2.2.5. Working Solution for JC-1

JC-1(T-3168, Molecular Probes Inc.) is dissolved in dimethyl sulfoxide, 0.15 mM (*see* **Note 1**).

2.2.6. Working Solution for PI

PI (P-4170, Sigma Chemical Co., St. Louis, MO) is dissolved in deionized water, 2.4 mM

3. Methods

3.1. Determination of Sperm Concentration

1. A 20 μL aliquot of either fresh boar semen (diluted with one volume of BTS) or a washed sperm suspension (150–400 × 10^6 sperm/mL) in TYR or BTS is pipetted into a micro-cuvette (catalog # 12300/1100, Minitube of America, Verona, WI).

2. The micro-cuvette is placed into the instrument micro-cuvette holder and inserted into a photometer (SpermaCue™, catalog # 12300/0500, Minitube of America, Verona, WI) and analyzed for optical density calibrated to record concentration as 10^6 sperm/mL.

3.2. Removal of Seminal Plasma and Semen Extenders by 2-Step Discontinuous Percoll Gradient Centrifugation

1. Sperm are extended to a concentration of 150×10^6/mL in 3 mL of BTS or TYR.
2. On day of use, the tubes containing 35% and 70% Percoll are warmed to room temperature and 70% Percoll (2 mL) is layered beneath 4 mL of 35% Percoll in 15 mL conical centrifuge tube.
3. The sperm suspension (3 mL) is layered on top of the 35% Percoll layer and then sperm are isolated from semen constituents and extenders by centrifugation at room temperature for 5 min at 200 g followed by 15 min at 1000 g.
4. The medium is aspirated from the loosely pelleted sperm cells and washed once in the TYR with centrifugation at 350 g for 10 min.
5. Sperms in the pellet of each sample are resuspended and diluted to 60×10^6/mL in TYR.

3.3. Menadione Induction of Reactive Oxygen Species

1. Aliquots containing 0, 2, 20, and 60 μM of menadione are removed from the 200 mM working solution, transferred into 0.5 mL TYR in 15 mL conical centrifuge tubes, and pre-warmed to 37°C in a heating block for 5 min.
2. After warming, a 0.5 mL aliquot containing 30×10^6 sperm is removed from the Percoll-washed sperm suspension and added to each pre-warmed treatment tube above and incubated aerobically at 37°C for sampling at 5, 30, 60, and 120 min for motion analysis and JC-1/PI flow cytometry.
3. The formation of ROS was confirmed by oxidation of hydroethidine measured by flow cytometry at 30 min *(4)*.

3.4. Staining with Cells Removed at Scheduled Times During Incubation with JC-1 and PI

1. A set of 12 × 75 mm polypropylene tubes each containing 467 μL of TYR are prepared for each sample to contain JC-1 and PI at 0.3 μM and 9.6 μM, respectively.
2. After the incubation treatments described in **Sectioon 3.3**, aliquots of 1×10^6 sperm (33 μL) are transferred to tubes containing JC-1 and PI.
3. The staining time and temperature for JC-1 and PI prior to flow cytometry are 10 min at 37°C (*see* **Notes 2-5** for an alternative JC-1/PI cell loading procedure).

3.5. Fluorescence Activated Flow cytometry

1. These instructions assume the use of a Beckman-Coulter Epics XL-MCL Analyzer (Beckman-Coulter, Hialeah, FL), or equivalent instrument, equipped with a single 488 nm excitation source and at least three fluorescence detectors. A forward and side light scatter gate was used to select single sperm from clumps and debris.
2. The standard filter configuration of the flow cytometer is altered for analysis of JC-1 fluorescence. The wave length of light directed to fluorescence detector 1 (FL1) with a 525 nm band pass (BP) filter to detect "green fluorescence" is not changed. However, the 575 BP and the 600 nm dichroic filters at FL2 are replaced with a 595 BP and a 645 nm dichroic filter to reflect wavelengths shorter than 645 nm to FL2 (JC-1 aggregate yellow-orange fluorescence) and pass wavelengths longer than 645 nm (PI red fluorescence) to FL4 (675 BP filter). The dichroic filter position at FL3 is left empty.
3. The collection of PI fluorescence by FL4 is electronically gated to distinguish between the viable, low-intensity PI fluorescence (PI−), and nonviable, high-intensity PI fluorescence (PI+), sperm populations as shown in **Fig. 8.1**, panel **A**. The percentage of viable sperm in panel a was 84.1%.
4. The J_{agg} fluorescence data (FL2, 595 BP) of the viable sperm population, derived by gating electronically on the PI- peak (panel **A**, **Fig. 8.1**), were plotted on a log scale against the log of JC-1 monomer fluorescence (from FL1) as illustrated in panel b. Similarly the J_{agg} fluorescence data of the non-viable sperm population, derived by gating on the PI+ peak (panel **A**),

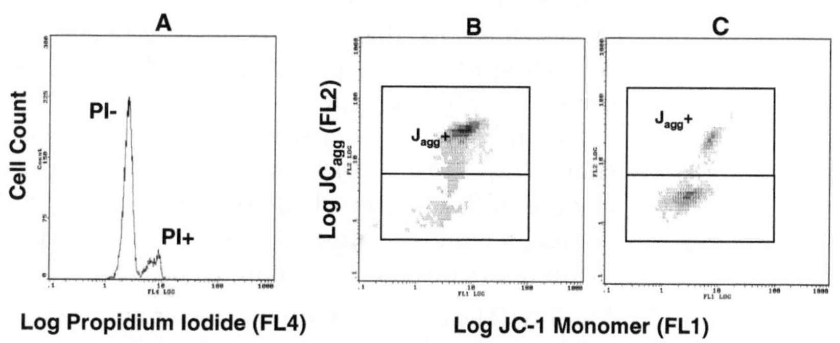

Fig. 8.1. Flow cytometric histograms and two dimensional dot plots of fluorescence intensities of boar sperm stained with 5,5′,6,6′-tetrachloro-1,1′,3,3′-tetraethylbenzimidazolylcarbocyanine iodide (JC-1) and propidium iodide (PI). The PI^- and PI^+ peaks (panel **A**) represent viable (dim PI fluorescence) and nonviable (bright PI fluorescence) sperm, respectively. The dot plots (panels **B** and **C**) show the JC-1 aggregate (J_{agg}) fluorescence intensity (vertical axis) of viable sperm gated on the PI^- population in panel **A** and of nonviable sperm gated on the PI^+ population in panel **A**, respectively. The fluorescence of JC-1 monomer, J_{agg}, and PI were collected in fluorescence detectors 1 (FL1), 2 (FL2), and 4 (FL4), respectively.

are plotted against the log of JC-1 monomer fluorescence as illustrated panel c. The energized cells in the upper cluster of cells in the top box in panels b and a, marked as J_{agg+}, are identified by the line set just above the lower cluster of dimly fluorescent cells in the bottom box in panel c. The percentage of viable sperm with J_{agg} fluorescence and the J_{agg} fluorescence intensity/sperm, expressed as channel number (CN) on a log scale from 0.1 to 1000, is recorded from the FL2 output. The percentage of sperm with J_{agg} fluorescence shown in panels b and c was 99 and 30, respectively.

3.6. Computer-assisted Motion Analysis

1. Sperm were incubated at a concentration of 30×10^6/mL at 37°C in a temperature controlled water bath, and aliquots were removed at specific times for motion analysis.
2. A Hamilton Thorne IVOS version-12 motion analysis system (Hamilton Thorne Biosciences, Beverly, MA) was set for boar sperm analysis *(9, 10)*.
3. A warmed 2 μL aliquot of each sperm suspension was transferred to one chamber of a warmed 20-micron four-chamber glass counting slide (SC 20-01 FA, Leja Products, Nieuw-Vennep, Netherlands). The counting slide was placed in a slide holder and the slide was scanned at 37°C using a video camera with a 10X negative phase objective to determine the proportion of motile and immotile sperm cells, and motion parameter values in eight pre-selected fields, which evaluated approximately 900 sperm cells.

3.7. Results of Menadione Treatment

1. ROS formation and viability: During the first 30 min of incubation, menadione caused an increase ($P < 0.01$) in ROS formation as measured by the intracellular oxidation of hydroethidine to ethidium *(4)*. The percentage of viable sperm containing ROS was low for 0 and 1 μM menadione, approximately 4%. However, increasing the menadione concentration to 10 and 30 μM increased the percentage of viable sperm containing ROS to 46.6% and 86.5%, respectively. The amount of ROS formed had no effect on sperm viability with a mean for the experiment of 93%, a result typical for boar sperm under this short exposure to ROS formation.
2. Motility: The key biological parameter in this experiment was sperm motion. The percentage of motile sperm was approximately 80% in the absence of menadione and was not significantly reduced during incubation for 120 min (**Table 8.1**). However, in the presence of menadione, the percentage of motile sperm decreased in a concentration and time related pattern. Incubation with 30 μM for 120 min decreased the percentage of motile sperm to essentially zero.

Table 8.1
Effect of menadione concentration and incubation time on the percent motility of boar sperm

	Incubation time (min)			
Menadione (μM)	5	30	60	120
0	84.2 ± 2.4[a]	82.5 ± 3.2[ab]	84.8 ± 2.7[a]	79.3 ± 4.8[ab]
1	78.5 ± 2.8[ab]	78.2 ± 4.2[ab]	72.8 ± 3.8[abc]	55.2 ± 6.4[d]
10	77.8 ± 3.4[ab]	60.0 ± 5.9[cd]	51.2 ± 7.8[de]	10.0 ± 2.8[f]
30	69.8 ± 4.9[bc]	40.0 ± 8.8[e]	6.8 ± 2.2[f]	0.7 ± 0.3[g]

Values are means ± SEM for ejaculates from six boars.

[a–g] Means within columns and rows with no superscript letter in common differ ($P < 0.05$).

3. Mitochondrial transmembrane potential: Most viable sperm were energized, with > 95% having high $\Delta\Psi_m$, i.e. J_{agg} fluorescence (**Table 8.2**). The only exception was in the sperm treated with 30 μM menadione for 120 min which decreased ($P < 0.05$) the percentage of viable energized sperm to 65%. However, while most viable sperm were energized exhibiting J_{agg} fluorescence, J_{agg} fluorescence intensity decreased ($P < 0.05$) in a menadione concentration and incubation time related fashion from a CN of 39 with 0 μM menadione at 5 min to a CN of 24, 19, and 9 at 30, 60, and 120 min, respectively, treated with 30 μM of menadione (**Table 8.3**).

Table 8.2
Effect of menadione concentration and incubation time on percentage of sperm cells with 5,5′,6,6′-tetrachloro-1,1′,3,3′-tetraethylbenzimidazolyl-carbocyanine iodide (JC-1) aggregates in sperm mitochondria

	Incubation time (min)			
Menadione (μM)	5	30	60	120
0	97.7 ± 0.6[a]	98.4 ± 0.5[a]	98.7 ± 0.3[a]	98.6 ± 0.3[a]
1	97.9 ± 0.5[a]	98.4 ± 0.4[a]	98.8 ± 0.3[a]	99.1 ± 0.2[a]
10	98.0 ± 0.6[a]	98.2 ± 0.5[a]	98.6 ± 0.4[a]	94.9 ± 3.7[a]
30	98.2 ± 0.5[a]	98.4 ± 0.4[a]	98.7 ± 0.3[a]	64.9 ± 13.4[b]

Values are means ± SEM for ejaculates from six boars.

[a,b] Means within columns and rows with no superscript letter in common differ ($P < 0.05$).

Table 8.3
Effect of menadione concentration and incubation time on 5,5′,6,6′-tetrachloro-1,1′,3,3′-tetraethylbenzimidazolyl-carbocyanine iodide (JC-1) aggregate red fluorescence intensity (channel number) in sperm mitochondria

	Incubation time (min)			
Menadione (μM)	5	30	60	120
0	38.9 ± 2.0^{acd}	38.2 ± 1.5^{abcd}	36.5 ± 2.1^{bc}	24.9 ± 2.3^{f}
1	41.0 ± 2.8^{ab}	39.0 ± 2.1^{ab}	35.2 ± 2.5^{cde}	23.1 ± 2.3^{fg}
10	38.9 ± 2.3^{abc}	31.3 ± 1.4^{e}	27.0 ± 1.7^{f}	16.4 ± 1.7^{h}
30	33.3 ± 1.2^{de}	24.5 ± 1.6^{f}	19.0 ± 1.2^{gh}	9.3 ± 1.0^{i}

Values are means ± SEM for ejaculates from six boars.
$^{a-i}$ Means within columns and rows with no superscript letter in common differ ($P < 0.05$).

The linear correlation between percent motile sperm and J_{agg} fluorescence intensity was high at 30, 60, and 120 min with Pearson correlation coefficients of 0.504 (P = 0.0120), 0.673 (P = 0.0003), and 0.761 ($P < 0.0001$); respectively. Therefore, the mean J_{agg} fluorescence intensity for sperm in a semen sample may be a useful marker for energized sperm cells and of semen quality following storage.

4. Notes

1. Addition of JC-1 from the 5 mM working solution in dimethyl sulfoxide to the aqueous suspension of the sperm sample must be accompanied by a brief, vigorous mixing on a vortex mixer to disrupt striations of the JC-1 to prevent precipitation.
2. A set of 12 × 75 mm polypropylene tubes are prepared to contain 30 × 10^6 sperm from the Percoll washed populations.
3. Experimental treatments such as menadione are added followed by JC-1 staining at 2.5 μM (0.5 μL of 5 mM working solution) and PI at 24 μM (10 μL of working 2.4 mM solution) in a final volume of 1 mL.
4. The cells are incubated aerobically at 37°C
5. At predetermined intervals during the incubation (treatments described in **Section 3.2**) after a brief mixing, 10,000 cells sperm cells are aspirated directly from the incubation tubes into the flow cytometer for analysis.

References

1. Halliwell B, Gutteridge JMC. Free radicals in biology and medicine, 2nd edn. Oxford University Press, Oxford. 1999.
2. Cramer WA, Knaff DB. Energy Transduction in Biological Membranes. Springer-Verlag, New York. 1990.
3. Floreani M, Napoli E, Palatini P. Role of antioxidant defenses in the species-specific response of isolated atria to menadione. Comp Biochem Physiol C Toxicol Pharmacol 2002; 132:143–151.
4. Guthrie HD, Welch GR. Determination of intracellular reactive oxygen species and high mitochondrial membrane potential in Percoll-treated viable boar sperm using fluorescence-activated flow cytometry. J Anim Sci 2006; 84: 2089–2100.
5. Cossarizza A, Baccarani-Contri M, Kalashnikova G, Franceschi C. A new method for the cytofluorimetric analysis of mitochondrial membrane potential using the J-aggregate forming lipophilic cation 5,5′,6,6′-tetrachloro-1,1′,3,3′-tetraethylbenzimidazol-carbocyanine iodide (JC-1). Biochem Biophys Res Commun 1993; 197:40–45.
6. Garner DL, Thomas CA, Joerg HW, DeJarnette JM, Marshall CE. Fluorometric assessments of mitochondrial function and viability in cryopreserved bovine spermatozoa. Biol Reprod 1997; 57:1401–1406.
7. Gravance CG, Garner DL, Miller MG, Berger T. Fluorescent probes and flow cytometry to assess rat sperm integrity and mitochondrial function. Reprod Toxicol 2001; 15:5–10.
8. Love CC, Thompson JA, Brinsko SP, Rigby SL, Blanchard TL, Lowry VK, Varner DD. Relationship between stallion sperm motility and viability as detected by two fluorescence staining techniques using flow cytometry. Theriogenology 2003; 60:1127–1138.
9. Vyt P, Maes D, Rijsselaere T, Dejonckheere E, Castryck F, Van Soom F. Motility assessment of porcine spermatozoa: A comparison of methods. Reprod Domest Anim 2004; 39:447–453.
10. Douglas-Hamilton DH, Smith NG, Kuster CE, Vermeiden JP, Althouse GC. Capillary-loaded particle fluid dynamics: effect on estimation of sperm concentration. J Androl 2005; 26:115–122.

Chapter 9

2,2,6,6-Tetramethylpiperidin-1-Oxyl Probes for Evaluating Oxidative Stress on the Cell Membrane and Mitochondria

Hidehiko Nakagawa and Naoki Miyata

Abstract

Oxidative stress is recognized to be involved in many pathological conditions, such as inflammation, arteriosclerosis, and neurodegenerative diseases. Reactive oxygen species (ROS), which induce oxidative stress responses, appear to induce different biochemical and cellular changes in each pathological condition. To elucidate the roles of ROS in these diverse pathological conditions, the measurement and evaluation of oxidative stress at subcellular levels would be very effective. We have developed EPR probes for oxidative stress in organelles, which are tagged with a fluorescent function, and assessed oxidative stress of the membrane and mitochondria using chemical probes for the corresponding organelles. These probes are localized to the expected cellular regions confirmed by confocal fluorescent microscopy, and for demonstration, the probes are employed to detect oxidative stress in RAW264.7 cells stimulated with LPS/IFN-γ. From the radical clearance rate of the probes localized in both the membrane and mitochondria, oxidative stress has been found to be consistently more severe in the membrane region, where NADPH oxidase is known to be upregulated by LPS/IFN-γ treatment. These specific probes have depicted each organelle under different oxidative stress conditions by a specific exogenous oxidative stimulus.

Key words: Membrane, mitochondria, nitroxide, spin probe, electron paramagnetic resonance, fluorescence, confocal microscopy.

1. Introduction

Oxidative stress is recognized to be involved in many pathological conditions, such as inflammation, arteriosclerosis, and neurodegenerative diseases *(1, 2)*. Reactive oxygen species (ROS),

From: *Methods in Molecular Biology, vol. 477: Advanced Protocols in Oxidative Stress I,*
Edited by: D. Armstrong, DOI: 10.1007/978-1-60327-517-0_9, © Humana Press, New York, NY

which induce oxidative stress responses, appear to induce different biochemical and cellular changes under each pathological condition *(3, 4, 5, 6, 7, 8, 9, 10, 11, 12)*. To elucidate the roles of ROS in those diverse pathological conditions, the measurement and evaluation of oxidative stress at sub-cellular levels would be very effective.

There are many reports indicating that ROS can be measured in cells indirectly by means of their reaction with stable radical probes, through which radial probes are readily reduced to non-radical species *(13)*. We focused on 2,2,6,6-tetramethylpiperidin-1-oxyl (TEMPO), a typical stable radical, as an EPR-detectable probe moiety. Under physiological conditions, TEMPO gradually converts to the non-radical form by reduction *(14, 15)*. When ROS are upregulated and cells are in a relatively oxidative environment, cellular reduction will be downregulated.

We have previously reported EPR probes for organelle-specific oxidative stress, assessing oxidative stress at the membrane and mitochondria using these probes for the corresponding organelles by confirming measurement areas with a fluorescent tag *(16, 17)*. These specific probes depicted each organelle under different oxidative stress conditions by a specific exogenous oxidative stimulus. These region-specific oxidative stress probes are promising to reveal the detailed mechanisms of oxidative stress in cells.

2. Materials

2.1. Equipment

1. A confocal microscope (LSM510, Carl Zeiss Co., Ltd.)
2. An EPR spectrometer (RE2X, JEOL, Tokyo, Japan) equipped with a personal computer and data analysis application software (WinRad, Radical Research Inc., Tokyo, Japan)
3. A quartz flat-type EPR cuvette (LLC-04B, sample volume = 130 μL; Labotec, Tokyo, Japan)
4. Conventional cell culture equipment: a 5% CO_2 humidified incubator, a clean bench, and so on.

2.2. Cell Culture

1. RAW264.7 cells, purchased from American Type Culture Collections (ATCC).
2. Dulbecco's minimum essential medium (Eagle's salt) (Sigma-Aldrich, St. Louis, MO, U.S.A.)
3. Fetal bovine serum (Biowest, S.A.S., Nuaillé, France), and antibiotics (penicillin and streptomycin, Invitrogen, Carlsbad, CA, U.S.A.)
4. Cell scraper (Sumitomo Bakelite, Co. Ltd., Tokyo, Japan)

5. Dulbecco-modified phosphate-buffered saline (D-PBS) (Invitrogen)
6. 100-mm diameter tissue culture dishes (Asahi Glass Co. Ltd., Funabashi, Japan or Corning Inc., Corning, NY, U.S.A.) to maintain the culture.
7. Glass-bottomed 30-mm diameter culture dishes (with a 10-mm diameter bottom glass) for confocal microscopy (Matsunami Glass Ind. Ltd., Osaka, Japan or MatTek Corp., Ashland, MA, U.S.A.)

2.3. Confocal Microscopy

1. Nucleus staining dye such as Hoecst33342 (Wako Pure Chemical Ind. Ltd., Osaka, Japan)
2. Mitochondria staining dye such as MitoRed® (PromoCell GmbH, Heidelberg, Germany).

2.4. Reagents for Cell Stimulation and Pharmacological Assessments

1. LPS (500 μg/mL in water) (Difco Laboratories, Detroit, MI, U.S.A.). Prepared solution is stored at −80°C until use.
2. Interferon-γ (IFN-γ; 150 kU/mL in water) (Sigma). Prepared solution is stored at −80°C until use.
3. Superoxide dismutase (SOD) and catalase (Sigma)

2.5. EPR Apparatus and Tools

1. EPR apparatus equipped to measure stable organic radicals in an aqueous solution can be used. We used a JEOL RE2X EPR spectrometer with an X-band microwave transmitter (JEOL, Tokyo, Japan). It is highly recommended to use apparatus connected to a computer system with a measurement application, such as WINRAD (Radical Research Inc., Tokyo, Japan). EPR apparatus should be turned on at least 30 min before measurement for reproductive signal intensity.
2. A manganese marker (Mn^{2+} in MnO) as an external standard for quantitative analysis.
3. A quartz flat-type cuvette (sample volume, 130 μL; Labotec, Tokyo, Japan)

3. Methods

3.1. Preparation and Storage of Oxidative Stress Probes

3.1.1. Synthesis of the Probes

The probes described here are not currently commercially available. To perform experiments using these probes, they should be synthesized. All the substrates and reagents for synthesis are commercially available, and the synthetic procedures are all conventional. For the synthesis details, see references 16, 17.

As novel organelle-specific oxidative stress probes, we have designed and synthesized two TEMPO derivatives: a

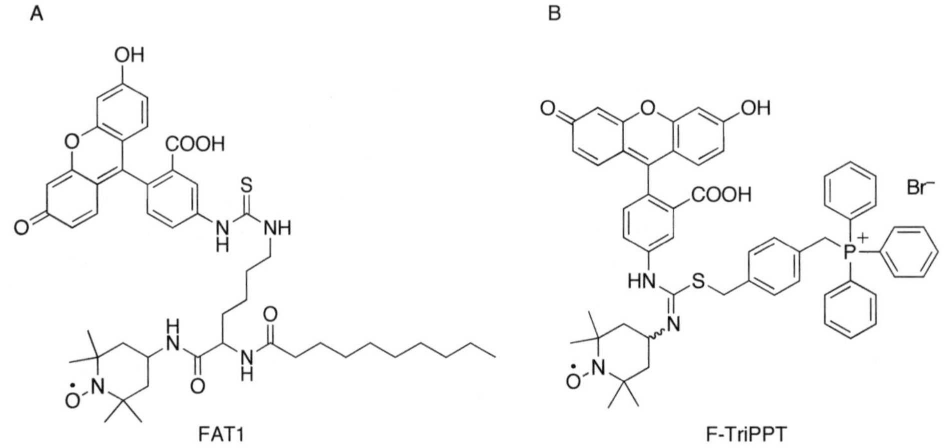

Fig. 9.1. Structures of membrane-specific (A, FAT1) and mitochondria-specific (B, F-TriPPT) oxidative stress probes.

membrane-localizing derivative with a long alkyl chain and a mitochondria-localizing derivative with a triphenylphosphonium moiety. These derivatives are tagged with a fluorescein moiety (**Fig. 9.1**).

3.1.2. Storage of Probes

These probe compounds have a fluorescent moiety and nitroxide radical moiety. To avoid decomposition during storage, the probes are kept under dark conditions at −30°C after flushing with inert gas (e.g. N_2 or Ar).

1. For powder, it is recommended that the compounds are placed in a sealed and screw-capped plastic or glass tube, flushed gently with Ar gas immediately before storage, and stored in a container in a freezer below −20°C.
2. For the solution of dimethylsulfoxide (DMSO), it is recommended that the compound solution is placed in a sealed and screw-capped plastic tube and stored in a freezer below −20°C. Before use, the solution is defrosted at room temperature and vortexed. It should be confirmed that the defrosted solution does not contain any precipitate.

3.2. Cell Culture and Stimulation

3.2.1. Cell Culture

Mouse RAW264.7 cells are cultured in DMEM culture medium containing penicillin and streptomycin, supplemented with fetal bovine serum according to ATCC instructions. Normal cell culture conditions are applicable for this method.[a] The cells are maintained at 37°C in a humidified 5% (v/v) CO_2 incubator under sub-confluent conditions.

3.2.2. Cell Stimulation

1. For the EPR experiment, cells are plated onto 10-cm culture dishes at 1.5 × 10^7 cells per dish with 15 mL of DMEM culture medium (Day 0). For confocal microscopy, cells are plated onto 3-cm glass-bottomed culture dishes. Alternatively, an aliquot of the cell suspension for the EPR experiment can be used for confocal observation.

2. The cells are subsequently incubated at 37°C in a humidified 5% (v/v) CO_2 incubator for 2 days.
3. On Day 2, the cells are stimulated with appropriate agents and duration. In our case, the culture medium was replaced with 5 mL of serum-free DMEM, and the cells were treated with LPS (E. coli, final concentration of 0.5 μg/mL) and IFN-γ (human recombinant, final 150 U/mL).
4. The treated cells are subsequently cultured for the designated time (5 h in our case).
5. The resulting cell culture is used for EPR experiments or confocal microscopy.

3.3. Loading Probes onto Cells

Here we describe the probe-loading and observation procedures when using an inflammatory cell model of mouse RAW264.7 cells prepared as shown above. The same procedures are applicable for other models or stimulations with appropriate modifications, such as heat-shock treatment and nephritis models.

3.3.1. Preparation of Probe Solution in DMSO

1. Probe compounds are weighed and dissolved in DMSO (biochemical or analytical grade) to yield 10 mM solution.
2. The prepared probe solution in DMSO is kept at room temperature during the probe-loading process under dark conditions.
3. The probe solution is relatively stable without light illumination. Stored at −30°C in the dark, it can be kept for weeks, but it is better to avoid repeated freezing and thawing.

3.3.2. Loading Probes

1. The appropriately stimulated cell culture is prepared as shown in **Section 3.2.**
2. The medium is replaced with 2 mL of D-PBS, and the cells are carefully scraped.
3. The cells are carefully washed once with 1 mL of D-PBS using a microcentrifuge.
4. The cells are then treated with 50–100 μM of the probe, by adding 5–10 μL of 10 mM DMSO solution to 1 mL of cell suspension, for 15 min under dark conditions.[b]
5. The medium containing the probes is removed, and the cells are washed three times with 1 mL of D-PBS.
6. Alternatively, without scraping the cells, they are washed once with D-PBS, and exposed to 50–100 μM of the probe by adding 10–20 μL of 10 mM DMSO solution to 2 mL of the medium (D-PBS) for 15 min under dark conditions.[b] The cells are then scraped into 1 mL of D-PBS.
7. For the membrane-specific probe, it is recommended that cells are co-stained with a nucleus marker dye, such as

Hoechst33342 (at a final concentration of 10 μg/mL) for confocal microscopy.

8. For the mitochondria-specific probe, it is recommended that cells are co-stained with a mitochondrial marker dye, such as MitoRed for confocal microscopy.

3.4. Confirmation of Cellular Localization of Probes by Confocal Microscopy

It is assessed by confocal microscopy that the probes are distributed to the appropriate region in probe-loaded cells. This assessment should be performed before using a new strain of cells, a new lot of probes, or new stimulation protocols.

3.4.1. Microscope Settings

The usual settings for fluorescein are used with an inverted fluorescence confocal microscope. We used an LSM510 confocal microscope (Carl Zeiss Co., Ltd.) with the following filter settings: DAPI/FITC (or EX405/488) for the membrane-specific probe co-stained with Hoechst 33342, or FITC/Cy3 (or EX488/543) for the mitochondria-specific probe co-stained with MitoRed.

3.4.2. Observation of Cells

Membrane-Specific Probe

Probe-loaded cells, co-stained with Hoechst33342, are observed under confocal fluorescence microscopy, and it is confirmed that fluorescence from the fluorescein moiety of the probe is localized in the very peripheral regions of a cell, whereas fluorescence from Hoechst33342 appears round and is placed in a relatively central region of a cell (**Fig. 9.2**).[c]

Mitochondria-Specific Probe

Probe-loaded cells, co-stained with MitoRed, are observed under confocal fluorescence microscopy, and it is confirmed that fluorescence from the fluorescein moiety is localized around the

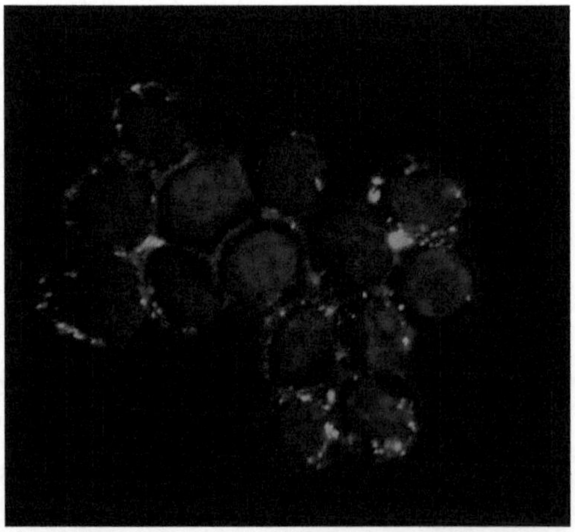

Fig. 9.2. RAW264.7 cells were stained simultaneously with Hoechst33342 and the membrane-specific probe (FAT1), and observed by confocal fluorescence microscopy. Hoechst33342 in the nuclei is shown as gray round shapes placed in a relatively central region of cells, and FAT1 is shown as gray peripheral circles around cells. (See the color figure in reference 17 for clear staining with blue and green, respectively. Reproduced form reference 17 with permission from Elsevier Science).

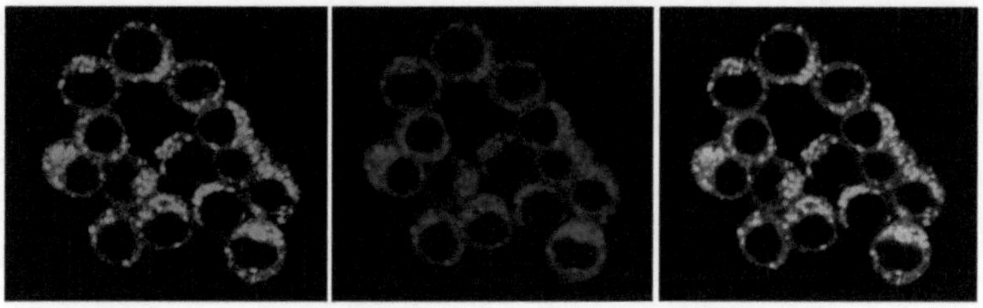

Fig. 9.3. RAW264.7 cells were stained simultaneously with MitoRed and the mitochondria-specific probe (F-TriPPT), and observed by confocal fluorescence microscopy. Distribution of F-TriPPT (left), distribution of MitoRed (center), and a merged image (right) in the same field are shown. (See the color figure in reference 16 for clear staining with green (left), red (center), and merged color image, respectively. Reproduced from reference 16 with permission from Elsevier Science).

nucleus in dotted forms, and co-localized with a mitochondria marker dye, MitoRed (**Fig. 9.3**).

3.5. EPR Measurement of Oxidative Stress Probes in Cells

Probe-loaded cells are subjected to EPR measurement after appropriate stimulations.[d] The observation periods and frequency depend on the EPR signal disappearing rate (this rate is referred to as the "spin clearance rate"). When the spin clearance rate is higher, EPR signals should be recorded more frequently for accurate evaluation of oxidative stress.

3.5.1. EPR Settings

The settings depend on the apparatus. Typical settings for organic nitroxide radicals in aqueous solution are usually sufficient. For our JEOL RE2X EPR spectrometer, the settings were as follows: center field, 336.5 mT; sweep width, ±7.5 mT; frequency, 9.4 GHz; microwave power, 10 mW; sweep rate, 0.067 min/mT (1 min/15 mT); modulation width 0.064; response, 0.03 sec.

3.5.2. Measurement of Probe Signals

For details of the basic EPR operation for measuring stable organic radicals in an aqueous solution, textbooks and handbooks should be consulted.

1. A suspension of cells in D-PBS, which are stimulated and treated with one of the probes, is introduced into a quartz flat-type EPR cuvette. For the best signal resolution, the suspension is leveled when the flat part of the cuvette is filled.
2. The cuvette is capped, and placed gently into the sample holder of the resonator. An MnO stick is set to the resonator as an external standard for quantification.
3. After tuning the resonator and transmitter, EPR signals are recorded.
4. The cell suspension in the flat-type cuvette is discarded, and the cuvette is washed with D-PBS. Then a new aliquot

of the cell suspension is placed in the cuvette for the next recording. EPR spectra are measured at the designated intervals.[e]

5. Within the observation range of the magnetic field, a three-line nitroxide signal of the probe is observed around the center. Two Mn^{2+} signals from a manganese marker (Mn^{2+} in MnO) are concomitantly observed on both sides of the nitroxide signal (**Fig. 9.4**).
6. Probe signal intensity (I) is quantified by calculating the 2nd integral (*i.e.* peak area) of the left-side peak of the three-line signal of the probe. When measurement application software is available, such as WINRAD (see Materials), the integrated signal areas will be automatically calculated by selecting the integration command. Alternatively, the signal height of one of the three peaks may be used as the signal intensity instead of the integrated value.
7. The signal intensity of one of the external standard (Mn^{2+} in MnO) signal peaks (I_0) is also quantified (left-side peak in the observation range is usually used).
8. The ratio ($I_R = I/I_0$) of the signal intensity of the probe against that of the external standard is calculated as the "relative signal intensity" for each time point. This ratio (relative signal intensity) is used to calculate the spin clearance rate.

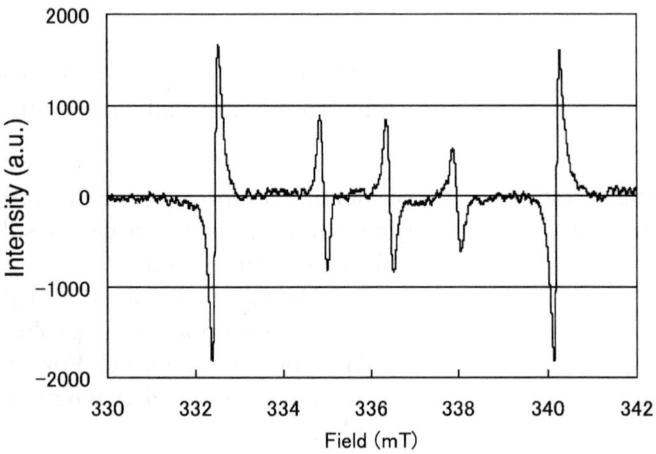

Fig. 9.4. EPR spectra of F-TriPPT in RAW264.7 cells were recorded with a JES-RE 2X spectrometer (JEOL Co. Ltd., Tokyo, Japan). The settings were as follows: microwave power, 10 mW; frequency, 9.42 GHz; field, 336.5 mT; sweep width, 7.5 mT; sweep time, 1 min; modulation width, 0.063 mT; gain, 2500; and time constant, 0.03. The signal intensity of the probe (I_R) was calculated from the second integral of the signal trace and the intensity of the standard Mn^{2+} signal.

3.6. Data Analysis and Evaluation of Regional Oxidative Stress

3.6.1. Calculation for Pseudo-First Order Rate of Spin Clearance

1. The relative signal intensity (I_R) at each time point is plotted on a natural logarithmic scale ($ln(I_R)$) for the function of time lapsed after the probe is loaded (t).
2. The decreasing rate of relative signal intensity (i.e. spin clearance rate) is calculated as the pseudo first order. The trend of the linear regression of the plot ($ln(I_R) = at + b$; a, trend; b, intercept) is the value of the pseudo-first order rate of spin clearance (**Fig. 9.5A** and **9.6A**).
3. The average spin clearance rate and its standard deviation are obtained from at least three independent determinations of the rate (i.e. three independent observations of signal disappearance). A reliable spin clearance rate may be obtained from six or more independent determinations.
4. To normalize the EPR results, the number of cells is counted in the suspension. The spin clearance rate obtained from each EPR experiment is normalized by the cell concentration. We normalized the spin clearance rate to 10^7 cells/mL (**Table 9.1**). Alternatively, the cell concentration may be determined and adjusted before the EPR experiment.

3.6.2. Data Interpretation

The EPR signal of the probes is based on the TEMPO moiety of the probe. The TEMPO signal is known to be gradually decreased in cells. The relative signal intensity (I_R) of the probe is also gradually decreased in control cells. Under our experiment conditions, the spin clearance rate in the control cells is about 0.0080 ± 0.0004 min^{-1} at 10^7 cells/mL for both membrane-specific and mitochondria-specific probes.[f]

When the probe concentration is sufficiently small compared with that of cellular reductants, probe signal disappearance is mainly due to the reducing reaction by cellular reductants. In a living cell, cellular reductants are consumed and reproduced, so that the concentration of the total reductants is balanced at a certain level in a normal cell by homeostasis. Based on this consideration, probe radical clearance (k_{obs}) is expressed as the pseudo-first

Table 9.1
Cell numbers in the measurement suspension for EPR detection

	Cell number
Control	2.47 ± 0.38
LPS/IFN-γ	2.52 ± 0.95
LPS/IFN-γ + SOD/catalase	2.13 ± 0.59

($\times 10^7$ mL^{-1}, Average ± SD).

order rate (Eq. 9.1). Here, T_{rad} is the probe radical concentration (observed as relative signal intensity), and R_{norm} is the total reductant concentration in a control cell.

$$\frac{d[T_{\text{rad}}]}{dt} = -k_{\text{obs}} \times [T_{\text{rad}}]\ (k_{\text{obs}} = \text{k} \times [R_{\text{norm}}]) \qquad (9.1)$$

Under the condition of oxidative stress, oxidants such as exogenous or intracellular ROS are upregulated and scavenged with cellular reductants, indicating increasing consumption of cellular reductants, so that the total reductant concentration is shifted, and balanced at a lower level. Under this condition, the probe radical clearance rate (k'_{obs}) is reduced due to the lower concentration of the reductants (R_{ox}) (Eqs. 9.2 and 9.3).

$$\frac{d[T_{\text{rad}}]}{dt} = -k'_{\text{obs}} \times [T_{\text{rad}}]\ (k'_{\text{obs}} = \text{k} \times [R_{\text{ox}}]) \qquad (9.2)$$

$$R_{\text{ox}} < R_{\text{norm}} \qquad (9.3)$$

In the experiments, when the observed spin clearance rate in the stimulated cells is lower than that in control cells, this indicates that the stimulated cells are under oxidative stress conditions.

In the *in vitro* LPS/IFN-γ-stimulated inflammatory model demonstrated, the spin clearance rates of both membrane-specific and mitochondria-specific probes were downregulated, meaning that the membrane and mitochondria region of the cell have fallen into oxidative stress (**Figs. 9.5 and 9.6**). Furthermore, the spin clearance rate in the membrane region (the rate of membrane-specific probe) was found to be lower than that at mitochondria (the rate of mitochondria-specific probe). Because the rate, k_{obs}, is not a function of the probe radical concentration, $[T_{rad}]$, the rate in the membrane region and mitochondria can be compared without considering the substantial concentration at each region. The lower spin clearance at the membrane suggests that oxidative stress at the membrane is more severe than at mitochondria. Consistently, it is known that LPS/IFN-γ induces NADPH oxidase, which is a membrane protein complex and produces superoxide efficiently in a cell via NF-κB activation.

3.7. Pharmacological Assessment of Oxidative Stress

It is necessary to confirm that the effects on spin clearance rates are actually due to the increase of oxidative stress by ROS in the selected model. For this purpose, pharmacological assessments are effective. Radical scavengers, inhibitors of oxidative stress enzymes, or antioxidants may be used. First, superoxide dismutase (SOD) and catalase are recommended to downregulate both superoxide and hydrogen peroxide. If nitric oxide (NO) is suspected to be involved in oxidative stress, nitric oxide synthase (NOS) inhibitors will be useful.

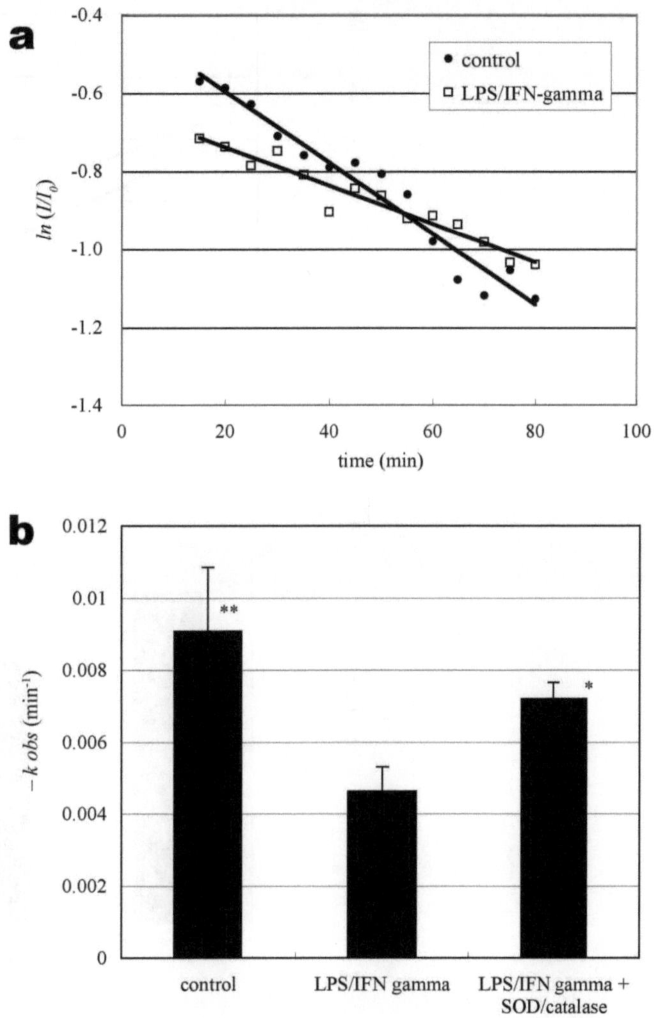

Fig. 9.5. Signal decay rate of organelle-specific oxidative stress probes in RAW264.7 cells with the membrane-specific probe (FAT1). (**A**) The time course of relative signal intensity measured at 5-min intervals, control cells (closed circle), LPS/IFN-γ-treated cells (open square). *I*, FAT1 signal intensity; I_0, Mn^{2+} external standard signal intensity. (**B**) Signal decay rate of FAT1 in RAW264.7 cells. Signal decay rates were calculated from EPR signal intensities of the probe in RAW264.7 cells treated with vehicle, or with LPS/IFN-γ, in the presence or absence of SOD/catalase. Values are presented as the means $\pm$ SD of 3–4 experiments. ANOVA and Bonferroni-type multiple *t* test indicated significant differences between LPS/IFN-γ and the control (**$P < 0.01$), and LPS/IFN-γ + SOD/catalase (*$P < 0.05$).

3.7.1. Assessment of the Involvement of Superoxide/Hydrogen Peroxide

1. SOD and catalase are dissolved in 1 mL of D-PBS at each concentration of 100 U/mL and 10 U/mL, respectively.
2. Stimulated and probe-loaded cells are precipitated by microcentrifuge, and resuspended in 1 mL of D-PBS containing SOD/catalase prepared in the previous step.
3. The cell suspension in D-PBS containing SOD/catalase is subjected to EPR measurement.

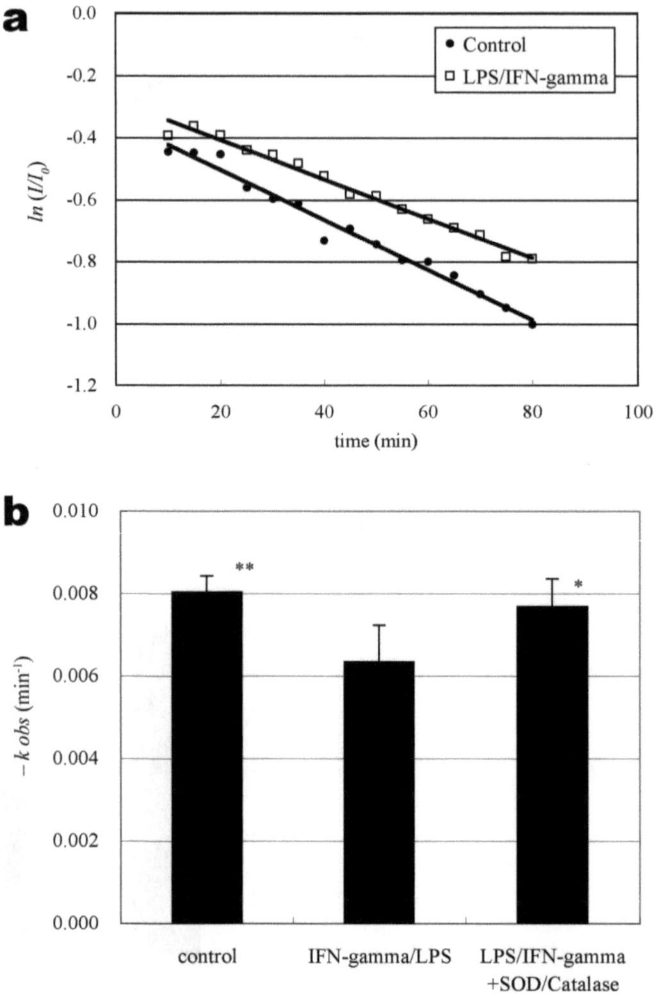

Fig. 9.6. Signal decay rate of organelle-specific oxidative stress probes in RAW264.7 cells with the mitochondria-specific probe (F-TriPPT). (**A**) The time course of relative signal intensity measured at 5-min intervals, control cells (closed circle), LPS/IFN-γ-treated cells (open square). I, F-TriPPT signal intensity; I_0, Mn^{2+} external standard signal intensity. (**B**) Signal decay rate of F-TriPPT in RAW264.7 cells. Signal decay rates were calculated from EPR signal intensities of the probe in RAW264.7 cells treated with vehicle, or with LPS/IFN-γ, in the presence or absence of SOD/catalase. Values are presented as the means $\pm$ SD of 4–7 experiments. ANOVA and Bonferroni-type multiple t test indicated significant differences between LPS/IFN-γ and the control (**$P < 0.01$), and LPS/IFN-γ + SOD/catalase (*$P < 0.05$).

3.7.2. Assessment of the Involvement of Other ROS or ROS-Related Enzymes

To confirm the involvement of other ROS, nitric oxide, or related enzymes, inhibitors or scavengers may be employed. Selection of inhibitors/scavengers depends on the stimulation and cell types. Cells can be treated with these inhibitors/scavengers by preparing D-PBS solution containing a focused inhibitor/scavenger, as described for SOD/catalase. The following reagents may be used for assessment.

1. DPI (diphenyleneiodonium, NADPH oxidase inhibitor, Sigma) *(18)* or Apocynin (4′-hydroxy-3′-methoxyacetophenone, NADPH inhibitor, Sigma) *(19)*.
2. L-NNA (N^{ω}-nitro-L-arginine, NOS inhibitor, Wako Pure Chemical Ind.) or L-NAME (N^{ω}-nitro-L-arginine methyl ester, NOS inhibitor, Tokyo Chemical Ind. Co. Ltd., Tokyo, Japan) *(20)*.

4. Notes

a. Note that excess ascorbate, iron ion, and other EPR active metals in the medium may interfere with the probe signals. Control EPR spectra should be checked without probes before experiments.

b. After addition of the probe solution, the medium (D-PBS) is immediately but gently shaken to disperse probe compounds to cells.

c. When the duration of staining with the membrane-specific probe is extended, fluorescence becomes observed in the inner membrane structures of the cell.

d. For EPR experiments, probe-loaded cells are not necessarily co-stained with organelle marker dyes such as Hoechst33342 or MitoRed.

e. Cells are precipitated and packed into the bottom in a flat-type cuvette when the cuvette containing the cell suspension is placed vertically into the resonator. Even after the cells are precipitated, EPR spectra can be obtained; however, it is recommended that a new aliquot of the suspension is introduced into the cuvette each time. When the measurement intervals are sufficiently short (for example, several minutes), two or three measurements can be performed before the cells are precipitated.

f. The quartz flat-type cuvette used holds 130 μL of cell suspension, so the control rate is also expressed as $0.0615 \pm 0.0031/\text{min}/10^7$ cells.

References

1. Irani K, Goldschmidt-Clermont PJ. Ras, superoxide and signal transduction. Biochem Pharmacol 1998;55(9):1339–1346.
2. Irani K, Xia Y, Zweier JL, et al. Mitogenic signaling mediated by oxidants in Ras-transformed fibroblasts. Science 1997;275 (5306):1649–1652.
3. Darley-Usmar V, Halliwell B. Blood radicals: Reactive nitrogen species, reactive oxygen species, transition metal ions, and the vascular system. Pharm Res 1996; 13(5):649–662.
4. Eckert A, Keil U, Marques CA, et al. Mitochondrial dysfunction, apoptotic cell death, and Alzheimer's disease.

Biochem Pharmacol 2003;66(8):1627–1634.

5. Emerit J, Edeas M, Bricaire F. Neurodegenerative diseases and oxidative stress. Biomed Pharmacother 2004;58 (1):39–46.
6. Fiskum G, Starkov A, Polster BM, Chinopoulos C. Mitochondrial mechanisms of neural cell death and neuroprotective interventions in Parkinson's disease. Ann NY Acad Sci 2003;991:111–119.
7. Halliwell B. Oxidative stress and neurodegeneration: Where are we now? J Neurochem 2006;97(6):1634–1658.
8. Jenner P. Oxidative stress in Parkinson's disease. Ann Neurol 2003;53 (Suppl. 3):S26–S36; discussion S-8.
9. Lenaz G. Role of mitochondria in oxidative stress and ageing. Biochem Biophys Acta 1998;1366(1–2):53–67.
10. Liu Y, Fiskum G, Schubert D. Generation of reactive oxygen species by the mitochondrial electron transport chain. J Neurochem 2002;80(5):780–787.
11. Nunomura A, Castellani RJ, Zhu X, Moreira PI, Perry G, Smith MA. Involvement of oxidative stress in Alzheimer disease. J Neuropathol Exp Neurol 2006;65(7): 631–641.
12. Zhao K, Luo G, Giannelli S, Szeto HH. Mitochondria-targeted peptide prevents mitochondrial depolarization and apoptosis induced by tert-butyl hydroperoxide in neuronal cell lines. Biochem Pharmacol 2005;70(12):1796–1806.
13. Nakagawa H, Moritake T, Tsuboi K, Ikota N, Ozawa T. Induction of superoxide in glioma cell line U87 stimulated with lipopolysaccharide and interferon-gamma: ESR using a new flow-type quartz cell. FEBS Lett 2000;471(2–3):187–190.
14. May JM, Qu ZC, Juliao S, Cobb CE. Ascorbic acid decreases oxidant stress in endothelial cells caused by the nitroxide tempol. Free Radic Res 2005;39(2): 195–202.
15. Iannone A, Bini A, Swartz HM, Tomasi A, Vannini V. Metabolism in rat liver microsomes of the nitroxide spin probe tempol. Biochem Pharmacol 1989; 38(16): 2581–2586.
16. Ban S, Nakagawa H, Suzuki T, Miyata N. Novel mitochondria-localizing TEMPO derivative for measurement of cellular oxidative stress in mitochondria. Bioorg Med Chem Lett 2007;17(7):2055–2058.
17. Ban S, Nakagawa H, Suzuki T, Miyata N. Novel membrane-localizing TEMPO derivatives for measurement of cellular oxidative stress at the cell membrane. Bioorg Med Chem Lett 2007;17(5): 1451–1454.
18. O'Donnell BV, Tew DG, Jones OT, England PJ. Studies on the inhibitory mechanism of iodonium compounds with special reference to neutrophil NADPH oxidase. Biochem J 1993;290 (Pt 1):41–49.
19. Stolk J, Hiltermann TJ, Dijkman JH, Verhoeven AJ. Characteristics of the inhibition of NADPH oxidase activation in neutrophils by apocynin, a methoxy-substituted catechol. Am J Respir Cell Mol Biol 1994;11(1):95–102.
20. Salerno L, Sorrenti V, Di Giacomo C, Romeo G, Siracusa MA. Progress in the development of selective nitric oxide synthase (NOS) inhibitors. Curr Pharm Des 2002;8(3):177–200.

Chapter 10

Zymographical Techniques for Detection of Matrix Metalloproteinases

Shinichi Iwai, Takako Nakanishi-Ueda, Donald Armstrong, and Katsuji Oguchi

Abstract

Matrix metalloproteinases (MMPs) are a family of zinc-dependent proteinases associated with extracellular matrix degradation, cellular migration, tissue remodeling, and angiogenesis. The activity of MMPs is regulated by the tissue inhibitors of metalloproteinases (TIMPs). Zymography and reverse zymography are useful to detect MMPs and TIMPs activities from various samples, for example vitreous, retina, plasma, and so on. Sample proteins are separated in substrate containing polyacrylamide gel by electrophoresis. The gel is incubated and then stained with Coomassie Blue. MMPs' activities are detected as clear bands.

Key words: Matrix metalloproteinase, tissue inhibitors of metalloproteinase, zymography, reverse zymography, plasma.

1. Introduction

Matrix metalloproteinases (MMPs) are a family of zinc-dependent proteinases associated with extracellular matrix (ECM) degradation, cellular migration, tissue remodeling, and angiogenesis *(1)*. MMPs are increased in various diseases, such as proliferative vitreoretinopathy *(2)*, corneal neovascularization *(3)*, and age – related macular degeneration (AMD) *(4, 5)*. MMPs are secreted by many types of cells as proenzymes (inactive form). After activation by proteolytic cleavage, the enzymes are capable of degrading many ECM components *(6)*. MMPs are believed to be the major enzymes responsible for ECM degradation *(7)*. MMP-2 and MMP-9, in particular, are

From: *Methods in Molecular Biology, vol. 477: Advanced Protocols in Oxidative Stress I,*
Edited by: D. Armstrong, DOI: 10.1007/978-1-60327-517-0_10, © Humana Press, New York, NY

involved in angiogenesis *(8, 9)*. In oxidative stress induced neovascularization upon exposure to lipid hydroperoxide, MMP-9 is upregulated *(10)*. The activity of MMPs is regulated by several types of inhibitors, of which the tissue inhibitors of metalloproteinases (TIMPs) are the most important *(11)*. The balance between MMPs and TIMPs is critical for the eventual remodeling of the ECM in tissue *(12)*.

1.1. Concept for Zymography

Zymography and reverse zymography are useful to detect MMPs and TIMPs in various samples, for example vitreous, retina, plasma, and so on. All types of substrate zymography originate from gelatin zymography *(13)*. Methods for all zymography use almost the same techniques except for substrate. Zymogram gel is polyacrylamide gel containing a specific substrate *(14, 15)*. Sample proteins are separated by electrophoresis. During electrophoresis, the sodium dodecyl sulfate (SDS) causes the MMPs to denature and become inactive. After electrophoresis the gel is washed, which causes the exchange of the SDS with Triton X-100. The latent MMPs are autoactivated without cleavage. The activation is believed to involve the "cysteine switch" due to the dissociation of Cys^{73} from the zinc molecule (**Fig. 10.1**) *(16, 17)*. Subsequently, the gel is incubated in an appropriate activation

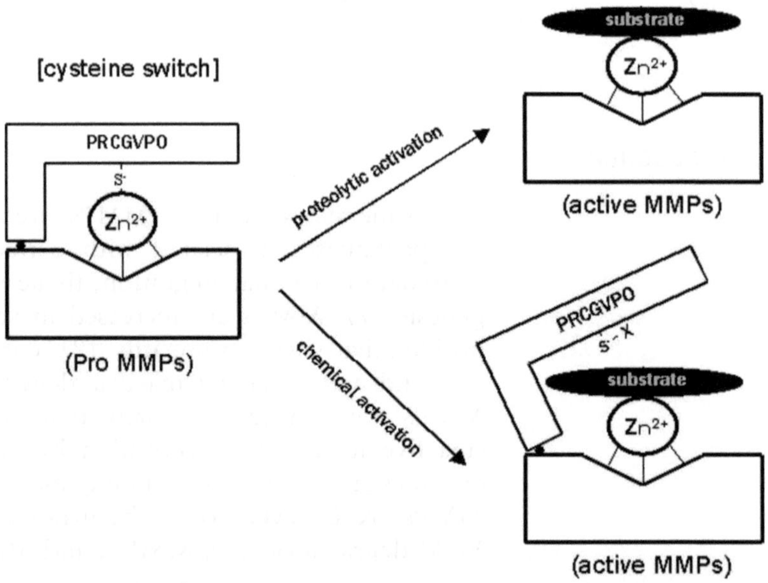

Fig. 10.1. Activation mechanisms of pro-MMPs (cysteine switch theory). All MMPs are released as inactive precursor forms (pro-MMPs). The proteolytic activation transfers pro-MMPs to active MMPs by MMP-activating cascade, plasmin, trypsin, and so on. The chemical activation is a disruption of the Cys^{73}-Zn^{2+} bond. Zymography makes use of the chemical activation mechanism.

buffer. During this incubation, the concentrated, renatured MMPs in the gel will digest the substrate. After incubation, the gel is stained with Coomassie Blue, and MMPs are detected as clear bands against a blue background of undegraded substrate *(15, 18)*. The clear band in the gel can be quantified by densitometry *(19)*.

2. Materials

2.1. Equipment and Supplies

1. The Mini-PROTEAN 3 system (Bio-Rad Laboratories, Inc., Hercules, CA)
2. Power supply (200 V, 500 mA)
3. Incubator (37°C)
4. Rotary shaker
5. Container for gel (a small plastic box for kitchen utensil etc.) (*see* **Note 1**)
6. ATTO imaging system (ATTO Co, Tokyo, Japan)
7. NIH Image software (Scion Co, Frederick, MD) (*see* **Note 2**)

2.2. Reagents and Stock Solutions

1. Acrylamide/Bis [30% (w/v), 2.67% (w/v)]: 146 g acrylamid and 4 g N′N′-Bis-methylene-acrylamide in 500 mL ddH_2O is filtered and Stored at 4°C in the dark (*see* **Note 3**).
2. 4x separating gel buffer 1.5 M Tris-HCl, pH 8.8: Consists of 18.15 g Tris base in ~70 mL ddH_2O. Adjust to pH 8.8 with HCl. Make up to 100 mL with ddH_2O and add 0.4 g gelatin (*see* **Note 4**). To dissolve gelatin use a microwave oven or a water bath.
3. 4x stacking gel buffer 0.5 M Tris-HCl, pH 6.8: Consists of 6 g Tris base in ~70 mL ddH_2O. Adjust to pH 6.8 with HCl. Make up to 100 mL with ddH_2O and store at 4°C.
4. 10% SDS (w/v): 10 g SDS with gentle stirring and bring up to 100 mL with ddH_2O. Store at room temperature.
5. 2x Sample Buffer: 4.4 mL ddH_2O, 1.0 mL 0.5 M Tris-HCl, pH 6.8, 0.8 mL Glycerol, 1.6 mL 10% (w/v) SDS, and 0.05% (w/v) 0.2 mL Bromophenol blue. Store at room temperature.
6. 5x Electrode Buffer consists of 15 g Tris base, 72 g Glycine, and 5 g SDS taken up to 1 L with ddH_2O. This should be at pH 8.3 (do no adjust) (*see* **Note 5**). Store at room temperature. Dilute to 1x (0.25 M Tris, 1.92 M Glycine and 1% SDS (w/v)) with ddH_2O before use. For one electrophoresis, 1x Electrode Buffer needs to be about 400 mL.

7. 10% (w/v) ammonium persulfate (APS): 100 mg APS in 1 mL ddH_2O made fresh daily.
8. TEMED (N, N, N′, N′-tetramethylethylenediamine): Store at 4°C and protect from light.
9. *n*-butanol.
10. 10% (v/v) ammonia water for glass cleaning (*see* **Note 6**). Store at room temperature.
11. 2.5% (v/v) Triton X-100: 2.5 mL Triton X-100 in 97.5 mL ddH_2O made fresh daily.
12. Incubation Buffer, 0.1 M Tris-HCl, 6.8 mM $CaCl_2$, 0.38 mM NaAzide consists of 2.4 g. Tris base, 0.3 g $CaCl_2$, 0.1 g NaAzide in ~300 mL ddH_2O. Adjust to pH 7.5 with in HCl. Make up to 400 mL with ddH_2O. Prepare fresh daily.
13. Staining solution consists of 2 g Coomassie Brilliant Blue [0.2% (w/v)], 500 mL Methanol [50% (v/v)], 100 mL Acetic Acid [10% (v/v)], and 400 mL ddH_2O. Store at room temperature.
14. Destaining solution consists of 400 mL Methanol [40% (v/v)], 100 mL Acetic Acid [10% (v/v)], and 500 mL ddH_2O. Store at room temperature.

2.3. MMP Standards and Molecular Weight

1. MMP standards: MMP-2; PF037 and MMP-9; PF038 (EMD Biosciences Inc, San Diego, CA)
2. Molecular weight markers are obtained from Dual Precision Plus Standards (Bio-Rad) and MultiMark Multi-Colored MW standards (EMD) *(18).*

3. Methods (*see* Note 7)

3.1. Sample Preparation

3.1.1. Rat Plasma

Blood specimens are taken from the inferior vena cava of rat and mixed with 3.2% sodium citrate solution in a volume ratio of 9:1. After 15 min of centrifugation at 3000 rpm, the supernatant as citrated plasma is used for the specimens. Plasma samples are diluted 100 times and 10 times in ddH_2O for measurement of MMP-2 and MMP-9, respectively *(20, 21).*

3.1.2. Positive Control

The 72-kDa purified proenzyme MMP-2 (0.8 ng protein per lane) and the 92-kDa purified proenzyme MMP-9 (0.1 ng protein per lane) were used as positive controls and standards for MMP-2 and MMP-9 (EMD) *(20)* (*see* **Note 8**).

3.2. Preparation of Gels

1. Clean the inside of glass with 10% ammonia water. Assemble glass cassette and casting stand according to the manufacture's instructions.
2. Prepare monomer solution for the separating gel by combining all reagents except APS and TEMED (**Table 10.1**) (*see* **Note 9**). Degas the solution under vacuum for at least 15 min.
3. After APS and TEMED are added into the solution, pour immediately into the prepared cast (about 4 mL each, until 1 cm under upper short plate) (*see* **Note 10**).
4. Overlay with n-butanol up to top of gel. Do not disturb until polymerization is complete in about 30–60 min (*see* **Note 11**).
5. Remove butanol and wash with water. Remove water from inside of glass with kimwipe.
6. After adding APS and TEMED into stacking gel buffer, insert well-forming comb and pour the solution immediately into the prepared cast. Do not disturb until polymerization is complete in about 30–60 min.

3.3. Electrophresis

1. Assemble the Mini-PROTEAN 3 Electrophresis Module. Fill the inner chamber with ~125 mL of Electrode Buffer and add another 200 mL of the Buffer to lower buffer chamber.
2. Dilute sample without boiling 1:1 with 2x Sample Buffer.

Table 10.1
There are 10 mL of separating solution and 5 mL of stacking solution to make two minigel (8.0 cm × 7.3 cm × 0.75 mm)

10% separating gel		10% stacking gel		15% separating gel (for reverse zymography)	
30% Acrylamide/Bis	3.3 mL	30% Acrylamide/Bis	1.65 mL	30% Acrylamide/Bis	5 mL
4X separating gel buffer	2.5 mL	4X stacking gel buffer	1.25 mL	4X separating gel buffer	2.5 mL
10% SDS	100 μL	10% SDS	50 μL	10% SDS	100 μL
ddH_2O	4 mL	ddH_2O	2 mL	ddH_2O	2.3 mL
				recombinant human Pro MMP-9	50 μL
APS	100 μL	APS	50 μL	APS	100 μL
TEMED	5–15 μL	TEMED	2–7 μL	TEMED	5–15 μL

3. Load the samples into the wells with a pipette using gel loading tips.
4. Electrophorese samples at 150 V as determined by trial with colored molecular weight maker and tracking dye.

3.4. Development and Staining of Gels

1. After electrophoresis, transfer the gel to a clean container for the detection of proteinase activity. SDS is washed out of the gels with about 100 mL of 2.5% Triton X-100 for 30 min by gentle agitation on a rotary shaker at room temperature (*see* **Note 12**).
2. Discard Triton X-100 and add directly (without water wash) (*see* **Note 13**) about 100 mL of Incubation Buffer to container.
3. Transfer the container with gel to an incubator (37°C) for 16–48 h.
4. After incubation, stain the gel in Staining solution for at least 1 h and destain in Destaining solution for just 10 min (*see* **Note 14**). Transfer the gel to ddH_2O, wash twice, and keep in ddH_2O over night. The bands are clearly visible and contrast well with the background.
5. The zymography bands are digitized using the ATTO imaging system, and analyzed using NIH Image software.

3.5. Examples

Figure 10.2 shows stained gelatin zymograms in various samples from rat plasma.

Rat samples are obtained from male 8-month-old Sprague-Dawley (SD) rats. MMP-2 and -9 in blood are very abundant. Plasma samples are 10 times diluted for MMP-9 (**Fig. 10.2A**) and 100 times for MMP-2 (**Fig. 10.2B**). If plasma sample are not

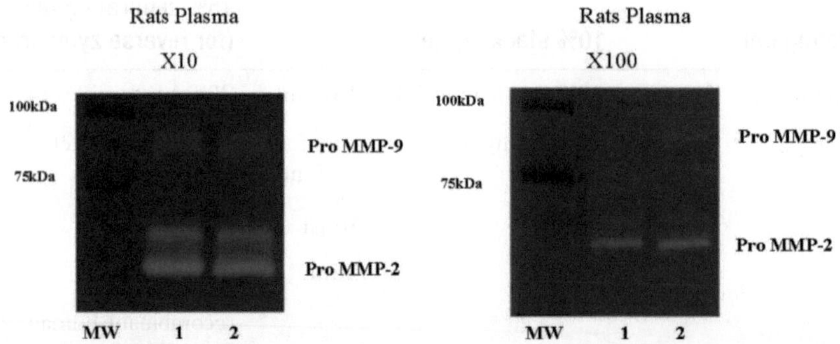

Fig. 10.2. Gelatin zymography for plasma in male 8-month-old SD rat. (**A**) Plasma samples were diluted 10 times in ddH_2O for measurement of MMP-9. Plasma samples need to diluted as MMPs in plasma were abundant. (MW: Molecular weight maker, Lane 1 and 2: male 8-month-old SD rats) (**B**) Plasma samples were diluted 100 times in ddH_2O for measurement of MMP-2. X100 sample gives a clearer expression of pro MMP-2 activity compared with X10. (MW: Molecular weight maker, Lane 1 and 2: male 8-month-old SD rats).

diluted, MMP-2 and -9 overexpression cannot be detected *(20)*. You can obtain MMP activities of other samples by the same methods when you have only the exchange of the plasma sample with another one. For example rabbit vitreous, retina, retinal pigment epithelium interphotoreceptor matrix (RPE-IPM) *(10)*, cornea *(22)* and rat vitreous *(20)* all have demonstrable MMP-2 and -9 activity.

4. Notes

1. Cannot use a container that has a projection at the bottom because gels are damaged and stained gels become mottled.
2. NIH Image software as Scion Image is downloaded at Scion Corporation home page as free software (http://www.scioncorp.com/index.htm).
3. Wear a mask and gloves while handling acrylamide powder and solution, because acrylamide is oncogenic and neurotoxic. Do not dispose of acrylamide powder and solution, and discard solid waste.
4. Casein zymography to detect for MMP-3 and -7 is 10% polyacrylamide gels containing 2 mg/mL β-casein instead of gelatin.
5. Make up to 1x Electrode Buffer, adjust usually to pH 8.3.
6. Cleaning the glass, especially inside, the bands in gels appear clear and beautiful.
7. This method of zymography is modified from Hawkes *et al. (18)*.
8. The MMP-9 activity per one molecule is stronger than MMP-2 activity (approximate 7 to 10 times).
9. We use human recombinant purified proenzyme MMP-9 for reverse zymography, because pro-MMP-9 has high affinity with TIMP-1 *(23)*. The cell culture medium can be used in place of purified proenzyme MMP-9 *(18)*.
10. Make a fine adjustment for amount of APS and TEMED.
11. When the border between separating and stacking gel is not straight, the band of MMP activity is disturbed.
12. Do not touch the separating gel because digital compression results. We control the stacking gels with extremitas when discarding solution. Thus, do not cut and put off a stacking gel until finish.
13. Remaining small volume (1–2 mL) of Triton X-100 is necessary for the exchange of the SDS with Triton X-100.
14. Do not extend the destaining time as the contrast of stained gel decreases and becomes unclear.

References

1. Nagase, H., and Woessner, J. F. (1999) Matrix metalloproteinases. *J. Biol. Chem.* **274**, 21491–21494.
2. Kon, C. H., Occleston, N. L., Charteris, D., Daniels, J., Aylward, G. W., and Khaw, P. T. (1998) A prospective study of matrix metalloproteinases in proliferative vitreoretinapathy. *Invest. Ophthalmol. Vis. Sci.* **39**, 1524–1529.
3. Kvanta, A., Sarman, S., Fagerholm, P., Seregard, S., and Steen, B. (2000) Expression of matrix metalloproteinase-2 (MMP-2) and vascular endothelial growth factor (VEGF) in inflammation-associated corneal neovascularization. *Exp. Eye Res.* **70**, 419–428.
4. Plantner, J. J., Jiang, C., and Smine, A. (1998) Increase in interphotoreceptor matrix gelatinase A (MMP-2) associated with age-related macular degeneration. *Exp. Eye Res.* **67**, 637–645.
5. Chau, K. Y., Sivaprasad, S., Patel, N., Donaldson, T. A., Luthert, P. J., and Chong, N. V. (2008) Plasma levels of matrix metalloproteinase-2 and -9 (MMP-2 and MMP-9) in age-related macular degeneration. *Eye 22*, 855–859.
6. Uemura, S., Matsushita, H., Li, W., Glassford, A. J., Asagami, T., Lee, K. H., Harrison, D. G., and Tsao P. S. (2001) Diabetes mellitus enhances vascular matrix metalloproteinase activity. *Circ. Res.* **88**, 1291–1298.
7. Visse, R., and Nagase, H. (2003) Matrix metalloproteinases and tissue inhibitors of metalloproteinases: structure, function, and biochemistry. *Circ. Res.* **92**, 827–839.
8. Vu, T. H., Shipley, J. M., Bergers, G., Berger, J. E., Helms, J. A., Hanahan, D., Shapiro, S. D., Senior, R. M., and Werb, Z. (1998) MMP-9/gelatinase B is a key regulator of growth plate angiogenesis and apoptosis of hypertrophic chondrocytes. *Cell* **93**, 411–422.
9. Bergers, G., Brekken, R., McMahon, G., Vu, T. H., Itoh, T., Tamaki, K., Tanzawa, K., Thorpe, P., Itohara, S., Werb, Z., and Hanahan, D. (2000) Matrix metalloproteinase-9 triggers the angiogenic switch during carcinogenesis. *Nat. Cell Biol.* **2**, 737–744.
10. Iwai, S., Aljada, A., Higa, A., Nakanishi-Ueda, T., Fukuda, S., Kamegawa, M., Iwabuchi, S., Ueda, T., Caballero, S., Browne, R., Afzal, A., Grant, M., Yasuhara, H., Koide, R., Oguchi, K., Dandona, P., and Armstrong, D. (2006) Activation of AP-1 and increased synthesis of MMP-9 in the rabbit retina induced by lipid hydroperoxide. *Curr. Eye Res.* **31**, 337–346.
11. Brew, K., Dinakarpandian, D., and Nagase, H. (2000) Tissue inhibitors of metalloproteinases: evolution, structure and function. *Biochem. Biophys. Acta.* **1477**, 267–283.
12. Nagase, H., Visse, R., and Murphy, G. (2006) Structure and function of matrix metalloproteinases and TIMPs. *Cardiovasc. Res.* **69**, 562–573.
13. Snoek-van Beurden, P. A., and Von den Hoff, J. W. (2005) Zymographic techniques for the analysis of matrix metalloproteinases and their inhibitors. *BioTechniques* **38**, 73–83.
14. Heussen, C., and Dowdle, E. B. (1980) Electrophoretic analysis of plasminogen activators in polyacrylamide gels containing sodium dodecyl sulfate and copolymerized substrates. *Anal. Biochem.* **102**, 196–202.
15. Fernández-Resa, P., Mira, E., Quesada, A. R. (1995) Enhanced detection of casein zymography of matrix metalloproteinases. *Anal. Biochem.* **224**, 434–435.
16. Springman, E. B., Angleton, E. L., Birkedal-Hansen, H., and Van Wart, H. E. (1990) Multiple modes of activation of latent human fibroblast collagenase: evidence for the role of a Cys73 active-site zinc complex in latency and a "cysteine switch" mechanism for activation. *Proc. Natl. Acad. Sci. USA* **87**, 364–368.
17. Okamoto, T., Akuta, T., Tamura, F., van Der Vliet, A., and Akaike, T. (2004) Molecular mechanism for activation and regulation of matrix metalloproteinases during bacterial infections and respiratory inflammation. *Biol. Chem.* **385**, 997–1006.
18. Hawkes, S. P., Li, H., and Taniguchi, G. T. (2001) Zymography and reverse zymography for detecting MMPs, and TIMPs, in *Matrix Metalloproteinase Protocols.* (Clark, I. M., ed.), Humana, Totowa, NJ, pp. 399–410.
19. Woessner, J. F. Jr. (1995) Quantification of matrix metalloproteinases in tissue samples. *Methods Enzymol.* **248**, 510–528.
20. Asano, Y., Iwai, S., Okazaki, M., Kumai, T., Munemasa, Y., Oonuma, S., Tadokoro, M., Kobayashi, S., and Oguchi, K. (2008) Matrix metalloproteinase-9 in spontaneously hypertensive hyperlipidemic rats. *Pathophysiology* In press.

21. Kamiya, Y., Iwai, S., Nara K., Okazaki, M., and Oguchi, K. (2005) Effects of green tea on matrix metalloproteinases in streptozotocin-induced diabetic rats. *J. Clin. Biochem. Nutr.* **37**, 77–85.

22. Ogura, H., Nakanishi-Ueda, T., Ueda, T., Iwai, S., Uchida, S., Saito, Y., Taguchi, Y., Yasuhara, H., Armstrong, D., Oguchi, K., and Koide, R. (2007) Effect of a dihydrobenzofuran derivative on lipid hydroperoxide-induced rabbit corneal neovascularization. *J. Pharmacol. Sci.* **103**, 234–240.

23. Vincenti, M. P. (2001) The matrix metalloproteinase (MMP) and tissue inhibitor of metalloproteinase (TIMP) genes, in *Matrix Metalloproteinase Protocols.* (Clark, I. M., ed.), Humana, Totowa, NJ, pp. 121–148.

Chapter 11

The Use of In Vivo Microdialysis Techniques to Detect Extracellular ROS in Resting and Contracting Skeletal Muscle

Graeme L. Close and Malcolm J. Jackson

Abstract

Reactive oxygen species (ROS) are constantly produced by skeletal muscle and this production is increased during contractile activity. Understanding the role that ROS play in skeletal muscle requires an understanding of the species of ROS produced, the subcellular site of production, the time-course of ROS production, and the effects of inhibiting these ROS using specific antioxidants or inhibitors. Unfortunately, due to the extremely short half-lives of ROS, many methods to study ROS have had to rely on downstream markers of ROS reactions which cannot provide specific information. In vivo microdialysis is one technique that allows access to a specific site of ROS production allowing the continuous measurement of ROS at rest and during contractile activity. This chapter will describe the technique of microdialysis to measure ROS in skeletal muscle as well as discussing specific methods to detect superoxide, nitric oxide, and hydroxyl radical activity using in vivo microdialysis in skeletal muscle at rest and during contractile activity.

Key words: Microdialysis, Cytochrome c, Salicylate, Superoxide, Hydroxyl, Exercise, Free radical, HPLC, Nitric oxide.

1. Introduction

There has been scientific interest in the generation of reactive oxygen and nitrogen species (ROS) in skeletal muscle for over 50 years, but the first convincing reports of muscular exercise-induced ROS production appeared in the late 1970s and early 1980s. Following this time, there has been a great expansion in scientific papers describing studies in this area. The field has been hampered by a lack of appropriate techniques to examine ROS in

From: *Methods in Molecular Biology, vol. 477: Advanced Protocols in Oxidative Stress I,*
Edited by: D. Armstrong, DOI: 10.1007/978-1-60327-517-0_11, © Humana Press, New York, NY

biological materials with accuracy and specificity, but despite this it is clear that ROS can induce oxidative damage to both proteins and lipids in muscle and they can modulate a number of cell signalling pathways and regulate the expression of multiple genes in this tissue.

This review will describe the background and application of one new approach to studying ROS in skeletal muscle, in vivo microdialysis. This approach offers the opportunity to study ROS activities in tissues in vivo in both humans and animals. Furthermore, the approach provides data in real time and can provide a dynamic picture of changes in physiological and pathological processes.

1.1. Microdialysis

The technique of microdialysis was first described in 1966 *(1)* and its theory and application have been comprehensively reviewed *(2)*. Microdialysis was initially used to monitor the concentration of low molecular weight substances in brain interstitial fluid although it is now used to study various tissues, including skeletal muscle. Analysis of the composition of the tissue extracellular (or interstitial) space is difficult due to a lack of access to that compartment. Microdialysis is one of the few techniques that allow the user to collect many substances from remote regions with only limited tissue trauma thus enabling the continuous monitoring of local biochemistry.

Once inserted into the tissue of interest, the microdialysis probe basically acts as a small blood capillary (**Fig. 11.1**). Before

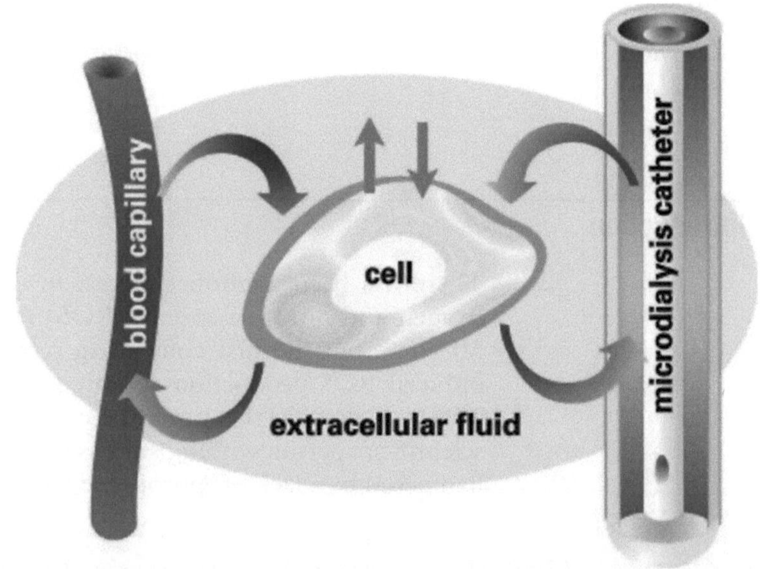

Fig. 11.1. Comparison of microdialysis probe with blood capillary. Image by courtesy of CMA Microdialysis AB.

a molecule enters the systemic circulation, it must transverse the extracellular space. Microdialysis allows the measurement of substances close to their site of origin before these reach the systemic circulation. This therefore allows the continuous monitoring of the substance of interest without dilution in the blood or the need to serially remove blood. This has major advantages for working with small animals and with highly reactive substances such as free radicals.

The microdialysis probe consists of a membrane permeable to water and small solutes. The probe is inserted into the interstitial space, and the membrane is continuously flushed with a solution devoid of the substance of interest. The probe is perfused on the inside with a physiological medium at a very slow flow rate and the outside of the membrane is in contact with the tissue extracellular fluid. This creates a concentration gradient causing bi-directional diffusion. The diffusion of substances will occur in the direction of the lowest concentration allowing water and low molecular weight compounds in the tissue extracellular fluid to perfuse across the dialysis membrane to be collected via the outlet port (**Fig. 11.2**). It is important to remember that the substance of interest must be present in the extracellular fluid. Furthermore, the constant flow results in incomplete equilibration of the dialysate with the extracellular fluid. This incomplete recovery will be discussed later in this chapter.

We have used this technique extensively for the detection of ROS in the extracellular fluid of rodent skeletal muscle *(3, 4, 5)*,

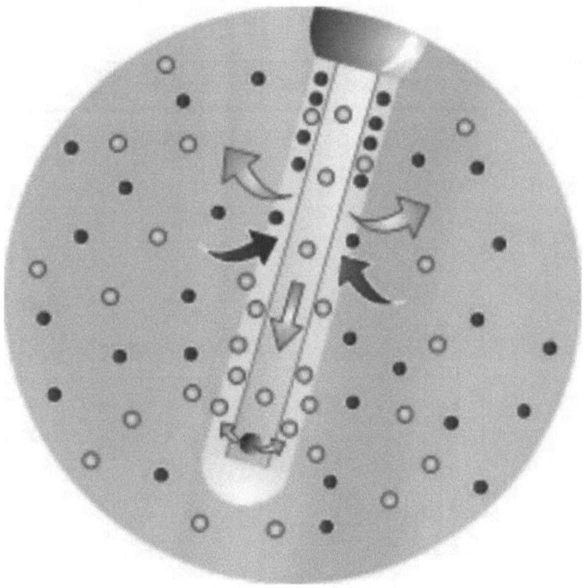

Fig. 11.2. Schematic representation of a microdialysis probe. Image by courtesy of CMA Microdialysis AB.

and recently this technique has been used by our laboratory to detect hydroxyl radical activity in the extracellular fluid of human muscle *(6)* and by other groups to detect superoxide activity in the extracellular fluid of human skeletal muscle *(7)*. This technique along with the use of transgenic mice has allowed us to gain a comprehensive understanding of the site and source of ROS production in quiescent and contracting skeletal muscle and this is summarised in **Fig. 11.3.**

1.2. Superoxide Detection Using Microdialysis

Superoxide is known to react with ferricytochrome c to form ferrocytochrome c *(8)* and this reduction results in an increased absorbance of cytochrome c when measured at 550 nm. It is therefore possible to measure this change in absorbance using a

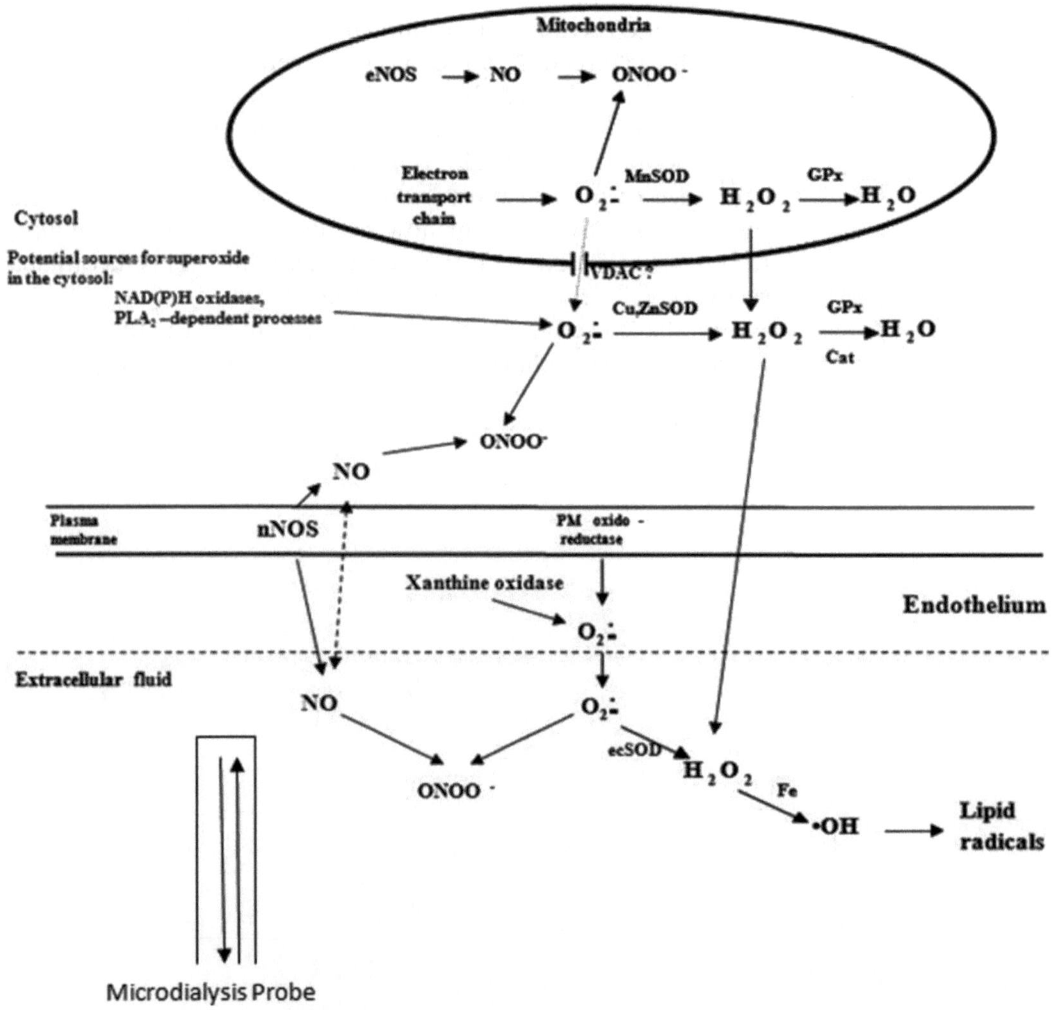

Fig. 11.3. Schematic representation of the sites of ROS production in skeletal muscle. Adapted from Jackson *et al.* (13). The microdialysis probe is inserted into the extracellular fluid allowing the measurement of several ROS close to the site of their production.

spectrophotometer with the change in absorbance representing the amount of reducing agent in the test medium *(8)*. It should be noted that cytochrome c is not capable of transmembrane movement and therefore this assay measures extracellular superoxide activity.

Due to the fact that cytochrome c can be reduced by both superoxide and nitric oxide, experiments have used specific inhibitors in an attempt to explore which radical species are causing the reduction of cytochrome c. For example, superoxide dismutase may be added to the test medium and any inhibition of cytochrome c reduction may then be attributed to superoxide. We have recently performed such experiments in conjunction with in vivo microdialysis and demonstrated that the predominant reductant of cytochrome c in skeletal muscle extracellular fluid is superoxide *(4)*. This study also demonstrated a marked increase in the reduction of cytochrome c in mouse extracellular fluid during a period of electrically stimulated isometric contractions (**Fig. 11.4**). We have reported approximately 0.2 nmol superoxide/15 min in the extracellular fluid of mouse skeletal muscle at rest and this can increase by approximately 100% during contractile activity.

1.3. Hydroxyl Radical Detection Using Microdialysis

This assay is based upon the reaction between the aromatic ring of salicylate and the hydroxyl radical. Hydroxyl radicals "attack" the phenolic ring of salicylate at either the 3 or the 5 position resulting in the formation of two stable metabolites, these being 2,3 and 2,5 dihydroxybenzoic acid (DHB) *(8)*. These metabolites

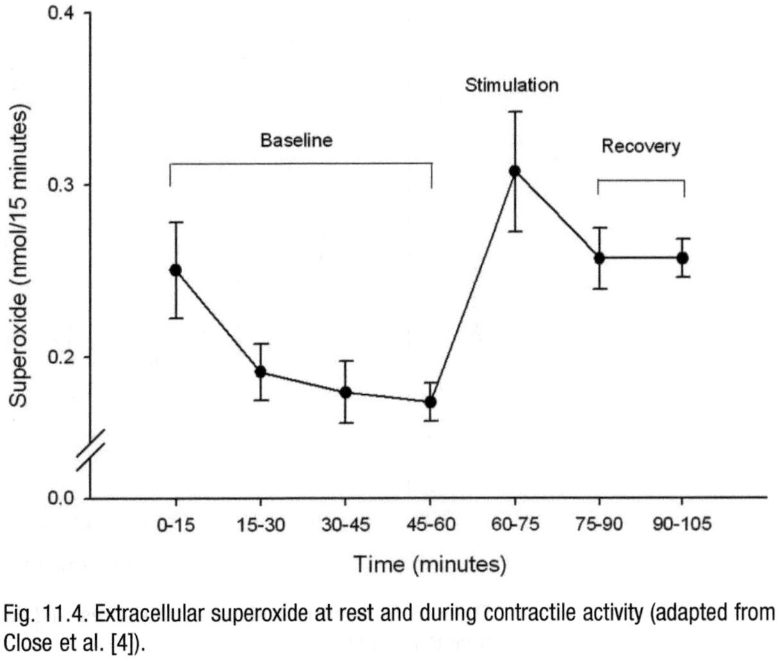

Fig. 11.4. Extracellular superoxide at rest and during contractile activity (adapted from Close et al. [4]).

can then be separated and measured using High Performance Liquid Chromatography with electro-chemical detection as an indicator of hydroxyl activity.

It is recognised that 2,5-DHB can also be produced enzymatically and may not be a specific assay for hydroxyl activity and therefore measurement of 2,3-DHB is now the assay of choice. This technique is frequently cited as a measure of hydroxyl activity although the specificity of this assay has also been questioned. It has been suggested that peroxynitrite may also contribute to the hydroxylation of salicylate and moreover that the presence of hydroxyl activity is not a prerequisite for hydroxylation reactions. However, we have recently demonstrated using specific ROS inhibitors such as L-NAME that hydroxyl is the predominant species causing the hydroxylation of salicylate in the extracellular fluid of mice that is detected by in vivo microdialysis *(4)*. We have reported the formation of approximately 5 pmol 2,3-DHB/15 min from mouse skeletal muscle at rest and we have observed a rise of approximately 100% during a 15 min contraction protocol (**Fig. 11.5**).

1.4. Nitric Oxide Detection Using Microdialysis

Some researchers have reported that microdialysates can be analysed for nitrite and nitrate content by the Griess reaction *(9)*, but we have found that use of a commercial fluorimetric assay (Cayman Chemical Co., USA) based on the method of Miles et al. *(10)* is required to ensure sufficient sensitivity with the very small volumes of sample available (60 μl/15 min collection).

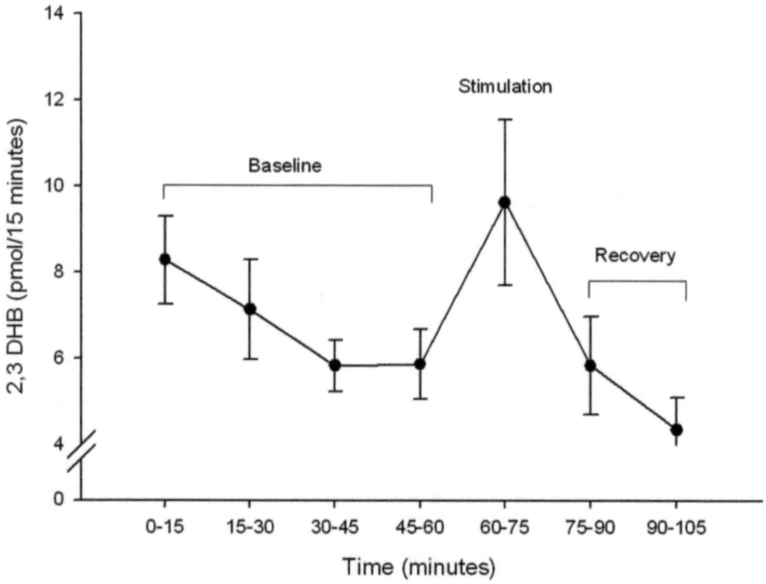

Fig. 11.5. Extracellular hydroxyl activity at rest and during contractile activity (adapted from Close et al. [4]).

This assay relies on the formation of nitrite and nitrate as the final products of Nitric Oxide (NO) reactions in vivo. Since the relative proportion of nitrite and nitrate is variable, the Cayman kit initially converts nitrate to nitrite using nitrate reductase. The nitrite is then converted to a fluorescent compound by adding DAN and NaOH and this compound can then be measured. We have observed resting total nitrite and nitrate values of approximately 500 pmoles/15 min in microdialysates from mouse skeletal muscle at rest with no major change following contractile activity *(11)*.

1.5. Comment and Limitations

There are a number of general limitations of this system, these include:

- It has been observed that inserting the microdialysis probe causes some local trauma to the tissue and this can have an effect on baseline ROS measures (see the initial values in **Figs. 11.4 and 11.5**). This is overcome by leaving the probe in place for 1 h to allow the baseline to stabilise prior to starting the experimental manipulation (*see* **Note 4**).
- Incomplete equilibration occurs across the dialysis membrane and for stable dialytes, this is corrected by assessing the recovery for a specific probe and experimental procedure. This is achieved by comparing the concentration of the substance in the microdialysis probe with the concentration of the test medium. Studies have shown that this recovery is dependent upon several factors including the length and diameter of the dialysis membrane, the flow rate, the temperature, and the molecular weight, shape, charge, and pH of the substance of interest. Recoveries of ROS across the microdialysis membrane have not been defined and are much more difficult to quantify than for relatively stable molecules. We therefore express the amount of ROS measured in terms of the amount recovered per 15 min. This approach does not allow us to calculate absolute values; however, our experiments are kinetic in nature where the main aim is to investigate the pattern of changes in ROS over time and hence it is not a crucial issue (*see* **Note 1**). For our studies, we ensure that the same microdialysis probes are used throughout to prevent variation due to differing probe recoveries. Caution should also be exerted when comparing data between different groups who may use different dialysis probes.
- The cellular source of ROS detected in the microdialysates cannot be defined since the skeletal muscle extracellular fluid is influenced by multiple cell types, including phagocytic cells, endothelial cells, and skeletal muscle cells.

- The slow flow rate requires a relatively long collection time to generate sufficient sample to be analysed (*see* **Note 1**). This means that immediate changes in local biochemistry cannot be measured and the samples represent what is occurring in the muscle over the 15 min collection period.

2. Materials

2.1. Microdialysis Probe and Introducer

Research groups have used two main types of probe, the "loop" dialysis membrane which is a short segment of a continuous tube, or "linear" probes that have a concentric tube arrangement whereby the perfusion fluid enters through an inner tube; flows to its distal end; exits the tube and enters the space between the inner tube and an outer dialysis membrane. The perfusion fluid then moves towards the proximal end of the probe and this is where the dialysis takes place. A wide variety of different probes with different molecular weight cut-offs are available. The "linear" probes are much more widely used than other types.

The microdialysis probes that we use are obtained from Microbiotech/se AB, Sweden and distributed in the UK by Royem Scientific Ltd. We have experience of only the "linear" probes and have regularly used 10, 4, or 2 mm membrane probes with 0.5 mm diameter and a molecular weight cut-off of 35,000 Da. The data presented in this chapter have been collected using a MAB 3.35.4 probe (Royem Scientific Ltd), which has a 35 mm shaft with a 4 mm dialysis membrane (**Fig. 11.6A**). The length of the inlet and outlet tubing can be customised with the manufacturer and specific information on the inlet and outlet tubing dead volume is provided. We currently use a 20 cm inlet and outlet tube that has 4.5 μl dead space in both the

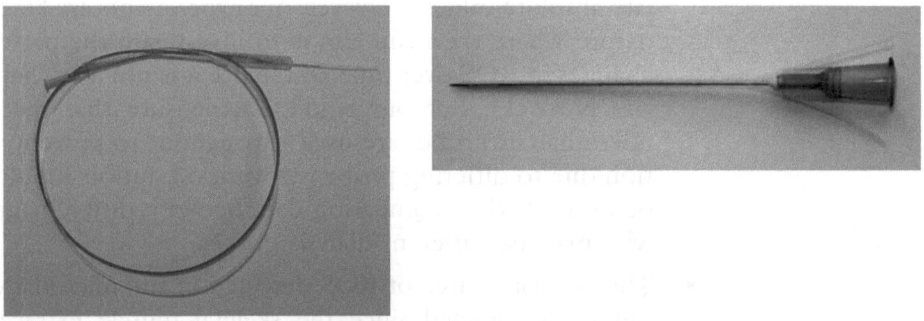

Fig. 11.6. **(A)** MAB3.35.4 microdialysis probe and **(B)** Plastic split introducer used for inserting the microdialysis probe. Reproduced with permission from Royem Scientific Ltd.

inlet and outlet tubing. The microdialysis probes are warranted for single use although if removed carefully they are suitable to be re-used in experimental models providing there has been no damage to the dialysis membrane (*see* **Note 3**). In our experience, the probes are best stored in saline between experiments to prevent the membrane from drying out. If the probe is stored dry, it is recommended to wet the membrane thoroughly prior to implantation. The membrane is made from polyether sulphone (PES) with polyurethane tubing, which means they can be autoclaved between experiments using saturated steam sterilisation at 121°C. Microdialysis probes are inserted into the muscle using a pre-split plastic introducer (**Fig. 11.6B**), also supplied by Royem Scientific Ltd.

2.2. Hamilton Syringe and Precision Pump

It is of vital importance that the flow rate is slow and consistent during microdialysis experiments and, therefore, a precision pump is necessary. There are many of these commercially available although the model we use is the KDS 220 Multi-Syringe Infusion Pump (KR Analytical). We also use glass Hamilton syringes to ensure a smooth delivery although other groups have used standard plastic syringes. Common flow rates used for this type of experiment are between 1 and 4 μl/min although slower rates can be used (*see* **Notes 1** and **2**). We also choose to use a multi-syringe pump to allow the infusion of several probes at the same time.

2.3. HPLC System

The HPLC system used to measure 2,3 and 2,5 DHB in our laboratory consists of a Rheodyne injector valve, HPLC pump and electrochemical detector (Gilson Model 303), Spherisorb 5 ODS column (HPLC technology) -25×4.6 mm with guard column and C-8 cartridge (BDH). The HPLC eluant consists of 34 mM sodium citrate, 27.7 mM acetate buffer (pH 4.75), and methanol 97.2:2.8, v/v. Standard solutions of 2,3-DHB and 2,5-DHB are prepared in HPLC-grade water. Twenty μl of samples or standards are eluted at a flow rate of 0.9 ml/min and monitored at +65 V with the electrochemical detector.

2.4. Spectrophotometers

For the measurement of NO using the Cayman kit, a fluorometric plate reader with the capacity to measure fluorescence using an excitation wavelength of 375 and an emission wavelength of 415 is required. We currently use a BMG Fluostar Optima (BMG Labtech). For the measurement of reduced cytochrome c, a spectrophotometer capable of measuring at 535–565 nm is required. We currently use a Bio-Tek Powerwave x340.

3. Methods

3.1. Insertion of the Microdialysis Probe

The microdialysis probes (**Fig. 11.6A**) are placed into the muscle using the split plastic introducer (**Fig. 11.6B**). The split plastic cannula is placed onto a 22G steel needle. The guide needle is then inserted into the muscle. Once the needle (and plastic cannula) are in the muscle, the needle is withdrawn leaving the plastic cannula in the muscle. The microdialysis probe can then be guided through the cannula and into the muscle (*see* **Note 5**). Once in position, the outlet tubing of the microdialysis probe is secured. The guide cannula is then removed from the muscle by slowly withdrawing the cannula whilst at the same time easing it apart. It is vital that the cannula is withdrawn at the same time as splitting to avoid damage to the muscle tissue or skin.

In mouse muscle, we routinely insert up to three microdialysis probes into a *gastrocnemius* muscle *(4)* and we have also inserted these probes into rat *anterior tibialis* muscle *(12)* and human *anterior tibialis* muscle *(6)*. In rodents, the probes are placed whilst the mice are fully anaesthetised and given suitable analgesia, whilst the probes are placed under local anaesthesia in humans (*see* **Note 6**). Probes can be left in place for several hours and we routinely leave the probes in mouse muscle for 3–4 h.

3.2. Perfusion of the Solution

Once the probes have been inserted into the muscle, the solution is then infused using the infusion pump. As previously stated, the recovery of the probe is dependent upon several factors including the flow rate and therefore an extremely slow flow rate is required to achieve optimum recovery. We routinely use 4 μl/min and collect samples for 15 min giving a total of 60 μl of dialysate per 15 min collection period. Samples should be collected onto ice during the 15 min collection period.

3.3. Reduction of Cytochrome c

- 50 μM cytochrome c (Sigma) is prepared in 0.9% saline.
- Cytochrome c is perfused through the microdialysis probe.
- Four, 15 min collections of microdialysates are initially made to allow the baseline to stabilise.
- Samples are collected and immediately stored on ice.
- Samples are then diluted 1:4 with distilled water in a clear microplate.
- Stock cytochrome c solution (50 μM) is diluted 1:4 with distilled water and added to the plate as a blank.
- Samples are analysed using scanning visible spectrometry at 535–565 nm.

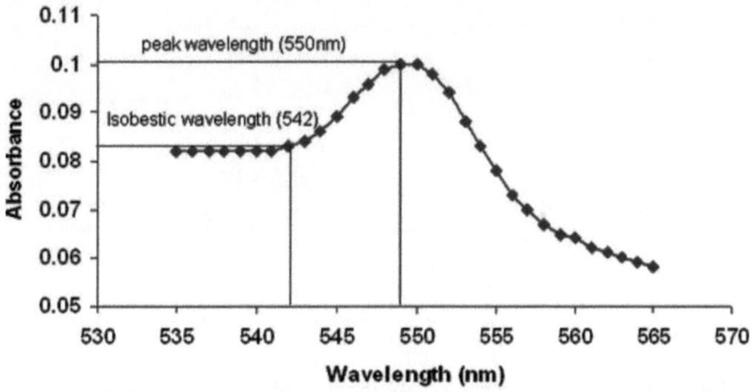

Fig. 11.7. Typical data from microdialysis experiment measuring the reduction of cytochrome c following electrical stimulation. The isobestic wavelength is subtracted from the peak wavelength and using a molar extinction coefficient of 21,000, the absorbance values can be converted to superoxide equivalents.

- The reduction of cytochrome c is calculated from the absorbance at 550 nm compared with that at the isobestic wavelength of 542 (**Fig. 11.7**).
- Results are expressed as superoxide equivalents using a molar extinction coefficient for reduced cytochrome c of 21,000 *(4)*.
- NB – Samples should not be frozen and should be analysed on the same day they are collected (*see* **Note** 7).

3.4. Hydroxylation of Salicylate

- 20 mM salicylate (Sigma) is prepared in normal saline.
- Salicylate is perfused through the microdialysis probe.
- Four, 15 min collections of microdialysates are initially made to allow the baseline to stabilise.
- Samples are collected and immediately stored on ice.
- Standard solutions of 2,3-DHB and 2,5-DHB are prepared ranging from 0.03 to 30 μM using HPLC grade water.
- 20 μl of sample/standard is eluted at a rate of 0.9 ml/min and monitored at +65 V with an electrochemical detector (**Fig. 11.8**).
- Samples can be stored at −40°C for up to 4 weeks prior to analysis.

3.5. Nitric Oxide

- 0.9% saline is prepared.
- Saline is perfused through the microdialysis probe.
- Four, 15 min collections of microdialysates are made.
- Samples are collected and immediately stored on ice.
- 20 μl samples are analysed using the Cayman chemical kit according to the manufacturer's instructions (*see* **Note 8**).
- Samples can be stored at −40°C for up to 4 weeks prior to analysis

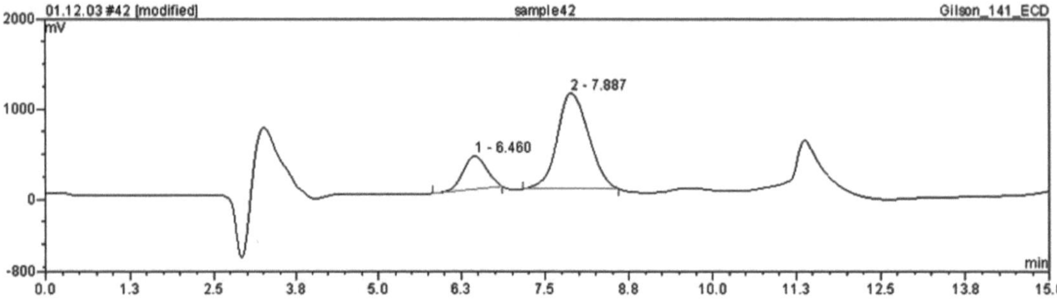

Fig. 11.8. Typical HPLC spectra for 2,3-DHB (peak 1 at 6.460 mins) and 2,5-DHB (peak 2 at 7.887 mins). X axis show a typical retention time for these peaks, which are generally run for 15 min.

3.6. Concluding Remarks

We have found that use of the microdialysis approach provides a novel means of accessing ROS within tissues in vivo. It is still relatively early in the development of this approach for monitoring ROS, but already significant advances have been achieved with this technique. A wider utilisation and evaluation of the technique is likely to lead to a greater understanding of the roles of extracellular ROS in normal physiology and in disease states.

4. Notes

We routinely monitor the following to ensure the reliability of the microdialysis approach:

1. As stressed throughout this chapter, a constant flow rate of the perfusate is essential for reliable microdialysis experiments. We monitor this by checking that the expected amount of dialysate is collected. For example, if we are perfusing at 4 μl/min for 15 min we always ensure that we have 60 μl of dialysate at the end of each 15 min collection period.
2. The length of the collection time also must be carefully monitored. Automated fraction collectors are now available that allow for precise timing of the collections.
3. Occasionally, a microdialysis probe can be damaged when inserting the probe. It is important to recognise when this has occurred. Typically this will result in a reduced volume or complete lack of any dialysate. It is vital that the initial dialysate is monitored and the probe is discarded if it is suspected that a probe has been damaged.
4. Occasionally, insertion of the microdialysis probe can result in a small bleed within the tissue and breakdown of the erythrocytes can lead to the dialysate appearing red in colour. This will significantly affect the data. It is important to monitor the colour of the dialysate and reject if there is any unusual colouring.

5. When placing the microdialysis probes into skeletal muscle of small rodents, it is possible to insert the probe too far and pass entirely through the muscle of interest. It is important to visually inspect the placement of the probe ensuring that it is placed into the muscle.
6. We routinely place up to three microdialysis probes into a single mouse *gastrocnemius* muscle. When placing multiple probes into the muscle, it is important to ensure that the "introducer" does not puncture a probe that is already in position. Insert the new probe away from the first probe or at a different angle to the first one.
7. When measuring superoxide production using the reduction of cytochrome c, it is necessary to regularly run a blank to check for any spurious reduction. We routinely test the perfusate reservoir from the Hamilton syringe to check for any abnormalities. There should be very little reduction of cytochrome c in the non-dialysed perfusate.
8. When measuring total nitrates and nitrites as an indicator of NO production, it is also advised to check the blank for any abnormalities. There should be very little nitrates and nitrites in the non-dialysed perfusate.

Acknowledgements

The authors would like to thank *The Wellcome Trust, Research into Ageing*, United States *National Institutes of Health, European Commission, Medical Research Council, The Food Standards Agency*, and *Biotechnology and Biological Sciences Research Council* for their financial support along with colleagues in the Cellular Pathophysiology Research group at the University of Liverpool.

References

1. Bito, L., H. Davson, E. Levin, M. Murray, and N. Snider, *The concentration of free amino acids and other electrolytes in cerebospinal fluid, in vivo dialysate of brain, and blood plasma of the dog.* J Neurochem, 1966. **13**(11): p. 1057–1067.
2. Benveniste, H. and P.C. Huttemeier, *Microdialysis – theory and application.* Prog Neurobiol, 1990. **35**(3): p. 195–215.
3. McArdle, A., J. van der Meulen, G.L. Close, D. Pattwell, H. Van Remmen, T.T. Huang, A.G. Richardson, C.J. Epstein, J.A. Faulkner, and M.J. Jackson, *Role of mitochondrial superoxide dismutase in contraction-induced generation of reactive oxygen species in skeletal muscle extracellular space.* Am J Physiol Cell Physiol, 2004. **286**(5): p. C1152–C1158.
4. Close, G.L., T. Ashton, A. McArdle, and M.J. Jackson, *Microdialysis studies of extracellular reactive oxygen species in skeletal muscle: factors influencing the reduction of cytochrome c and hydroxylation of salicylate.* Free Radic Biol Med, 2005. **39**(11): p. 1460–1467.
5. Pattwell, D., A. McArdle, R.D. Griffiths, and M.J. Jackson, *Measurement of free radical production by in vivo microdialysis*

during ischemia/reperfusion injury to skeletal muscle. Free Radic Biol Med, 2001. **30**(9): p. 979–985.

6. Crowe, A.V., A. McArdle, F. McArdle, D.M. Pattwell, G.M. Bell, G.J. Kemp, J.M. Bone, R.D. Griffiths, and M.J. Jackson, *Markers of oxidative stress in the skeletal muscle of patients on haemodialysis.* Nephrol Dial Transplant, 2007. **22**(4): p. 1177–1183.
7. Hellsten, Y., J.J. Nielsen, J. Lykkesfeldt, M. Bruhn, L. Silveira, H. Pilegaard, and J. Bangsbo, *Antioxidant supplementation enhances the exercise-induced increase in mitochondrial uncoupling protein 3 and endothelial nitric oxide synthase mRNA content in human skeletal muscle.* Free Radic Biol Med, 2007. **43**(3): p. 353–361.
8. Murrant, C.L. and M.B. Reid, *Detection of reactive oxygen and reactive nitrogen species in skeletal muscle.* Microsc Res Tech, 2001. **55**(4): p. 236–248.
9. Shintani, F., S. Kanba, T. Nakaki, K. Sato, G. Yagi, R. Kato, and M. Asai, *Measurement by in vivo brain microdialysis of nitric oxide release in the rat cerebellum.* J Psychiatry Neurosci, 1994. **19**(3): p. 217–221.
10. Miles, A., Y. Chen, M. Owens, and M. Grisham, *Fluorimetric determination of nitric oxide.* Methods, 1995. 7(1): p. 40–47.
11. Vasilaki, A., A. Mansouri, H. Remmen, J.H. van der Meulen, L. Larkin, A.G. Richardson, A. McArdle, J.A. Faulkner, and M.J. Jackson, *Free radical generation by skeletal muscle of adult and old mice: effect of contractile activity.* Aging Cell, 2006. **5**(2): p. 109–117.
12. Pattwell, D., T. Ashton, A. McArdle, R.D. Griffiths, and M.J. Jackson, *Ischemia and reperfusion of skeletal muscle lead to the appearance of a stable lipid free radical in the circulation.* Am J Physiol Heart Circ Physiol, 2003. **284**(6): p. H2400–H2404.
13. Jackson, M.J., D. Pye, and J. Palomero, *The production of reactive oxygen and nitrogen species by skeletal muscle.* J Appl Physiol, 2007. **102**(4): p. 1664–1670.

Chapter 12

Cell-Free Antibody Capture Method for Analysis of Detergent-Resistant Membrane Rafts

Anil Bamezai and Colleen Kennedy

Abstract

Cholesterol-rich microdomains present on the plasma membrane appear to play an important role in spatio-temporal regulation of cell signaling and cell adhesion processes. Compositional heterogeneity of these microdomains and their coalescence during cell–cell interactions may provide one mechanism for triggering and/or regulating signaling cascades from the plasma membrane to the cell interior. Biochemical analyses of distinct lipid microdomain subpopulations and single-rafts obtained from unstimulated and ligand-stimulated cells are critical for deciphering functional role of lipid rafts. We have designed a cell-free assay that captures detergent-resistant lipid rafts with an antibody against a raft-resident molecule and detects the presence of another lipid raft molecule. Moreover, this cell-free assay provides a simple and quick way to examine the simultaneous presence of two proteins in the lipid rafts, and has the potential to estimate trafficking of molecules in and out of the lipid microdomains during cell signaling on a single lipid raft-basis.

Key words: Lipid rafts, Detergent-resistant microdomains, ELISA, Electron microscopy, GPI-anchored protein, Cholesterol;, Ly-6, Thy-1.

1. Introduction

The fluid Mosaic model proposed by Singer and Nicholson suggests that the constituents of the plasma membrane (phospholipids, cholesterol, and proteins) are homogenously distributed and move in the plane of the membrane in an unrestrained manner *(1)*. Inconsistent with this proposed model was the observation that Immunoglobulin G and Immunoglobulin M receptors on the surface of mature B lymphocytes co-capped after binding to a specific antigen *(2)*. In addition, the presence

From: *Methods in Molecular Biology, vol. 477: Advanced Protocols in Oxidative Stress I,*
Edited by: D. Armstrong, DOI: 10.1007/978-1-60327-517-0_12, © Humana Press, New York, NY

of supermolecular protein complexes in the plasma membrane poses restriction on the movement of some proteins in the plane of the membrane *(3, 4)*. Heterogeneity of the plasma membrane, with respect to lipid composition, also has been recognized. The outer leaflet of the bilayer is enriched in phosphatidylcholine and sphingolipids whereas the inner leaflet mostly contains phosphatidylserine and phosphatidylethanolamine *(5, 6)*. This nonhomogenous distribution of lipids across the membrane bilayer points to an asymmetrical nature of the plasma membrane. Membrane asymmetry is more pronounced in epithelial cells in which lipids differentially segregate to apical and basolateral side of the plasma membrane *(7)*. A more recent model of the cell membrane suggests that a phase separation in the membrane exists where sphingolipids with their saturated acyl side chains in association with cholesterol form tight clusters (also known as lipid rafts) that segregate from other unsaturated phospholipids in the same plane of the membrane, further contributing to the asymmetric nature of the membrane *(8)*.

Tight packing of sphingolipids and cholesterol imparts thermodynamic stability to lipid rafts more than the liquid-disordered phase of the membrane constituted of unsaturated phospholipids and largely devoid of cholesterol *(9)*. The liquid-ordered property of the lipid rafts imparts them insolubility in nonionic detergents at 4°C. The insoluble microdomians have been given a variety of acronyms, for example, detergent-resistant membranes (DRMs), glycolipid-enriched microdomains (GEMs), or detergent-resistant lipid rafts (DRLRs), *(10, 11, 12)*. In addition to sphingolipids and cholesterol, lipid microdomains are composed of glycosylphosphatidylinositol (GPI)-anchored proteins, which tether to the outer leaflet of the membrane with a lipid anchor *(13)*. Other cytoplasmic and transmembrane proteins including $p56^{lck}$, $p59^{fyn}$, linker of activated T cells (LAT), calcium channel protein, p21ras, heterotrimeric/monomeric G proteins, and gangliosides (GM1) with lipid modifications are also present in lipid rafts *(14, 15, 16)*. A number of other signaling molecules traffic through these microdomains during cell signaling *(17, 18)*. Lipid rafts have been implicated in signaling with functional consequences that range from cell survival and proliferation to oxidative stress and cell death *(19, 20, 21)*.

1.1. Size and Compositional Heterogeneity of Lipid Rafts

The existence of lipid rafts on normal cells has been questioned and hotly debated *(22)*. A body of published data suggests that these membrane microdomains exist on live cells and their reported size varies from 5 to 150 nm in diameter, depending on the method used and/or the cell type studied *(23, 24)*. A number of reports suggest heterogeneous lipid rafts, with respect to protein and lipid composition *(25, 26, 27, 28, 29)*. Size and compositional heterogeneity of these microdomains may be an integral part of the regulation of signaling. These initial observations have led a number of us to propose that lipid rafts play an important role in cell signaling *(30)*. In the absence of cell stimulation, the lipid rafts are small and dynamic, as much as these membrane microdomains are incomplete signaling units. The heterogenous nature of lipid rafts allows compartmentalization of signaling molecules, thereby preventing an initial trigger for a signaling cascade from the plasma membrane. Engagement of a receptor with an appropriate ligand initiates coalescing of lipid rafts, bringing together different signaling molecules that are sequestered in different microdomains. Promotion of molecular interaction through coalescing of lipid rafts may provide one possible mechanism by which cell signaling is regulated from the plasma membrane to the cell interior. To be able to understand whether this mechanism is operational, it is critical to study the size and compositional heterogeneity of lipid rafts. Methods and experimental approaches that would dissect lipid rafts into distinct subsets, as well as allow their analysis on a single raft basis, will be critical to elucidate the role of these microdomains in cell signaling. Both the biochemical characterization and direct visualization of these membrane entities will provide much needed functional understanding of lipid rafts.

2. Materials

2.1. Equipment

1. SpectraMax 190 ELISA plate reader. (Molecular Devices, Sunnyvale, CA, USA).
2. Sorval Legend RT centrifuge (ThermoFischer Scientific, Walthan, MA, USA).
3. Beckman Coulter Avanti J-E ultracentrifuges (Beckman Coulter, Fullerton, CA, USA).
4. Beckman L8-M ultracentrifuge (Beckman Coulter, Fullerton, CA, USA).
5. Hitachi H-7600 transmission electron microscope (Hitachi, Gaithersburg, MD, USA).

2.2. Cell Culture

Dulbecco's Modified Eagle Medium (DMEM) with 4.5 g/ml of glucose/liter (GIBCO/BRL, Carlsbad, CA, USA) supplemented with 10% heat inactivated fetal bovine serum (Atlanta Biologicals, Atlanta, GA, USA), L-glutamine, sodium pyruvate, and streptomycin, fungizone (GIBCO/BRL, Carlsbad, CA, USA).

2.3. Generation of Detergent-resistant Lipid Rafts

1. MBS cell lysis buffer: 25 mM MES, 150 mM NaCl (pH 6.5), 1% Triton X-100, 1 mM Na_3VO_4, 2 mM EDTA, 1 mM PMSF, and 1 ug/ml aprotinin (Sigma-Aldrich, St. Louis, MO, USA).
2. 85% Sucrose solution: 85% sucrose (w/v) in MBS.
3. 35% Sucrose solution: 35% sucrose (w/v) in MBS.
4. 5% Sucrose solution: 5% sucrose (w/v) in MBS containing 1 mM Na_3VO_4.
5. SW40 centrifuge tubes (Beckman Instruments, Palo Alto, CA, USA).

2.4. Lipid Raft Elisa

1. 96-well, flat bottom, high binding, enzyme immunoassay/radioimmuno assay (EIA/RIA) (Costar, New York, NY, USA).
2. Capture antibody, e.g., immuno-purified anti-mouse CD90 (BD Biosciences, San Jose, CA, USA) is resuspended in carbonate/bicarbonate buffer (Sigma-Aldrich, St. Louis, MO, USA) at 2 μg/ml concentration.
3. Wash buffer: Phosphate-buffered saline with Tween 20 (PBS-T); Prepare from 10× stock solution with 137 mm NaCl, 2.7 mm KCl, 4.3 mm Na2HPO4-7H20, 1.4 mm KH2PO4, pH 7.3. Dilute 100 ml with 900 ml double distilled water (DDW) and add 0.05% Tween 20 (Sigma, St. Louis, MO, USA) and stir it before use.
4. Blocking Buffer: Phosphate-buffered saline supplemented with 1% (w/v) fraction V bovine serum albumin (BSA) (Fisher Scientific, Pittsburg, PA, USA).
5. Dilution buffer: PBS-T/BSA; Prepare PBS-T supplemented with 1% (w/v) BSA.
6. Detection Antibody: Biotin anti-mouse Ly6A/E (Sca-1) (D7) (BD Biosciences, San Jose, CA, USA) at 1 μg/ml in PBS-T/BSA.
7. Avidin-HRP: PBS-T/BSA supplemented with 1:250 dilution from avidin-horseradish peroxidase stock (BD Biosciences, San Jose, CA, USA).
8. ABTS Peroxidase Substrate system: ABTS is 2,2′-azino-di[3-ethyl-benzthiazoline 6-sulphonate] (Solution A) and Peroxidase Substrate is 0.02% solution of H_2O_2 in Citric acid buffer (Solution B) (KPL, Gaithersburg, MD, USA).
9. ABTS stop solution: Dilute this solution 1:5 in ddH2O prior to use (KPL, Gaithersburg, MD, USA).
10. Spectrophotometer (Spectramax 190 platereader) (Molecular Devices, Sunnyvale, CA, USA).

2.5. Electron Microscopy (EM): Formvar Coated Grids

1. Nickel grids (Electron Microscopy Sciences, Fort Washington, PA, USA).
2. Formvar Solution in ethylene dichloride (Electron Microscopy Sciences, Hatfield, PA, USA) in chloroform (Fisher scientific, Pittsburg, PA, USA).
3. Standard microscope slides (Fisher Scientific, Pittsburgh, PA, USA).

2.6. Immunogold Labeling

1. Incubation buffer: Prepare a 10× PBS stock with 137 mM NaCl, 2.7 mM KCl, 4.3 mM Na_2HPO_4-$7H_2O$, 1.4 mM KH_2PO4, pH 7.3. Dilute 100 ml with 900 ml water, add 1% of 10% BSA-C (Aurion, Costerweg, The Netherlands).
2. Capture antibody: Purified anti-mouse CD90 (Thy-1) (G7) (BD Biosciences, San Jose, CA, USA) at 4 μg/ml in carbonate/bicarbonate buffer (Sigma-Aldrich, St. Louis, MO, USA).
3. Blocking buffer: 1× PBS supplemented with 0.05% (w/v) fraction V bovine serum albumin (BSA) (Fisher Scientific, Pittsburg, PA, USA).
4. Detection antibody: Biotin conjugated anti-mouse Ly6A/E (Sca-1) (D7) (BD Biosciences, San Jose, CA, USA) at 3 μg/ml in incubation buffer.
5. Colloidal gold: Goat anti-biotin antibody conjugated to 10 nm gold particles (Aurion, Costerweg, Netherlands) diluted from stock at a 1:20 dilution in incubation buffer.
6. Gluteraldehyde solution: 1% gluteraldehyde (Electron Microscopy Sciences, Hatfield, PA, USA) in ddH_2O.
7. Osmium Tetra-oxide solution: 1% OsO_4 (Electron Microscopy Sciences, Hatfield, PA, USA) in ddH_2O.
8. Tannic acid solution: 1% (w/v) tannic acid (Electron Microscopy Sciences, Hatfield, PA, USA) in ddH_2O.
9. Uranyl acetate solution: 2% uranyl acetate (Electron Microscopy Sciences, Hatfield, PA, USA) in ddH_2O.
10. Ten humid chambers: 100 mm Petri dishes overlaid with wet Whatman filter paper and overlaid with parafilm.

3. Methods

Nonionic detergents render lipid rafts insoluble aiding in their separation from other sub-cellular organelles on a sucrose-gradient following ultracentrifugation based on their low density. Biochemical analyses of detergent-resistant membrane domains have been instrumental in developing the "Lipid raft" hypothesis *(8, 12)*. Use of SDS/PAGE coupled with Western blot analysis has allowed examination of the composition of lipid microdomains with specific antibodies directed against them *(17, 18)*. This technique has been used to examine alterations in the composition that might occur during a cellular process *(17, 18)*. Isolating lipid rafts from the membrane and subjecting them to SDS/PAGE coupled with Western blotting does not completely examine trafficking of signaling proteins in and out of lipid rafts, simply because of their heterogeneous nature that we all have come to recognize *(25, 26, 27, 28, 29, 30)*. The vesicular shape and the compositional heterogeneous nature of lipid rafts have allowed us to devise methods to capture and detect lipid rafts and examine them on subpopulation as well as on single raft-basis.

3.1. Preparation of Samples for Analyses

1. Cell culture: Mouse $CD4^+$ T-T hybrid cell line, YH16.33, *(31)* are passaged twice a week by transferring 100 μl from parent culture flask into new 25 m^2 flask with 10 ml of fresh DMEM media cocktail and maintained at 37°C in 10% CO_2 incubator. For generating 40×10^6 cells the cultures are proportionally expanded in 125 cm^2 flasks supplemented with 100 ml DMEM media cocktail per flask.
2. Washing cells: Cells are harvested by transferring cells into sterile 50 ml tubes and centrifuged at 1000 rpm for 10 min, at 4°C. Cell pellet is resuspended in 1 × PBS and washed three times with chilled 1 × PBS to remove extraneous proteins from media and fetal bovine serum.

3.2. Detergent-Resistant Lipid Rafts

1. Harvest 40×10^6 cells in culture media and place in 50 ml centrifuge tube.
2. Centrifuge cells at 1000 rpm for 2 min at 4°C.
3. Decant supernatant and resuspend cells in 10 ml of 1× PBS.
4. Centrifuge cells at 1000 rpm for 3 min at 4°C.
5. Decant supernatant and resuspend in 1 ml of ice-cold MBS for 30 min on ice.
6. Homogenize lysates with 10 strokes of a loose-fitting dounce homogenizer.
7. Mix homogenized cell lysates with 1 ml 85% sucrose and place in SW40 centrifuge tube.
8. Overlay sample with 6 ml of 35% sucrose, followed with 4 ml of 5% sucrose.
9. Ultracentrifuge for 18 h at 200,000 g at 4°C.
10. Fractionate samples by collecting 1 ml fractions from the bottom of the gradient tube.

3.3. Lipid Raft Elisa (see Notes 1 and 2)

1. Coat the 96-well plate with 50 μl of capture antibody. Cover the plate with saran wrap and incubate overnight at 4°C.
2. Wash the microwells four times with PBS-T. For each washing step pipette 100 μl of PBS-T in each well. Remove the buffer by dumping the fluid in the sink and then gently pat the plate on a stack of paper towels for removing buffer sticking to the edges of the wells.
3. Add 100 μl of blocking buffer to each well and incubate at room temperature for 30 min. This step will block nonspecific binding sites in the microwells.
4. Wash the microwells three times with 100 μl of PBS-T.
5. Add 50 μl of lipid raft fractions to appropriate wells after diluting the raft fractions at 1:5 in PBS-T/BSA solution.
6. Wash the microwells nine times with 100 μl PBS-T buffer.
7. Add 50 μl of biotinylated detection antibody to appropriate wells. Cover the plate with saran wrap and incubate at room temperature for 60 min.
8. Wash the microwells six times with 100 μl PBS-T buffer.
9. Add 50 μl of avidin-HRP to appropriate wells. Cover the microtitre plate with saran wrap and incubate at room temperature for 30 min.
10. Wash the microwells eight times with 100 μl PBS-T buffer.
11. Add 100 μl of ABTS substrate to appropriate microwells. The A and B solutions should be mixed together in 1:1 ratio. Cover the microtitre plate with aluminum foil and incubate at room temperature for 10–30 min, or until color develops.
12. Add 100 μl of ABTS stop solution.
13. Read absorbance at 405 nm with spectramax plate reader.

3.4. EM: Formvar Coated Grids

1. Clean nickel grids by sonicating three times in ethanol.
2. Clean standard microscope glass slides with ethanol and air dry them.
3. Using a kimwipe, gently wipe dishsoap and layer it over the glass slide (*see* **Note 3**).
4. Wipe the glass slide with kimwipe.
5. Dip the glass slide in formvar and let sit for a few seconds. Carefully take the slide out and let sit in the fumes of the formvar for a few more seconds (*see* **Note 4**).
6. Score around the edges of the glass slide to be able to release formvar.
7. In a bowl of clean ddH_2O, put formvar coated glass slide at a 45° angle to allow formvar to slide off into water (*see* **Note 5**).

8. Carefully arrange nickel grids on floating formvar.
9. Using a different microscope slide wrapped in parafilm, carefully scoop the formvar with the grids up from the water bowl and dry grids on the parafilm wrapped slide (*see* **Note 6**).

3.5. EM: Immunogold Labeling (see Notes 7–9)

1. Pipette 30 μl drops of capture antibody on the parafilm of humid chamber and incubate grids on top of each drop with formvar coated side facing down for overnight at 4°C.
2. Pipette 75 μl drops of incubation buffer on parafilm of the 2nd set of humid chambers and wash grids on these drops, four times, for 5 min each at room temperature.
3. Pipette 30 μl drops of blocking buffer in 3rd set of humid chamber and incubate grids on the drop for 20 min at room temperature.
4. Wash grids on 75 μl drops of incubation buffer two times, for 5 min each at room temperature.
5. Incubate grids on 30 μl drops of sample overnight at 4°C.
6. Wash grids on 75 μl drops of incubation buffer six times, for 5 min each at room temperature.
7. Incubate grids on 30 μl drops of detection antibody for 60 min at room temperature.
8. Wash grids on 75 μl drops of incubation buffer four times, for 5 min each at room temperature.
9. Incubate grids on 30 μl drops of blocking buffer for 15 min at room temperature.
10. Wash grids on 75 μl drops of incubation buffer two times, for 5 min at room temperature.
11. Incubate grids on 30 μl drops of goat anti-biotin gold for 60 min at room temperature.
12. Wash grids on 75 μl drops of incubation buffer six times, for 5 min at room temperature.
13. Wash grids on 75 μl drops of ddH_2O two times, for 5 min at room temperature.
14. Incubate grids on 30 μl drops of 1% glutraldehyde in ddH_2O for 5 min at room temperature (*see* **Note 10**).
15. Wash grids on 75 μl drops of ddH_2O two times, for 5 min at room temperature.
16. Dry grids, specimen side up, on dry Whatman filter paper.
17. Incubate grids on 50 μl drops of 1% osimum in ddH_2O for 10 min at room temperature (*see* **Note 10**).
18. Wash grids on 75 μl drops of ddH_2O two times, for 5 min at room temperature.
19. Incubate grids on 50 μl drops of 1% tannic acid in ddH_2O for 30 min at room temperature.

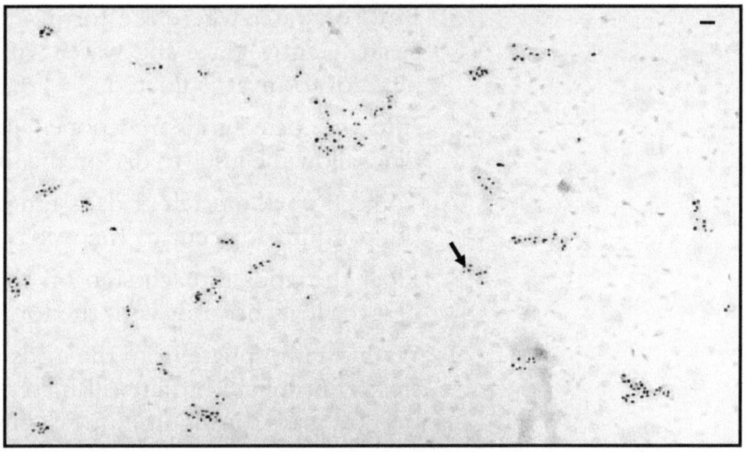

Fig. 12.1. Immunogold labeling of detergent-resistant lipid rafts. Lipid rafts were captured on a nickel grid by anti-Thy1 monoclonal antibody and detected with biotinylated anti-Ly-6A/E monoclonal antibody followed by incubation with anti-biotin antibody conjugated with colloidal gold (10 nm). The captured detergent-resistant lipid rafts were first fixed with 1% $Os0_4$ and then counterstained with 1% tannic acid and 2% uranyl acetate before their analysis by transmission electron microscopy. Arrow indicates one of many detergent-resistant lipid rafts captured with anti-Thy-1 antibody and detected with anti-Ly-6A/E. Many detergent-resistant lipid rafts captured with anti-Thy-1 antibody do not show the presence of anti-Ly-6A/E molecule in them (dark entities without colloidal gold). The bar drawn on top right of the figure = 50 nm.

20. Wash grids on 75 μl drops of ddH_2O two times, for 5 min at room temperature.
21. Incubate grids on 50 μl drops of 2% uranyl acetate in ddH_2O for 30 min at room temperature (*see* **Notes 10** and **11**).
22. Wash grids on 75 μl drops of ddH_2O two times, for 5 min at room temperature.
23. Dry grids, specimen side up, on dry Whatman filter paper.
24. Analyze grids on a Transmission Electron Microscope. An example of the results produced is shown in **Fig. 12.1.**

4. Notes

1. The microwells should not dry out during the washing steps.
2. Air bubbles in the microwells are undesirable at each ELISA step and will interfere in the reading of the microwell plate on spectrophotometer plate reader. Slow and careful pipetting prevents the introduction of air bubbles.
3. High humidity makes it harder for the formvar to solidify and stick to the grids. Carrying out coating of grids on a day of low humidity will alleviate this problem.
4. Dishwashing soap allows the formvar to stick to the glass slide.

5. Bowl of water where the formvar is floated should be very clean. Gently grace the water with a kimwipe before each piece of formvar is floated.
6. After the coated grids are scooped up with the parafilm covered slide, allow the grids to dry on the slide in a covered Petri dish.
7. To keep track of each grid, assign it a number by writing on the parafilm adjacent to the grid.
8. Float the grids in each step on the same side. It would be best to float the grids with the formvar coat facing the drop.
9. Wash off phosphate from the grids prior to immuno-staining steps. Phosphate buffer will increase chances of contamination on grids during immuno-staining.
10. Remove excess water/buffer from the grid after each washing/incubation step before beginning the next staining step by touching the edge of the grid with a filter paper.
11. Limit exposure to light when staining with uranyl acetate and air dry the grids completely before placing them in the grid box.
12. Gluteraldehyde, Osmium tetroxide, tannic acid, and uranyl acetate are toxic and generate considerable unpleasant smell in the laboratory. Use these chemicals with utmost care in a chemical fume hood.

Acknowledgements

The authors would like to thank Dr. Norman Dollahon, Mr. Matthew D. Nelson, and Ms. Sally Shrom for helpful discussions. This work was supported by Department of Biology, Villanova University.

References

1. Singer, S.J. and Nicolson, G.L. (1972) Fluid mosaic model of the structure of cell membranes. *Science* **175**, 720–731.
2. Goding, J.W. and Layton, J.E. (1976) Antigen-induced co-capping of IgM and IgD-like receptors on murine B cells. *J Exp Med* **144**, 852–857.
3. Damjanovich, S., Matko, J., Matyus, G., Szabo, G., Szollosi, J., Pieri, J., Farkas, T., and Gaspar, R. (1998) Supramolecular receptor structures in the plasma membrane of lymphocytes revealed by flow cytometric energy transfer, scanning force- and transmission electron-microscopic analyses. *Cytometry* **33**, 225–233.
4. Vereb, G., Szollosi, J., Matko, J., Nagy, P., Farkas, L., Vigh, L., Matyus, L., Waldmann, T., and Damjanovich, S. (2003) Dynamic, yet structured: The cell membrane three decades after the Singer-Nicolson model. *Proc Natl Acad Sci USA* **100**, 8053–8058.
5. Renooij, W., Van Golde, L.M., Zwaal, R.F., and Van Deenen, L.L. (1976) Topological asymmetry of phospholipids metabolism in rat erythrocyte membranes: Evidence for flip-flop of lecithin. *Eur J Biochem* **61**, 53–58.
6. Rothman, J.E. and Lenard, J. (1977) Membrane asymmetry. *Science* **195**, 743–753.

7. Keller, P., Toomre, D., Diaz, E., White, J., and Simons, K. (2001) Multicolour imaging of post-Golgi sorting and trafficking in live cells. *Nat Cell Biol* **2**, 140–149.
8. Simons, K and Ikonen, E. (1997) Functional rafts in cell membranes. *Nature* **387**, 569–572.
9. Almeida, P.F., Pokorny, A., and Hinderliter, A. (2005) Thermodynamics of membrane domains. *Biochim Biophys Acta* **1720**, 1–13.
10. Hoessli, D. and Rungger-Brandle, E. (1985) Association of specific cell-surface glycoproteins with a Triton X-100-resistant complex of plasma membrane proteins isolated from T-lymphoma cells. *Exp Cell Res* **156**, 239–250.
11. Hoessli, D. and Rungger-Brandle, E. (1983) Isolation of plasma membrane domains from murine T lymphocytes. *Proc Natl Acad Sci USA* **80**, 439–443.
12. Brown, D.A. (1992) Interactions between GPI-anchored proteins and membrane lipids. *Trends Cell Biol* **2**, 338–343.
13. Simons, K. and van Meer, G. (1988) Lipid sorting in epithelial cells. *Biochemistry* **27**, 6197–6202.
14. Janes, P.W., Ley S.C., and Magee, M.I. (1999) Aggregation of lipid rafts accompanies signaling via the T cell antigen receptor. *J Cell Biol* **147**, 447–461.
15. Prior, I.A., Harding, A., Yan, J., Sluimer, J., Parton, R.G., and Hancock, J.F. (2001) GTP-dependent segregation of H-ras from lipid rafts is required for biological activity. *Nat Cell Biol* **3**, 368–375.
16. Moffett, S. Brown, D.A., and Linder, M.E. (2000) Lipid-dependent targeting of G proteins into rafts. *J Biol Chem* **275**, 2191–2198.
17. Xavier, R., Brennan, T., Li, Q., McCormack, C., and Seed, B. (1998) Membrane compartmentation is required for efficient T cell activation. *Immunity* **8**, 723–732.
18. Montixi, C., Langlet, C., Bernard, A.M., Thimonier, J., Dubois, C., Wurbel, M.A., Chauvin, J.P., Pierres, M., and He, H.T. (1998) Engagement of T cell receptor triggers its recruitment to low-density detergent-insoluble membrane domains. *EMBO J* **17**, 5334–5348.
19. Saarma, M. (2002) GDNF recruits the signaling crew into lipid rafts. *Trends Neurosci* **24**, 427–429.
20. Scheel-Toellner, D., Wang, K., Singh, R., Majeed, S., Raza, K., Curnow, S.J., Salmon, M., and Lord, J.M. (2002) The death-inducing signaling complex is recruited to lipid rafts in Fas-induced apoptosis. *Biochem Biophy Res Commun* **297**, 876–879.
21. Morgan, M.J., Kim, Y.S., and Liu, Z. (2007) Lipid rafts and oxidative stress-induced cell death. *Antioxid Redox Signal* **9**, 1471–1483.
22. Munro, S. (2003) Lipid rafts: Elusive of illusive? *Cell* **115**, 377–388.
23. Simons, K. and Toomre, D. (2000) Lipid rafts and signal transduction. *Nat Rev Mol Cell Biol* **1**, 31–39.
24. Brown, D.A. and London, E. (1998) Functions of lipid rafts in biological membranes. *Annu Rev Cell Dev Biol* **14**, 111–116.
25. Vyas, K.A., Patel, H.V., Vyas, A.A., and Schnaar, R.L. (2001) Segregation of gangliosides GM1 and GD3 on cell membranes, isolated membrane rafts, and defined supported lipid monolayers. *Biol Chem* **382**, 241–250.
26. Schade, A. and Levine, A. (2002) Lipid raft heterogeneity in human peripheral blood T lymphoblasts: A mechanism for regulating the initiation of TCR signal transduction. *J Immunol* **168**, 2233–2239.
27. Pike, L.J. (2004) Lipid rafts: Heterogeneity on the high seas. *Biochem J* **378**, 281–292.
28. Kiyokawa, E., Baba, T., Otsuka, N., Makino, A., Ohno, S., and Kobayashi, T. (2005) Spatial and functional heterogeneity of sphingolipid-rich membrane domains. *J Biol Chem* **280**, 24072–24084.
29. George, S., Nelson, M., Dollahan, N., and Bamezai, A. (2006) A novel approach to examining compositional heterogeneity of detergent-resistant lipid rafts. *Immunol Cell Biol* **84**, 192–202.
30. Golub, T., Wacha, S., and Caroni, P. (2004) Spatial and temporal control of signaling through lipid rafts. *Curr Opin Neurobiol* **14**, 542–550.
31. Yeh, E.T., Reiser, H., Bamezai, A., and Rock, K.L. (1988) TAP transcription and phosphatidylinositol linkage mutants are defective in activation through the T cell receptor. *Cell* **52**, 665–674.

Chapter 13

Determination of Acrolein by High-Voltage Capillary Electrophoresis from Oxidized Fatty Acids

Rafael Medina-Navarro

Abstract

Acrolein is an α,β-unsaturated aldehyde with enormous capacity of reaction, occurs in the air like a pollutant, but it is (we know now) an important lipid peroxidation product as well. The compound is one of the several aldehydes produced from fatty acid oxidation, although it is particularly important because it constitutes the major electrophyle aldehyde derived from lipid oxidation. Acrolein can be formed actively from oxidized fatty acids and undergo aldolic condensation in alkaline pH; this is a particular characteristic that we have used in its process of separation with capillary electrophoresis. We have shown that the oxidation of unsaturated fatty acids forms acrolein, and that the use of capillary electrophoresis to be a powerful, sensitive, and attractive method for separation, identification, and quantization of this and other aldehydes from in vitro lipid peroxidation.

Key words: Aldehydes, Acrolein, High-voltage capillary electrophoresis, Unsaturated fatty acids, Lipid peroxidation, Malondialdehyde, Thiobarbituric acid.

1. Introduction

Acrolein is an atmospheric pollutant, which is a product of lipid peroxidation and a compound that can be formed endogenously from the oxidized biological constituents *(1)*. Acrolein is a lipoperoxidation product; in addition, like other similar compounds it is prone to generate free radicals by itself, even though the mechanisms by which this happens are still unknown. The exposure of acrolein leads to time- and dose-dependent ROS generation and lipid peroxidation in some tissues *(2, 3)*. In the same way, antioxidants can reduce membrane damage and cell death produced by the acrolein's effects. Acrolein can probably interact with mitochondria

From: *Methods in Molecular Biology, vol. 477: Advanced Protocols in Oxidative Stress I*,
Edited by: D. Armstrong, DOI: 10.1007/978-1-60327-517-0_13, © Humana Press, New York, NY

leading to overproduction of reactive oxygen species. Acrolein induces oxidative stress in brain mitochondria *(3)*.

The contribution of α,β-unsaturated aldehydes such as acrolein to various disease processes has prompted the development of methodologies designed for their separation and identification. Aldehydes like acrolein and 4-hydroxy-nonenal can each react with amino acid residues on proteins and produce stable derivatives. One of these aminoacids is lysine, which is converted into the chemically stable end product, FDP-Lysine (N^{α}-acetyl N^{ϵ}-(3-formyl-3, 4-dehydro piperidino) lysine) *(1)*. This end product represents an epitope of the antibodies against acrolein-modified proteins, and has been used as a marker for the formation of acrolein adducts in vivo *(4, 5, 6)*. Immunochemical analysis then represents the procedure of choice for evaluating acrolein-protein interaction. Some methods, such as HPLC using fluorescence detection, have been used in the determination of acrolein when it is present in high levels in urine, by liquid chromatography of its quinoline derivative, essentially to monitor acrolein pharmacokinetics in patients after cyclophosphamide treatment *(7)*. Other similar methodologies have been implemented for quantitative determination of acrolein in plasma (cyclophosphamide pharmacokinetics) *(8)*, and acrolein and other low-aliphatic aldehydes in human saliva such as high-performance liquid chromatography (HPLC) and capillary electrophoresis *(9)*. Free acrolein from environmental polluted air has been quantitated using HPLC with UV detection and derivatized with hydrazine and its derivatives *(10)*. The standard method for the determination of gaseous carbonyls is to collect these onto 2, 4-dinitrophenyl hydrazine (DNPH) coated solid sorbent followed by solvent extraction and analysis of the derivatives using HPLC. Robust methods that involve gas chromatography (GC) have been developed essentially to determine atmospheric acrolein, using carbonyl group capture by both bisulfite and pentafluorophenyl hydrazine derivatives *(11, 12)*. These technologies have not been used with quantitative purposes for free lipid peroxidation derivative products in vivo, principally due to the low concentration that we expect to obtain from biological tissues and due to the enormous reactivity of aldehydes with macromolecules like proteins, nucleic acids and all kinds of lipids.

We found that it is possible to "capture" acrolein and malondialdehyde (MDA) produced from unsaturated fatty acids peroxidation at alkaline pH, taking advantage of the property of α,β-unsaturated aldehydes to condense (aldolic condensation). Later, we designed a procedure using capillary electrophoresis (CE) in their zonal mode (CZE) and examined aldehydes using the high-resolution separation capability of this technology. The fractions collected at the end of the procedure allows one to characterize the components obtained. In addition, in the present chapter, we will show the ozone procedure for the simple, clean, and efficient aldehyde generation from fatty acid oxidation.

2. Materials

2.1. Instruments and Devices

1. Capillary electrophoresis system P/ACE TM series 5000 and System Gold software©1991 from Beckman Instruments, Inc., Fullerton, CA, USA.
2. Diode Array detector for spectral analysis.
3. P/ACE cartridge for capillary cooling, Beckman Instruments, Inc., Fullerton, CA, USA.
4. Free capillary columns (polymide-coated fused silica), 57-cm long (50 cm to the detector window), 75 μm of internal diameter, Beckman Instruments, Inc., Fullerton, CA, USA (*see* **Note 1**).
5. Microfiltration tubes (Microcon, Millipore, Bedford, MA, USA) with a 10-kDa molecular mass cut-off.
6. Scanning spectrophotometer.
7. Vortex mixer.

2.2. Reagents and Solvents

1. 1,1,3,3-tetramethoxypropane malondialdehyde bis [dimethyl acetal] (Sigma Chemical Co., St. Louis, MO 63103 USA).
2. Acrolein 90% (GC) (Sigma Chemical Co., St. Louis, MO 63103 USA).
3. Thiobarbituric acid (TBA) (Sigma Chemical Co., St. Louis, MO 63103 USA).
4. 9,12-octadecadienoic acid (linoleic acid) (Sigma Chemical Co., St. Louis, MO 63103 USA).
5. 5,8,11,14-eicosatetraenoic acid (arachidonic acid) (Sigma Chemical Co., St. Louis, MO 63103 USA).
6. Tris (2-amino-2-hydroxymethylpropane-1,3-diol) (Sigma Chemical Co., St. Louis, MO 63103 USA).
7. Ferric chloride (anhydrous) (Sigma Chemical Co., St. Louis, MO 63103 USA).
8. Sodium tetraborate (Sigma Chemical Co., St. Louis, MO 63103 USA).
9. Capillary regenerator solution (Beckman Instruments Inc., Fullerton, CA, USA).
10. HPLC grade water.

2.3. Buffers and Solutions

1. Running buffer: 0.1 M borate, pH 8.2. Dissolve 20.12 g of sodium tetraborate with water and complete to 1l (*see* **Notes 2** and **3**).
2. 5p8 (phosphate-buffered saline): 150 mM NaCl, 5 mM sodium phosphate, pH 8.0.
3. Thiobarbituric acid solution (TBA): 15% w/v trichloroacetic acid, 0.375% w/v thiobarbituric acid, and 0.25 N hydrochloric acid. The solution can be mildly heated to assist in the dissolution of the TBA.
4. Sodium hydroxide 1% w/v.
5. Tris/HCl, 100 mM, pH 8.0.
6. Nitrogen of 99.9999% purity, BIP® (Built-in-Purifier).

3. Methods

3.1. Standards Preparation

A malondialdehyde (MDA) stock solution must be prepared from the acid hydrolysis of malondialdehyde bis [dimethyl acetal] previously distilled. Approximately 164.2 mg of the compound is removed and combined with 100 ml of 1% sulphuric acid (v/v) (Stock). After resting at room temperature for 2 h, 1 ml of the solution is taken and adjusted to a 100-ml volume with more of 1% sulphuric acid (v/v). In the spectrophotometric assessment, the stock solution should be of an approximate concentration of 10 mM, and 0.1 mM of the final solution, if we use an extinction coefficient of 13700 for non-adducted, and of 153000 for the MDA-TBA adduct respectively, with a variation of no more than 2% *(13)*. Approximately 100 μl of the final solution is combined with the same volume of 1% NaOH. The solution is allowed to rest in the dark for 30 min. The volume is adjusted to 1 ml using borate buffer, 0.01 M, pH 8.2. The final concentration of the solution is 10 μM. The MDA condensation product formation can be verified at 266–267 nm. For acrolein, a Stock solution was prepared taking 20 μL of the bottle (90%, d = 0.849), poured on 20 mL of HPLC water and keeping chilled on ice (*see* **Note 4**). From this solution, 100 μl is taken to which an equal volume of 1% NaOH is added. After adjusting to 1 ml with borate buffer 0.01 M, pH 8.2, the final acrolein concentration should be 1.4 mM. The aldolic condensation is monitored at 266 nm. Before the standards are injected in the CE system, the identity of the free aldehydes is verified with a spectral analysis of the thiobarbituric acid reaction (**Fig. 13.1**). Then, the runs are conducted for a mix of standards previously prepared (**Fig. 13.3**). The absorption of the mixture and of the erythrocyte membranes in the inserted graph is proportional to the degree of oxidation.

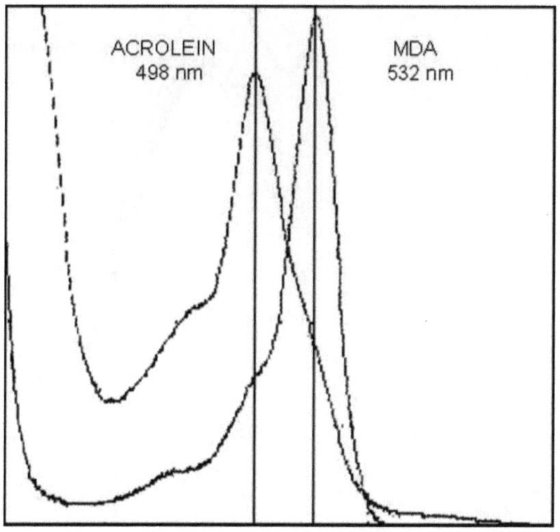

Fig. 13.1. Details of the spectrum runs generated from the products of the reaction of acrolein and malondialdehyde (MDA) standards with the thiobarbituric acid (TBA). Acrolein and MDA are both reactive products of the TBA (TBARS), but with different maximum absorption peaks. (Reproduced with permission of SAGE publications from reference 16).

3.2. Preparation and Analysis of Fatty Acid Vesicles

Mix 40 mg of linoleic acid and 0.40 mg of arachidonic acid in 4 ml of Tris/HCl, 100 mM, pH 8.0. A suspension is formed by intense agitation in a vortex mixer for a period of 30 min. Later on, the mixture is taken to a final concentration of 7.2 mM using the same Tris/HCl buffer *(14)*. Vesicles must be always freshly prepared. Immediately before the assays, 200 μl of the fatty acid suspension is taken and the volume adjusted to 1000 μl with run buffer 0.1 M borate, pH 8.2, for a final concentration of 1.44 μM. Ensure aldolic condensation with the addition of 5 μl 1% NaOH solution just before oxidation. Then, the polar phase containing the aldehydes is obtained after the lipid extraction, with the addition of 2 ml chloroform–methanol (2:1), intense agitation (vortex mixer) for 10 min, and final centrifugation at 2500 × *g* for 15 min at 4°C. Under these conditions, it is possible to verify the spectral analysis of the samples (**Fig. 13.2**) Samples prepared for capillary electrophoresis can be injected directly and the previous chloroform/methanol residue evaporated under a stream of nitrogen (*see* **Section 2.3**) for 1 min at 27°C (**Fig. 13.3**).

3.3. Open Erythrocyte Membrane Preparation and Analysis

Open erythrocyte membranes used for experiments described here are prepared by hypotonic lysis obtained from adult volunteers' complete blood samples, collected in sterile test tubes containing heparin as an anticoagulant, as originally described by Steck *(15)*. The entire process is verified at 4°C, and on the same day the

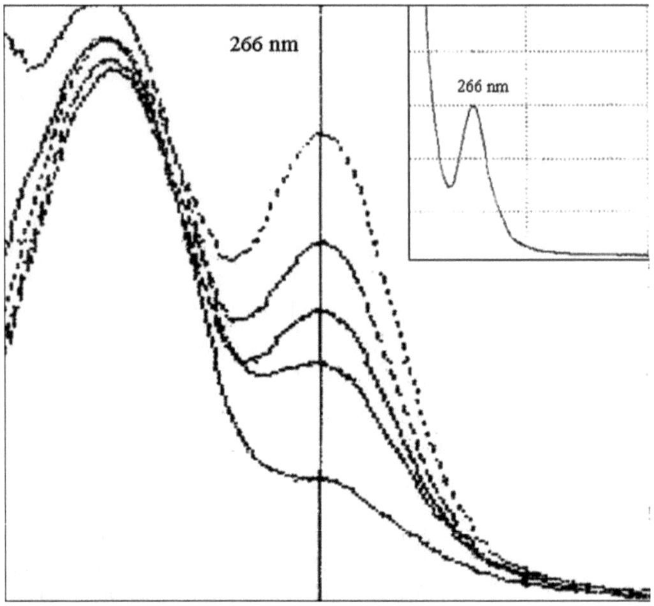

Fig. 13.2. Scanning of the polar fractions from fatty acids (arachidonic and linoleic) exposed to increasing amounts of ozone (1.37, 2.75, 4.13, and 5.51 μg of ozone/nmol of fatty acids). In the inserted graph, the polar phase of a sample of open erythrocyte membranes is oxidized with 3.8 μg of ozone/μg of protein. The exposure takes place as described in **Section 3.3(1)** and the polar extraction is described in **Sections 3.2** and **3.3** for erythrocyte membranes. (Reproduced partially with permission of SAGE publications from reference 16).

experiment is to be performed. Heparin, EDTA, or citrate may be used for anticoagulation. Red blood cells are washed three times at 2300 g for 10 min, re-suspending the pellet in five volumes of PBS. Hemolysis is initiated by rapidly and thoroughly mixing 1 ml of packed cells with 40 ml of 5P8 [*see* **Section 2.3**(**2**)]. The ghosts are pelleted by centrifugation at 22,000 × *g* for 10 min in an angle-head rotor. The supernatant is removed and the tube is tipped and rotated on its axis so that the loosely packed ghost slides away from a small hard button, rich in proteases, which may then be aspirated. Repeat the wash two more times. At the end of the procedure, the membranes' average protein concentration is around 3.5 μg/μl. The oxidation takes place using a volume equivalent to 70 μg of membrane protein on a volume corresponding to 1000 μl of run buffer. Ensure condensation with the addition of 5 μl 1% NaOH solution just before oxidation. Similar to the case of fatty acid vesicles, after the erythrocyte membrane suspension is oxidized, the polar phase containing the aldehydes is obtained with 2 ml of chloroform-methanol (2:1) extraction, intense agitation for 10 min, and final centrifugation at 2500 × *g* for 15 min at 4°C. Then, the same procedure and instructions described for fatty acid vesicles are fallowed (**Fig. 13.2**).

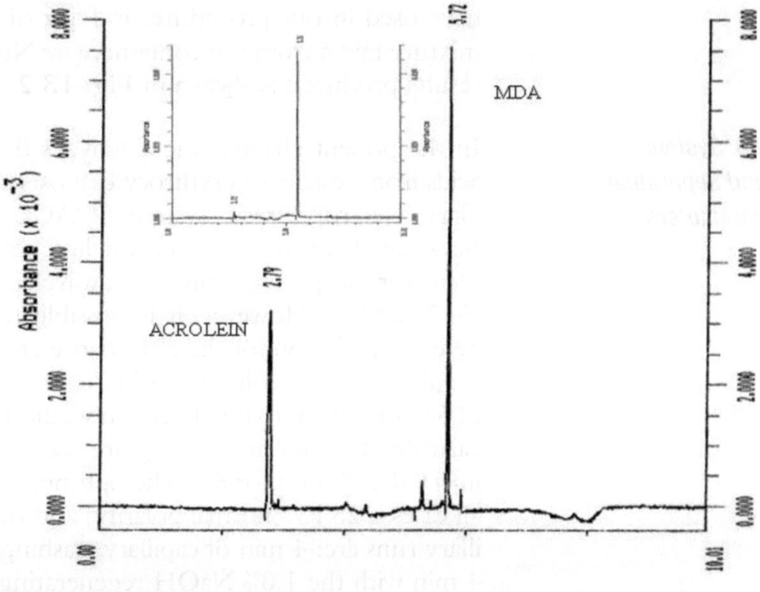

Fig. 13.3. Electropherograms of the acrolein and malondialdehyde standards. Run conditions: open capillary of fused-silica, 0.75 μm of internal diameter × 50 cm of distance from the detector; potential applied of 20 kV/40 μA; borate buffer 0.1 M, pH 8.2 at stable temperature of 27°C; maximum absorbance at 266 nm and a bandwidth of 10. In the inserted electropherogram, identical conditions for the oxidation product of fatty acids (arachidonic/linoleic); the ozone concentration in this case is 4.2 μg/nmol of fatty acids. (Reproduced with permission of SAGE publications from reference 16).

3.4. Fatty Acid Oxidation Modes

The ways by which fatty acids can be oxidized are several. In fact, fatty acids can undergo auto-oxidation upon exposure to air. However, when the objective is to obtain a highly efficient oxidation, we must use an oxygen or nitrogen reactive species source, generating over controlled conditions. At the end of chapter in **Section 4** (*see* **Note 8**), we will show another possible alternative.

3.5. Fatty Acid Oxidation with Ozone

There are trade mark devices for ozone production for diverse purposes and some of them with automatic ozone concentration measurement. The simplest way is the ozone production from an electric arc, from a water purification apparatus connected to an ultra pure oxygen source. The ozone mixture must be circulated constantly and poured into a container. The oxygen-ozone mixture can be extracted with a needle and syringe, through a rubber stopper on the container. The needle is sealed and the oxidation is carried inside the syringe, agitated for 15 min using a vortex mixer. If ozone concentration needs to be determined, standard curves should be obtained carrying out an experiment measuring the iodide concentration using an extinction coefficient of 25,200/350 nm, as a product of the stochiometric reaction of ozone with potassium iodide. The ozone concentration that we have used in our procedures is 1 μg of ozone per ml of the gas mixture taken from the container (*see* **Note 5**). An example of the results produced is shown in **Fig. 13.2**.

3.6. System and Separation Parameters

In the present chapter, the aldehydes derived from oxidized fatty acids from vesicles or erythrocyte ghosts are analysed using the capillary electrophoresis system P/ACE TM series 5000 from Beckman Instruments, Inc., on line with a Diode Arrangement Detector for spectral analysis at a wavelength range of 190–600 nm (512 diodes). However, it is possible to use a conventional UV detector and monitor the absorbance around 266 nm. In all cases, the free capillary columns of id 75 μm, od 375 μm, and a length of 57 cm (50 cm to the detection window) built within the P/ACE cartridge for capillary cooling are used. The samples are pressure-injected at 0.5 psi for 5 s. The optimum separation voltage used in all cases is 25 kV (normal polarity) and the treatment between capillary runs are: 4 min of capillary washing with the running buffer; 4 min with the 1.0% NaOH regenerating solution; 6 min washing with water and 6 min of draining previous to the injection of the sample with the running buffer at higher pressure: 15 psi. In all cases, the running buffer should be 0.1 M borate, pH 8.2. For the electropherograms presented here, detection is set at a wavelength of 266 nm with a maximum bandwidth of 10. Examples of the results produced are shown in **Figs. 13.3** and **13.4**.

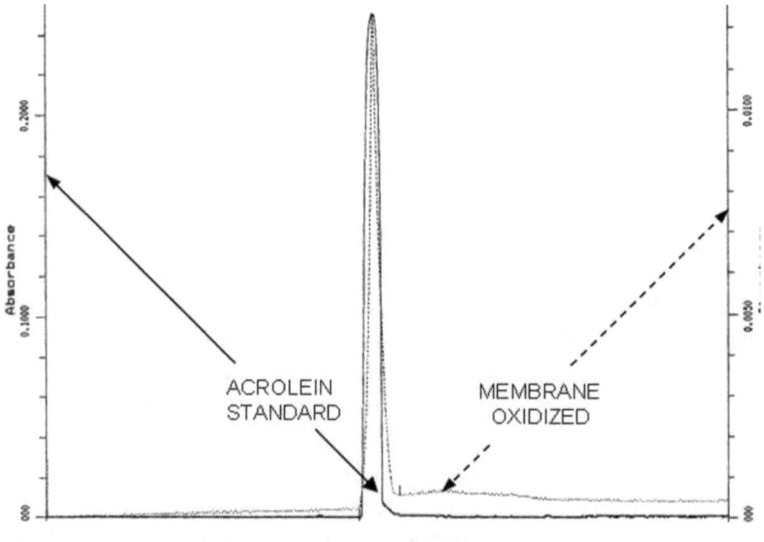

Fig. 13.4. Superposition of two electropherograms corresponding to an acrolein standard and the products of the oxidation of erythrocyte membranes with ozone; the former corresponding to the discontinuous line. In this selected electropherogram, both peaks had precisely a retention time of 2.9 min. The variation between two consecutive runs is usually less than 1.5%. Other conditions are as described in **Fig. 13.3** and in **Sections 3.3** and **3.5**. (Reproduced with permission of SAGE publications from reference 16).

3.7. Fractions Collected (see Notes 6 and 7)

The total estimated capillary volume at 27°C and 0.798×10^{-2} P is 2544 nl. Taking into account that the first separated compound appears at approximately three minutes after having started the run, the system must be programmed to stop the separation three minutes after the run. The system is programmed to inject 312.4 nl of the sample at low grade pressure (0.5 psi) for 1 min; the volume corresponding to 7 cm of capillary. This sequence is repeated 50 times and these correspond to the first fraction. The capillary must be drained between each run. Every five runs, the instrument will be programmed to carry out another four consecutive injections (in their corresponding vials) equivalent to an additional four fractions containing 312.4 nl each. The main rationale behind this is that the first fraction would contain the first compound given that it includes the volume equivalent to the distance to the window detector, that is to say 7 cm. Given that the concentration of the first compound is expected to be approximately 9.5 times lower, four injections of the same fraction are done for each. The volumes are adjusted at the end of the procedure. The fraction is collected in micro-vials containing 0.375% TBA solution. An example of the results produced is shown in **Fig. 13.5**.

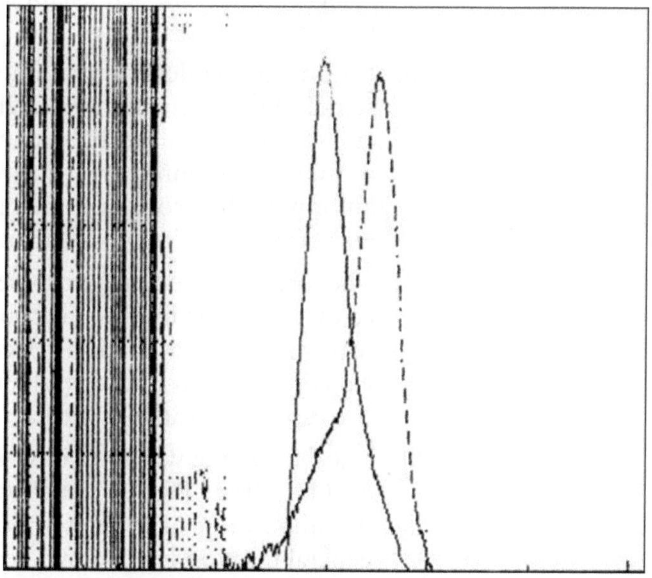

Fig. 13.5. Detail of the spectral scanners corresponds to the fractions collected. The sample corresponds to fatty acids (arachidonic/linoleic) after being oxidized with ozone. The maximums of the fractions correspond to acrolein (498 nm) and malondialdehyde (532 nm). In the oxidation procedure, ozone concentration in this case is 9 μg of ozone/nmol of free fatty acids. The separation procedure is described in **Sections 3.2** and **3.5**. (Reproduced with permission of SAGE publications from reference 16).

4. Notes

1. A new capillary must be conditioned before it can be used. Pretreat the capillary for 10 min with 0.1 M sodium hydroxide, 5 min with water, and 10 min with run buffer.
2. All buffers and solutions should be prepared in water presented a resistivity of 18.2 MΩ × cm;we refer to this type of water in the text in some instances
3. Buffers should be filtered through 0.45-μm filters prior to use they are stored in a refrigerator and discarded after 4 days.
4. Acrolein is an extremely toxic and volatile compound and this property must be taken into account. Open the container and flask in the extraction hood. Always use gloves and store the flask at 4°C with a visible warning label. It is convenient to manipulate in the cold and make a stock solution from the compound flask; it needs a special stopper from which it is possible to draw in the aldehyde with a Hamilton syringe.
5. Other alternatives to iodide measurement should be the addition of starch and titrate with a standardized thiosulfate solution. There are different procedures for an acceptable good determination of ozone from an air mixed sample, and some of them are scholar practices described in chemistry text books.
6. For fraction collection, we must know some basic capillary standard parameters: total length, capillary volume (nl), and an injection volume by second at defined pressure. The parameters required are usually available with the capillary kit or with the provider. In the present chapter, these parameters are shown in the procedure description.
7. The maximum spectrophotometric absorption of the adduct acrolein-TBA is used to corroborate its identity (**Fig. 13.5**) with the use of a fraction collector. However, we have previously demonstrated with the use of GC/MS that this peak corresponds positively with acrolein *(17)*.
8. Another simple way to oxidize fatty acids is through a source of OH. The Fenton reaction is an iron salt-dependent decomposition of hydrogen peroxide to produce OH. The general reaction is as follow: $Fe^{2+} + H_2O_2 \rightarrow Fe^{3+} + OH + OH^-$. Addition of a reducing agent such as ascorbate exacerbates the oxidative effect. There are several modes of Fenton reaction. For the oxidation of 1 mM of unsaturated fatty acids with good result, we have used the formula consisting of 10 μM ferrous sulphate, 100 μM H_2O_2, and 1 mM ascorbate in 0.1 M sodium phosphate buffer, pH 7.4. Samples that must stay in the dark for at least 6 h need to be

partitioned (chloroform/methanol) and the upper layer evaporated to dryness with nitrogen and re-suspended in run buffer. Prior to injections, samples should be dialysed or filtered using centrifugal filtration devices (*see* **Section 2.1**). Ensure aldolic condensation as described in **Section 3.2**.

Acknowledgments

The author would like to thank Sr Antonio Lopez de Silanes and Laboratorios Silanes for the financial support in the realization of some of the content described in the present chapter.

References

1. Uchida, K., Kanematsu, M., Sakai, K., Matsuda, T., Hattori, N., Mizuno, Y., et al. (1999) Protein-bound acrolein: potential markers for oxidative stress. *Proc Natl Acad Sci USA* **95**, 4882–4887.
2. Luo, J., and Shi, R. (2004) Acrolien induces axolemmal disruption, oxidative stress, and mitochondrial impairment in spinal cord tissue. *Neurochem Int* **44**, 475–486.
3. Luo, J., and Shi, R. (2004) Acrolien induces oxidative stress in brain mitochondria. *Neurochem Int* **46**, 443–452.
4. Noiri, E., Yamada, S., Nakao, A., Tsuchiya, M., Masaki, I., Fujing, K., et al. (2002) Serum protein acrolein adducts: utility in detecting oxidant stress in hemodialysis patients and reversal using a vitamin E-bonded hemodialyzer. *Free Radic Biol Med* **33**, 1651–1656.
5. Arikketh, D., Niranjali, S., and Devaraj, H. (2004) Detection of acrolein-lysine adducts in plasma low-density lipoprotein and in aorta of cyclophosphamide-administered rats. *Arch Toxicol* **78**, 397–401.
6. Calingasan, N.Y., Uchida, K., and Gibson, G.E. (1999). Protein-bound acrolein: a novel marker of oxidative stress in Alzheimer's disease. *J Neurochem* **72**, 751–756.
7. al-Rawithi, S., el-Yazigi, A., and Nicholls, P.J. (1993) Determination of acrolein in urine by liquid chromatography and fluorescence detection of its quinoline derivative. *Pharm Res* **10**, 1587–1590.
8. Paci, A., Rieutord, A., Guillaume, D., Traoré, F., Ropenga, J., Husson, H.P., et al. (2000) Quantitative high-performance liquid chromatographic determination of acrolein in plasma after derivatization with Luminarin 3. *J Chromatogr B Biomed Sci Appl* **739**, 239–246.
9. Annovazzi, L., Cattaneo, V., Viglio, S., Perani, E., Zanone, C., Rota, C., et al. (2004) High-performance liquid chromatography and capillary electrophoresis: methodological challenges for the determination of biologically relevant low-aliphatic aldehydes in human saliva. *Electrophoresis* **25**, 1255–1263.
10. Herrington, J., Zhang, L., Whitaker, D., Sheldon, L., and Zhang, J.J. (2005) Optimizing a dansylhydrazine (DNSH) based method for measuring airborne acrolein and other unsaturated carbonyls. *J Environ Monit* 7, 969–976.
11. Seaman, V.Y., Charles, M.J., and Cahill, T.M. (2006) A sensitive method for the quantification of acrolein and other volatile carbonyls in ambient air. *Anal Chem* **78**, 2405–2412.
12. Ho, S.S., and Yu, J.Z. (2004) Determination of airborne carbonyls: comparison of a thermal desorption/GC method with the standard DNPH/HPLC method. *Environ Sci Technol* **38**, 862–870.
13. Slater, T.F. (1984) Overview of methods used for detecting lipid peroxidation. *Meth Enzymol* **105**, 283–376.
14. Hicks, M., Delbridge, L., Yue, D.K., and Reeve, T.S. (1988) Catalysis of lipid peroxidation by glucose and glycosylated collagen. *Biochem Biophys Res Commun* **151**, 649–655.

15. Steck, T.L., and Kant, J.A. (1974) Preparation of impermeable ghosts and inside out vesicles from human erythrocyte membranes. *Meth Enzymol* **31**, 72–180.
16. Medina-Navarro, R., Mercado-Pichardo, E., Hernandez-Perez, O., and Hicks, J.J. (1999) Identification of acrolein from the ozone oxidation of unsaturated fatty acids. *Hum Exp Toxicol* **18**, 677–682.
17. Medina-Navarro, R., Duran-Reyes, G., Diaz-Flores, M., Hicks, J.J., and Kumate, J. (2004) Glucose-stimulated acrolein production from unsaturated fatty acids. *Hum Exp Toxicol* **23**, 101–105.

Part II
Antioxidant Technology and Application

Chapter 14

Cupric Ion Reducing Antioxidant Capacity Assay for Food Antioxidants: Vitamins, Polyphenolics, and Flavonoids in Food Extracts

Reşat Apak, Kubilay Güçlü, Mustafa Özyürek, Burcu Bektaşoğlu, and Mustafa Bener

Abstract

Antioxidants are health beneficial compounds through their combat with reactive oxygen and nitrogen species and free radicals that may cause tissue damage leading to various diseases. This work reports the development of a simple and widely applicable antioxidant capacity index for dietary polyphenols, vitamins C and E, and plasma antioxidants utilizing the copper(II)-neocuproine (Cu(II)-Nc) reagent as the chromogenic oxidizing agent. This novel method based on an electron-transfer mechanism was named by our research group as 'cupric reducing antioxidant capacity', abbreviated as the CUPRAC method. The method is comprised of mixing the antioxidant solution with aqueous copper(II) chloride, alcoholic neocuproine, and ammonium acetate aqueous buffer at pH 7, and subsequently measuring the developed absorbance at 450 nm after 30 min. Since the color development is fast for compounds like ascorbic acid, gallic acid, and quercetin but slow for naringin and naringenin, the latter compounds are assayed after incubation at 50°C on a water bath for 20 min. The flavonoid glycosides are hydrolyzed to their corresponding aglycones by refluxing in 1.2 M HCl-containing 50% MeOH so as to exert maximal reducing power towards Cu(II)-Nc. The CUPRAC antioxidant capacities of synthetic mixtures are equal to the sum of individual capacities of antioxidant constituents, indicating lack of chemical deviations from Beer's law. Tests on antioxidant polyphenols demonstrate that the highest CUPRAC capacities are observed for epicatechin gallate, epigallocatechin gallate, quercetin, fisetin, epigallocatechin, catechin, and caffeic acid in this order, in accord with the number and position of the –OH groups as well the conjugation level of the molecule. The parallelism of the linear calibration curves of pure antioxidants in water and in a given complex matrix (plant extract) demonstrates that there are no chemical interactions of interferent nature among the solution constituents, and that the antioxidant capacities of the tested antioxidants are additive, in conformity to the Beer's law. For individual determination of ascorbic acid in fruit juices with a modified CUPRAC procedure, flavonoids are pre-extracted as their La(III) complexes prior to assay. For apricot extracts, a modified version of the CUPRAC assay based on anion exchange separation at pH 3 is applied, since sulfited-dried sample extracts contain the hydrosulfite anion interfering with the determination. For herbal tea infusions, the standard CUPRAC protocol is applied. The CUPRAC reagent is stable, easily accessible, low-cost, and is sensitive

From: *Methods in Molecular Biology, vol. 477: Advanced Protocols in Oxidative Stress I,*
Edited by: D. Armstrong, DOI: 10.1007/978-1-60327-517-0_14, © Humana Press, New York, NY

toward thiol-type antioxidants unlike FRAP. The reaction is carried out at nearly physiological pH as opposed to the acidic pH of FRAP or to the alkaline pH of Folin methods, constituting a basic advantage for the realistic assay of biological fluids.

Key words: CUPRAC antioxidant capacity assay, Vitamins; Polyphenolics; Flavonoids, Spectrophotometry.

1. Introduction

Antioxidants are health beneficial compounds through their combat with reactive oxygen and nitrogen species and free radicals that may cause tissue damage leading to various diseases. It is important to measure the antioxidant potency of food material and human plasma for food quality estimation and for diagnosis and treatment of diseases, respectively. Current literature clearly states that there is no single 'total antioxidant capacity' index available for food labeling because of the lack of standard quantitation methods. This work reports the development of a simple and widely applicable antioxidant capacity index for dietary polyphenols and vitamins C and E. This novel method based on an electron-transfer mechanism was named by our research group as 'cupric reducing antioxidant capacity', abbreviated as the CUPRAC method. This method is advantageous over FRAP (ferric reducing antioxidant power) since the redox chemistry of copper(II) involves more favourable kinetics and selectivity (*see* **Note 1**). The chromogenic redox reagent used for the CUPRAC assay is bis(neocuproine)copper(II) chelate (**Fig. 14.1**). This reagent is useful at pH 7, and the absorbance of the Cu(I)-chelate formed as a result of redox reaction with reducing polyphenols is measured at 450 nm. The color is due to the Cu(I)-Nc chelate formed (*see* **Fig. 14.2**, for Cu(I)-Nc spectra obtained with reacting varying concentrations of ascorbic acid with the CUPRAC reagent). The reaction conditions such as reagent concentration, pH, and oxidation time at room and elevated temperatures are optimized as shown in **Section 3**, derived from other sources *(1, 2, 3, 4, 5)*.

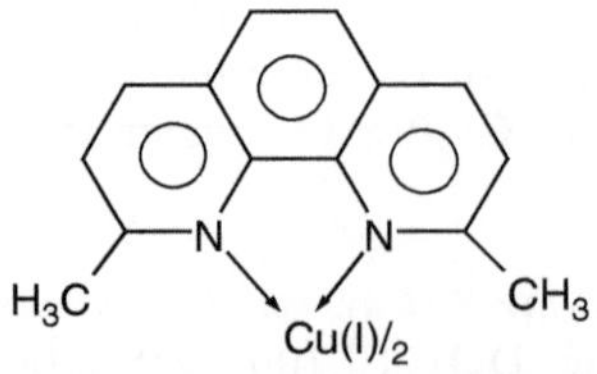

Fig. 14.1. The CUPRAC chromophore: Bis(neocuproine)copper(I) chelate cation.

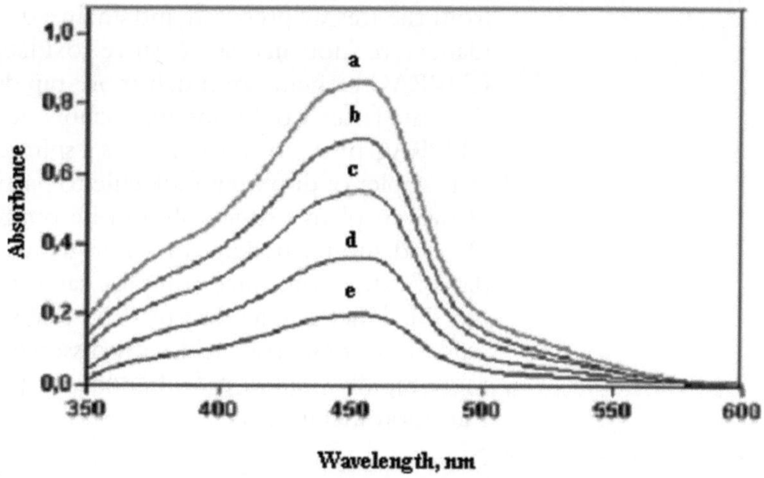

Fig. 14.2. Visible spectra of Cu(I)-Nc chelate produced as a result of CUPRAC reaction with varying concentrations of ascorbic acid: **(A)** 6.10×10^{-5} M, **(B)** 4.88×10^{-5} M, **(C)** 3.66×10^{-5} M, **(D)** 2.44×10^{-5} M, **(E)** 1.22×10^{-5} M.

The chromogenic oxidizing reagent of the developed CUPRAC method, i.e., bis(neocuproine)copper(II) chloride (Cu(II)-Nc), reacts with n-electron reductant antioxidants (AO) in the following manner:

$$\text{n Cu(Nc)}_2^{2+} + \text{n-electron reductant (AO)} \leftrightarrow \text{n Cu(Nc)}_2^{+} + \text{n-electron oxidized product} + \text{n H}^{+}. \quad (14.1)$$

In this reaction, the reactive Ar-OH groups of polyphenolic antioxidants are oxidized to the corresponding quinones (Ar = O) and Cu(II)-Nc is reduced to the highly colored $Cu(Nc)_2{}^{+}$ chelate showing maximum absorption at 450 nm. Although the concentration of Cu^{2+} ions is in stoichiometric excess of that of neocuproine in the CUPRAC reagent for driving the redox equilibrium reaction represented by (Eq. 14.1) to the right, the actual oxidant is the $Cu(Nc)_2{}^{2+}$ species and not the sole Cu^{2+}, because the standard redox potential of the Cu(II/I)-neocuproine is 0.6 V, much higher that of the Cu^{2+}/Cu^{+} couple (0.17 V) *(4)* (*see* **Note 2**). As a result, polyphenols are oxidized much more rapidly and efficiently with Cu(II)-Nc than with Cu^{2+}, and the amount of colored product (i.e., Cu(I)-Nc chelate) emerging at the end of the redox reaction is equivalent to that of reacted Cu(II)-Nc. The liberated protons are buffered in ammonium acetate medium. In the normal CUPRAC method ($CUPRAC_N$), the oxidation reactions are essentially complete within 30 min. Flavonoid glycosides require acid hydrolysis to their corresponding aglycons for fully exhibiting their antioxidant potency. Slow reacting antioxidants need elevated temperature incubation so as to complete their oxidation with the CUPRAC reagent *(1, 2)*. Special precautions to exclude oxygen from the freshly prepared and analyzed solutions of pure antioxidants are not necessary since oxidation reactions with the CUPRAC reagent are much more rapid than with dissolved O_2 (i.e., the latter would not appreciably occur during the period of CUPRAC protocol since there is a spin restriction for the ground state triplet of dioxygen molecule to participate in fast reactions). However, plant extracts should be purged with N_2 to drive off O_2, and should be kept in a refrigerator if not analyzed on the day of extraction, since complex catalyzed reactions with unpredictable kinetics may take place in real systems. Additionally, the oxidation of ascorbic acid with dissolved oxygen may take place more rapidly than of polyphenolics, especially in the presence of transition metal salts.

2. Materials

2.1. Equipment

1. The absorbances are measured with a Varian (Australia) Cary 1-E UV-Vis. spectrophotometer using a pair of matched quartz cuvettes of 1 cm thickness.
2. The pH measurements are made with the aid of an E512 Metrohm-Herisau (Switzerland) pH-meter equipped with a glass electrode.
3. Ultra-Torrax CAT X-620 model (M. Zipperer GmbH, Germany) homogenizer apparatus is used for extraction.
4. An Adams Dynac centrifuge apparatus (NJ, USA) is used for separation of the clear fractions of pharmaceutical tablet solutions and fruit juices.
5. The chromatograph is from Perkin Elmer (Shelton, CT, USA), and consists of a pump (Perkin Elmer Series 200 HPLC pump), an injection valve (Model 7725i; Rheodyne, Cotati, CA, USA), a Hamilton 25-μL syringe (Reno, NV, USA), an analytical stainless-steel column packed with Agilen brand Zorbax Eclipse XDB-C18 sorbent (250 mm × 4.6 mm, 5 μm) from Hichrom (Berkshire, UK), and a variable-wavelength UV/Vis detector (Perkin Elmer Series 200).

2.2. Reagents and Buffers

1. Fisetin (FS), quercetin (QR), rutin (RT), naringin (N), naringenin (NG), (-)epicatechin (EC), (-) epigallocatechin (EGC), (-)epicatechin gallate (ECG), (-)epigallocatechin gallate (EGCG), morin (MR), vanilic acid (VA), syringic acid (SA), chlorogenic acid (CGA), hesperetin, and meta-phosphoric acid are purchased from Sigma Chemical Co., Steinheim, Germany.

2. (-)Catechin (CT), gallic acid (GA), kaempherol (KP), sinapic acid (SNA), and hesperidin (HD) are from Fluka Chemicals., Steinheim, Germany.
3. Ferulic acid (FRA), *p*-coumaric acid (CMA), caffeic acid (CFA), L-ascorbic acid (AA), and trolox (6-hydroxy-2,5,7, 8-tetramethylchroman-2-carboxylic acid, abbreviated as TR) are supplied from Aldrich Chemicals Co., Steinheim, Germany.
4. α-Tocopherol (TP), methanol (MeOH), sodium hydroxide, lanthanum(III) chloride heptahydrate, $NaH_2PO_4.2H_2O$, H_3PO_4 (concentrated), sodium acetate, and 96% ethanol (EtOH) are from E. Merck, Darmstadt, Germany.
5. Ethylacetate, hydrochloric acid (concentrated), acetic acid, and absolute ethanol are from Riedel-de Haen, Steinheim, Germany.
6. CUPRAC assay reagents: Neocuproine (2,9-dimethyl-1,10-phenanthroline) is from Sigma Chemical Co., Steinheim, Germany; ammonium acetate (NH_4Ac) and copper(II) chloride dihydrate are from E. Merck, Darmstadt, Germany.
7. ABTS assay reagents: ABTS (2,2′-azinobis(3-ethylbenzothiazoline-6-sulfonic acid) diammonium salt) radical reagent is purchased from Fluka, Steinheim, Germany (kept at +4°C); potassium persulfate, from E. Merck, Darmstadt, Germany.
8. Folin assay reagents: Folin-Ciocalteu's phenol reagent (stored at +4°C) is supplied from Sigma Chemical Co., Steinheim, Germany; copper(II) sulfate, sodium carbonate, and sodium potassium tartarate are from E. Merck, Darmstadt, Germany.

2.3. Supplies

1. Grape juice is supplied from Kavaklidere Şaraplari AS, Ankara, Turkey.
2. The fruit juice commercial samples are purchased from the Turkish market, Istanbul; grape juice from Taskobirlik, Nevsehir, Turkey, and 100% orange juice (gold label) from Cappy, The Coca Cola Co. of Turkey. The pharmaceuticals and vitamin supplements are purchased from pharmacies in Istanbul: (i) GNC vitamin C 1000 dietary supplement with bioflavonoids and rose hips (containing vitamin C (as ascorbic acid) 1000 mg, and 100 mg citrus bioflavonoids complex); (ii) Redoxon Vitamin C tablets (containing 500 mg ascorbic acid); (iii) Boots chewable vitamin C with bioflavonoids (containing 500 mg vitamin C (as ascorbic acid), and 15 mg bioflavonoids).
3. Apricots: Fresh apricots as five different species (Hacihaliloglu, Cologlu, Kabaasi, Soganci, and Zerdali) are collected from the gardens of Malatya Bahce Bitkileri Arastirma Enstitusu (Malatya Garden Plants Research Institute, Malatya, Turkey).

4. Herbal teas: Scarlet pimpernel (*Anagallis arvensis*), everlasting (immortelle) (*Helichrysum italicum*), buckthorn (*Rhamnus cathartica*), fumitory (*Fumaria officinalis*), plantain (*Plantago major*), common mallow (*Malva sylvestris*), French lavender (*Lavandula stoechas*), shrubby germander (*Teucrium fruticans*), yarrow (*Achillea millefolium*), puncture vine (*Tribulus terrestris*), coriander (*Coriandrum sativum*), sweet basil (*Ocimum basilicum*), lemon balm (*Melissa officinalis*), coltsfoot (*Tussilago farfara*), thyme (*Thymus vulgaris*), marshmallow (*Althea officinalis*), and green tea (*Camellia sinensis*) are donated by Malatya Pazari AS (Istanbul, Turkey) as dried herbal tea material. Turkish Rize black tea is purchased from Caykur AS, Rize, Turkey and Turkish blended black tea from Balkupu AS, Kayseri, Turkey.
5. Tea bags: Lipton English breakfast tea is purchased from Unilever San. Tic. Turk AS, Ceylon blend tea from Ece Gida Sanayi ve Ticaret AS, Izmir, Turkey, sage (*Salvia officinalis*) from Botany AS, Izmir, Turkey, green tea with lemon from Doga Bitkisel Urunler Sanayi ve Ticaret AS, Istanbul, Turkey, green tea from Dogadan AS, Ankara, Turkey, blackberry (*Rubus fruticosus*) with various additives and aroma from TEMA Vakfi, Istanbul, Turkey, Caykur black tea bag from Altincay, stinging nettle (*Urtica dioica*) and linden flower (*Tilia europaea)* from TEMA Vakfi, Istanbul, Turkey.

3. Methods

3.1. Original CUPRAC Method Applied to Pure Antioxidant Solutions and Their Mixtures

1. Preparation of CUPRAC assay solutions: $CuCl_2$ solution, 1.0×10^{-2} M Cu(II), is prepared by dissolving 0.4262 g $CuCl_2.2H_2O$ in water, and diluting to 250 mL. Ammonium acetate (NH_4Ac) buffer at pH = 7.0, 1.0 M, is prepared by dissolving 19.27 g NH_4Ac in water and diluting to 250 mL. Neocuproine (Nc) solution, 7.5×10^{-3} M, is prepared daily by dissolving 0.039 g Nc in 96% ethanol, and diluting to 25 mL with ethanol. Trolox, 1.0×10^{-3} M, is prepared in 96% ethanol. The CUPRAC reagent is low-cost, stable, and easily accessible (*see* **Note 3**). The reagent is not affected by air or sunlight (*see* **Note 4**).
2. Preparation of polyphenolics and vitamins solutions: All polyphenolic compounds and vitamins solutions are freshly prepared in 96% EtOH at 1 mM (1.0×10^{-3} M) concentration prior to measurement.
3. Normal sample measurement: To a test tube are added 1 mL each of Cu(II), Nc, and NH_4Ac buffer solutions.

Antioxidant sample (or standard) solution (*x* mL) and H_2O (1.1 − *x*) mL are added to the initial mixture so as to make the final volume = 4.1 mL. The tubes are stoppered, and after 1 h, the absorbance at 450 nm (A_{450}) is recorded against a reagent blank. Since the analytical wavelength falls in the visible range, standard colorimeters may also be used for measurement (*see* **Note 5**). The standard calibration curves of each antioxidant compound is constructed in this manner as absorbance *vs* molar concentration., and the molar absorptivitiy of the CUPRAC method for each antioxidant is found from the slope of the calibration line concerned *(1)*. The scheme for normal measurement is summarized as:

1 mL Cu(II) + 1 mL Nc + 1 mL buffer + *x* mL antioxidant solution + (1.1 − *x*) mL H_2O; total vol. = 4.1 mL, measure A_{450} against a reagent blank after 30 min of reagent addition.

4. Incubated sample measurement: The mixture solutions containing sample and reagents are prepared as described in 'normal measurement'; the tubes are stoppered and incubated for 20 min in a water bath at a temperature of 50°C. The tubes are cooled to room temperature under running water, and their A_{450} values are measured.

5. Hydrolyzed sample measurement: A suitable mass of the polyphenol or vitamin standard is weighed such that the final antioxidant concentration of the methanolic solution would be 1.0 mM. Each standard is dissolved in a suitable volume of 50% MeOH. In a 100 mL flask, 5 mL of 1.2 M aqueous HCl is added to each solution and diluted to the mark with 50% MeOH. This solution is decanted to a distillation flask into which a few pieces of boiling stone are added, and refluxed at 80°C for 2 h. The flask is cooled to room temperature under running tap water. The hydrolyzate is neutralized with approx. 6.5 mL of 1 M NaOH. The neutralized hydrolyzate is then subjected to 'normal measurement'.

6. Hydrolyzed and incubated sample measurement: The neutralized hydrolyzate is subjected to incubation at 50°C in a water bath for 20 min. The A_{450} of running water-cooled samples are 'normally measured'.

7. Measurement of ternary antioxidant mixture solutions: Individual 1.0 mM solutions of the antioxidant compounds are prepared in 96% EtOH. Ternary mixtures of the antioxidants are prepared in suitable volume ratios such that the final absorbance of the mixture does not exceed 0.80 using the CUPRAC method. To the mixtures are added 1 mL each of Cu(II), Nc, and NH_4Ac buffer in this order. Water is added for dilution to a final volume of 4.1 mL. The ternary

mixture solutions are subjected to both 'normal' and 'incubated measurement' so as to test the hypothesis of the additivity of absorbances due to each antioxidant, and the theoretically computed CUPRAC antioxidant capacities of the mixtures are compared to those experimentally found.

8. Method of standard antioxidant additions to real mixtures (fruit juices): The commercial grape juice kept in a glass bottle is used as such. The grape juice is appropriately diluted with water such that its original CUPRAC absorbance at 450 nm would lie between 0.2 and 0.4 abs. units. The suitable dilution ratio selected is 1:10 for grape juice. The standard calibration curve of trolox is redrawn in this real solution so as to observe the parallelism between the calibration line (trolox and individually) in water and in real solution (**Fig. 14.3**) (*see* **Notes 6** and 7).

3.2. Individual Spectrophotometric Determination of Ascorbic acid with a Modified CUPRAC Assay in the Presence of Flavonoids

1. Preparation of solutions: Two distinct ascorbic acid (AA) solutions at an equimolar concentration of 1.0×10^{-3} M are prepared, one in de-aerated distilled water by bubbling N_2, and the other in 80 mL glacial acetic acid + 30 g metaphosphoric acid + water mixture (further called 'metaphosphoric acid–glacial acetic acid mixture') in a final solution volume of 1000 mL, as recommended in literature for stability *(6)*. It is

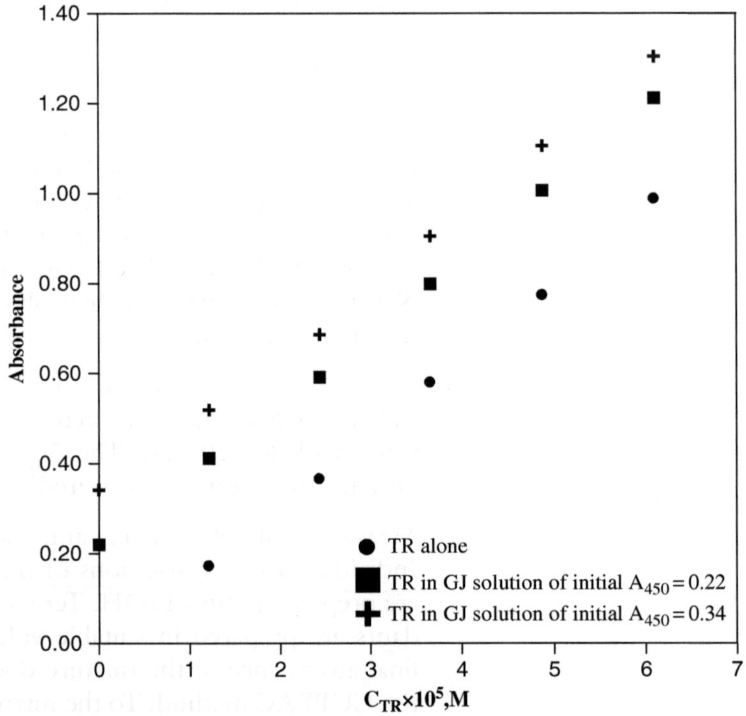

Fig. 14.3. Calibration curve of trolox (TR) in grape juice (GJ).

later shown that daily prepared AA solution has no stability problem in the CUPRAC assay, as compared to the standard AA solution prepared with respect to the metaphosphoric acid procedure, and daily prepared fresh solutions of AA are used thereafter. La(III) aqueous stock solution at 3.0×10^{-3} M concentration (for chelation of phenolics and flavonoids) is prepared from $LaCl_3.7H_2O$. The stock solutions of other antioxidants like QR, CAT, FRA, NG, MR, and FS are prepared daily as 1.0×10^{-2} or 1.0×10^{-3} M (as required) solutions in EtOH. As for the commercial fruit juice samples, the grape juice (Taskobirlik, Nevsehir, Turkey) is used as such, and the orange juice (Cappy gold, 100%) is diluted at a ratio of 1:10 with distilled water or with 'metaphosphoric acid–glacial acetic acid mixture' for HPLC analysis after filtration. The vitamin tablets assayed are first crushed to fine powder in an agat mortar, and then dissolved with a 1:1 (v/v) mixture of EtOH-H_2O, and diluted with the same solvent mixture to either 250 or 500 mL.

2. Preparation of HPLC assay solutions: The mobile phase for HPLC analysis with isocratic elution is prepared from phosphate-acetate buffer, as recommended in Lykkesfeldt's modified version of analysis *(7)*. This buffer is prepared by dissolving 7.8 g of sodiumdihydrogen phosphate dihydrate and 6.8 g of sodium acetate in 490 mL of bidistilled water, and adding 10 mL MeOH *(7)*. The pH of the mobile phase thus prepared is adjusted to pH 5.4 with phosphoric acid. The flow rate of the mobile phase is 0.8 mL min^{-1}, and the detection wavelength is 240 nm.

3. Ascorbic acid determination in the absence and presence of flavonoids:

Method applied to pure AA solution: One millilitre of 3.0×10^{-3} M La(III) is added to 0.5 mL of 1.0×10^{-3} M pure AA solution, the mixture is diluted to 5 mL with distilled water, and extracted for 1 min with two successive portions of 5 mL EtAc *(5)*. After phase separation, 0.5 mL withdrawn from the lower aqueous phase is analyzed for AA with the CUPRAC method in a total aqueous phase volume of 4.0 mL, as described in **Section 3.1**.

Method applied to (AA + flavonoids) mixture solution: To 0.5 mL of 1.0×10^{-3} M AA solution are added different proportions of flavonoid stock solutions (at 10^{-2} or 10^{-3} M concentration) so as to prepare binary, ternary, quaternary or more complex mixtures of AA + flavonoids. La(III) stock solution twice the volume of the flavonoid is added, and the final mixture is diluted to 5 mL with H_2O. The aqueous phase is extracted with two successive 5-mL volumes of EtAc, the phases are left to separate, and a suitable aliquot of the lower aqueous phase – separated from flavonoids – is analyzed with the CUPRAC method, as described in **Section 3.1**.

4. Recommended procedure for ascorbic acid determination in complex samples: One millilitre of ascorbic acid analyte solution (at weakly acidic-to-neutral pH; $2 \leq pH \leq 7$) containing flavonoids is taken, and 2 mL of 3.0×10^{-3} M La(III) solution is added, and well mixed. The final volume is diluted to 5 mL with H_2O, and extracted with two successive 5-mL volumes of EtAc. The aqueous raffinate phase is kept for the CUPRAC assay.

One millilitre of Cu(II) solution is placed in a separate test tube, and then 1.0 mL of Nc, 1 mL of ammonium acetate buffer, 1.0 mL of aqueous raffinate (above) solution containing AA are added by mixing in this order. After 1 min, the absorbance of the solution – of 4.0 mL total volume – at 450 nm is recorded against a reagent blank.

For AA aqueous solutions prepared in the mixture of glacial acetic acid–metaphosphoric acid, it is necessary to add approximately 0.5 mL of 3 M NaOH to the solution mentioned above so as to bring its final pH around 7.

5. Application to pharmaceutical tablets: The crushed and powdered pharmaceutical tablets containing vitamin C are dissolved in 1:1 (v/v) EtOH-H_2O, and the turbid solution is centrifuged at 5000 rpm (1500 g) for 5 min. The centrifugate is filtered through a glass fiber/polyester filter (Chromafil) of 0.45-μm pore size, diluted at a ratio of 1:10 with H_2O, and the final solution is analyzed for AA simultaneously with the modified CUPRAC method and with HPLC for validation.
6. Application to fruit juices: The grape juice is assayed as such, and orange juice after dilution with H_2O at a ratio of 1:10. The turbid solution is centrifuged and filtered, as described above for pharmaceutical tablet solutions. To (x) mL of fruit juice in final form, ($2x$) mL La(III) and ($10.0 - 3x$) mL H_2O are added, and extracted with two successive portions of 10 mL EtAc (x was usually taken as 2.0 mL). After the separation of phases, 1.0 mL of the lower aqueous phase is taken for analysis with the modified CUPRAC procedure and with HPLC for validation.

3.3. CUPRAC Assay of Fresh, Sun-dried and Sulfited-dried Malatya Apricot (Prunus Armeniaca) in Comparison with Other Methods

1. Standard CUPRAC assay for total antioxidant capacity of apricot extracts: The 'normal sample measurement' protocol is applied to (x) mL of apricot (fresh, sun-dried, sulfited-dried, and desulfited samples) extract, as described in **Section 3.1**. Since the molar absorptivity of trolox in the CUPRAC method is $\varepsilon = 1.67 \times 10^4$ L mol^{-1} cm^{-1}, and the calibration curve for pure trolox is a line passing through

the origin, the trolox equivalent molar concentration of the apricot extract sample in final solution may be found by dividing the observed absorbance to the ε for trolox (optical cuvette thickness = 1 cm) *(8)*. The trolox equivalent antioxidant capacity may be traced back to the original extract considering all dilutions, and proportionated to the initial mass of apricot sample taken to find a capacity in the units of micromoles TR/g dry matter. When the technique of standard additions is used for apricot extracts, ($x = 0.1 - 0.6$) mL of extract are taken such that the initial CUPRAC absorbance of the solution is approximately 0.2. Varying volumes between 50 and 250 μL of 1.0×10^{-3} M trolox stock solution and enough water are added to make the final total volume = 4.1 mL, and the absorbance reading is made as described for direct measurements. The standard additions technique is performed for confirming the validity of the principle of additivity in absorbance measurements, i.e., for showing the parallelism of calibration lines of trolox in pure solution and in trolox standard-added extracts, similar to the situation in **Fig. 14.3** (*see* **Notes 6** and 7). Naturally, this parallelism is also indicative of the additivity of individual antioxidant capacities of constituents in the mixture (*see* **Note** 7).

2. Preparation of ABTS assay solutions: The chromogenic radical reagent ABTS, at 7.0 mM concentration, is prepared by dissolving 0.1920 g of the compound in water, and diluting to 50 mL. To this solution is added 0.0331 g $K_2S_2O_8$ such that the final persulfate concentration in the mixture be 2.45 mM. The resulting chromophore containing ABTS radical cation solution is left to mature at room temperature in the dark for 12–16 h, and then used for TEAC assays.

3. ABTS assay of total antioxidant capacity: The matured ABTS radical solution of blue–green color is diluted with 96% ethanol at a ratio of 1:10. The absorbance of the 1:10 diluted $ABTS^+$ radical cation solution is 1.28 ± 0.04 at 734 nm. To 1 mL of the radical cation solution, 4 mL of ethanol are added, and the absorbance at 734 nm is read at the end of the first and sixth minutes. The procedure is repeated for the unknown extract by adding 1 mL of the radical cation solution to (x) mL of apricot extract (x varied between 0.1 and 0.5 mL) and ($4.0 - x$) mL of ethanol, and recording the absorbance readings at the end of first and sixth minutes. The absorbance difference (ΔA) is found by subtracting the extract absorbance from that of the reagent blank (pure radical solution), and this is correlated to trolox equivalent antioxidant concentration with the aid of a linear calibration curve (usually the absorbance decrease at the sixth minute is used for calculations).

4. Preparation of Folin assay solutions: The solutions used in the Folin assay of polyphenolics are prepared as follows: Lowry A: 2% aqueous Na_2CO_3 in 0.1 M NaOH; Lowry B: 0.5% $CuSO_4$ aqueous solution in 1% $NaKC_4H_4O_6$ solution; Lowry C: prepared freshly as mixture (50 mL Lowry A + 1 mL Lowry B); Folin-Ciocalteau reagent is diluted with H_2O at a volume ratio of 1:3 prior to use. All percentages are given as (w/v), and distilled and deaerated (N_2-bubbled) water is used throughout.
5. Folin method of total phenolics assay: To (x) mL of the apricot extract (such that x varied between 0.1 and 0.5 mL) is added ($2.0 - x$) mL H_2O. An aliquot of 2.5 mL of Lowry C solution is added, and the mixture is let to stand for 10 min. At the end of this period, 0.25 mL of Folin reagent is added, and 30 more minutes is allowed for stabilization of the blue color formed. The absorbance against a reagent blank is measured at 750 nm. Trolox is used as the reference standard. Folin method works at the distinctly alkaline pH of 10 (*see* **Note 8**) so as to enable the Folin reagent to oxidize the phenolate anions of polyphenolic antioxidants.
6. Calculation of total antioxidant capacity and phenolics content: The molar absorptivity of trolox in the above methods are as follows:

   $\varepsilon_{TR} = 1.67 \times 10^4$ L mol^{-1} cm^{-1} (CUPRAC method); $\varepsilon_{TR} = 2.6 \times 10^4$ L mol^{-1} cm^{-1} (ABTS method); $\varepsilon_{TR} = 4.65 \times 10^3$ L mol^{-1} cm^{-1} (Folin method).

   If a fruit extract obtained from fresh, sun-dried, sulfited-dried, and desulfited apricot samples (initial volume = V_i) prepared from (m) grams of solid matter is diluted (r) times prior to analysis, and a sample volume of (V_s) is taken for analysis from the diluted extract, and color development (after addition of reagents) is made in a final volume of (V_f) to yield an absorbance of (A_f), then the trolox equivalent antioxidant capacity of the apricot sample (in µmol TR per gram of solid matter, or simply µmol TR/g) is found using the equation:

   Antioxidant capacity (in µmol TR/g) = $(A_f/\varepsilon_{TR})(V_f/V_s)\, r\,(V_i/m) \times 10^3$
7. Treatment of apricot samples: Fresh apricots are supplied as mentioned in **Section 2.3**. The collected samples are subjected to pretreatment procedures according to the traditional practice of the region under the supervision of the Apricot Institute, and classified into three forms as fresh, sun-dried, and sulfited-dried samples. Fresh samples are suitably packaged (in 1-kg packs), and immediately brought to our laboratories (in Istanbul University) by air transport, and deep-frozen at −25°C prior to analysis. A part of the fresh samples are deseeded and sun-dried on concrete floor.

Another part of the fresh samples are fumigated with sulfur dioxide in sulfiting rooms, then sun-dried and deseeded. Both sun-dried (without additive) and sulfited-dried samples in 1-kg packs are deep-freezed as described.

8. Solvent extraction of apricot samples and anion exchange separation/analysis of sulfite: The described procedure can be applied to all kinds of dried fruits previously treated with sulfur dioxide for safe drying (*see* **Note 9**). Three different weighings (each approximately 4 g) of each form of the apricots are taken and extracted; three measurements are made for each extract. The extraction procedure was as described by Garcia-Alonso et al. *(9)*. Briefly, the deep-frozen samples are homogenized in cold methanol (3 × 25 mL), and centrifuged at 10.000 rpm for 10 min; the centrifugates are filtered through an ordinary filter paper into a 100-mL flask, and the settled pulp at the bottom of the centrifuge tube is taken up with methanol containing some water, filtered into the same flask, and diluted to a final volume of 100 mL. CUPRAC, ABTS/TEAC, and Folin assays are conducted immediately on these extracts. For desulfiting the sulfited-dried sample extracts, 25 mL of the extract adjusted to pH 3.0 (by adding 0.1 and 0.3 mL of 1 M HCl) is passed thrice through a 5-g resin bed of Dowex-X8 (Cl^- form) strongly basic anion exchanger (thoroughly prewashed with water) placed in a 25-mL burette. The bed height of the anionic resin is approximately 10 cm, and the flow rate is 1.6 mL min^{-1}. The resin-retained sulfite (as HSO_3^-) is eluted with 0.1 M NaOH, and analyzed with the same CUPRAC reagent (at pH 7); the molar absorptivity of the CUPRAC reagent for sulfite is $\varepsilon_{sulfite} = 4.92 \times 10^3$ L mol^{-1} cm^{-1}, and the equation for the linear calibration of sulfite is: $A_{CUPRAC} = 4.92 \times 10^3 C_{sulfite} + 0.0223$. The desulfited apricot extracts are brought to a pH of 7 by the addition of 0.08–0.2 mL of 1.2 M NaOH, and analyzed for their antioxidant content using the standard CUPRAC protocol.

3.4. CUPRAC and Polyphenolic Content of Herbal Teas

1. Steeping method applied to herbal teas to obtain infusions: A mass of 1.5 g of dry herbal matter is wrapped in a purse of standard membrane filter material, the purse is transfered to a beaker, 250 mL of freshly boiled water is added, and the beaker is covered with a watch glass, let to stand for 5 min. The purse is then removed, and the partly turbid solution (let to cool to room temperature) is filtered through a black band Whatman quantitative filter paper. If the method of standard additions is to be applied for analyzing this infusion, the clear extract is properly diluted so as to give a final absorbance (at 450 nm) around 0.2 in the CUPRAC measurement;

otherwise, the infusion is diluted to yield a final absorbance of about 0.43 for highest photometric precision (i.e., the relative indeterminate error with respect to measured absorbance is minimized around an absorbance of 0.43 units) *(10)*. The diluted sample solutions are analyzed for their antioxidant capacities with the use of CUPRAC *(1)* and ABTS *(11)* methods; and for their phenolic content with the aid of Folin method *(12)*.

2. Infusion preparation from tea bags: Tea bags supplied from the market are weighed as such to yield masses ranging between 1.40 and 2.25 g. The tea bags are dipped into and pulled out of beakers containing 250 mL of freshly boiled water for the first 2 min, and let to steep for the remaining 3 min in the covered beakers (total steeping time is 5 min, same as that of the herbal teas). The bags are removed, and the partly turbid solutions are filtered through a black band Whatman quantitative filter paper after cooling to room temperature. The clear extract is properly diluted for CUPRAC and ABTS assays of total antioxidant capacity, and for Folin assay of polyphenolics content. All infusions are analyzed as fresh as possible for reliability of the results; however, the assay results do not change even after 48 h storage of the tea extracts at +4°C.
3. Antioxidant capacity assay of herbal tea infusions: The infusions are assayed by standard CUPRAC (*see* **Section 3.1**), and ABTS and Folin methods (*see* **Section 3.3**), for comparative evaluation. The trolox (TR) equivalent antioxidant capacity is calculated as described in **Section 3.3** in the units of mmol TR/g dry herbal material.

3.5. Results

1. Original CUPRAC method, modifications and applications to food antioxidants: The trolox equivalent antioxidant capacities (TEAC coefficients) of pure antioxidant compounds are calculated. The TEAC coefficients (i.e., the reducing potency – in trolox mM equivalents – of 1 mM antioxidant solution under investigation) of various antioxidant compounds found with the developed CUPRAC method and compared to those measured with the reference methods are tabulated in **Table 14.1**. The linear calibration curves of the tested antioxidants as absorbance *vs* concentration with respect to the CUPRAC method (figures not shown) generally give correlation coefficients close to unity ($r \geq 0.999$) within the useful absorbance range of 0.1–1.1. The highest antioxidant capacities in the CUPRAC method are observed for epicatechin gallate, epigallocatechin gallate, quercetin, fisetin, epigallocatechin, catechin, caffeic acid, epicatechin, gallic acid, rutin, and chlorogenic acid in this order, in accordance with theoretical expectations, because the number and position of the hydroxyl groups as well

Table 14.1
Antioxidant capacities of the polyphenolic compounds (in the units of TEAC: trolox equivalent antioxidant capacity) as measured by CUPRAC, ABTS/TEAC, and Folin assays

	$TEAC_{CUPRAC}$				$TEAC_{ABTS}$	
Antioxidant name	$TEAC_N$	$TEAC_I$	$TEAC_H$	$TEAC_{H\&I}$	$TEAC_{6min}$	$TEAC_{Folin}$
Flavonoids						
Epicatechin gallate (ECG)	5.32	5.65			3.51	4.35
Epigallocatechingallate (EGCG)	4.89	5.49			3.15	2.78
Quercetin (QR)	4.38				2.77	5.17
Fisetin (FS)	3.90	4.18			2.62	3.90
Epigallocatechin (EGC)	3.35	3.60				
Catechin (CT)	3.09	3.56	3.08	3.49	3.14	4.09
Epicatechin (EC)	2.77	2.89			2.69	3.22
Rutin (RT)	2.56			3.80	1.15	6.75
Morin (MR)	1.88	3.32			1.79	3.37
Kaempherol (KP)	1.58	1.87			0.90	2.01
Hesperetin (HT)	0.99	1.05	0.85	0.98	1.11	4.50
Hesperidin (HD)	0.97	1.11	0.79	0.95	1.40	3.29
Naringenin (NG)	0.05	2.28		3.03	0.64	5.52
Naringin (N)	0.02	0.13			0.62	1.12
Hydroxycinnamic acids						
Caffeic acid (CFA)	2.89	2.96	2.87	3.22	1.39	3.27
Chlorogenic acid (CGA)	2.47	2.72	1.20	1.42	1.21	2.84
Ferulic acid (FRA)	1.20	1.23	1.18	1.34	2.16	3.49
p-Coumaric acid (CMA)	0.55	1.00	0.53	1.15	1.63	2.54
Vitamins						
α-tocopherol (TP)	1.10	1.02	0.99	0.87	1.02	
Ascorbic acid (AA)	0.96				1.03	
Benzoic acids						
Gallic acid (GA)	2.62				3.48	1.23
Sinapic acid (SNA)	1.24	2.17			1.11	3.39
Vanillic acid (VA)	1.24	1.52	1.32	1.57	1.25	3.05
Syringic acid (SA)	1.12	1.64	1.13	1.67	1.50	2.49

as the degree of conjugation of the whole molecule are important for easy electron transfer. The results obtained with the Folin method are generally higher than with others, because the essential component of the Folin reagent, i.e., molybdophospho-tungstate heteropoly acid, has an indefinite but much higher redox potential than those of the other reagents which lie in the range of $E^0 = 0.6–0.7$ V (*see* **Note 8**). The Folin reagent may be considered as a nonspecific oxidant for polyphenols, in accord with relevant literature *(13)*. Flavonoid glycosides normally yielding much lower TEAC values in other electron transfer-based methods are hydrolyzed to their corresponding aglycons, and exhibit their full antioxidant capacity.

2. Structure-acitivity relationships in the CUPRAC method: A novel antioxidant capacity assay eligible for standardization (i.e., to be used in determining the antioxidant status of food and biological fluids) has to consider some structural requirements of antioxidant potency. For example, the presence of 5-hydroxy-4-keto group in A & C rings in flavonols, the 2,3-double bond connecting the two ring systems of flavonol *via* conjugation, and the 3′,4′-dihydroxy substitution of the B ring (*see* **Fig. 14.4**) are considered as important structural

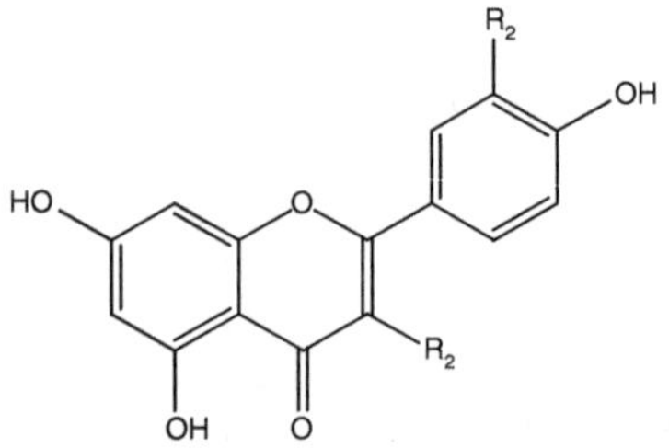

Quercetin (QR): $R_1 = R_2 = –OH$
Fisetin (FS): $R_1 = –OH$, $R_2 = –H$
Rutin (RT): $R_1 = –O\text{-Rutinase}$, $R_2 = –OH$

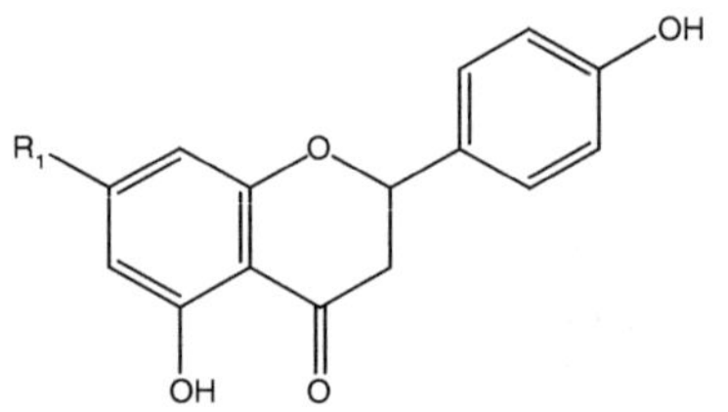

Naringenin (NG): R_1–OH
Naringenin (N): R_1–O-Neohesperidase

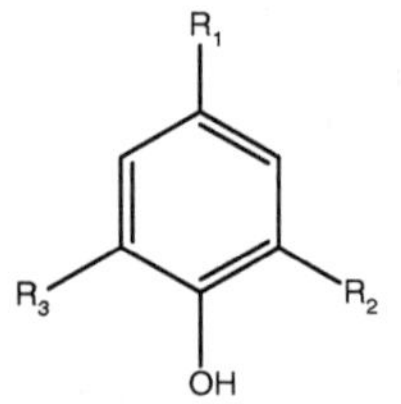

Gallic acid (GA): R_1= -COOH, R_2=R_3= -OH
Caffeic acid (CFA): R_1= -CH=CH-COOH, R_2= -OH, R_3= -H
Ferulic acid (FRA): R_1= -CH=CH-COOH, R_2= -OCH_3, R_3= -H
p-Coumaric acid (CMA): R_1= -CH=CH-COOH, R_2= R_3= -H

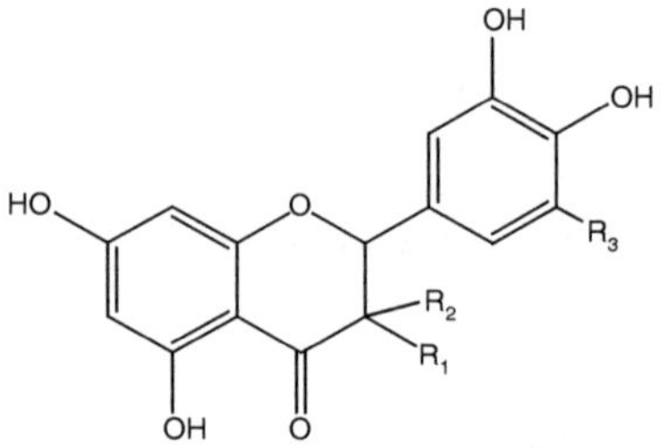

Catechin: R_1= -OH, R_2= R_3= -H,
Epicatechin: R_1= R_3= -H, R_2= -OH
Epicatechin gallate: R_1= R_3= -H, R_2= -O-galloyl
Epigallocatechin: R_1= -H, R_2=R_3= -OH
Epigallocatechin gallate: R_1= -H, R_2= -O-galloyl R_3= -OH

Galloyl:

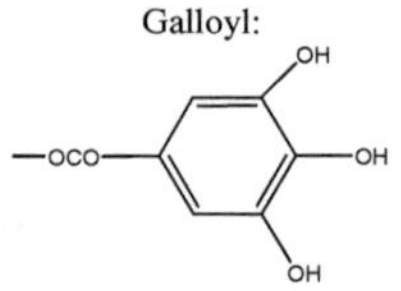

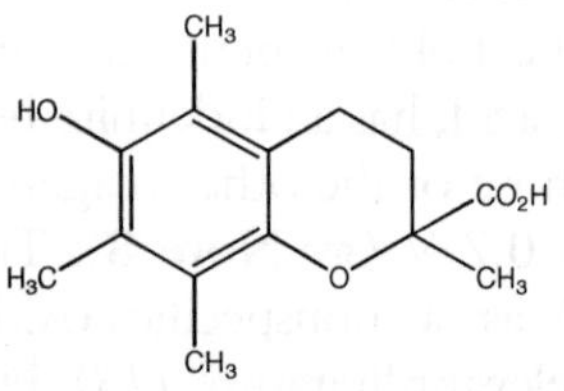

Trolox

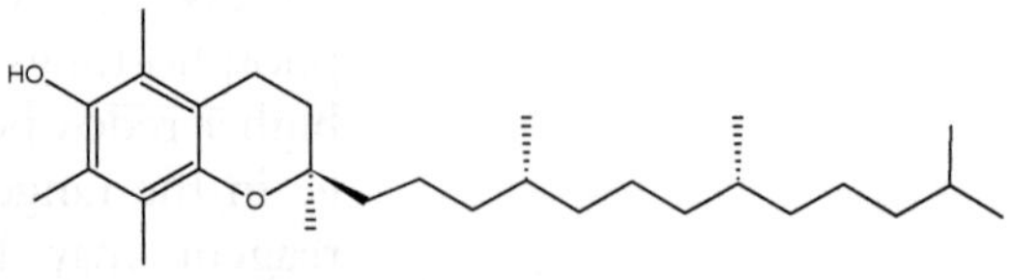

?-tocopherol (vitamin E)

Fig. 14.4. The formulas of the hydrophilic and lipophilic antioxidants tested with the CUPRAC assay.

characteristics for antioxidant potency *(14, 15)*, all three of which are combined in quercetin. As a result, the TEAC coefficient in the CUPRAC assay is highest among flavonols for quercetin. Fisetin has one –OH group less than quercetin, and therefore gives the lower TEAC value (**Table 14.1**). Rutin, having an O-rutinase substituent instead of –OH in the 3-position (**Fig. 14.4**), shows the lower capacity. In general, when flavonoid glycosides are hydrolyzed to the corresponding aglycons (i.e., O-sugar substituent being converted to –OH), their CUPRAC antioxidant capacities significantly improve.

As for hydroxycinnamic acids, which are almost the most abundant phenolic components in the citrus family and in some other fruits, the TEAC coefficients with respect to the CUPRAC method (and with respect to the ABTS assay, as shown in parantheses) are as follows: caffeic acid 2.9 (1.4), chlorogenic acid 2.5 (1.2), ferulic acid 1.2 (2.2), and *p*-coumaric acid 0.6 (1.6). The trolox equivalent capacity order for these phenolic acids is just the opposite of that of the most widely used ABTS assays *(15)*. Structural properties of hydroxycinnamic acids would normally dictate that two –OH bearing caffeic and chlorogenic acids should exhibit higher TEAC coefficients than one –OH bearing ferulic and *p*-coumaric acids (*see* **Fig. 14.4** for formulas). Furthermore, ferulic acid having an electron-donating methoxy group in *ortho*-position relative to the phenolic –OH, thereby allowing increased stabilization of the resulting aryloxyl radical through electron delocalization after H-atom donation by the hydroxyl group, should show a higher TEAC coefficient than *p*-coumaric acid which lacks such a group. Thus, structural requirements dictate that hydroxycinnamic acids should have a TEAC order as measured by the CUPRAC and not by the ABTS assay. Moreover, the order of peroxyl radical scavenging ability of hydroxycinnamic acids, and thus the order for their ability to enhance the resistance of LDL to oxidation, is measured as caffeic acid > chlorogenic acid > ferulic acid > *p*-coumaric acid *(15, 16, 17)*, again entirely consistent with the results of the CUPRAC method. Although the antioxidant activity of rosmarinic acid, another powerful antioxidant hydroxycinnamic acid, is reported to be much higher than those of trolox *(18)*, α-tocopherol, caffeic acid, ferulic acid, BHT *(19)* and other plant phenolics *(20)*, the ABTS/TEAC and DPPH methods applied by different researchers report low TEAC values within the interval of 1.5–2.0 for rosmarinic acid *(21, 22)* while the CUPRAC method finds a TEAC value close to that of quercetin. The TEAC order of hydroxycinnamic acids (**Table 14.1**) clearly reflects the superiority of CUPRAC over other similar ET-based methods. Gallic acid has one more –OH group than chlorogenic acid, and therefore shows the higher capacity. The catechin group, also known as 'tea antioxidants', gives a capacity order in accord with the number and position of their –OH groups, together with the overall extent of conjugation in the molecule. Inspection of **Table 14.1** reveals that all these structural requirements for antioxidant potency are met by the TEAC results of the CUPRAC assay. The TEAC coefficients of individual antioxidants found with CUPRAC correlate linearly ($r = 0.8$) to those of ABTS, but the correlation of both assays to Folin is poor (**Table 14.1**), probably because both CUPRAC and ABTS are similar ET-based antioxidant assays with close reduction potentials while the exact potential of the Folin reagent with the presumably higher potential is not definitely known (*see* **Note 8**). The phenols are essentially dissociated to phenolates in the Folin assay carried out in alkaline medium, facilitating oxidation. As a result, the high TEAC values listed in **Table 14.1** achieved with the Folin assay reveal that the Folin reagent should have oxidized the tested phenolics to a greater extent than either CUPRAC or ABTS/TEAC.

Among the antioxidants tested by the CUPRAC assay, those of which the oxidation potentials were measured with cyclic voltammetry (CV) by Firuzi et al. *(23)* are quercetin (0.39 V), fisetin (0.39 V), catechin (0.45 V), rutin (0.46 V), and naringenin (0.89 V). The standard potentials and TEAC values (with respect to the normal CUPRAC method) of these antioxidants correlate linearly with a high correlation coefficient (in absolute value), i.e., $r = -0.986$, meaning that antioxidants within a given class (flavonoids, hydroxycinnamic acids, etc.) with the lowest oxidation potentials are the most active compounds in the CUPRAC assay, showing the highest TEAC coefficients *(3)*. Briefly, oxidizability of an antioxidant correlates well with its CUPRAC capacity (in TEAC units), confirming that CUPRAC is a genuine electron transfer-based assay.

3. CUPRAC assay of synthetic and real mixtures: The experimentally measured capacities are generally within ± 5% interval of the theoretically computed values using the formula:

$$\text{Capacity}_{\text{total}} = \text{TEAC}_1\,\text{concn}_1 + \text{TEAC}_2\,\text{concn}_2 + \text{TEAC}_3\,\text{concn}_3 + \ldots + \text{TEAC}_n\,\text{concn}_n. \quad (14.2)$$

where 1,2, ... ,*n* denote the corresponding constituents of the synthetic mixture. The comparison of expected (using Eq. 14.2) and experimentally found antioxidant capacities of synthetic mixture solutions (as mM trolox-equivalents) found with CUPRAC are generally in accord with each other (**Table 14.2**) (*see* **Note 7**). The accordance of theoretical and

Table 14.2
The comparison of expected and found CUPRAC antioxidant capacities of synthetic mixture solutions (as mM trolox equivalents) (*see* Note 7)

Method	Composition of mixture	Capacity expected* (as mM TR-equivalent)	Capacity found experimentally (as mM TR-equivalent)
Normal measurement	10 μL of 1 mM QR 10 μL of 1 mM CT 10 μL of 1 mM RT	2.45×10^{-2} mM	2.55×10^{-2} mM
Incubated measurement	10 μL of 1 mM QR 10 μL of 1 mM CT 10 μL of 1 mM RT	2.56×10^{-2} mM	2.97×10^{-2} mM
Normal measurement	30 μL of 1 mM GA 20 μL of 1 mM CFA 50 μL of 1 mM CMA	3.99×10^{-2} mM	3.96×10^{-2} mM
Incubated measurement	30 μL of 1 mM GA 20 μL of 1 mM CFA 50 μL of 1 mM CMA	4.58×10^{-2} mM	4.25×10^{-2} mM
Normal measurement	20 μL of 1 mM GA 20 μL of 1 mM FRA 20 μL of 1 mM CFA	3.27×10^{-2} mM	3.09×10^{-2} mM
Incubated measurement	20 μL of 1 mM GA 20 μL of 1 mM FRA 20 μL of 1 mM CFA	3.32×10^{-2} mM	3.11×10^{-2} mM
Normal measurement	20 μL of 1 mM FRA 50 μL of 1 mM CMA 20 μL of 1 mM CFA	2.66×10^{-2} mM	2.62×10^{-2} mM
Incubated measurement	20 μL of 1 mM FRA 50 μL of 1 mM CMA 20 μL of 1 mM CFA	3.26×10^{-2} mM	3.10×10^{-2} mM
Normal measurement	20 μL of 1 mM FRA 50 μL of 1 mM CMA 20 μL of 1 mM GA	2.53×10^{-2} mM	2.45×10^{-2} mM
Incubated measurement	20 μL of 1 mM FRA 50 μL of 1 mM CMA 20 μL of 1 mM GA	3.09×10^{-2} mM	3.21×10^{-2} mM
Normal measurement	10 μL of 1 mM QR 10 μL of 1 mM CT 50 μL of 1 mM TR	3.04×10^{-2} mM	3.08×10^{-2} mM
Incubated measurement	10 μL of 1 mM QR 10 μL of 1 mM CT 50 μL of 1 mM TR	3.15×10^{-2} mM	3.09×10^{-2} mM

*Found by means of the equation: $Capacity_{mixture} = TEAC_1 \cdot C_1 + TEAC_2 \cdot C_2 + TEAC_3 \cdot C_3$

experimental findings, combined with the parallelism of the linear calibration curves of each antioxidant compound tested in the presence of the other compound, effectively demonstrate that there are no chemical interactions of intereferent nature among the synthetic solution constituents, and that the antioxidant capacities of the tested antioxidants are additive *(1)* (*see* **Notes 6** and 7). Thus the proposed CUPRAC method may be effectively used for the antioxidant capacity assay of synthetic mixtures and real solutions.

4. Individual spectrophotometric determination of ascorbic acid (AA) with a modified CUPRAC assay in the presence of flavonoids: When a La(III) chloride solution is added to an AA solution and extracted with ethyl acetate (EtAc), it is observed that the aqueous phase gives the expected CUPRAC absorbance while the organic phase is not colored with the CUPRAC reagent. This means that AA remains in aqueous medium when EtAc extraction in the presence of excess La(III) is carried out (*see* **Fig. 14.5**). The spectra of Cu(I)-Nc solutions obtained from AA (in 1.25×10^{-2} mM solution) alone, AA + La(III), and AA + QR + La(III) after extraction (i.e., remaining in the aqueous phase) almost overlap with each other (**Fig. 14.5**).

The calibration equation for HPLC determination of ascorbic acid (AA) is $A = 5.6 \times 10^{9}\ C_{AA} + 4.5 \times 10^{4}$ ($r = 0.9999$), where A is the integrated peak area. HPLC analysis for method validation is also applied to pharmaceutical tablets. The retention time for AA under the described conditions is 3.2 min (in pure solution). The chromatogram for 1.0×10^{-3} M standard AA solution is given in **Fig. 14.6.**

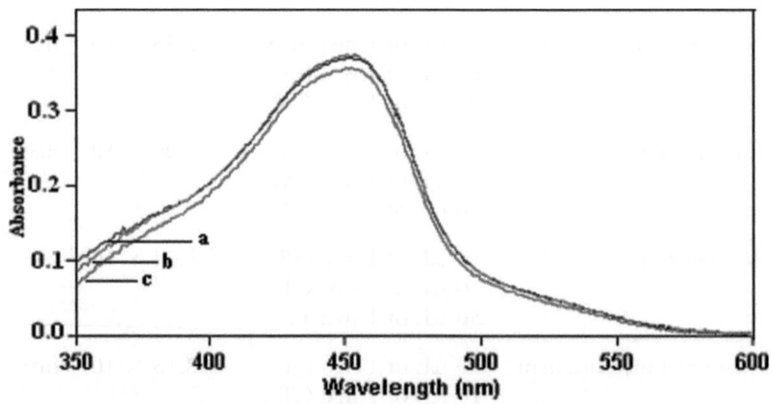

Fig. 14.5. The spectra of Cu(I)-Nc solutions obtained from **(A)** AA (in 1.25×10^{-2} mM solution) alone, **(B)** AA + La(III), and **(C)** AA + QR + La(III) after extraction (i.e., remaining in the aqueous phase).

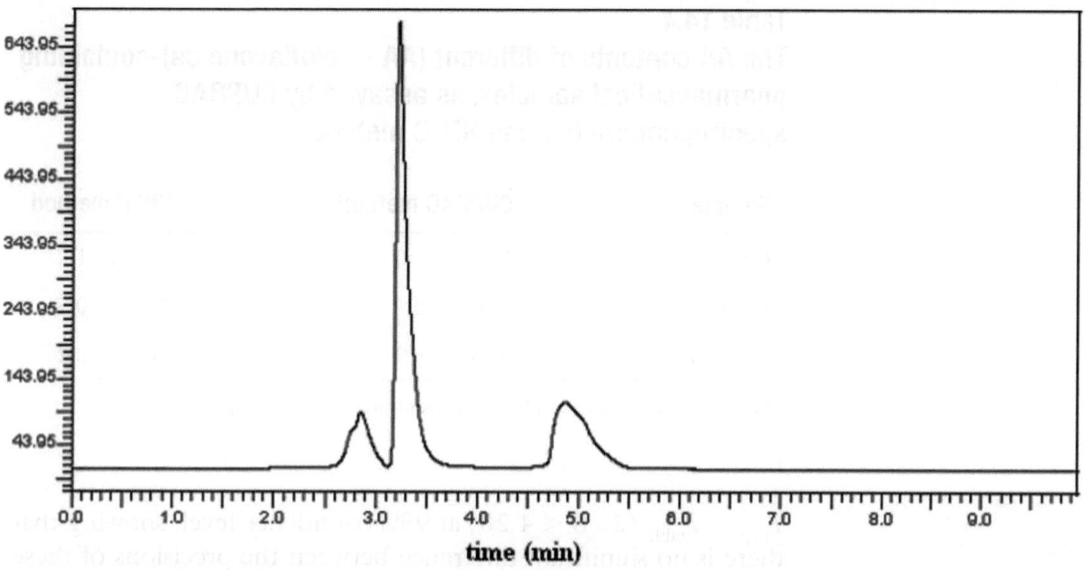

Fig. 14.6. The chromatogram for 1.0×10^{-3} M standard AA solution.

Table 14.3 depicts the CUPRAC absorbance results of raw (untreated) grape and orange juices and the absorbances after La(III)-EtAc extraction. The AA contents of these fruit juices are calculated from the final aqueous phase absorbance values after extraction, and these values correspond to those found by HPLC (**Table 14.3**).

The analytical results for different (AA + bioflavonoids)-containing pharmaceutical samples are depicted in **Table 14.4**. The results for the preliminary extractive CUPRAC method are in good agreement with those of HPLC. The CUPRAC findings for the GNC dietary supplement sample containing AA + bioflavonoids are statistically compared with those of HPLC using Student's t- and F-tests. For $N = 7$, $t_{exp} < t_{table}$ ($0.694 < 2.179$) at 95% confidence level, that the means of the two methods are essentially alike is confirmed. Similarly,

Table 14.3
The AA contents of grape and orange juices calculated from the final aqueous phase absorbance values after La(III)-EtAc extraction

Sample	CUPRAC method	La-CUPRAC method	HPLC method
1 mL grape juice	0.4352	0.2764 (C_{AA}:4.20×10^{-4} M)	(C_{AA}:3.98×10^{-4} M)
1 mL (1:10) orange juice	0.1851	0.1301 (C_{AA}:2.37×10^{-3} M)	(C_{AA}:2.00×10^{-3} M)

The corresponding AA concentrations found with CUPRAC spectrophotometric and HPLC methods are given in parantheses.

Table 14.4
The AA contents of different (AA + bioflavonoids)-containing pharmaceutical samples, as assayed by CUPRAC spectrophotometric and HPLC methods

Sample	CUPRAC method	HPLC method
GNC	974 ± 2.0	973 ± 1.2
Boots	486 ± 2.9	492 ± 0.7
Redoxon	464 ± 2.0	467 ± 3.1

The AA concentrations of pharmaceuticals in mg/tablet.

$F_{exp} < F_{table}$ (2.78 < 4.28) at 95% confidence level, showing that there is no significant difference between the precisions of these two methods. The results of the *t*- and *F*-tests for Redoxon tablet are 0.677 and 2.115; and for Boots tablet 2.098 and 17.84, respectively.

5. CUPRAC assay of fresh, sun-dried and sulfited-dried Malatya apricot (*Prunus Armeniaca*) in comparison with other methods: **Table 14.5** shows the CUPRAC, ABTS/TEAC, and Folin assay results of five varieties of Malatya apricots in four different forms (fresh, sun-dried, sulfited-dried, and desulfited). The antioxidant capacities are given as trolox (TR) equivalents, in the units of μmol TR/g solid matter. Since other reducing agents as well as phenolics react with the Folin reagent in a molybdenum blue-type heteropoly acid reaction *(24)*, the results of the Folin assay may also be reported as trolox equivalents, because trolox is a reference antioxidant compound in these assays. However, the Folin results are significantly greater than those of CUPRAC and ABTS, because polyphenolics are included along with reducing compounds (antioxidants).

When analyzed with the aid of linear regression, our CUPRAC assay correlates well with Folin phenolics content (**Fig. 14.7**) (*see* **Note 10**). Thus the higher antioxidant capacity of Malatya apricots may result from their high total phenolics and not from their ascorbic acid content. The trolox-equivalent antioxidant capacities (as stoichiometric TEAC coefficients in the original CUPRAC method) of most phenolic compounds (quercetin, catechins, gallic acid, etc.) are greater than 2 *(1)*, meaning that each phenolic or flavonoid behaves as a reducing agent in the redox reaction of concern capable of donating electrons greater than or equal to two molecules of trolox or ascorbic acid, contributing with a greater share to the overall TEAC antioxidant capacity.

Table 14.5
CUPRAC, ABTS/TEAC, and Folin assay TEAC results of five varieties of Malatya apricots in four different forms

Type of apricot	Fresh apricot (μmol TR/g)			Sun-dried apricot (μmol TR/g)			Desulphited Apricot (μmolTR/g)			Sulphited-Dried Apricot (μmolTR/g)		
	$TEAC_{CUPRAC}$	$TEAC_{ABTS/TEAC}$	$TEAC_{Folin}$	$TEAC_{CUPRAC}$	$TEAC_{ABTS/TEAC}$	$TEAC_{Folin}$	$TEAC_{CUPRAC}$	$TEAC_{ABTS/TEAC}$	$TEAC_{Folin}$	$TEAC_{CUPRAC}$	$TEAC_{ABTS/TEAC}$	$TEAC_{Folin}$
Hacihaliloglu	3.92 ± 0.54	3.50 ± 1.14	11.38 ± 1.47	15.30 ± 0.70	16.16 ± 0.13	35.20 ± 2.37	13.52 ± 0.56	14.45 ± 0.54	40.59 ± 0.47	37.84 ± 5.24	22.97 ± 0.80	83.35 ± 5.72
Cöloglu	4.23 ± 0.24	4.46 ± 0.93	10.73 ± 2.61	15.14 ± 0.42	16.46 ± 0.59	40.37 ± 5.25	14.84 ± 0.29	15.58 ± 0.30	43.16 ± 1.70	52.47 ± 0.49	31.41 ± 0.47	99.77 ± 3.98
Kabaasi	3.22 ± 0.17	3.18 ± 0.29	9.04 ± 1.24	13.55 ± 0.17	12.29 ± 0.45	33.82 ± 4.25	12.78 ± 1.72	12.45 ± 0.69	39.99 ± 3.85	38.83 ± 2.31	25.45 ± 0.69	87.69 ± 6.67
Soganci	2.67 ± 0.08	2.86 ± 0.16	8.47 ± 1.93	9.32 ± 0.19	9.45 ± 1.56	28.38 ± 8.09	9.68 ± 0.47	10.27 ± 1.50	34.48 ± 1.15	31.25 ± 0.62	15.38 ± 0.83	78.52 ± 0.35
Zerdali	4.06 ± 0.14	3.37 ± 0.35	10.90 ± 1.49	17.92 ± 3.05	15.81 ± 2.44	43.33 ± 0.87	17.00 ± 0.90	16.24 ± 0.73	43.13 ± 0.58	35.22 ± 3.52	18.66 ± 0.59	84.39 ± 0.47

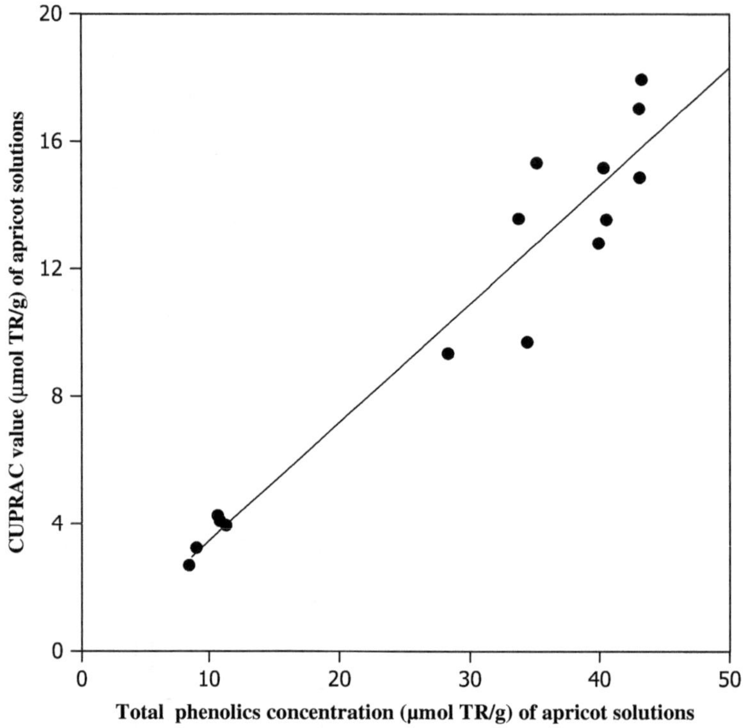

Fig. 14.7. The correlation of CUPRAC assay results with Folin total phenolics content.

The optimal pH of sulfite removal is chosen as pH 3.0, because this pH is well above the pK_{a1} of sulfurous acid (H_2SO_3), i.e., 1.8, where the predominant form of sulfite is HSO_3^-, which can easily be retained by the strongly basic anion exchanger (*see* **Note 9**). This is also a suitable pH for ascorbic acid, since this antioxidant compound is in molecular form at this pH, and therefore would not be retained by the anionic resin, thereby preventing a loss in observed antioxidant capacity of the sulfited-dried apricot extract. The CUPRAC reagent reacts easily with sulfite ($\varepsilon_{sulfite} = 4.92 \times 10^3$ L $mol^{-1}cm^{-1}$), and therefore this reagent can simultaneously assay the sulfite content and antioxidant capacity of sulfited-dried fruit extracts (*see* **Note 9**). The sulfite content of Zerdali, Soganci, Kabaasi, Hacihaliloglu, and Cologlu varieties are found as 479, 602, 744, 732, and 948 ppm SO_3^{2-}, respectively, well below the maximum residual sulfur dioxide limit of 2000 ppm set by FDA for dried fruits. The sulfite contents of sulfited-dried apricots correlate well with the difference in CUPRAC antioxidant capacity between the sulfited-dried and desulfited (by anion-exchange) apricot extracts ($r = 0.97$). It is remarkable that the sun-dried apricots, locally named as 'gun-kurusu kayisi', retain their antioxidant values nearly as much as sulfited-dried apticots (**Table 14.5**). As Asma described for Malatya apricots *(25)*, sulfite-drying causes a weight reduction factor of about 3.75, which more or less explains our increased antioxidant capacities as a result of drying (*see* **Table 14.5**). However, the work of Halvorsen et al. *(26)* reports extremely high FRAP capacities for dried apricots, where the capacity seems to increase by more than six times upon drying of the fresh fruit. To our understanding, a capacity increase at this level could not have arisen merely from drying, and the sulfite content may have also played part in the observed overall capacity, as sulfite is a reducing agent which may react with the FRAP reagent to produce the colored Fe(II)-tripyridyltriazine chelate (*see* **Note 9**). In further tests of antioxidant assay with dried fruits, this point should always be considered.

6. CUPRAC assay and polyphenolic content of herbal teas: The molar absorptivity of trolox in the CUPRAC, ABTS, and Folin methods are given in **Section 3.3**.

If a herbal infusion (initial volume = V_{CUPRAC}) prepared from (m) g of dry matter is diluted (r) times prior to analysis, and a sample volume of (V_s) is taken for analysis from the diluted extract, and color development (after addition of reagents) is made in a final volume of (V_f) to yield an absorbance of (A_f), then the trolox equivalent antioxidant capacity of the herb (in mmol TR per gram of dry matter, or simply mmol TR/g) is found using the equation:

Antioxidant capacity (in mmol TR/g) = $(A_f/\varepsilon_{TR})\,(V_f/V_s)\,r\,(V_{CUPRAC}/m)$

Example calculation: 1.5465 g of lemon balm (dry herbal tea material) is weighed, and prepared into a 250-mL infusion; 8 mL of this infusion is diluted to 100 mL prior to analysis (dilution ratio = r = 12.5). The volume of sample solution taken for analysis is $V_s = 0.2$ mL, and the total volume of final solution (in which color development is made) in the CUPRAC method is $V_f = 4.1$ mL. The trolox equivalent capacity of lemon balm using the above equation is $(0.401/1.67 \times 10^4)\,(4.1/0.2)\,12.5\,(250/1.5465) = 0.99$ mmol TR/g.

Different herbal tea infusions prepared from dried plants give CUPRAC values of total antioxidant capacity ranging between 0.1 and 1.6 mmol trolox equivalent per gram of dry matter, i.e., mmol TR/g (*see* **Table 14.6**). The inter-assay coefficient of variation of measurements is less than 1.7% (*see* **Note 6**). The results are generally in good agreement with either ABTS assay values or total phenolics contents (*see* **Note 10**). The highest antioxidant capacities are observed for scarlet pimpernel, sweet basil, green tea, and lemon balm in this order (1.63, 1.18, 1.07, and 0.99 mmol TR/g, respectively). For dried leaves, the capacity of green tea, i.e., 1.1 mmol TR/g, is higher than that of ordinary or blended black tea, mean 0.4 mmol TR/g. For infusions prepared from ready-to-use tea bags, the CUPRAC values (as mmol TR/g) are highest for

Table 14.6
The total antioxidant capacities of herbal infusions, either prepared from dried plant material or manufactured tea bags, as measured by the CUPRAC and ABTS/TEAC assays, combined with their total phenolics content as measured by the Folin assay

Herbal tea				
Botanic name	English name	CUPRAC value (mmol TR/g)	ABTS/TEAC value (mmol TR/g)	Total phenolics value (mmol TR/g)
Anagallis arvensis	Scarlet pimpernel	1.63	1.02	1.47
Helichrysum italicum (H. Arenarium)	Everlasting (immortelle)	0.37	0.35	0.35
Rhamnus cathartica	Common buckthorn	0.25	0.24	0.51
Fumaria officinalis	Fumitory herb	0.25	0.26	0.56
Plantago major	Plantain	0.31	0.10	0.38
Malva sylvestris	Common mallow	0.23	0.21	0.47
Lavandula stoechas	French lavender	0.30	0.16	0.29
Teucrium fruticans	Shrubby germander	0.32	0.13	0.29
Achillea millefolium	Yarrow (Achilea herba)	0.09	0.07	0.23
Tribulus terrestris	Puncture vine (land caltrops)	0.08	0.07	0.18
Coriandrum sativum	Coriander seed	0.49	0.50	0.83
Ocimum basilicum	Sweet basil	1.18	0.77	1.10
Melissa officinalis	Lemon balm	0.99	0.63	1.04
Tussilago farfara	Coltsfoot	0.48	0.46	0.71
Thymus vulgaris	Domesticated thyme	0.54	0.52	0.61
Althaea officinalis	Marshmallow	0.05	0.05	0.10
Camellia sinensis	Green tea	1.07	0.86	1.06
Camellia sinensis	Turkish Rize ordinary tea	0.42	0.37	0.47
Camellia sinensis	Turkish blended ordinary tea	0.38	0.34	0.42
Urtica dioica	Stinging nettle	0.18	0.15	0.19
Tilia europaea (cordata)	Linden flower (lime)	0.18	0.11	0.33

Rubus fruticocus	Blackberry	0.44	0.14	0.82
Foeniculum vulgare	Fennel	0.05	0.04	0.16
Salvia officinalis	Common sage	0.56	0.41	0.85
Camellia sinensis	Green tea	0.94	0.62	0.97
Camellia sinensis	Green tea with lemon	1.61	1.74	1.38
Camellia sinensis	Ceylon blended ordinary tea	4.41	4.05	4.06
Camellia sinensis	English breakfast ordinary tea	1.26	1.20	1.09
Camellia sinensis	Turkish çaykur ordinary tea	0.83	0.51	0.61

Ceylon blended ordinary tea (4.41), green tea with lemon (1.61), English breakfast ordinary tea (1.26), and green tea (0.94), all of which are manufactured types of *Camellia sinensis.*

Figure 14.8 shows the correlation of total phenolics content (as measured by the Folin assay) with the CUPRAC total antioxidant capacities of herbal teas. The linear curve in **Fig. 14.8** has

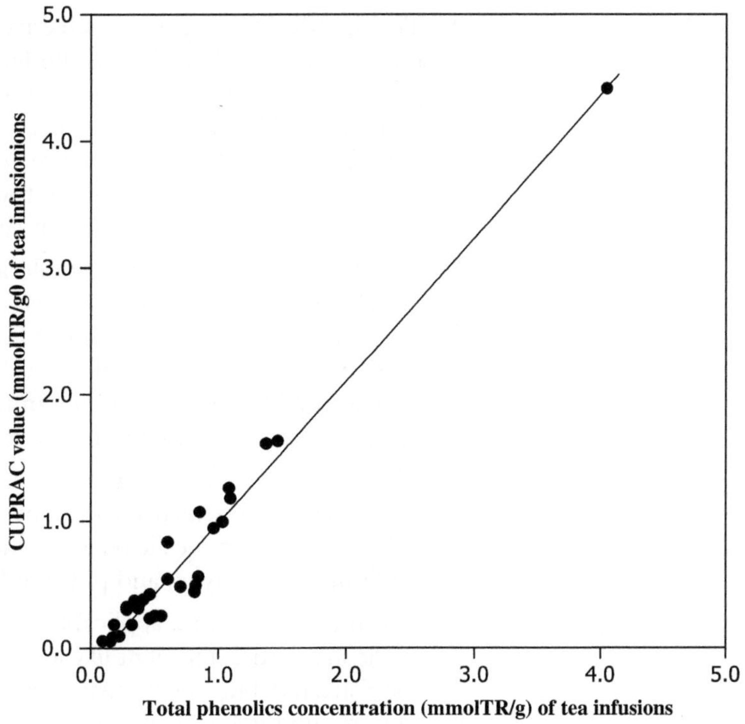

Fig. 14.8. The correlation of CUPRAC assay results with total phenolics content.

a high correlation coefficient (r) of 0.966, showing that the CUPRAC assay results correlate well with total phenolics content of herbal teas (*see* **Note 10**), constituting a significant improvement over other antioxidant assay methods in literature since most of the observed antioxidant capacity of plant extracts is a direct outcome of polyphenolic compounds.

4. Notes

The advantages of the CUPRAC method over other similar assays are summarized below:

1. The CUPRAC reagent is fast enough to oxidize thiol-type antioxidants *(2, 27)*, whereas according to the protocol developed by Benzie and Strain *(28)*, the FRAP method may only measure with serious negative error certain thiol-type antioxidants like glutathione (i.e., the major low molecular-weight thiol compound of the living cell). The CUPRAC assay also responds much faster than FRAP to certain hydroxycinnamic acids. The possible reason for this with respect to electronic configurations is the kinetic inertness of high-spin d^5-Fe(III) having half-filled d-orbitals, while CUPRAC utilizing d^9-Cu(II) oxidant involves faster kinetics.
2. The CUPRAC reagent is selective, because it has a lower redox potential than that of the ferric–ferrous couple in the presence of phenanthroline or other Fe(II)-stabilizing similar ligands. The standard potential of the Cu(II,I)-Nc redox couple is about 0.6 V (4), close to that of $ABTS^{+}$/ABTS, i.e., 0.68 V. Simple sugars and citric acid, which are not true antioxidants, are not oxidized in the CUPRAC method.
3. The reagent is much more stable and easily accessible than the chromogenic radical reagents (e.g., ABTS, DPPH). The cupric reducing ability measured for a complex sample (such as a food extract) may indirectly but efficiently reflect the total antioxidant power of the sample even though no major radicalic species is utilized in the assay.
4. The redox reaction giving rise to a colored chelate of Cu(I)-Nc is relatively insensitive to a number of parameters adversely affecting certain radicalic reagents such as DPPH *(29)*, i.e., air, sunlight, solvent type, and pH, the latter to a certain extent.
5. The method is easily applicable to conventional laboratories using standard equipment like a colorimeter rather than more sophisticated but costly instrumental techniques of analysis. The method involves minimal sample preparation, the experimental procedure is flexible, and suitable for automation.

6. The CUPRAC absorbance *vs* concentration curves are perfectly linear over a wide concentration range, unlike those of other methods yielding polynomial curves. The method perfectly complies with dilution, as the absorbance *vs* concentration curves of diluted extracts pass through the origin *(10)*. The molar absorptivity (e.g., $\varepsilon = 7.3 \times 10^4$ L $mol^{-1}cm^{-1}$ for quercetin) is sufficiently high to sensitively determine most phenolic antioxidants. Preconcentration of the CUPRAC chromophore using a weakly acidic cation-exchange resin is also possible for very sensitive applications *(30)*. The within-run and between-run coefficients of variation (CV) of the CUPRAC method (0.7 and 1.5%) are much lower than those of most methods that find wide use in total antioxidant assays *(2)*.
7. The total antioxidant capacity (TAC) values of antioxidants found with CUPRAC are perfectly additive, i.e., the TAC of a phenolic mixture is equal to the sum of TAC values of its constituent polyphenols. The parallelism of the linear calibration curves of pure antioxidants in water and in a given complex matrix (such as a food extract) demonstrates that there are no chemical interactions of interferent nature among the solution constituents, and that the antioxidant capacities of the tested antioxidants are additive, in conformity to the Beer's law.
8. The CUPRAC redox reaction producing colored species is carried out at nearly physiological pH (pH 7 of ammonium acetate buffer) as opposed to the acidic conditions (pH 3.6) of FRAP, or to the basic conditions (pH 10) of Folin-Ciocalteu (FCR) assay. At more acidic conditions than the physiological pH, the reducing capacity may be suppressed due to protonation on phenolics, whereas at more basic conditions, proton dissociation of phenolics (converted into phenolates) would enhance a sample's reducing capacity. Thus, the CUPRAC assay working at a pH close to that of physiological fluids gives a realistic estimate of antioxidants in a sample.
9. The CUPRAC procedure can be applied to all kinds of dried fruits previously treated with sulfur dioxide for safe drying, where SO_2 is an antibacterial agent and antioxidant. Sulfite residues remaining in dried fruits interfere with most antioxidant assays since the redox potential of the CUPRAC and similar (e.g., ABTS, Folin, FRAP) oxidizing reagents is sufficient to oxidize sulfite to sulfate (*see* **Note 2**). Thus the hydrosulfite (HSO_3^-) anion should first be separated from the antioxidant solution at pH 3 on an anion exchanger column, and then the standard CUPRAC assay should be applied to the column eluate *(8)*.

10. The method proves to correlate well with ABTS or Folin-Ciocalteu assays in herbal plant infusions *(10)* and apricot extracts *(8)* since all these assays work with the same mechanism of electron-transfer.

Acknowledgments

The authors would like to express their gratitude to Istanbul University Research Fund, Bilimsel Arastirma Projeleri Yurutucu Sekreterligi, for the funding of Project YOP-4/27052004, and to State Planning Organization of Turkey for the Advanced Research Project of Istanbul University (2005K120430). The authors would like to extend their gratitude to TUBITAK (Turkish Scientific and Technical Research Council) for the Research Projects 105T402 and 106T514.

References

1. Apak, R., Güçlü, K., Özyürek, M., and Karademir, S. E. (2004) A novel total antioxidant capacity index for dietary polyphenols, vitamins c and e, using their cupric ion reducing capability in the presence of neocuproine: CUPRAC method. *J. Agric. Food Chem.* **52**, 7970–7981.
2. Apak, R., Güçlü, K., Özyürek, M., Karademir, S. E., and Altun, M. (2005) total antioxidant capacity assay of human serum using copper(II)-neocuproine as chromogenic oxidant: The CUPRAC method. *Free Radic. Res.* **39**, 949–961.
3. Apak, R., Güçlü, K., Demirata, B., Özyürek, M., Çelik, S. E., Bektaşoğlu, B., Berker, K. I., and Özyurt, D. (2007) Comparative evaluation of total antioxidant capacity assays applied to phenolic compounds, and the CUPRAC assay. *Molecules* **12**, 1496–1547.
4. Tütem, E., Apak, R., and Baykut, F. (1991) Spectrophotometric determination of trace amounts of copper(I) and reducing agents with neocuproine in the presence of copper(II). *Analyst* **116**, 89–94.
5. Özyürek, M., Güçlü, K., Bektaşoğlu, B., and Apak, R. (2007) Spectrophotometric determination of ascorbic acid by the modified CUPRAC method with extractive separation of flavonoids-La(III) complexes. *Anal. Chim. Acta* **588**, 88–95.
6. Association of Official Analytical Chemists (1990) K. Helrick, *Official Methods of Analysis* (15th edn.), AOAC, Food Composition, Additives, Natural Contaminants, Washington, DC, **2**, p. 1059.
7. Lykkesfeldt, J. (2000) Determination of ascorbic acid and dehydroascorbic acid in biological samples by high-performance liquid chromatography using subtraction methods: reliable reduction with tris[2-carboxyethyl]phosphine hydrochloride. *Anal. Biochem.* **282**, 89–93.
8. Güçlü, K., Altun, M., Özyürek, M., Karademir, S. E., and Apak, R. (2006) Antioxidant capacity of fresh, sun- and sulfited-dried Malatya apricot (*Prunus armeniaca*) assayed by CUPRAC, ABTS/TEAC and Folin methods. *Int. J. Food Sci. Technol.* **41**, 76–85.
9. Garcia-Alonso, M., Pascual-Teresa, S., Santos-Buelga, C., and Rivas-Gonzalo, J. C. (2004) Evaluation of the antioxidant properties of fruits. *Food Chem.* **84**, 13–18.
10. Apak, R., Güçlü, K., Özyürek, M., Karademir, S. E., and Erçağ, E. (2006) The cupric ion reducing antioxidant capacity (CUPRAC) and polyphenolic content of some herbal teas. *Int. J. Food Sci. Nutr.* **57**, 292–304.
11. Re, R., Pellegrini, N., Proteggente, A., Pannala, A., Yang, M., and Rice-Evans, C. (1999) Antioxidant activity applying an improved ABTS radical cation decolorization assay. *Free Radic. Biol. Med.* **26**, 1231–1237.
12. Singleton, V. L., Orthofer, R., and Lamuela-Raventos, R. M. (1999) Analysis of total phenols and other oxidation substrates and

antioxidants by means of Folin-Ciocalteu reagent. *Meth. Enzymol.* **299**, 152–178.

13. Singleton, V. L., and Rossi, J. A. (1965) Colorimetry of total phenolics with phosphomolybdic-phosphotungstic acid reagent. *Am. J. Enol. Vitic.* **16**, 144–158.
14. Bors, W., Hellers, W., Michel, C., and Saran, M. (1990) Radical Chemistry of Flavonoid Antioxidants. In: I. Emerit, L. Packer and C. Auclair (Eds.), *Antioxidants in Therapy and Preventive Medicine*, Plenum Press, New York, **1**, 165–170.
15. Rice-Evans, C. A., Miller, N. J., and Paganga, G. (1997) Antioxidant properties of phenolic compounds. *Trends Plant Sci.* **2**, 152–159.
16. Castelluccio, C., Bolwell, G. P., Gerrish C., and Rice-Evans, C. A. (1996) Differential distribution of ferulic acid to the major plasma constituents in relation to its potential as an antioxidant. *Biochem. J.* **316**, 691–694.
17. Kanski, J., Aksenova, M., Stoyanova, A., and Butterfield, D. A. (2002) Ferulic acid antioxidant protection against hydroxyl and peroxyl radical oxidation in synaptosomal and neuronal cell culture systems *in vitro*: structure-activity studies. *J. Nutr. Biochem.* **13**, 273–281.
18. Tepe, B., Eminagaoglu, O., Akpulat, H. A., and Aydin, E. (2007) Antioxidant potentials of rosmarinic acid levels of methanolic extracts of *Salyvia verticillata* (L.) subsp. *verticillata* and *S. verticillata* (L.) subsp. *amasiaca* (Freyn & Bornm.) Bornm. *Food Chem.* **100**, 985–989.
19. Chen, J. H., and Ho, C. -T. (1997) Antioxidant activities of caffeic acid and its related hydroxycinnamic acid compounds. *J. Agric. Food Chem.* **45**, 2374–2378.
20. Cervellati, R., Renzulli, C., Guerra, M. C., and Speroni, E. (2002) Evaluation of antioxidant activity of some natural polyphenolic compounds using the briggs-rauscher reaction method. *J. Agric. Food Chem.* **50**, 7504–7509.
21. Kim, D. -O., and Lee, C. Y. (2004) Comprehensive study on vitamin C equivalent antioxidant capacity (VCEAC) of various polyphenolics in scavenging a free radical and its structural relationship. *Critic. Rev. Food Sci. Nutr.* **44**, 253–273.
22. Miliauskas, G., van Beek, T. A., Venskutonis, P. R., Linssen, J. P. H., and de Waard, P. (2004) Antioxidative activity of *Geranium Macrorrhizum. Eur. Food Res. Technol.* **218**, 253–261.
23. Firuzi, O., Lacanna, A., Petrucci, R., Marrosu, G., and Saso, L. (2005) Evaluation of the antioxidant activity of flavonoids by "ferric reducing antioxidant power" assay and cyclic voltammetry. *Biochim. Biophys. Acta.* **1721**, 174–184.
24. Santos-Buelga, C., and Scalbert, A. (2000) Proanthocyanidins and tannin-like compounds: nature, occurence, dietary intake and effects on nutrition and health (review). *J. Sci. Food Agric.* **80**, 1094–1117.
25. Asma, B. M. (2000) Kayisi yetistiriciligi (Apricot Growing). Malatya, Turkey: Evin Publishers.
26. Halvorsen, B. L., Holte, K., Myhrstad, M. C. W., Barikmo, I., Hvattum, E., Remberg, S. V., et al. (2002). A systematic screening of total antioxidants in dietary plants. *J. Nutr.* **132**, 461–471.
27. Tütem, E., and Apak, R. (1991) Simultaneous spectrophotometric determination of cystine and cysteine in amino acid mixtures using copper(II)-neocuproine reagent. *Anal. Chim. Acta* **255**, 121–125.
28. Benzie, I. F. F., and Strain, J. J. (1996) The ferric reducing ability of plasma (FRAP) as a measure of "antioxidant power": the FRAP assay. *Anal. Biochem.* **239**, 70–76.
29. Özçelik, B., Lee, J. H., and Min, D. B. (2003) Effects of light, oxygen, and pH on the absorbance of 2,2-diphenyl-1-picrylhydrazyl. *J. Food Sci.* **68**, 487–490.
30. Özyürek, M., Çelik, S. E., Berker, K. I., Güçlü, K., Tor, I., and Apak, R. (2007) Sensitivity enhancement of CUPRAC and iron(III)-phenanthroline antioxidant assays by preconcentration of colored reaction products on a weakly acidic cation exchanger. *React. Func. Polym.* **67**, 1478–1486.

Chapter 15

Redox Property of Ribonucleotide Reductase Small Subunit M2 and p53R2

Xiyong Liu, Lijun Xue, and Yun Yen

Abstract

Human ribonucleotide reductase (RR) small subunits, M2 and P53R2, play key roles in forming RR holoenzyme and supplying nucleotide precursors for DNA replication and repair. Currently, we are studying the redox property, structure, and function of hRRM2 and p53R2. In the cell-free system, p53R2 did not oxidize a reactive oxygen species (ROS) indicator Carboxy-H_2DCFDA, but hRRM2 did. Further studies demonstrated that purified recombinant p53R2 protein has the catalase activity to scavenge H_2O_2. Over-expression of p53R2 reduced intracellular ROS and protected the mitochondrial membrane potential against oxidative stress, whereas over-expression of hRRM2 did not result in the collapse of mitochondrial membrane potential. Our findings suggest that p53R2 may play a key role in defending oxidative stress by scavenging ROS, and this antioxidant property is also important for its enzymatic activity.

Key words: Ribonucleotide reductase, p53R2, Antioxidant.

1. Introduction

It is well known that the basic structure and function of ribonucleotide reductase (RR) is closely linked to its redox state. In *E. Coli*, active B2 (RR small subunit) harbors the iron center and the radical in oxidized states *(1)*. However, oxido-reduction studies in *E. Coli* and mouse showed the existence of a stable, fully reduced form of the B2 protein *(2, 3)*. The B2 protein shows dynamic carboxylate, radical, and water shifts in different redox forms *(4)*. Similarly, cytochrome C oxidase, which has redox active tyrosine in the binuclear center, participates in reducing oxygen in respiration *(5, 6)*, suggesting that proteins

From: *Methods in Molecular Biology, vol. 477: Advanced Protocols in Oxidative Stress I,*
Edited by: D. Armstrong, DOI: 10.1007/978-1-60327-517-0_15, © Humana Press, New York, NY

possessing tyrosyl radicals with a binuclear center may function as reducing or oxidizing reagents.

hRRM2 and p53R2, two small subunits of human RR, belong to class I RR, which have a functional motif: a tyrosyl radical coupled with μ-oxo-diiron (III) cluster *(4)*. Overexpression of hRRM2 can have significant effects on the biological properties of cells, tumor development, metastasis, and drug resistance *(7)*, but hydroxyurea, an inhibitor of hRRM2, can protect the cell from H_2O_2-induced DNA damage through free radical scavenging activity *(8)*. Conversely, p53R2, induced by tumor suppressor p53, may be important for supplying deoxyribonucleotides to the DNA damage repair system. For example, p53R2-deficient mice showed more susceptibility to oxidative stress than wild-type mice *(9)*. Therefore, p53R2 may have a role that is distinct from the role of hRRM2 in regulating a cellular redox state, but the redox property and functions of p53R2 remain unclear. Here we compare the redox properties of p53R2 and hRRM2 proteins, and clarify the structure and function of p53R2 relative to its redox properties.

2. Materials

2.1. Equipment

1. Vertical Laminar flow hoods: SterilGARD Hood, The Baker company, Inc. (Sanford, ME).
2. CO_2 Incubator: Model 3158 S/N 32083-3069 Forma Scientific, Inc. (Marietta, OH).
3. Inverted phase contrast microscope: Leitz Diavert (New York, NY).
4. Liquid nitrogen tank: CRYO Biological Storage System 4753/ Locator 8 Therolyne (Lambertville, NJ).
5. Incubator shaker series: New Brunswick Scientific/Excella E25 (Edison, NJ).
6. Eppendorf Multiporator (Westbury, NY).
7. PCR: MJ Research/DTC-100™ Programmable Thermal Controller (Waltham, MA).
8. *f* Max microplate reader (Molecular Devices, Sunnyvale, CA).
9. Invitrogen Novex Mini-cell (Carlsbad, CA).
10. Eppendorf centrifuge 5417R (Westbury, NY).
11. Beckman J2-21M/E centrifuge (Fullerton, CA).
12. Precision water bath (Waltham, MA).

2.2. Reagents and Buffers

2.2.1. Cell Culture and Lysis

1. RPMI 1640 Medium (Gibco/BRL, Bethesda, MD) supplemented with 10% fetal bovine serum (FBS Hyclone, Ogden, UT) and 1% penicillin/streptomycin (Gibco/BRL, Bethesda, MD).
2. Trypsin solution, 0.25%, and ehtylaenediamine tetraacetic acid (EDTA), 1 mM, are from Gibco/BRL.
3. Geneticine (Gibco/BRL, Bethesda, MD) is dissolved in tissue culture water at 100 mg/ml and stored at −20°C, and then added 300 μg/ml for stable clone selection in KB cells.
4. Cell lysis buffer: 75 mM Tris-HCl, pH 6.8, 1.5%, w/v, sodium dodecyl sulfate (SDS), 7.5%, w/v, glycerol, 200 mM β-mercaptotethanol, 0.03%, w/v, bromophenol blue, 0.003%, w/v, pyronin-Y. Store in aliquots at −20°C.
5. Teflon cell scrapers (Fisher, Pittsburg, PA).

2.2.2. Mammalian Expression Plasmid Construction

1. QIAconnect RNA to cDNA Kit (Qiagen, Valencia, CA) is used for RNA extraction and production of cDNA, which includes RNeasy Mini Kit (Qiagen, Valencia, CA), and QuantiTect Reverse Transcription Kit (Qiagen, Valencia, CA).
2. QIAGEN Plasmid Maxi Kit (Qiagen, Valencia, CA) is used for plasmid isolation.
3. QIAGEN PCR Cloning Kit contains: 2× Ligation Master Mix (50 μl), pDrive Cloning Vector (0.5 μg) and distilled water (1.7 ml).
4. pcDNA 3.1 (+) vector is commercially available from Invitrogene (Carlsbad, CA).
5. Prokaryotic expression vector pET28 (Novagen, Madison, WI).
6. Primers were designed and synthesized by City of Hope Core-facility:
   a. hRRM2: 5′-ATCCGGATCCACTATGCTCTCCCTC-CGTGT
      3′-GCTTAAGCTTATTTAGAAGTCAGCATCCAAG.
   b. p53R2: 5′-TCGGATCCATGGGCGACCCGGAAAGG
      3′-GCGGCCGCTTAAAAATCTGCATCCAA.
7. Luria broth (LB) medium containing 30 μg/ml kanamycin is used to grow transformed bacteria.

2.2.3. Protein Expression and Purification

1. Recombinant human RRM2 and p53R2 proteins were expressed using the bacterial strain BL21 (DE3) (Stratagene, La Jolla, CA).
2. Ni^{2+}-resin (Qiagen, Valencia, CA) affinity chromatography column was used to purify the recombinant proteins.

3. Luria broth medium containing 30 μg/ml kanamycin is used for transformed bacterial growth.
4. Isopropyl-1-thio-β-D-galactopyranoside (TPTG), 1 mM, is used for inducing protein expression.
5. Benzonase nuclease (Qiagen, Valencia, CA).
6. Washing buffer: 50 mM NaH_2PO_4, 800 mM NaCl, 50 mM imidazole, pH 7.0, 0.1% Triton X-100, and 10 mM 2-mercaptoethanol.
7. Dilution buffer containing 50 mm NaH_2PO_4, 300 mm NaCl, and 125 mm imidazole, pH 7.0.
8. Dialysis buffer: 25 mM Tris-HCl, pH 7.4.

2.2.4. Fluorescence Assays of Carboxy-H_2DCFDA Oxidation

1. Carboxy-H_2DCFDA (Molecular Probes, Eugene, OR) is an indicator for reactive oxygen species (ROS).
2. Esterases (purified from porcine liver) (Sigma, St. Louis, MO) was used as control.
3. Mitochondrial lysate was extracted from KB cell by using Mitochondrial extract kit (Imgenex, San Diego, CA).
4. 5× reaction buffer containing 0.5 M Tris-HCl, pH 7.5.
5. Hank's balanced salt solution (HBSS).

2.2.5. Catalase Activity

Catalase activity was determined using Amplex® Red Catalase Assay Kit (Molecular Probes, Eugene, OR).

1. Amplex Red reagent (MW = 257, Component A). Just prior to use, dissolve the contents of the vial of Amplex Red reagent (0.26 mg) in 100 μl DMSO (*see* **Note 1**).
2. Dimethylsulfoxide (DMSO), anhydrous (Component B), 500 μl.
3. Horseradish peroxidase (Component C), 20 U, where 1 unit is defined as the amount of enzyme that will generate 1.0 mg purpurogallin from pyrogallol within 20 s at 20°C, pH 6.0. Prepare a 100-U/ml solution of horseradish peroxidase (HRP) by dissolving the contents of the vial containing HRP (Component C) in 200 μl of 1× reaction buffer, 0.1 M Tris-HCl, pH 7.5 (*see* **Note 2**).
4. Hydrogen peroxide (H_2O_2) (MW = 34, Component D), 500 μl of a stabilized ~3% solution; the actual concentration is indicated on the component label.
5. Prepare a 1× working solution of reaction buffer by adding 4 ml of 5× reaction buffer, 0.5 M Tris-HCl, pH 7.5 (Component E) to 16 ml of deionized water (dH_2O).
6. Catalase (Component F), 100 U, where 1 unit is defined as the amount of enzyme that will decompose 1.0 μmole of H_2O_2 per minute at 25°C, pH 7.0. Prepare a 1000-U/ml

solution of catalase by dissolving the contents of the vial containing catalase (Component F) in 100 μl of dH_2O.

2.2.6. Inner Mitochondrial Membrane Potential Analysis

1. JC-1 dye (Molecular Probes, Eugene, OR) was dissolved in DMSO to produce a 1-mg/ml stock solution in our study.
2. Cell culture material is same as **Section 2.1**.

3. Methods

3.1. RRM2 and p53R2 Expression Vector Construction, Protein Expression and Purification

1. Total RNA was isolated from human oropharyngeal carcinoma KB cells using QIAGEN RNeasy Mini Kit, and converted into cDNA by using QuantiTect Reverse Transcription Kit. The coding sequences of human p53R2 and hRRM2 proteins were obtained by reverse transcription-PCR using the corresponding primer pairs (indicated in **Section 2.2 4**).
2. The PCR products were purified and inserted into the pDrive cloning vector (Qiagen) via U-A ligation.
3. After restriction digestion, the coding sequence of each RR subunit was then cloned in-frame with an NH2-terminal 6× His-tag into the prokaryotic expression vector pET28 (Novagen).
4. All constructs (pET28a-p53R2 and pET28a-hRRM2) were verified by DNA sequencing.
5. An overnight culture of the transformed BL21 (DE3) bacteria was diluted 50-fold in 1 l of LB medium containing 30 μg/ml kanamycin.
6. BL21 (DE3) was transfected and grown at 37°C for 4 h.
7. Induced by 1 mM IPTG for an additional 3 h at 30°C.
8. To harvest, pellet cell via centrifugation.
9. Cell pellets were disrupted by incubating with Bugbuster and benzonase nuclease (Qiagen) at 4°C for 30 min with vigorous agitation, and the lysate was centrifuged at 16,000 × *g* for 30 min at 4°C.
10. The supernatant was incubated with Ni^{2+}-resin at 4°C for 30 min, loaded onto a column, washed with at least 30-fold bed volume of a buffer containing 50 mM NaH_2PO_4, 800 mM NaCl, 50 mM imidazole, pH 7.0, 0.1% Triton X-100, and 10 mM 2-mercaptoethanol.
11. Elute with a buffer containing 50 mM NaH_2PO_4, 300 mM NaCl, and 125 mM imidazole, pH 7.0.

12. The eluates were thoroughly dialyzed against 2000 sample volumes of a buffer containing 25 mM Tris-HCl, pH 7.4, at 4°C overnight.
13. Protein concentration was determined using the Bio-Rad Protein Assay kit. Protein purity was determined by densitometric scanning of the Coomassie-stained SDS–PAGE gel. For immunoblotting, protein samples were analyzed using goat anti-p53R2 and anti-hRRM2 antibodies (Santa Cruz Biotechnology).
14. The purified proteins aliquoted and stored at −70°C (*see* **Note 3**).

3.2. Determining the Ability of RR Small Subunits to Oxidize Carboxy-H2DCFDA In vitro

The recombinant human p53R2 and hRRM2 proteins were purified and immediately used to evaluate their ROS generating activity in a cell-free system containing esterase or mitochondrial extract. Carboxy-H_2DCFDA (Molecular Probes) is an indicator for reactive oxygen species (ROS) that does not fluoresce until hydrolyzed by esterase and oxidized. Without esterase, Carboxy-H_2DCFDA is not sensitive to oxidants *(9)*. For functional assays in a cell-free system, esterases inherent in mitochondrial extracts or commercially purchased esterase (Sigma) were added. The fluorescence intensity (FI) of Carboxy-H_2DCF in the presence of p53R2 or hRRM2 without adding the esterase was considered background (**Fig. 15.1A**).

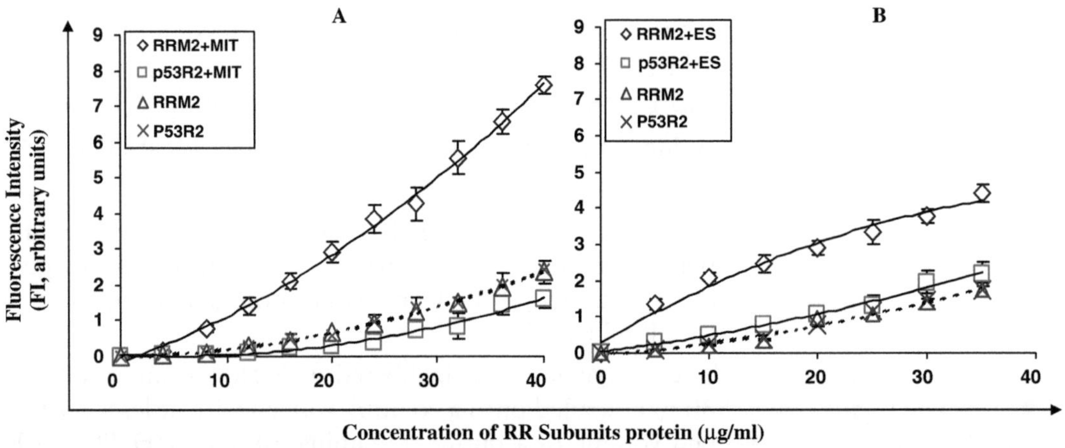

Fig. 15.1. Recombinant purified p53R2 lacked ability to oxidize Carboxy-H_2DCFDA. Carboxy-H_2DCFDA (10 μM) was incubated at 37°C with mitochondrial lysate of KB cells (MIT) **(A)** or esterases purchased (ES) **(B)** for 30 min in the presence of purified proteins hRRM2 or p53R2. Control incubations contained hRRM2 or p53R2 but no MIT (in **A**) or no ES (in **B**). Carboxy-DCF generation was monitored fluorometrically. Data shown are the mean based on three independent experiments, each carried out in triplicate.

1. Add 40 μl 5× reaction buffer containing 0.5 M Tris-HCl, pH 7.5.
2. Add 10 μM carboxy-H_2DCFDA.
3. Add 240 μg/ml of mitochondrial lysate (**Fig. 15.1A**) or 75 μg/ml of esterase enzyme(**Fig. 15.1B**).
4. Add 0, 5, 10, 15, 20, 25, 30, 35 μg/ml recombinant ribonucleotide reductase small subunits p53R2 or M2.
5. Add H_2O to make total volume at 200 μl.
6. Incubate at 37°C for 30 min.
7. The fluorescence of the oxidized form of carboxy-H_2DCFDA was measured using a *f*Max microplate reader (Molecular Devices) with a fluorescence excitation at 485 nm and emission at 538 nm.

After adding the mitochondrial extract, FI of p53R2 was unchanged at 4–40 μg/ml range; whereas the FI of hRRM2 significantly increased. Furthermore, the FI of p53R2 was much lower than that of hRRM2 at 6% and 20% of hRRM2 at 8 μg/ml and 40 μg/ml, respectively (**Fig. 15.1A**). These results suggest that hRRM2 is able to generate ROS in the presence of mitochondrial extract, but p53R2 is not. Similarly, when mitochondrial extract was replaced with pure esterase, the FI of p53R2 was consistently close to background and lower than that of hRRM2 (**Fig. 15.1B**), suggesting that p53R2 does not have the intrinsic ability to oxidize Carboxy-H_2DCF. However, larger differences in FI were observed between hRRM2 and p53R2 in Assay 1 than in Assay 2, suggesting the mitochondrial environment may influence redox properties of p53R2 and hRRM2.

3.3. Determine Scavenge H_2O_2 Ability of p53R2

The inability of p53R2 to oxidize carboxy-H2DCFDA indicates that p53R2 might be an antioxidant. To explore this hypothesis, experiments were designed to determine the catalase activity of purified p53R2 in breaking down H_2O_2. The specific catalase activity of p53R2 was derived from the standard curve of catalase activity and increased in a dose-dependent manner (**Fig. 15.2**).

1. Prepare a catalase standard curve: Dilute an appropriate amount of the 1000 U/ml catalase solution into 1× reaction buffer to produce catalase concentrations of 0–4.0 U/ml. Use 1× reaction buffer without catalase as a negative control. A volume of 25 μl will be used for each reaction.
2. Dilute the catalase-containing samples in 1× reaction buffer. A total volume of 25 μl will be used for each reaction (*see* **Note 4**).
3. Pipette 25 μl of the diluted experimental samples, standard curve samples, and controls into separate wells of a 96-well microplate.

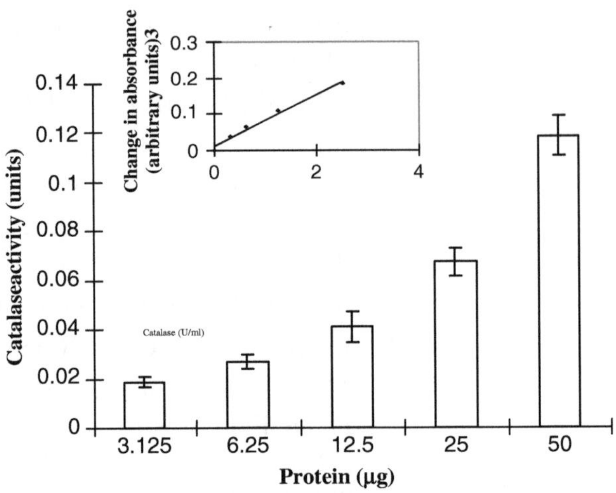

Fig. 15.2. P53R2 scavenged H_2O_2. Insert, catalase standard curve. Catalase was detected using the Amplex Red reagent-based assay. Initially, each reaction contained the indicated amounts of catalase and 20 μM H_2O_2 in 1× reaction buffer and was incubated for 30 min. The final reaction contained 50 μM Amplex Red reagent and 0.2 U/ml HRP and was incubated at 37°C. After 30 min, absorbance was measured in a microplate reader using absorbance at 550 nm. Change in absorbance is reported as the observed absorbance intensity subtracted from that of a no-catalase control. The absorbance intensity of the indicated amount of p53R2 was detected as for a catalase standard curve. The absorbance intensity of p53R2 was then changed to catalase-specific activity according to a catalase standard curve. Catalase activity, where 1 unit is defined as the amount of enzyme that will decompose 1.0 μmole of H_2O_2 per minute at pH 7.0 at 25°C. Data shown are the mean ± SE based on three independent experiments, each carried out in quadruplicate.

4. Prepare a 40-μM H_2O_2 solution by adding 10 μl of the 20 mM H_2O_2 solution to 4.99 ml 1× reaction buffer.
5. Pipette 25 μl of the 40 μM H_2O_2 solution to each microplate well containing the samples and controls.
6. Incubate the reaction for 30 min at room temperature.
7. Prepare a working solution of 100 μM Amplex Red reagent containing 0.4 U/ml HRP by adding 50 μl of the Amplex Red reagent stock solution and 20 μl of the HRP stock solution to 4.93 ml 1× Reaction Buffer. This 5 ml volume is sufficient for ~100 assays (*see* **Note 5**).
8. Begin the second phase of the reaction by adding 50 μl of the Amplex Red/HRP working solution to each microplate well containing the samples and controls.
9. Incubate the reaction for 30 min or longer at 37°C, protected from light.
10. The fluorescence of the oxidized form of Carboxy-H2DCFDA was measured using a *f*Max microplate reader

(Molecular Devices) with a fluorescence excitation of 485 nm and emission at 538 nm (*see* **Note 6**).

11. Report the change in fluorescence or absorbance by subtracting the sample value from that of the no-catalase control.

The result yielded from this study (in **Fig. 15.2**) suggests that p53R2 can directly break down H_2O_2.

3.4. Inner Mitochondrial Membrane Potential Analysis

Because mitochondria is particularly susceptible to oxidative damage, leading to decrease of mitochondrial membrane potential ($\Delta\Psi_{mt}$) (10), we investigated the influence of both p53R2 and hRRM2 on $\Delta\Psi_{mt}$ in response to H_2O_2 in KB-wt (parental KB cell), KB-p53R2S (p53R2 sense transfectant), KB-M2S (hRRM2 sense transfectant), KB-p53R2AS (p53R2 antisense transfectant), and KB-M2AS (hRRM2 antisense transfectant) clones with JC-1. JC-1 is $\Delta\Psi_{mt}$ sensitive probe, which forms monomers (green fluorescence) at a low $\Delta\Psi_{mt}$ and J-aggregates (red fluorescence) at a higher $\Delta\Psi_{mt}$ (11). Expression of p53R2 increased three-fold with KB-p53R2S compared to the KB-wt or KB-vector, whereas expression of p53R2 decreased with KB-p53R2AS compared to the control (**Fig. 15.3A**). hRRM2 protein decreased under IPTG induction in KB-M2AS clone (**Fig. 15.3B**). KB-M2S clone demonstrated over-expression of the hRRM2 RNA and protein (data not shown). The change of $\Delta\Psi_{mt}$ was indicated as the ratio of red/green fluorescence.

1. JC-1 dye (Molecular Probes) was dissolved in DMSO to prepare a 1-mg/ml stock solution.
2. Cells were incubated with 1 μg/ml JC-1 for 30 min at 37°C (*see* **Note 7**).
3. Trypsinized, resuspended in medium at a density of approximately 10^6 cells/ml, and transferred on ice to a flow cytometer (MoFlo MLS, Dako, Carpinteria, CA).
4. JC-1 was excited at 488 nm and the monomer emission signal (green) was analyzed at 525 nm. Simultaneously, the aggregate signal (red) was analyzed at 590 nm. The distribution of red and green fluorescence from JC-1 was displayed in two-color contour plots. The ratio between red and green signals was measured.

In KB-wt cells, treatment with H_2O_2 (250 μM) resulted in shifting of the ratios toward lower values when compared with untreated cells (**Fig. 15.3C**, KB-wt). A threshold ratio of 100 severs to divide the cells into two distinct groups, one representing the untreated cells with >100 ratios and the other representing cells with injured mitochondria with <100 ratios. On the basis of this distinction, we designated a threshold ratio of 100 to distinguish between cells with intact (high ratio, designated as R2 cells) and injured mitochondria (low ratio, designated as R1

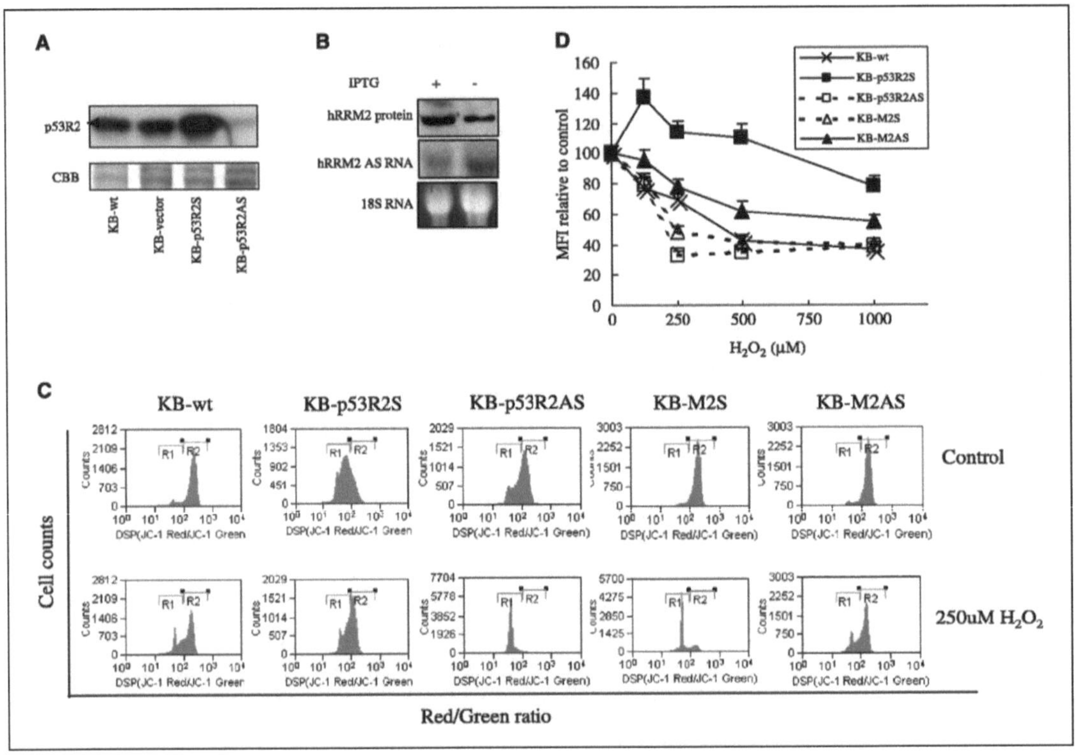

Fig. 15.3. P53R2 protected mitochondrial membrane from H_2O_2 attack. **(A)** Expression of p53R2 protein in KB-wt cell and KB-vector (pcDNA3.1 vector), KB-p53R2S, and KB-p53R2AS clones. **(B)** Expression of antisense hRRM2 was induced by 5 mM IPTG. The hRRM2 protein and hRRM2 AS RNA level were determined here with 18S RNA control. **(C)** Histogram distribution of red/green fluorescence ratios in KB-wt, KB-p53R2S, KB-p53R2AS, KB-M2S, and KB-M2AS cells after treatment with 250 μM of H_2O_2 for 24 h. Cells were stained with 1 μg/ml JC-1 for 30 min at 37°C before harvest. Histogram distribution showed two populations (R1 and R2) of cells with intact (high ratio, designated as R2 cells) and injured mitochondria (low ratio, designated as R1 cells). **(D)** Variations of the mean red/green fluorescence intensity ratio (MFI) as a function of the H_2O_2 concentration. Data shown are the mean ± SE based on three independent experiments.

cells). Treatment with H_2O_2 (250 μM) resulted in the cellular shift from R2 to R1 when compared to the respective untreated clones, in contrast to KB-p53R2S cells, which shifted from R1 to R2 (**Fig. 15.3C**). To clearly demonstrate the effect of p53R2 and hRRM2 on $\Delta\Psi_{mt}$ induced by H_2O_2, the mean fluorescence intensity ratio (MFI) of the red/green ratio was plotted as a function of H_2O_2 concentration (**Fig. 15.3D**). KB-p53R2S and KB-M2AS cells were more resistant to the H_2O_2 stress than KB-wt cells as indicated by the increase in MFI, whereas KB-p53R2AS and KB-M2S cells were more sensitive to H_2O_2 stress than KB-wt cells. The results showed that p53R2 protects the mitochondrial membrane against oxidative stress. It is also consistent with the study using p53R2 knockout mouse embryonic fibroblasts (MEFs) in which MEFs containing p53R2 (p532R2-wt) had higher viability against H_2O_2 stress than mutant MEFs (p53R2-KO) *(12)*, although the difference of viability between p53R2-wt and p53R2-KO cells was much higher than the difference of their DNA repair ability. Our results may explain this variance, as p53R2 protects cells against the loss of $\Delta\psi$ from H_2O_2 stress by decreasing ROS. This idea is supported by our results of catalase activity displayed by recombinant p53R2 protein (**Fig. 15.2A**), and intracellular ROS was reduced in p53R2 over-expressing cells compared to control cells in response to H_2O_2 (**Fig. 15.2B**). These findings are in agreement with another study demonstrating the catalase activity of B2 protein from *E. Coli (13)*.

ROS are potentially dangerous by-products of cellular metabolism that have direct effects on cell growth and development and cell survival, and have a significant role in the pathogenesis of cancer. The antioxidant ability of p53R2 implies that p53R2 plays an important role in preventing cancer because, not only can p53R2 supply deoxyribonucleotides to the DNA repair system, but it can also scavenge ROS (this report) to maintain genomic integrity in response to oxidative stress.

4. Notes

1. This stock solution should be stored at −20°C, protected from light.
2. After use, the remaining solution should be divided into small aliquots and stored frozen at −20°C.
3. Do not defrost protein until use. p53R2 and hRRM2 have an identical tyrosyl radical; X-band electron paramagnetic resonance (EPR) spectra of purified hRRM2 and p53R2 were determined to measure the activity of proteins.
4. The samples' catalase concentrations will be four-fold lower in the final reaction volume.
5. The final concentrations for the Amplex Red reagent and the HRP will be two-fold lower in the final reaction volume.
6. Because the Amplex Red reaction is continuous (not terminated), fluorescence or absorbance may be measured at multiple time points to follow the kinetics of the reactions.
7. It is not necessary to change culture medium.

Acknowledgments

Grant support: NCI R01 Grant CA72767.

References

1. Reichard, P. and A. Ehrenberg (1983). Ribonucleotide reductase – a radical enzyme. *Science* **221**(4610): 514–9.
2. Sahlin, M., A. Graslund, et al. (1989). Reduced forms of the iron-containing small subunit of ribonucleotide reductase from *Escherichia coli*. *Biochemistry* **28**(6): 2618–25.
3. Davydov, A., P. P. Schmidt, et al. (1996). Reversible red-ox reactions of the diiron site in the mouse ribonucleotide reductase R2 protein. *Biochem Biophys Res Commun* **219**(1): 213–8.
4. Kolberg, M., K. R. Strand, et al. (2004). Structure, function, and mechanism of ribonucleotide reductases. *Biochim Biophys Acta* **1699**(1–2): 1–34.
5. Proshlyakov, D. A., M. A. Pressler, et al. (1998). Dioxygen activation and bond cleavage by mixed-valence cytochrome c oxidase. *Proc Natl Acad Sci U S A* **95**(14): 8020–5.
6. Proshlyakov, D. A., M. A. Pressler, et al. (2000). Oxygen activation and reduction in respiration: involvement of redox-active tyrosine 244. *Science* **290**(5496): 1588–91.
7. Fan, H., C. Villegas, et al. (1998). The mammalian ribonucleotide reductase R2 component cooperates with a variety of oncogenes in mechanisms of cellular transformation. *Cancer Res* **58**(8): 1650–3.
8. Rauko, P., D. Romanova, et al. (1997). DNA-protective activity of new ribonucleotide reductase inhibitors. *Anticancer Res* **17**(5A): 3437–40.
9. Kimura, T., S. Takeda, et al. (2003). Impaired function of p53R2 in Rrm2b-null mice causes severe renal failure through attenuation of dNTP pools. *Nat Genet* **34**(4): 440–5.
10. Sastre, J., F. V. Pallardo, et al. (2003). The role of mitochondrial oxidative stress in aging. *Free Radic Biol Med* **35**(1): 1–8.
11. Smiley, S. T., M. Reers, et al. (1991). Intracellular heterogeneity in mitochondrial membrane potentials revealed by a J-aggregate-forming lipophilic cation JC-1. *Proc Natl Acad Sci U S A* **88**(9): 3671–5.
12. Tanaka, H., H. Arakawa, et al. (2000). A ribonucleotide reductase gene involved in a p53-dependent cell-cycle checkpoint for DNA damage. *Nature* **404**(6773): 42–9.
13. Sahlin, M., B. M. Sjoberg, et al. (1990). Activation of the iron-containing B2 protein of ribonucleotide reductase by hydrogen peroxide. *Biochem Biophys Res Commun* **167**(2): 813–8.

Chapter 16

Antioxidant QSAR Modeling as Exemplified on Polyphenols

Bono Lučić, Dragan Amić, and Nenad Trinajstić

Abstract

Methodology for deriving quantitative structure-activity relationship (QSAR) models based on computed molecular descriptors, representing numerically structural features of polyphenols, and applicable to the antioxidant activity of polyphenols is delineated. The application of this methodology is illustrated on a data set of 100 polyphenols. Prior to the computation of molecular descriptors, molecular structures are coded in the SMILES form, a computer-acceptable version of structure, and then converted to the 3D form by the CORINA program. Using 3D structures, molecular descriptors can be calculated by one of several programs developed (we used the DRAGON program in this study). Finally, using computer program for selection of most important descriptors in the model, a two-descriptor model is selected and its use is illustrated.

Key words: Polyphenols, Flavonoids, Antioxidant activity, Free radical scavenging, QSAR, Quantitative structure-activity relationship, Molecular descriptors, TEAC assay, ABTS.

1. Introduction

Oxidative stress induced by reactive oxygen species (ROS) such as $O_2^{\bullet -}$, $OH^{\bullet}$, or lipid peroxyl radicals $LOO^{\bullet}$, can result in damaging of cellular proteins, nucleic acids, and lipids, which has been implicated in the pathogenesis of various diseases, including coronary heart disease and some form of cancer *(1)*. Flavonoids are a class of natural polyphenolic antioxidants capable of combating ROS by scavenging free radicals, chelating metal ions, inhibiting prooxidant enzymes, and activating antioxidant and detoxifying enzymes *(2, 3)*. Due to protective effects of flavonoids, they are recognized as potential candidates for use as drugs in illnesses such as cancer, atherosclerosis, cardiovascular, and coronary heart diseases and many other age-related diseases *(4)*.

From: *Methods in Molecular Biology, vol. 477: Advanced Protocols in Oxidative Stress I,*
Edited by: D. Armstrong, DOI: 10.1007/978-1-60327-517-0_16, © Humana Press, New York, NY

QSAR and QSPR (quantitative structure-property relationships) are theoretical models used to estimate or predict the physicochemical and biologically important activities or properties of molecules. Since 1964 and the paper by Hansch and Fujita *(5)*, this field has been developing into a science highly useful in the process of designing new compounds having desirable properties by QSAR driven modification of basic chemical structure. Forty years later, a set of rules that lead to standardization of QSAR protocols was defined by the scientific community working in this field through the work on the project supported by European Commission *(6)*.

The structural features (characterized by various kinds of molecular descriptors) of flavonoids (polyphenols) important for antioxidant activities can be used to build appropriate QSAR models in order to predict the activities of many other untested flavonoids, and to direct the synthesis of flavonoid compounds with higher potency for potential clinical application *(7, 8)*. Recent overview of QSAR models for flavonoids, a sub-class of polyphenolic compounds, is given *(9)*. Amić et al. emphasized the problem discussed recently by other authors. All published QSARs were developed using in vitro experimental antioxidant activity, and transferability of such models to in vivo situations is questionable, and should be investigated in future *(9, 10, 11)*.

It should be emphasized that the above-mentioned in vitro studies are much simpler and controllable compared to in vivo studies for screening antioxidants for the structure-activity relationships (SARs) elucidation. Detailed investigations should be done on the actual fate of flavonoid compounds in vivo (adsorption, distribution, metabolism, and excretion, ADME) to be able to predict their effects in living cells *(10, 11)*.

Because of that, models that can be obtained by presented method should be used with caution, taking into account the fact that antioxidant activities estimated or predicted by a model contain all limitations that were included in experimental values used for the model development, as well as those related to the shortcomings of modeling process and methodology, that includes many simplifications and approximations. The main shortcomings of the modeling process are related to errors (a) of the program used to produce molecular structures (that are not the same as the real ones); (b) due to the imperfection of programs for computation of molecular descriptors; and (c) related to the use of improper methods for validation of models. In this study, we will use the simplest form of QSAR models based on the multivariate linear regression (MLR).

2. Materials

The TEAC (Trolox equivalent antioxidant capacity) assay is frequently used for constructing SARs of antioxidant activity of flavonoids *(12, 13, 14)*. The TEAC assay is based on the scavenging of the ABTS radical ($ABTS^{\bullet +}$) converting it into a colorless product *(15)*. The TEAC is defined as the millimolar concentration of a Trolox solution with equivalent antioxidant potential to a 1 mM solution of the compound under investigation. It has been reported that TEAC value of the series of flavonoids correlate well with the number of aromatic hydoxyl groups *(16)*. However, recently it has been found that the possible contribution of reaction products to the TEAC of a compound limits the applicability of this assay in constructing of SAR *(17, 18)*. A new, vitamin C equivalent antioxidant capacity (VCEAC) assay, on a weight basis (mg/L), using the ABTS radical, was applied to ascertain SAR of polyphenols. A good linear correlation between VCEAC and the number of OH groups on the flavonoid core was found *(19)*.

2.1. Data set of polyphenols and experimental TEAC values

1. Experimental TEAC values of compounds related to their antioxidant activity that will be used in modeling can be obtained by direct measurement in one laboratory, or collected from results of different measurements from literature or internet sources. Recently, it is mentioned that, among them, TEAC values obtained by the method that uses ABTS as a target radical appears to be the one applicable to a wide range of foods and beverages *(15)*, despite some protocol failures that were mentioned – (*see* **Note 1**).
2. Taking into account Notes 1 and 2, we selected for illustration of described methodology data set of 100 polyphenols from Cai et al. *(20)*, because the authors performed two measurements (improved $ABTS^{\bullet +}$ and DPPH assays) of the antioxidant capacities of polyphenols, and obtained high correlation between these two assays ($R = 0.995$). High correlation between two measurements of antioxidant activities is an additional confirmation of consistency of obtained TEAC values, which is also suggested by Yoo et al. *(15)*.
3. All experimental details related to $ABTS^{\bullet +}$ and DPPH assays were described in reference 20.
4. Determinations of all samples were carried out in triplicate for both $ABTS^{\bullet +}$ and DPPH assays. All results were calculated as mean $\pm$ standard deviation (SD) *(20)*.
5. For illustration of QSAR methodology, only TEAC experimental values obtained by $ABTS^{\bullet +}$ assay from reference 20 will be used in the modeling.

3. Methods

3.1. Generation of Appropriate Forms of Molecular Structures

1. Initial structures of molecules were prepared in SMILES (Simplified Molecular Input Line Entry Specification) form *(21, 22)*, according to the SMILES rules (more details about SMILES and some examples are given in **Note 3**).
2. The SMILES strings of molecules were converted to 3D structures using the CORINA program, a fast and reliable rule-based SMILES to the 3D structure model generator, developed by Gasteiger et al. *(23, 24)* (*see* **Note 4**). We have been using version 1.1 of the CORINA program implemented on the Online SMILES Translator server *(25)* to convert SMILES strings to 3D Structure Data Format (SDF). There are several ways to convert/obtain 3D structure from SMILES, but according to our experience, for large data sets of diverse compounds, CORINA works very fast and gives reasonably good 3D structures, with small failures, i.e. 3D structures that are incorrect (*see* **Note 5**).

3.2. Computation of Numerical Parameters that Describe Properties of Molecular Structures

1. Molecular descriptors are numerical representations of molecular structures. For a set of molecules, it is possible to design a new set of specific molecular descriptors, but usually they have been calculated by some of the computer programs like Dragon or CODESSA *(26, 27, 28)* (*see* **Note 6**). In this study, we selected and calculated only (about) 350 simpler molecular descriptors by the Dragon 5.4 program *(26, 27)*.
2. Based on statistical criteria, only 110 molecular descriptors having single correlation coefficient with corresponding TEAC values for selected training set of 21 polyphenols higher than 0.37 were taken into account in the process of selecting the models.

3.3. Training and Test Sets, and Model Validation Procedures

1. For a selected set of 100 experimental TEAC values, we have to perform partition of data into the training and test sets. We selected 21 compounds in the training set (distribution in **Fig. 16.1**) and remaining 79 compounds in the test set (*see* **Note 7**). Performing partition of data, we found that distribution of all 100 experimental values is far from being normal (**Fig. 16.2**).
2. Three main validation procedures have been usually done for estimating the quality of a model: fit, internal validation (or cross-validation), and prediction (*see* **Note 8** and reference 29).
3. Before the modeling, we have to select statistical parameters that will be used for judging the quality of the model. There

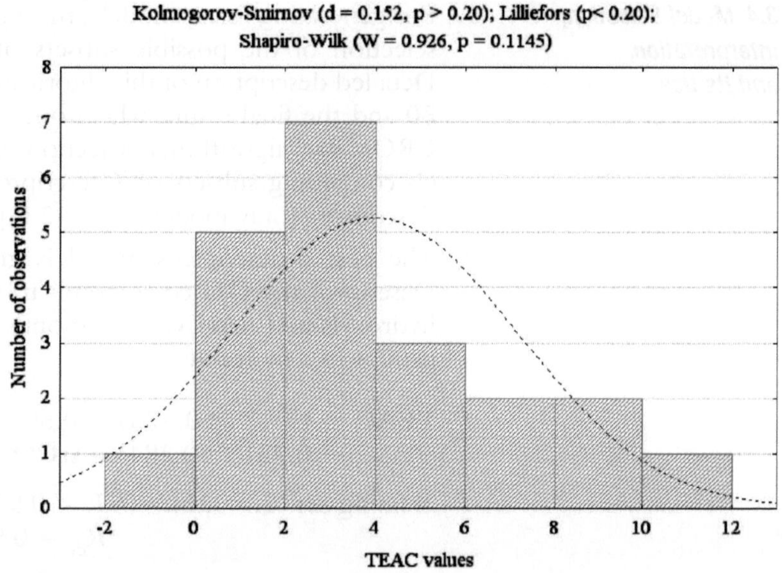

Fig. 16.1. Normality tests of distribution of 21 TEAC values (training set) used for development of the model out of 100 TEAC values taken from reference 20 (note that interval (-2,0) contain one polyphenol with the TEAC value equal to 0.0).

are many statistical parameters in use, but the simplest parameters of standard errors of fit (S), cross-validation (S_{cv}) and prediction (S_{pred}), as well as corresponding correlation coefficients (R, R_{cv}. R_{pred}) are sufficient, if they are calculated correctly for each procedure (*see* **Note 9**).

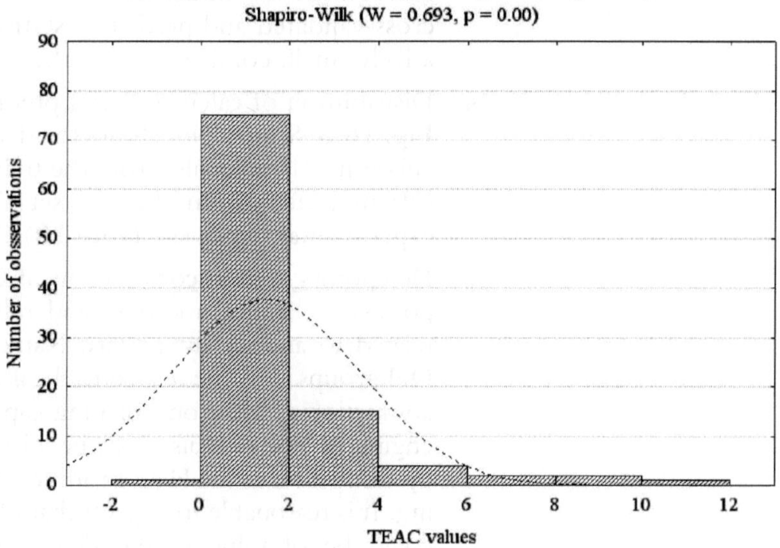

Fig. 16.2. Normality tests of distribution of all 100 TEAC values (training and test sets together) from reference 20 (note that interval (-2,0) contain one polyphenol with the TEAC value equal to 0.0).

3.4. Model Selection, Interpretation, and Its Use

1. For generating linear models, the algorithm CROMRsel for selection of the possible subsets of descriptors was used. Detailed description of this algorithm was given in reference 30 and the final result, which one can obtain by using this CROMRsel algorithm, is selection of the best possible models containing subsets of I descriptors among the set of K descriptors (only model for $I = 2$ is presented in this study).
2. The best 2-descriptors model is given below (Eq. 16.1). Descriptor **nArOH** corresponds to the number of aromatic hydroxyls and **nArCO** corresponds to the number of keto groups in a molecule.

$$\mathbf{TEAC} = 1.043\ (\pm 0.378) + 0.593\ (\pm 0.042)\ \mathbf{nArOH} - 0.921\ (\pm 0.284)\ \mathbf{nArCO} \qquad (16.1)$$

training set ($N = 21$): $R = 0.975$, $S = 0.692$; $R_{cv} = 0.967$, $S_{cv} = 0.796$

test set ($N_{tst} = 79$): $R_{pred}^2 = 0.906$, $R_{pred} = 0.952$; $S_{pred} = 1.016$

To estimate TEAC antioxidant activity for a new polyphenolic compound, preferably similar to those used in the training set, one should simply count the total numbers of aromatic OH and keto groups and then insert them into Eq. 16.1.

3. One can see that each regression coefficient is three times (or more for descriptor **nArOH**) higher than the corresponding error of regression coefficient (given in brackets), confirming significance of related descriptor. This is also a prerequisite for model stability. Differences between fit and cross-validated and predictive statistical parameters are relatively small, confirming that the model is of good quality.
4. Distribution of calculated and predicted values is shown in **Fig. 16.3**. Scatter plot shows the distribution of error in calculation (21 molecules from the training set) and prediction (79 molecules from the test set) in the range of TEAC experimental values (*see* **Note 10**).
5. Descriptors used in construction of QSAR model (Eq. 16.1) possess a chemical meaning and reflect some driving forces related to antioxidant activity. Namely, increased number of OH groups could be related to the increased ability of H atom abstraction or electron donating capacity and increased scavenging of free radicals. Polyphenol phenoxyl radicals formed by abstraction of the H atom are stabilized by hydrogen bonding. It is reasonable to suggest that QSAR models of that type could be of value in the designing of new and efficient polyphenolic antioxidants. They may also shed some light on the mechanism of antioxidant action.

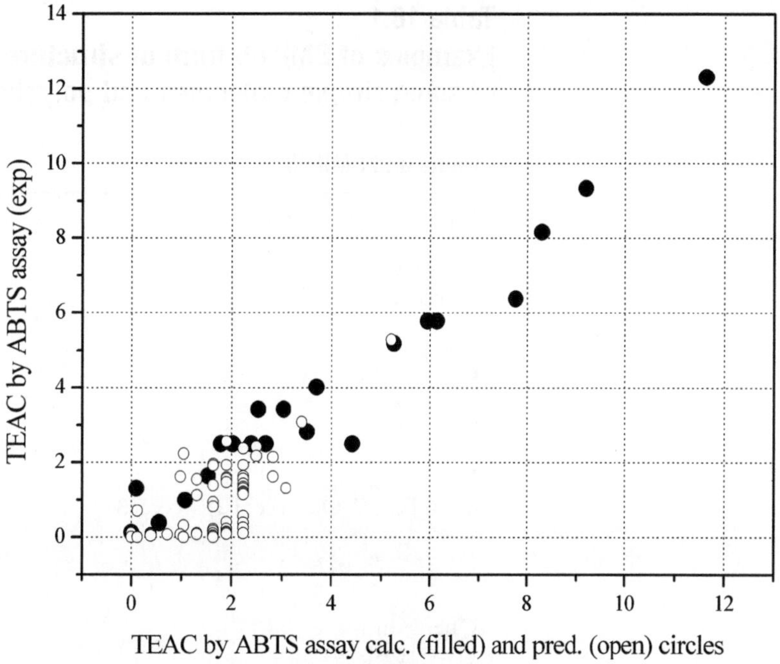

Fig. 16.3. Sater plot of observed *vs* calculated (for 21 molecules from the training set – filled circles) and observed *vs* predicted (for 79 molecules from the test set – open circles) TEAC values.

4. Notes

1. Disadvantages of ABTS method mentioned in reference 15 are: (a) very sensitive with temperature and light; (b) extra step to thermally generate free radical from ABTS; (c) not standardized, hence hard to compare experimental values obtained by different laboratories. People working in QSAR field rarely control the quality of experimental data. QSAR models can help in detecting some weakness of experimental procedures, but besides the correctness and the quality of performed parts of modeling processes, they are primarily dependent on the quality of experimental data used in modeling.
2. In order to measure antioxidant activities accurately, one must consider many variables associated with the targeted free radicals and variable reaction mechanisms on individual antioxidants. It is recommended that various additional factors associated with the reactions, such as extraction solvents, temperature, light, and time, should be considered in order to have a reliable result of antioxidant measurement. In addition, a strong suggestion is to perform more than two antioxidant

Table 16.1
Examples of SMILES form of structure (SMILES string) of some simple molecules and polyphenols

Name and SMILES	Molecular structure
Ethene: C=C; Ethyne: C#C	
Benzene c1ccccc1	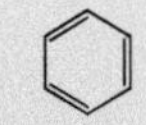
Gallic acid C(=O)(O)c1cc(O)c(O)c(O)c1	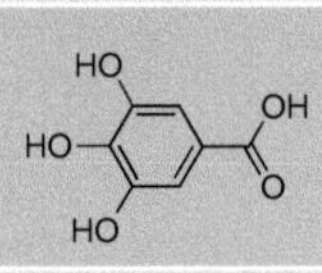
Flavone O=C1C=C(Oc2ccccc12)c3ccccc3	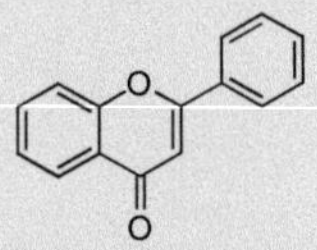
Quercetin O=C1C(O)= C(Oc2cc(O)cc(O)c12)c3ccc(O)c(O)c3	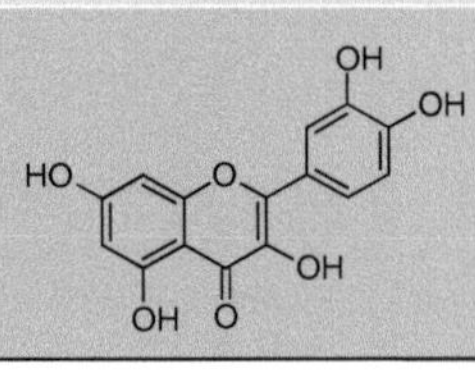

assays *(15)*, but after an intensive overview of literature we found two measurements only in reference 20 (i.e. in almost all studies we found only one assay of measurement).

3. In **Table 16.1** several examples of SMILESes are given. There are many programs used for generation of SMILESes from 2D charts of the molecule, and also, there are several SMILES notations. Obtained SMILESes should be checked by visual inspection to be sure that we have correct molecular structures coded in the SMILES structure form.
4. One of advantages of the CORINA program is that we can quickly and simply generate good 3D structures of a huge number of molecules. Structure parameters in CORINA are derived after analysis of experimental bond lengths and bond angles for different bonds and bond types appearing in databases containing experimental structures of molecules.
5. It is possible to perform structure optimization at a higher level of computation. However, the problem is selection of an appropriate conformation for a polyphenol from a huge number of conformations that are possible, and additional difficulties appear when a large number of structures have to be optimized.
6. There are many programs for calculating several hundreds of molecular descriptors, but the Dragon program, together

with CODESSA *(28)* is the most popular in the last 10 years. These programs used SMILES or 3D molecular files (SDF) as inputs. There are several successful attempts in designing simple descriptors for describing the antioxidant activities of polyphenols that account for the number and specificity in position of OH groups. One simple example was presented in reference 31.

7. In reference 20, as well as in some other studies (see, for example, references 19 and 32) that contain measurements of antioxidant activity of polyphenols, the distribution of experimental TEAC values of polyphenols is not normal. Usually, a population of lower activity is more abundant than that of middle or higher activity. TEAC experimental values of selected data set containing 21 compounds pass all three normality tests implemented in the Statistica 7.1 program *(33)*. Kolmogorov-Smirnov one-sample test for normality *(33)* is based on the maximum difference between the sample cumulative distribution and the hypothesized cumulative distribution. In addition, we also used Statistica 7.1 for performing normality testing of TEAC values by the Shapiro-Wilk W test and the Lilliefors test. All three tests of normality show, at the level of significance (at least) of 0.05, that our sample of TEAC values for the training set has (approximately) normal distribution. Taken all (100) experimental TEAC values from reference 20 (both from ABTS and DPPH assays) one can see that they do not pass the tests of normality (*see* **Fig. 16.2**).

8. **Fit procedure** – when all data from the training set were used to perform the least square fit between selected descriptors and experimental values on the training set. **Cross-validation** – in this paper only leave-one-out (LOO) cross-validation (CV) was done. LOO CV procedure is a procedure usually used for evaluating the model stability, and is performed on the training set. During LOO CV procedure, each of the N molecules is taken away only once. Using the remaining $N-1$ molecules, the multivariate regression model was generated. Applying the generated model, in each step, the TEAC value for excluded compound was calculated. Finally, we have prediction (in LOO CV sense) of TEAC values for N molecules. **Prediction** is a procedure done to test real predictive capabilities of developed model on the test set compounds (79 molecules) not used in model selection procedure.

9. We expressed statistical performance of fit, and of leave-one-out (LOO) cross-validation (CV) for each developed model on the training set by the correlation coefficient of fit (R) and cross-validation (R_{cv}). We also calculated the standard error of fit (S) and of LOO CV (S_{cv}). Predictive performance of developed models are evaluated on the test set (79 compounds),

and expressed by the predictive correlation coefficient calculated by Eq. 16.2

$$R_{\text{pred}} = \sqrt{1 - \sum_{i=1}^{79} (X_{\text{exp},i} - X_{\text{pred},i})^2 / \sum_{j=1}^{21} (X_{\text{exp-train},j} - \bar{X}_{\text{exp-train}})^2} \quad (16.2)$$

and by the standard error of prediction (S_{pred}) (Eq. 16.3)

$$S_{\text{pred}} = \sqrt{\sum_{i=1}^{N=79} (X_{\text{exp},i} - X_{\text{pred},i})^2 / N} \quad (16.3)$$

In Eqs. 16.2 and 16.3, $X_{\text{exp},i}$ and $X_{\text{pred},i}$ are experimental and predicted TEAC values for the test set, respectively. In Eq. 16.2, $X_{\text{exp-train},j}$ are experimental TEAC values for the training set, and $\bar{X}_{\text{exp-train}}$ is their corresponding mean value.

10. Such a plot is usually used for illustration and comparison of errors in calculation (fit) of the training set and prediction of the test set, as well as in order to obtain insight of distribution of errors in the whole range of experimental values.

Acknowledgments

This research was supported by Ministry of Science, Education and Sports of the Republic of Croatia through grant 098-1770495-2919 (BL and NT) and grant 079-3211 (DA and BL).

References

1. Heim, K. E., Tagliaferro, A. R., and Bobilya, D. J. (2002) Flavonoid antioxidants: chemistry, metabolism and structure-activity relationships. *J. Nutr. Biochem.* **13**, 572–584.
2. Bors, W., Heller, W., Michel, K., and Stettmaier, K. (1996) Flavonoids and Polyphenols: Chemistry and Biology, in *Handbook of Antioxidants* (Cadenas, E., and Packer L., eds.), Marcel Dekker, Inc., New York, NY, pp. 409–466.
3. Soobrattee, M. A., Neergheen, V. S., Luximon-Ramma, A., Arouma, O. I., and Bahorun T. (2005) Phenolics as potential antioxidant therapeutic agents: mechanism of actions. *Mutat. Res.* **579**, 200–213.
4. Havsteen, B. H. (2002) The biochemistry and medical significance of the flavonoids. *Pharmacol. Ther.* **96**, 67–202.
5. Hansch, C., and Fujita, T. (1964) ρ-σ-π Analysis. A method for the correlation of biological activity and chemical structure. *J. Am. Chem. Soc.* **86**, 1616–1626.
6. Netzeva, T. I., Worth, A. P., Aldenberg, T., Benigni, R., Cronin, M. T. D., Gramatica , P., Jaworska J. S., Kahn, S., Klopman, G., Marchant, C. A., Myatt, G., Nikolova-Jeliazkova, N., Patlewicz, G. Y., Perkins, R., Roberts, D. W., Schultz, T. W., Stanton, D. T., van de Sandt, J. J. M., Tong, W., Veith, G., and Yang, C. (2005) Current Status of Methods for Defining the Applicability Domain of (Quantitative) Structure–Activity Relationships. The Report and Recommendations of ECVAM Workshop 52. *ATLA* **33**, 155–173, http://ecb.jrc.it/qsar/publications/(accessed December 17, 2007).

7. Zhang, H.-Y. (2005) Structure-activity relationships and rational design strategies for radical-scavenging antioxidants. *Curr. Comp. Aided Drug Des.* **1**, 257–273.
8. Zhang, H.-Y., Yang, D.-P., and Tang, G.-Y. (2006) Multipotent antioxidants: from screening to design. *Drug Discov. Today* **11**, 749–754.
9. Amić, D., Davidović-Amić, D., Bešlo, D., Rastija, V. Lučić, B., and Trinajstić, N. (2007) SAR and QSAR of the antioxidant activity of flavonoids. *Curr. Med. Chem.* **14**, 827–845.
10. Haenen, G. R. M. M., Arts, M. J. T. J., Bast, A., and Coleman, M. D. (2006) Structure and activity in assessing antioxidant activity in vitro and in vivo. A critical appraisal illustrated with the flavonoids. *Environ. Toxicol. Pharmacol.* **21**, 191–198.
11. Hermans, N., Cos, P., Maes, L., De Bruyne, T., Vanden Berghe, D., Vlietinck, A. J., and Pieters, L. (2007) Challenges and pitfalls in antioxidant research. *Curr. Med. Chem.* **14**, 417–430.
12. Rice-Evans, C. A., Miller, N. J., and Paganga G. (1996) Structure-antioxidant activity relationships of flavonoids and phenolic acids. *Free Radic. Biol. Med.* **20**, 933–956.
13. Rice-Evans, C. A., and Miller, N. J. (1998) Structure-Antioxidant Activity Relationships of Flavonoids and Isoflavonoids, in *Flavonoids in Health and Disease* (Rice-Evans, C. A. and Packer, L., eds.), Marcel Dekker, Inc., New York, NY, pp. 199–219.
14. Lien, E. J., Ren, S., Bui, H.-H., and Wang, R. (1999) Quantitative structure-activity relationship analysis of phenolic antioxidants. *Free Radic. Biol. Med.* **26**, 285–294.
15. Yoo, K. M., Kim, D.-O., and Lee C. Y. (2007) Evaluation of different methods of antioxidant measurement. *Food. Sci. Biotechnol.* **16**, 177–182.
16. van den Berg, R., Haenen, G. R. M. M., van den Berg, H., van der Vijgh, W., and Bast, A. (2000) The predictive value of the antioxidant capacity of structurally related flavonoids using the Trolox equivalent antioxidant capacity (TEAC) assay. *Food Chem.* **70**, 391–395.
17. Arts, M. J. T. J., Dallinga, J. S., Voss, H.-P., Haenen, G. R. M. M., and Bast, A. (2003) A critical appraisal of the use of the antioxidant capacity (TEAC) assay in defining optimal antioxidant structures. *Food Chem.* **80**, 409–414.
18. Arts, M. J. T. J., Haenen, G. R. M. M., Voss, H. -P., and Bast, A. (2004) Antioxidant capacity of reaction products limits the applicability of the Trolox equivalent antioxidant capacity (TEAC) assay. *Food Chem. Toxicol.* **42**, 45–49.
19. Kim, D.-O., and Lee, C. Y. (2004) Comprehensive study on vitamin C equivalent antioxidant capacity (VCEAC) of various polyphenolics in scavenging a free radical and its structural relationship. *Crit. Rev. Food Sci. Nutr.* **44**, 253–273.
20. Cai, Y.-Z., Sun, M., Xing, J., Luo, Q., and Corke, H. (2006) Structure-radical scavenging activity relationships of phenolic compounds from traditional Chinese medicinal plants. *Life Sci.* **78**, 2872–2888.
21. Weininger, D. (1988) SMILES, a chemical language for information systems. 1. Introduction to methodology and encoding rules. *J. Chem. Inf. Comput. Sci.* **28**, 31–36.
22. SMILES notation on Daylight web site http://www.daylight.com/dayhtml/doc/theory/theory.smiles.html or http://cactus.nci.nih.gov/services/smiles.html (accessed December 10, 2007).
23. Gasteiger, J., Sadowski, J., Schuur, J., Selzer, P., Steinhauer, L., and Steinhauer, V. (1996) Chemical information in 3D space. *J. Chem. Inf. Comput. Sci.* **36**, 1030–1037.
24. CORINA Program (2006) Molecular Networks GmbH Computerchemie, Erlangen, Germany (http://www.molecular-networks.com) and http://www2.chemie.uni-erlangen.de/software/ (accessed December 10, 2007).
25. Oellien, F., and Nicklaus, M. C. (2004) Online SMILES Translator and Structure File Generator, http://cactus.nci.nih.gov/services/translate/(accessed December 10, 2007).
26. TALETE srl, (2006) DRAGON Professional 5.4 program, Milano, Italy (http://www.talete.mi.it/ dragon.htm).
27. Todeschini, R. and Consonni, V. (2000) *Handbook of Molecular Descriptors*, Wiley-VCH, Weinheim.
28. Katritzky, A. R., Lobanov, V., and Karelson, M. (1995) CODESSA 2.21 Program, University of Florida, Gainesville, USA (http://ark.chem.ufl.edu).
29. Gramatica, P. (2007) Principles of QSAR models validation: internal and external. *QSAR Comb. Sci.* **26**, 694–701.
30. Lučić, B., and Trinajstić, N. (1999) Multivariate regression outperforms several robust architectures of neural networks in

QSAR modeling. *J. Chem. Inf. Comput. Sci.* **39**, 121–132.

31. Amić, D., Davidović-Amić, D., Bešlo, D., and Trinajstić, N. (2003) Structure-radical scavenging activity relationships of flavonoids. *Croat. Chem. Acta* **76**, 55–61.
32. Miyahara, M., Ohtaka, H., Katayama, H., Tatsumi, Y., Miyaichi, Y., and Tomimori T. (1993) Structure-activity relationship of flavonoids in suppresing rat liver lipid peroxidation. *Yakugaku Zasshi* **113**, 133–154.
33. StatSoft, Inc. (2005) STATISTICA 7.1 (data analysis software system), version 7.1 (http://www.statsoft.com).

Chapter 17

Immunohistochemical Staining of Cyclooxygenases with Monoclonal Antibodies

Ghassan M. Saed

Abstract

Immunohistochemistry is an important tool that is often used for the diagnosis of several diseases in the pathology laboratory. The quality and sensitivity of immunohistochemical staining is affected by formalin fixation, which results in variable loss of antigenicity, known as a masking effect. While the sensitivity of immunohistochemistry is excellent for certain antigens, other antigens such as COX-1 and COX-2 are difficult to identify, especially in formalin-fixed, paraffin sections. Antigen retrieval is a technique that re-exposes epitopes and allows detection of masked antigens with standard immunohistochemical procedures. One common method involves partial, enzymatic pre-digestion with trypsin or pepsin while other, nonenzymatic procedures or heat-mediated antigen retrieval methods include pressure-cookers, hot plates, or microwave (MW) irradiation of tissue sections in water or a variety of antigen-retrieval solutions. In this chapter, we will describe a technique that provides a more reliable, much simpler approach for the demonstration of cyclooxygenase-1 and cyclooxygenase-2 expression in frozen, vibratome or paraffin sections, and/or cells in cultures.

Key words: COX-1, COX-2, Immunofluorescence, Post-operative adhesions, Ovarian cancer.

1. Introduction

Cyclooxygenases (COXs), also referred to as prostaglandin (PG) endoperoxide synthases, catalyze the rate-limiting step in PG synthesis. Mammalian cells contain at least two isozymes of COX: COX-1 and COX-2. Cyclooxygenase-1 is a well-characterized enzyme originally purified from ovine and bovine vesicular glands and platelets. The COX-1 gene is constitutively expressed in most tissues, and the protein levels do not fluctuate in response to stimuli such as cytokines or growth

From: *Methods in Molecular Biology, vol. 477: Advanced Protocols in Oxidative Stress I,*
Edited by: D. Armstrong, DOI: 10.1007/978-1-60327-517-0_17, © Humana Press, New York, NY

factors *(1)*. In contrast, expression of the COX-2 gene is induced by a variety of agents, including growth factors *(2)*, phorbol esters *(3)*, and cytokines *(4)*. Both COX-1 and COX-2 perform two enzymatic functions: as COXs, they convert arachidonic acid to PGG2, and as peroxidases, they convert PGG2 to PGH2. Recently, a third distinct COX isozyme, COX-3, as well as two smaller COX-1-derived proteins (partial COX-1 [PCOX-1] proteins), have been described elsewhere *(5)*. Cyclooxygenase-3 and one of the PCOX-1 proteins (PCOX-1a) are made from the COX-1 gene but retain intron 1 in their mRNAs. Partial COX-1 proteins additionally contain an in-frame deletion of exons 5–8 of the COX-1 mRNA. In humans, COX-3 mRNA is expressed as a 5.2-kb transcript and is most abundant in cerebral cortex and heart *(6)*.

Fluorescence microscopy is a rapidly expanding and invaluable tool of investigation. Immunofluorescence is a very important tool, which is widely used for visualization of specific molecules within a cell or tissue, allowing for the determination of their distribution and localization *(7)*. Typically, immunofluorescence requires two sets of antibodies: a primary antibody which is used against the antigen of interest and a subsequent, secondary, fluorescently-labeled antibody that recognizes the primary antibody *(8, 9)*. Immunofluorescence can be either direct or indirect. In direct immunofluorescence, a specific antibody that has been labeled with a fluorophore is added to microscope slide with cells or tissue. The amount of fluorescence detected is directly proportional to the amount of antigen present *(8)*. In indirect immunofluorescence, a primary unlabeled antibody is allowed to bind to the specific antigen and subsequently incubated with a fluorescently-labeled secondary antibody, which recognizes the primary antibody *(9)*. The advantage of the indirect method is that it reduces the need for a large stock of labeled antibodies, and also usually results in greater fluorescence intensity.

Immunofluorescence has been successfully utilized by numerous studies that deal with cytochemical and histochemical staining to identify chromosomes, DNA content, proteins, cellular structures, hormones, and vitamins *(10)*. Because of the extreme sensitivity of the fluorochromes and the fluorescence microscope, it is possible to quantitate small amounts of specific antigens within cells and tissues.

There are several fluorochromes commonly used in fluorescence microscopy. One being the popular nucleic acid stain 4′,6-diamidino-2-phenylindole (DAPI), which fluoresces in the blue region of the visible light spectrum. Other commonly used dyes are Texas Red and fluorescein isothiocyanate (FITC), which fluoresce in the red and green regions of the visible light spectrum, respectively. Texas Red is used in combination with DAPI and FITC for multiple stained specimens that can be observed due to the red, blue, and green fluorescence of the dyes.

2. Materials

2.1. Equipment

1. Biological Safety Cabinet
2. Chemical Fume Hood
3. Humidified Chamber or water bath
4. Microtome
5. Oven
6. Axiovert 25 inverted microscope (Zeiss, Thornwood, NY), or equivalent fluorescence capable microscope
7. Microscope mounted camera
8. Fluorescence filters for Texas Red, DAPI, and FITC (Zeiss)

2.2. Reagents and Buffers

2.2.1. Cell Culture

1. Dulbecco's modified eagle medium (DMEM) (Gibco, Invitrogen, Carlsbad, CA) supplemented with 10% Fetal Bovine Serum (FBS, USA Scientific, Ocala, FL) and 1% Antiboitic-Antimycotic (100 × liquid) (Gibco, Invitrogen).
2. Trypsin solution with ethylenediamine tetraacetic acid EDTA (0.05% Trypsin with EDTA 4Na) (Gibco, Invitrogen).
3. Phosphate-buffered saline 1×, pH 7.4 (PBS; Gibco, Invitrogen) for washing.
4. Paraformaldehyde (Sigma-Aldrich, St. Louis, MO): Prepare 4% solution in PBS fresh for each experiment using a hot plate in the fume hood.

2.2.2. Blocking and Staining

1. 0.2% Triton X-100 (Sigma-Aldrich, St. Louis, MO) diluted in PBS.
2. 0.05% Tween-20 (BioRad, Hercules, CA) diluted in PBS.
3. Xylenes (Sigma-Aldrich, Catalog number 247642).
4. 10 mM sodium citrate, 0.05% Tween-20, pH 6.0 diluted in deionized water. May be stored for 3 months at room temperature.
5. Ethanol, 100%, 95%, 80% (Sigma-Aldrich).
6. 3% Hydrogen peroxide.
7. 10% Normal Goat Serum (Santa Cruz, Santa Cruz, CA), diluted in PBS.

8. 10% Normal Donkey Serum (Santa Cruz), diluted in PBS.
9. Monoclonal COX-1 Antibody (Cayman Chemical, Ann Arbor, MI).
10. Monoclonal COX-2 Antibody (AbCam, Cambridge, MA).
11. Donkey anti-mouse Texas Red secondary antibody (Santa Cruz).
12. Goat anti-rabbit FITC secondary antibody (Santa Cruz).
13. Prolong Gold Antifade Reagent with DAPI (Invitrogen, Carlsbad, CA).

2.3. Supplies

1. Cryomolds (Fisher Scientific, Pittsburg, PA).
2. Coverslips (Fisher).
3. Microscope Slides (Fisher).
4. Axiovision 4 MOD Multi Channel Software (Zeiss), or equivalent software.

3. Methods

3.1. Preparation

Slides for immunohistochemical staining can be prepared from cultured cells, frozen tissue sections, or paraffin embedded tissues as follows.

3.2. Cell Culture

1. Cultured cells (5×10^6 cells) are grown on autoclaved glass cover slips overnight at 37°C to approximately 50–70% confluency.
2. Wash with PBS.
3. Fix the cells by submerging coverslips into a 4% paraformaldehyde solution for 12–15 min at room temperature. From this point forward, do not allow your coverslips to dry (*see* **Note 1**).
4. The paraformaldehyde solution should be discarded (into a hazardous waste container), and coverslips are washed three times, 5 min each with 0.05% Tween-20 diluted in PBS (*see* **Note 2**).
5. Permeabilize the cells by incubating coverslips in 0.2% Triton X-100 diluted in PBS for 10 min at room temperature.
6. Wash the coverslips, three times, 5 min each with 0.05% Tween-20 diluted in PBS (*see* **Note 3**).

3.3. Frozen Sections

1. Snap freeze fresh tissues in liquid nitrogen or isopentane pre-cooled in liquid nitrogen, embedded in OCT compound in cryomolds. Store frozen blocks at −80°C.

2. Cut 5–7 μm thick cryostat sections and mount on coated slides. Store slides at −80°C.
3. Before staining, warm slides to room temperature (30–45 min) and fix in 4% paraformaldehyde for 15–20 min at room temperature.
4. The paraformaldehyde solution should be discarded (into a hazardous waste container), and then wash the coverslips three times, 5 min each with 0.05% Tween-20 diluted in PBS.
5. Permeabilize the tissue using 0.2% Triton X-100 diluted in PBS for 15–20 min at room temperature.
6. Wash in 0.05% Tween-20 diluted in PBS three times, 5 min each.

3.4. Paraffin

1. To help tissues adhere to the slide, you must heat the slide in an oven at 65°C for at least 2 h up to 24 h.
2. Deparaffinize tissue sections by immersing them in Xylenes two times, 5 min each at room temperature.
3. Hydrate tissues with 100% ethanol two times, 3 min each at room temperature.
4. Hydrate with 95% ethanol two times, 2 min at room temperature.
5. Hydrate with 80% two times, 2 min at room temperature.
6. Wash in 0.05% Tween-20 diluted in PBS three times, 5 min each.
7. Block for endogenous peroxidase activity by submerging the slide in 3% hydrogen peroxide for 20 min (*see* **Note 4**).
8. Wash in 0.05% Tween-20 diluted in PBS three times, 5 min each.
9. Citrate Buffer antigen retrieval is done by submerging slide in the pre-heated citrate buffer (95–100°C) in a water bath for 20–40 min. Turn off water bath, remove container with slide, allow slides to cool for 20 min at room temperature. Dispose off citrate buffer in a hazardous waste container (*see* **Notes 5–6**).

3.5. Blocking with Serum

1. Sections are incubated with 10% normal serum diluted in PBS for 30–35 min at room temperature in the dark (*see* **Note 7**).
2. You may alternatively block with the 3–10% bovine serum albumin (BSA) diluted in PBS at room temperature in the dark (*see* **Note 8**).

3.6. Incubation

1. **Double Immunofluorescence**: Incubate sections with COX-1 and COX-2 primary antibodies diluted to 2.5 μg/ml, and diluted 100× respectively in 3% normal serum diluted in PBS for 1 h at room temperature in a humidified chamber in the dark (*see* **Notes 9–11**).
2. Wash three times, 5 min each with 0.05% Tween-20 diluted in PBS in the dark.

3.7. Secondary Antibody

1. Incubate sections with fluorescently conjugated secondary antibodies, diluted 100× in 3% BSA-PBS solution, for 45 min at room temperature in a humidified chamber in the dark (*see* **Note 12**).
2. Wash three times, 10 min each with 0.05% Tween-20 diluted in PBS at room temperature in the dark.

3.8. Mounting and Counterstaining

1. Slides are mounted with a few drops of Prolong Gold Antifade Reagent with DAPI.
2. Seal the edges of the coverslip after mounting using clear nail polish.
3. Store slides in the dark at 4°C (*see* **Note 13**).

3.9. Detection and Quantification

1. Slides are examined with the Axiovert 25 inverted microscope using DAPI, FITC, and Texas Red fluorescent filters (*see* **Note 14**).
2. DAPI is visualized using a filter that has excitation and emission wavelengths of 365 and 445 nm, respectively (*see* **Note 15**).
3. FITC is visualized using a filter that has excitation and emission wavelengths of 470 and 525 nm, respectively.
4. Texas Red is visualized using a filter that has excitation and emission wavelengths of 596 and 613 nm, respectively.
5. Images are taken using the Axiovision software and microscope mounted camera (*see* **Note 16**).
6. Quantification of data may be desired and is achieved by photographing the cells in black and white and then using the Axiovision software to analyze a section and measure the intensity of the fluorescence.
7. In **Fig. 17.1**, the slides were incubated with mouse-anti-COX-1 monoclonal primary antibody along with rabbit-anit-COX-2 monoclonal primary antibody followed by donkey-anti-mouse Texas Red secondary antibody and goat-anti-rabbit-FITC secondary antibody. Weak scattered COX-1 and COX-2 fluorescence signals were detected at 63 × magnification using an Axiovert 25 inverted microscope from Zeiss.

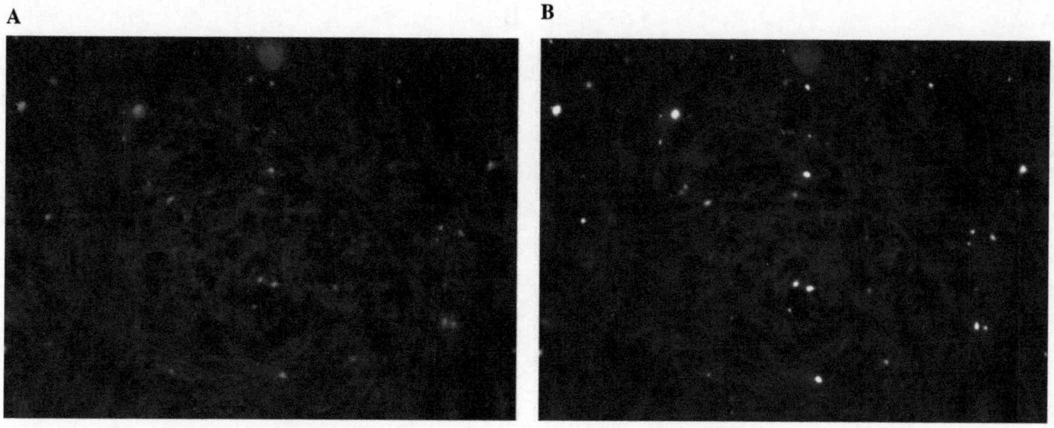

Fig. 17.1. Immunofluorescence staining of COX-1 **(A)** and COX-2 **(B)** in paraffin-embedded and formalin-fixed normal ovarian tissue as described in methods.

8. In **Fig. 17.2**, the slides were incubated in the same manner as in **Section 3.9.6**, followed by nuclei staining using Prolong Gold Antifade Reagent with DAPI (shown in color images only). Stronger COX-2 than COX-1 fluorescence signals were detected at 63× magnification using an Axiovert 25 inverted microscope from Zeiss.
9. In **Fig. 17.3**, the cells on coverslips were incubated in the same manner as in **Section 3.9.6**, followed by nuclei staining using Prolong Gold Antifade Reagent with DAPI (shown in color images only). A stronger COX-2 fluorescence signal was higher in adhesion (d) as compared to normal fibroblasts (b). In contrast, a stronger COX-1 fluorescence signal was higher in normal (a) as compared to

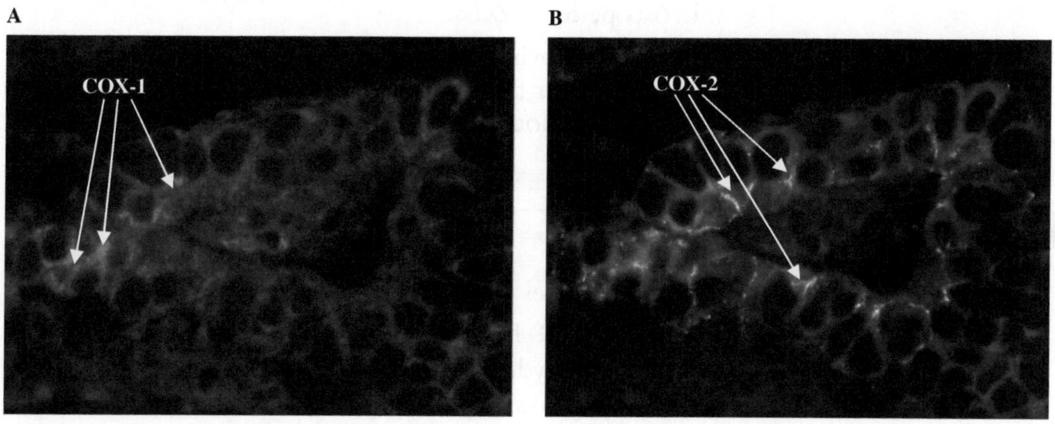

Fig. 17.2. Immunofluorescence staining of COX-1 **(A)** and COX-2 **(B)** in paraffin-embedded and formalin-fixed ovarian cancer tissue as described in methods.

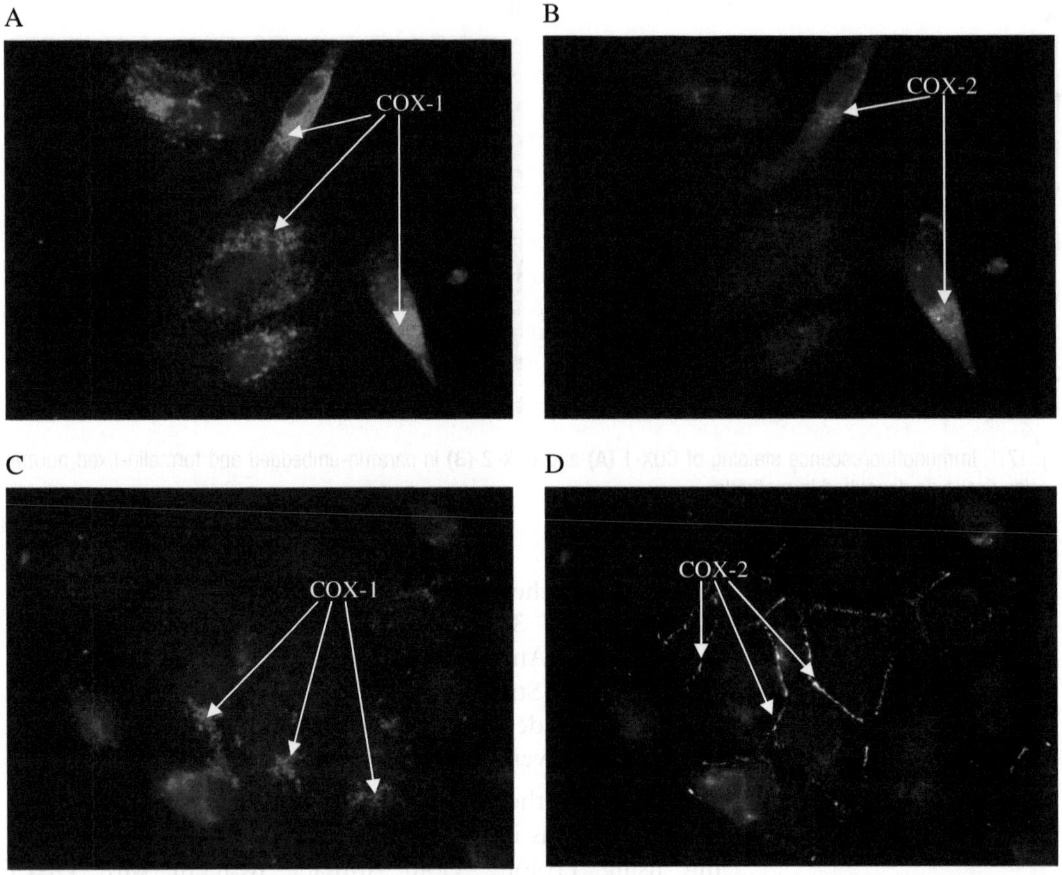

Fig. 17.3. Immunofluorescence staining of COX-1 **(A)** and COX-2 **(B)** in primary cultures of fibroblasts established from normal peritoneum, and immunofluorescence staining of COX-1 **(C)** and COX-2 **(D)** in adhesion fibroblasts established from adhesion tissues obtained from the same patient as described in methods.

adhesion fibroblasts (c). All fluorescence signals were detected at 63× magnification using an Axiovert 25 inverted microscope from Zeiss.

10. Using the Axiovision software, you can measure the average amount of your target of interest in a cell and compare it to the average amount of your other target.

4. Notes

1. Cells and tissues should remain wet after fixation throughout the staining process.
2. Paraformaldehyde should be handled in a chemical fume hood and always disposed off into a hazardous waste container for proper disposal.
3. 0.05% Tween-20 diluted in PBS helps to reduce background staining.

4. Some cells or tissues contain endogenous peroxidase. Using HRP conjugated antibody may result in high, nonspecific background staining. Pretreatment of your sample will reduce high background staining of tissues.
5. Formalin-fixation and paraffin-embedding forms protein cross-links that mask epitopes for COX-1 and COX-2; therefore, proper antigen retrieval such as heat, enzymatic digestion, or saponin should be utilized. However, do not use this pretreatment with frozen sections or cultured cells that are not paraffin-embedded.
6. Sodium Citrate Buffer antigen retrieval must be done at this point. The solution is designed to break the protein cross-links, unmasking the antigens and epitopes in formalin-fixed and paraffin embedded tissue sections, thus enhancing staining intensity of antibodies.
7. Normal serum should be of the same species as that of which your secondary antibody was raised. For example, if your primary antibody is mouse, and secondary antibody is goat anti-mouse, than normal goat serum should be used.
8. Nonspecific antibody staining may occur, besides poor knowledge of the antibody reactivity, due to factors such as endogenous enzymes or fluorochromes, endogenous antibody binding activity (Fc receptors), as well as cross reactivity of the secondary reagents with endogenous proteins. For this reason, it may be necessary to use normal serum of the same species as that of which your secondary antibody was raised instead of bovine serum albumin.
9. Single Immunofluorescence is performed by incubating with only one antibody. Incubate slides with either COX-1 or COX-2 primary antibody, diluted to 6.5 μg/ml and diluted 100× respectively, in 3% normal serum for 1 hour at room temperature in a humidified chamber in the dark.
10. Incubation with the primary antibody may also be done overnight at 4°C.
11. For double immunofluorescence, you must use antibodies from two different species to prevent cross-reactivity.
12. Many fluorophores are available for your conjugated secondary antibody. You may choose different colors based on your needs. Antibody dilution is determined by titration and should be performed by the end user to determine optimum concentration.
13. After slides have been sealed, you should protect them from light. Slides may be stored in the dark at 4°C for several months.
14. Axiovert 25 inverted microscope and filters were used (Zeiss), but any fluorescence capable microscope may be used.

15. We did stain the nuclei with DAPI, but it is not shown in the black and white photos.
16. Equivalent software and a microscope mounted camera can be used.

Acknowledgments

The author would like to thank Miss Nicole M. Fletcher, B.S. for patiently reviewing, editing, and optimizing the conditions to ensure reproducibility of this method.

References

1. Vane J. Towards a better aspirin. *Nature* 1994;367:215–6.
2. DuBois RN, Awad J, Morrow J, Roberts LJ, 2nd, Bishop PR. Regulation of eicosanoid production and mitogenesis in rat intestinal epithelial cells by transforming growth factor-alpha and phorbol ester. *J Clin Invest* 1994;93:493–8.
3. Kujubu DA, Fletcher BS, Varnum BC, Lim RW, Herschman HR. TIS10, a phorbol ester tumor promoter-inducible mRNA from Swiss 3T3 cells, encodes a novel prostaglandin synthase/cyclooxygenase homologue. *J Biol Chem* 1991;266:12866–72.
4. Chen G, Wilson R, McKillop JH, Walker JJ. The role of cytokines in the production of prostacyclin and thromboxane in human mononuclear cells. *Immunol Invest* 1994;23:269–79.
5. Schwab JM, Schluesener HJ, Laufer S. COX-3: just another COX or the solitary elusive target of paracetamol? *Lancet* 2003;361:981–2.
6. Shaftel SS, Olschowka JA, Hurley SD, Moore AH, O'Banion MK. COX-3: a splice variant of cyclooxygenase-1 in mouse neural tissue and cells. *Brain Res Mol Brain Res* 2003;119:213–5.
7. Ockleford CD. An atlas of antigens: fluorescence microscope localisation patterns in cells and tissues. New York, NY: Stockton Press, 1990.
8. Kupper H, Storz H. Double staining technique using a combination of indirect and direct immunofluorescence with monoclonal antibodies. *Acta Histochem* 1986;78:185–8.
9. Wheatley SP, Wang YL. Indirect immunofluorescence microscopy in cultured cells. *Methods Cell Biol* 1998;57:313–32.
10. Tornehave D, Hougaard DM, Larsson L. Microwaving for double indirect immunofluorescence with primary antibodies from the same species and for staining of mouse tissues with mouse monoclonal antibodies. *Histochem Cell Biol* 2000;113:19–23.

Chapter 18

Examining the Endogenous Antioxidant Response Through Immunofluorescent Analysis of Nrf2 in Tissue

Kathryn A. Lindl and Kelly L. Jordan-Sciutto

Abstract

As organisms designed to depend upon oxygen to sustain life, humans are necessarily and continually exposed to damaging oxidizing agents. As a vital protective measure, oxygen-dependent organisms have developed a highly evolutionarily conserved mechanism for preventing oxidative stress. NF-E2 (nuclear factor (erythroid-derived 2))-related factor-2 (Nrf2) is the primary regulator of this endogenous antioxidant response. Many diseases that plague human society, ranging from various cancers to neurodegenerative diseases, have oxidative stress as a component of their etiology, and thus, much disease research has focused on Nrf2, both as a potential point of biological failure and as a promising therapeutic target. As a transcription factor, Nrf2 is active only when it is nuclear, and is regulated largely by its subcellular distribution. Thus, Nrf2 protein levels and subcellular localization are both key pieces of information when studying the endogenous antioxidant response. Immunofluorescent analysis (IFA) of Nrf2 in human tissue is a particularly powerful tool in the study of Nrf2 in disease, because it allows examination of both of these regulatory mechanisms that modulate Nrf2 activity.

Key words: Nrf2, Immunofluorescence, Confocal microscopy, Endogenous antioxidant response.

1. Introduction

Oxidative stress is a constant presence in the cells that make up animal life forms and, thus, inevitably plays a significant role in the pathogenesis of many diseases. To combat this omnipresent oxidative stress, animal cells have evolved an endogenous antioxidant response. Nuclear factor (erythroid-derived 2) (NF-E2)-related factor 2 (Nrf2), is the key regulator of antioxidant response genes, making it an essential player in this endogenous cellular antioxidant response *(1, 2, 3, 4, 5, 6, 7)*.

From: *Methods in Molecular Biology, vol. 477: Advanced Protocols in Oxidative Stress I,*
Edited by: D. Armstrong, DOI: 10.1007/978-1-60327-517-0_18, © Humana Press, New York, NY

Nrf2 is a member of the Cap 'n' Collar family of transcription factors, which includes Nrf1, Nrf3, p45, Bach1, and Bach2 *(2)*. Nrf2 binds the antioxidant responsive element (ARE), also known as the electrophile response element (EpRE)), a *cis*-acting regulatory element found in the 5′-flanking region of genes encoding enzymes for antioxidant production and of other Nrf2 target genes *(2, 3, 8)*. Genes regulated by Nrf2 include heme oxygenase-1 (HO-1), ubiquitin/PKC-ζ-interacting protein A170, glutathione S-transferases (GST), NAD(P)H quinone oxidoreductase (NQO1), peroxiredoxin 1, catalase, glutathione peroxidase, superoxide dismutase (SOD), and thioredoxin *(1, 4, 9, 10, 11, 12)*. Additionally, Nrf2 regulates the expression of scavenger receptors, chaperone proteins, proteosome proteins, and transcriptional regulators *(9, 11)*.

Nrf2 is largely regulated through the control of its subcellular distribution, rather than through its induction, evidenced by the fact that Nrf2 mRNA levels remain relatively constant throughout treatment with oxidative stressors *(13, 14)*. Subcellular localization of Nrf2 is controlled by the actin-binding, cytoplasmic Kelch-like erythroid cell-derived protein with CNC homology (ECH)-associated protein 1 (Keap1), which binds the N-terminal Neh2 [Nrf2-ECH homology 2] domain of Nrf2, tethering it to the cytoplasm *(2, 14)* (**Fig. 18.1**). In addition to regulating Nrf2 subcellular localization, Keap1 also regulates Nrf2 protein levels, by mediating proteasomal degradation of Nrf2. Keap1 binds the ubiquitin ligase, Cullin 3 (Cul3) *(15)*. In effect, Keap1 acts as a bridge between Cul3 and Nrf2, resulting in rapid Nrf2 degradation *(16, 17)*. As shown in **Fig. 18.1**, in the presence of mediators of oxidative stress, such as reactive oxygen species (ROS), modification of Keap1 cysteines and

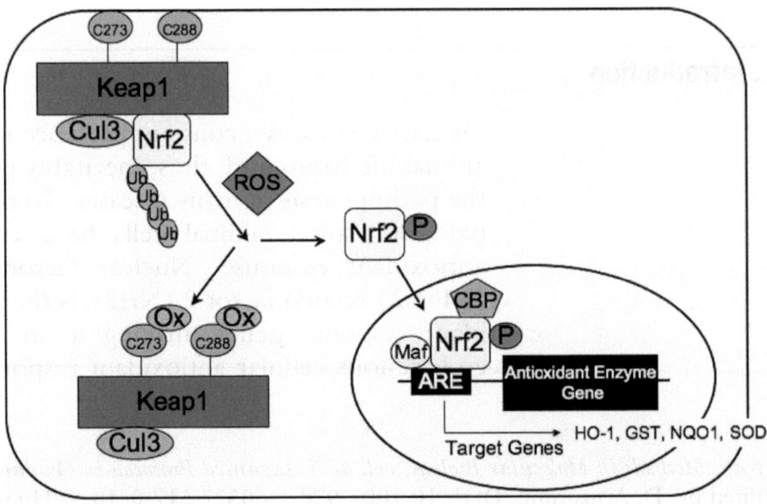

Fig. 18.1. Schematic for Nrf2 regulation and function.

phosphorylation of Nrf2 act cooperatively as regulatory mechanisms allowing Keap1 release of Nrf2, and consequently, nuclear translocation of Nrf2 *(2, 18, 19, 20, 21, 22, 23, 24, 25)*. This system of regulatory mechanisms provides precise control of Nrf2 activation.

While Nrf2 subcellular localization is a crucial determining factor of Nrf2 activity, several additional mechanisms modulate Nrf2 activity. Nrf2 DNA binding and transactivation both depend upon Nrf2 heterodimerazation with small Maf proteins *(1, 26)*. In addition to small Mafs, studies have suggested that c-Jun and activating transcription factor 4 (ATF4) also form heterodimers with Nrf2 and augment its transactivation activity *(27, 28)*. Further, Nrf2 has two transcriptional activation domains. These domains bind CBP (cyclic adenosine monophosphate (cAMP) response element binding protein (CREB) binding protein) or p300, which then mediate transcriptional activation of Nrf2 target genes *(29, 30)*. Synergistic activity of simultaneous CBP/p300 binding at both Nrf2 transactivation domains may explain the large potency of Nrf2 transactivation activity as compared with that of other CNC transcription factors *(9, 29)*.

The exquisite control over Nrf2 function (**Fig. 18.1**) implies an extreme importance of precise Nrf2 activity, and also provides a large number of potential points of malfunction and targets for therapeutic strategies in disease states affected by oxidative stress. Because Nrf2 is regulated by control of both its subcellular localization and its protein level, immunofluorescent analysis (IFA) is a particularly powerful tool for examining the endogenous antioxidant response as controlled by this transcription factor. In particular, IFA in human tissue allows study of the role Nrf2 plays in human disease. However, because Nrf2 is expressed at relatively low levels, particularly in certain tissues, such as brain, IFA of this protein is not straightforward. Further, IFA is only scientifically useful for semiquantitative analysis of protein levels and subcellular localization when it is performed with certain considerations to allow for distinction between slight variations in protein level and when it is analyzed with confocal laser microscopy.

2. Materials

2.1. Equipment

1. Microtome.
2. Biorad Radiance 2100 laser confocal microscope equipped with Argon, Green He/Ne, Red Diode, and Blue Diode lasers (Biorad, Hercules, CA, USA), as described previously *(31)*.

2.2. Reagents

1. Histoclear (National Diagnostics, Atlanta, GA, USA).
2. Ethanol (200 proof, Absolute, Anhydrous, ACS/USP Grade) in deionized H_2O (dI H_2O) at 70%, 90%, and 95%, and 100% ethanol.
3. A stable solution of 30% H_2O_2 (Sigma-Aldrich, St. Louis, MO, USA) is mixed with methanol (Thermo Fisher Scientific, Waltham, MA, USA) to a final concentration of 3% H_2O_2.
4. 10X Target Retrieval Solution (Dako, Carpinteria, CA, USA) in dI H_2O to its 1X concentration.
5. Phosphate buffered saline (PBS) (ScyTek Laboratories, Inc., Logan, UT, USA) purchased at 25X and diluted in dI H_2O to its final 1X concentration.
6. Normal goat serum (Chemicon (Millipore), Billerica, MA, USA) diluted in PBS to a final concentration of 5%.
7. Super Pap Pen (Ted Pella, Inc., Redding, CA, USA)
8. Normal Antibody Diluent (NAD) (ScyTek Laboratories, Inc., Logan, UT, USA).
9. Tyramide Signal Amplification (TSA Biotin System) (TSA Kit) (Perkin Elmer, Waltham, MA, USA).
10. Streptavidin FITC (Jackson ImmunoResearch Laboratories, Inc., West Grove, PA, USA).
11. Nucleic acid stains: DAPI (Molecular Probes), propidium iodide (Sigma-Aldrich), or Hoechst (Thermo Fisher Scientific) diluted in NAD according to manufacturers' recommendations.
12. RNase A (Roche Pharmaceuticals, Nutley, NJ, USA) diluted in NAD according to manufacturer's recommendation.
13. Citifluor AF1 mounting media (Citifluor, Ltd., London, UK).
14. Nail polish (any commercially available variety).
15. Primary antibodies: Nrf2 C-20, Nrf2 H-300 (Santa Cruz Biotechnology, Santa Cruz, CA, USA), Nrf2 MO1, Nrf2 MO3 (Abnova Corporation, Taipei City, Taiwan).
16. Secondary Antibodies (Jackson ImmunoResearch Laboratories Inc.): biotin-conjugated goat anti-rabbit antibody, biotin-conjugated goat anti-mouse antibody, fluorescein isothiocyanate (FITC)-conjugated goat anti-rabbit antibody, FITC-conjugated goat anti-mouse antibody.

2.3. Supplies

1. MetaMorph 6.0 image analysis software (Universal Imaging, Inc, Downingtown, PA, USA).

3. Methods

When the endogenous antioxidant response is active, the overall protein levels of Nrf2 increase and its subcellular localization is altered. Given the importance of these two regulatory mechanisms in Nrf2 activation, immunofluorescent analysis (IFA) is an especially powerful tool for the examination of Nrf2-mediated endogenous antioxidant activity in human tissues. However, Nrf2 is a relatively scarce protein, particularly in certain types of tissue, such as brain. Thus, to successfully visualize this protein using immunofluorescence, the signal must be amplified using tyramide signal amplification (TSA). This amplification method utilizes the enzymatic activity of horseradish peroxidase (HRP), which induces activation of short-lived, but highly reactive tyramide, which then covalently attaches to nucleophilic residues near the site to which the HRP is tethered. To start, Nrf2 is recognized by a primary antibody directed against it, which is then recognized by a secondary antibody that is conjugated to biotin. The biotin then binds HRP-conjugated streptavidin with high affinity, affixing the HRP in close vicinity to Nrf2 proteins. Then, biotinylated tyramide is added, and subsequently activated by the HRP causing it to covalently bind to nearby electrophilic residues. Finally, fluorescein isothiocyanate (FITC)-conjugated streptavidin binds to the biotinylated tyramide, allowing amplified visualization of Nrf2.

In addition to allowing visualization of scarce proteins, TSA also allows more flexibility in the choice of primary antibodies that can be used to triple- and quadruple-label tissue. Specifically, when TSA is not used, only one polyclonal antibody made in each type of animal may be used (e.g. only one rabbit antibody, not two, may be used in any given sample), or in the case of monoclonal antibodies, only one of each type of antibody isotype may be used. However, when using TSA, primary antibodies can be used at such a low dilution that they are not visible without the use of TSA. Thus, after TSA is used to amplify one antibody, even though a second secondary will recognize both primary antibodies made in the same animal, the TSA-amplified antibody will be such a small amount that it will not contribute to the signal visualized with the second secondary.

While TSA is a powerful tool for the visualization of Nrf2, it must be used appropriately if it is to allow for discrimination between protein levels in different experimental conditions (e.g. disease versus normal). Specifically, the appropriate dilution factor for the primary antibody must be empirically determined prior to data generation. Further, confocal laser microscopy must be used to visualize fluorescent staining of Nrf2, when attempting to accurately discern subcelluar localization of this protein.

3.1. Immunofluorescent labeling in paraffin-embedded tissue

1. Slice paraffin-embedded tissue into 7 μm sections using a microtome. Mount individual sections on (+) charged slides and allow them to dry, and, thus, they become fixed to the slide.
2. Heat paraffin-embedded sections to 55°C overnight. If necessary, this step can be shortened, but slides must be heated for at least 30 minutes. Then, deparaffinize slides by 3 incubations in Histoclear for 10 minutes each (*see* **Note 1**).
3. Rehydrate sections as follows: 2 incubations for 10 minutes each in 100% alcohol followed by 5-minute incubations in each of 95% alcohol, 90% alcohol, 70% alcohol, and deionized H_2O.
4. Incubate slides for 30 minutes in a solution of 3% H_2O_2 in methanol to quench endogenous peroxidase activity in the tissue, followed by a 5-minute wash in deionized H_2O.
5. Perform antigen unmasking by immersing slides for 1 hour in Target Retrieval Solution (DAKO) preheated to 95°C (*see* **Note 1**). After a 1-hour incubation in the Target Retrieval Solution at 95°C, remove the Coplan jar containing the slides from the water bath to the benchtop, and allow it to cool gradually to room temperature. After cooling, wash slides three times in PBS for 5 minutes each.
6. To minimize nonspecific binding of secondary antibodies, block slides by incubating them for 1 hour in a phosphate-buffered saline (PBS)-based solution of 10% normal serum of the animal in which the secondary antibody was produced. This incubation is done at room temperature. Unless otherwise indicated, all incubations and washes should be conducted at room temperature. After 1 hour of blocking, wash slides three times in PBS for 5 minutes each.
7. Remove slides from the PBS wash one at a time, dry the portions of the slide that have no tissue with a paper towel and encircle the tissue with the hydrophobic marking of a Pap Pen (*see* **Note 2**). Apply Nrf2 primary antibody to the area enclosed by the Pap Pen mark and incubate slides overnight at 4°C. Antibody should be diluted in Normal Antibody Diluent (NAD). To allow for discrimination between the levels of protein in different experimental conditions (e.g. diseased versus normal tissue), determination of the correct primary antibody dilution is critical, and must be empirically determined prior to use of the antibody for data generation. (*see* **Section 3.2** for a detailed procedure for empirically optimizing primary antibody dilution.) After overnight incubation, wash slides three times in PBS for 5 minutes each.

8. This step begins the TSA amplification process. Incubate slides for 30 minutes (*see* **Note 3**) in biotin-conjugated secondary antibody directed against immune fragments of the animal in which the primary antibody was made (e.g. use an antibody directed against rabbit IgG for a primary antibody made in a rabbit). For all secondary antibodies used throughout the course of the protocol, use antibodies made in the animal against which the tissue was blocked in step 6. Secondary antibodies should be diluted in NAD, following the manufacturer's recommendation (*see* **Note 4**). Following the 30-minute incubation, wash slides three times in PBS for 5 minutes each.
9. Incubate slides first in TNB (Tris NaCl Blocking)(TSA kit) Buffer for 30 minutes, then in horseradish peroxidase (HRP)-conjugated Streptavidin (TSA kit) diluted in TNB at a dilution factor of 1:400 for 30-minutes. Following these incubations, wash slides three times in PBS for 5 minutes each.
10. Incubate slides for 30 minutes in biotinylated tyramide, diluted 1:100 in amplification diluent (TSA kit). Following this incubation, wash slides three times in PBS for 5 minutes each (*see* **Note 5**).
11. Incubate slides for 30 minutes in Streptavidin conjugated to fluorescein isothiocyanate (FITC) or another fluorescent molecule whose excitation and emission frequencies are different from those of the nucleic acid markers to be used (e.g. DAPI or propidium iodide). Following this incubation, wash slides three times in PBS for 5 minutes each (*see* **Note 6**).
12. Incubate slides in second primary antibody for 1 hour at 37°C, 2 hours at room temperature, or overnight at 4°C (*see* **Note 7**). The second primary antibody may be a cell-type specific marker, used to allow determination of cell-type specific Nrf2 staining, or it may be another experimental protein of interest. However, the recognized protein must be sufficiently abundant so that its visualization does not require TSA amplification, as only one course of TSA amplification may be used in any staining protocol. Following this incubation, wash slides three times in PBS for 5 minutes each.
13. Incubate slides for 30 minutes in the second secondary antibody, which should be conjugated to cyanine 3 or another fluorescent molecule whose excitation and emission frequencies are different from the fluorescent molecule already used during TSA amplification and from the nucleic acid markers to be used. This antibody should be directed against immune fragments of the animal in which the second primary antibody was

made and should be diluted in NAD, following the manufacturer's recommendation. Following the 30-minute incubation, wash slides three times in PBS for 5 minutes each.

14. Steps 12 and 13 can be repeated for a third or fourth primary antibody. Please note, no two primary antibodies may be used that were made in the same type of animal, except for the host animal of the antibody amplified using TSA, which may be used twice (*see* **Section 3.3(3)** for necessary controls if this is to be the case). Additionally, each primary antibody must be visualized using a fluorescent molecule with distinct excitation and emission frequencies from others used in the same staining protocol.
15. If desired, incubate slides in a nucleic acid stain to label cell nuclei (*see* **Notes 8** and **9**). Dilute stain according to manufacturer's recommendation. If propidium iodide or Hoechst stains are to be used, slides must first be incubated in RNase A, diluted according to manufacturer's recommendation, for at least 30 minutes at room temperature.
16. Coverslip slide using three drops of Citifluor mounting medium (*see* **Note 10**). Seal the edges of the coverslip with nail polish to prevent leakage of the mounting medium and the resultant drying of the tissue.

3.2. Empirically optimizing primary antibody dilution

1. To allow for discrimination between the levels of protein in different experimental conditions (e.g. diseased versus normal tissue), use of the correct primary antibody dilution is critical, and must be empirically determined prior to use of the antibody for data generation. If too much antibody is used, the signal becomes saturated and any differences between experimental conditions are masked and cannot be seen, thereby resulting in a type II error (false positive) (*see* **Note 11**).
2. To empirically determine the appropriate primary antibody dilution, test different dilutions in one case from each experimental condition (e.g. disease versus normal), which will allow discernment of differences in levels of the target protein. For this process, it may be possible to avoid use of tissue that is highly experimentally valuable, in favor of tissue that is relevant but not as valuable. For example, when studying Alzheimer's disease, a disease that greatly affects hippocampal brain tissue, determination of appropriate primary antibody dilutions can often be conducted in a brain region that is more abundant, but still has pathology such as the anterior cingulate cortex, using both normal and diseased tissue.
3. Determination of the ideal dilution may require several rounds of empirical testing. To start, use dilutions that span

the range provided by the manufacturer, in a nonlinear distribution. For example, if the manufacturer recommends using a dilution between 1:100 and 1:1000, the testing dilutions should be 1:100, 1:300, and 1:1000. If the manufacturer does not provide recommendations and there are no previous publications using the antibody for IFA, start testing dilutions between 1:100 and 1:1000. Ideally, protein level differences will be visible with one of the dilutions used. However, if this is not the case, and if your lowest dilution factor does not show staining, then try lower dilution factors using a similar distribution scheme. Alternatively, if even the most dilute antibody results in very robust staining, then repeat the process using higher dilutions with a similar distribution (i.e. 1:1000, 1:3000, 1:10000). When one dilution shows potential as the optimal dilution, one more test should be conducted using dilutions closer to the chosen dilution. *See* **Fig. 18.2** for an example protocol scheme.

3.3. Controls

1. During empirical determination of the appropriate dilution factor for the primary antibody, two controls must be performed. Both of these controls need only be performed once for a given type of tissue (species and tissue type) and a given antibody. However, when a new lot of any antibody (primary or secondary) is purchased, these controls should be rerun, to ensure that the antibody has not changed from lot to lot.
2. The first control is conducted to ensure that the secondary antibody does not cause nonspecific background staining and must be conducted each time a new secondary antibody is used, regardless of the primary antibody being used. Prepare

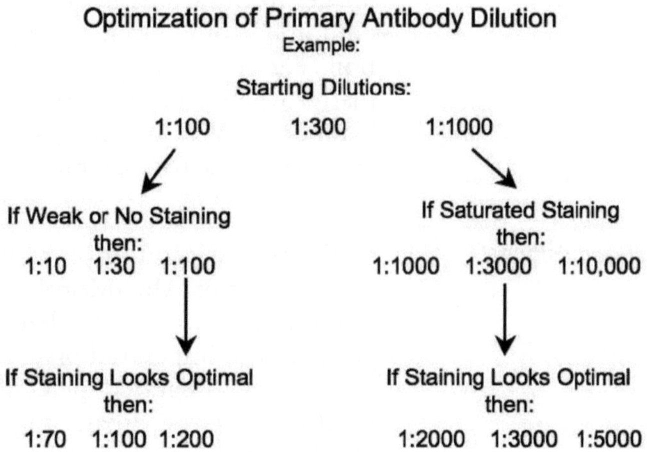

Fig. 18.2. Optimization of primary antibody dilution.

one slide that is not incubated in primary antibody. Instead, this slide should be incubated overnight in NAD as the other slides are incubating in primary antibody (*see* **Section 3.1.**(7)). Following this step, treat this control slide exactly as the other slides are treated, following each step of the protocol. This control slide should show no staining.

3. The second control acts to ensure that there is no visible Nrf2 signal if the primary antibody signal is not amplified using TSA. This control must be performed every time a new primary antibody is used with TSA, regardless of the secondary antibody being used, and is particularly important if a second primary antibody made in the same animal will be used (e.g. if using both a rabbit Nrf2 antibody and a rabbit cell-type marker antibody in the same slides). This control slide is prepared by skipping the TSA process. Thus, to prepare this slide, at step 8 (*see* **Section 3.1**(**8**)), which starts the TSA amplification, rather than using the biotin-conjugated secondary antibody, incubate your control slide with a secondary antibody conjugated to the same fluorescent molecule as that conjugated to the streptavidin in the final TSA step (*see* **Section 3.1**(**11**)). Do NOT treat this slide with the remaining TSA steps, but rather, continue with the protocol starting from incubation with the second primary (*see* **Section 3.1**(**12**)). This control slide should show no staining.

3.4. Confocal laser microscopy image capture

1. Images should be captured using confocal laser microscopy to allow determination of Nrf2 subcellular localization, specifically whether it is in the nucleus or the cytoplasm. When choosing which fluorescent molecules to use for visualization of antibody staining, consider the lasers included in available confocal microscopes. Collect 5–10 images from each tissue section. Images should be collected randomly from within the regions of interest in each section (e.g. in cortical brain tissue consider gray versus white matter when collecting images) and in a researcher-blinded manner. Collect all images at uniform settings to allow comparisons across samples.

3.5. Post-acquisitional analysis

1. Perform post-acquisition analysis for immunofluorescent staining using MetaMorph 6.0 image analysis software (*see* **Section 2.3**(**1**)). To determine total Nrf2 protein levels, measure integrated pixel intensity for Nrf2 per image, where the integrated pixel intensity is defined as total pixel intensity per image times the area of pixels positive for the signal. For colocalization of Nrf2 with specific cell types in each image, examine integrated pixel intensity for

Nrf2 that is overlapping with the pixels positive for your specific cell-type marker. Often when determining Nrf2 colocalization with specific cell types, it is necessary to normalize the values to the area of pixels positive for the specific phenotypic marker to account for variations in phenotypic marker expression in different experimental conditions. Additionally, Nrf2 subcellular localization can be examined in a similar manner by determining Nrf2 colocalization with nuclear markers, such as propidium iodide.

4. Results

1. AD is a region-specific neurodegenerative disease, affecting in particular the hippocampus and various parts of the cortex. Examination of Nrf2 in hippocampal autopsy tissue from AD patients reveals that Nrf2 subcelluar localization is altered as compared with control autopsy hippocampal tissue *(32)*. In this study by Ramsey *et al.*, phenotypic cell-type specific markers and nucleic acid stains demonstrate that Nrf2 is present in both neurons and astrocytes, and allow discrimination of different Nrf2 subcellular staining patterns in these two cell types (**Figure 2** from *(32)*). Specifically, the findings presented by Ramsey *et al.* show that in normal hippocampal CA1 neurons, Nrf2 is predominantly nuclear, while in AD hippocampal CA1 neurons, it is predominantly cytoplasmic, and therefore not transcriptionally active. These findings suggest that Nrf2 is unable to initiate the endogenous antioxidant response in neurons in the hippocampi of AD patients, and may thereby contribute to pathogenesis of this disease that tragically affects so many lives.
2. The protocol outlined in this chapter can be extended to examination of proteins other than Nrf2. As can be seen in Lindl et al *(33)*, the endoplasmic reticulum (ER) stress response protein, Binding Protein (BiP), can be visualized using IFA with TSA in human midfrontal cortical autopsy tissue from HAD patients. In this tissue, BiP total protein levels were quantified using Metamorph software and were seen to increase in gray matter. Additionally, astrocytic BiP increased as determined using Metamorph software to analyze BiP staining that colocalized with astrocytic marker, GFAP staining. These findings demonstrate that the ER stress response is activated in the midfrontal cortical gray matter in HAD autopsy tissue.

5. Notes

1. The antigen-unmasking step (*see* **Section 3.1(5)**) is usually done in a Coplan jar, which will crack if heated quickly. Thus, it is prudent to prepare the Target Retrieval Solution in a Coplan jar while the slides are in the first step of Histoclear, and place the jar in a water bath set to 37°C and allow the Target Retrieval Solution to heat as the water bath heats up to 95°C.
2. When applying Pap Pen, be careful not to mark over the tissue and do not allow the tissue to dry out.
3. As a general rule, all 30-minute incubations should be limited to 30 minutes. Specifically: do not allow slides to incubate in secondary antibodies for much longer than 30 minutes, as this may increase nonspecific binding of the secondary antibody; do not allow slides to incubate in Streptavidin-HRP or -FITC for much longer than 30 minutes (*see* **Section 3.1(9)**), as this may increase background staining; finally, **it is critical** that slides are not allowed to incubate in biotinyl-tyramide for longer than 30 minutes (*see* **Section 3.1(10)**), as this step utilizes an enzymatic activity and, thus, additional incubation time may significantly increase staining, resulting in an over-amplification of both specific and nonspecific antibody signal. Two exceptions to this general rule: 1) endogenous peroxidase quenching (*see* **Section 3.1(4)**), which can be extended to several hours and 2) the TNB incubation (*see* **Section 3.1(9)**), which can be greatly extended, but care should be taken that TNB does not remain on slides for a length of time that would allow contamination of the buffer.
4. Occasionally, the manufacturers, recommended dilutions for secondary antibodies are not optimal and result in high non-specific or background staining. If this is the case, apply the steps outlined in **Section 3.1(1)** for optimization of the primary antibody dilution to optimization of the secondary antibody dilution.
5. After step 10 (*see* **Section 3.1(10)**), slides should be kept in the dark as much as possible. Changing of solutions can be done in a light room, but during washes and incubations, slides should be kept in a dark place.
6. If the Nrf2 antibody is the only primary antibody that will be used, proceed directly from step 11 to step 16 (*see* **Sections 3.1(11)–3.1(16)**).
7. Incubations in primary antibody should be carried out for 1 hour at 37°C, 2 hours at room temperature, or overnight at 4°C. Ideally, slides should be incubated in the Nrf2 antibody overnight at 4°C. Further, any additional experimental proteins

of interest should be incubated overnight at 4°C, ideally. However, antibodies to cell-type specific markers often work well in any of the three incubation conditions, and do not result in nonspecific background staining or weak specific staining. In particular, antibodies to microtubule-associated protein 2 (MAP2) and glial fibrillary-associated protein (GFAP) are known to work at any of the incubation conditions.

8. While all nucleic acid stains thoroughly mark astrocytic and oligodendrocyte nuclei, propidium iodide most clearly marks neuronal nuclei.

9. To save time, RNase A and nucleic acid stains can be included in other incubations; in particular, secondary antibody incubations, since they utilize the same conditions, are particularly convenient. However, please note the RNase A step *must be completed prior* to adding propidium iodide or Hoechst stain.

10. When coverslipping, avoid trapping air bubbles on top of the tissue in the Citifluor, as this will distort the visualization of the tissue. To prevent air bubbles, roll the coverslip onto the tissue from one end of the slide, rather than dropping it directly onto the slide.

11. The ideal antibody dilution will result in a signal just stronger than that required to visualize the protein of interest, thus avoiding saturation of the signal.

12. This protocol can be used for immunohistochemistry with some slight modification. Specifically, rather than using fluorescent-tagged streptavidin in the final TSA step (*see* **Section 3.1**(**11**)), use Strep-HRP as used in step 9 (*see* **Section 3.1**(**9**)). To visualize Nrf2 staining in this case, use a kit such as Vector Red (Vector Laboratories, Burlingame, CA, USA) according to manufacturer's recommendation. Then, dehydrate slides by immersion for 5 minutes each in dI H_2O, 70% ethanol, 90% ethanol, 95% ethanol, and two times in 100% ethanol. Finally, coverslip the slides using Permount. Keep in mind that immunohistochemistry does not have the quantitative power of IFA and that it does not allow for the use of cell-type markers to distinguish Nrf2 expression in different cell types.

13. This protocol can be used for IFA in tissue culture with some slight modification. First, wash cells three times in PBS for 5 minutes each and then two times for 10 minutes each in PBS with 0.1% Tween-20. Then, fix cells using a 30-minute incubation period in 4% paraformadlehyde/2% sucrose in PBS. Skip the steps required for deparaffinization, and proceed directly to blocking. However, include 0.1% Triton-X 100 in the blocking solution, to ensure the cells are fully permeablized.

References

1. Itoh, K., Chiba, T., Takahashi, S., Ishii, T., Igarashi, K., Katoh, Y., Oyake, T., Hayashi, N., Satoh, K., Hatayama, I., Yamamoto, M. and Nabeshima, Y. (1997) An Nrf2/small Maf heterodimer mediates the induction of phase II detoxifying enzyme genes through antioxidant response elements. *Biochem Biophys Res Commun*, 236, 313–22.
2. Motohashi, H. and Yamamoto, M. (2004) Nrf2-Keap1 defines a physiologically important stress response mechanism. *Trends Mol Med*, 10, 549–57.
3. Lee, J.M. and Johnson, J.A. (2004) An important role of Nrf2-ARE pathway in the cellular defense mechanism. *J Biochem Mol Biol*, 37, 139–43.
4. McMahon, M., Itoh, K., Yamamoto, M., Chanas, S.A., Henderson, C.J., McLellan, L.I., Wolf, C.R., Cavin, C. and Hayes, J.D. (2001) The Cap'n'Collar basic leucine zipper transcription factor Nrf2 (NF-E2 p45-related factor 2) controls both constitutive and inducible expression of intestinal detoxification and glutathione biosynthetic enzymes. *Cancer Res*, 61, 3299–307.
5. Chan, K. and Kan, Y.W. (1999) Nrf2 is essential for protection against acute pulmonary injury in mice. *Proc Natl Acad Sci U S A*, 96, 12731–6.
6. Kraft, A.D., Johnson, D.A. and Johnson, J.A. (2004) Nuclear factor E2-related factor 2-dependent antioxidant response element activation by tert-butylhydroquinone and sulforaphane occurring preferentially in astrocytes conditions neurons against oxidative insult. *J Neurosci*, 24, 1101–12.
7. Leung, L., Kwong, M., Hou, S., Lee, C. and Chan, J.Y. (2003) Deficiency of the Nrfl and Nrf2 transcription factors results in early embryonic lethality and severe oxidative stress. *J Biol Chem*, 278, 48021–9.
8. Itoh, K., Tong, K.I. and Yamamoto, M. (2004) Molecular mechanism activating Nrf2-Keap1 pathway in regulation of adaptive response to electrophiles. *Free Radic Biol Med*, 36, 1208–13.
9. Kobayashi, M. and Yamamoto, M. (2005) Molecular mechanisms activating the Nrf2-Keap1 pathway of antioxidant gene regulation. *Antioxid Redox Signal*, 7, 385–94.
10. Kwak, M.K., Itoh, K., Yamamoto, M., Sutter, T.R. and Kensler, T.W. (2001) Role of transcription factor Nrf2 in the induction of hepatic phase 2 and antioxidative enzymes in vivo by the cancer chemoprotective agent, 3H-1, 2-dimethiole-3-thione. *Mol Med*, 7, 135–45.
11. Lee, J.M., Calkins, M.J., Chan, K., Kan, Y.W. and Johnson, J.A. (2003) Identification of the NF-E2-related factor-2-dependent genes conferring protection against oxidative stress in primary cortical astrocytes using oligonucleotide microarray analysis. *J Biol Chem*, 278, 12029–38.
12. Ishii, T., Itoh, K., Takahashi, S., Sato, H., Yanagawa, T., Katoh, Y., Bannai, S. and Yamamoto, M. (2000) Transcription factor Nrf2 coordinately regulates a group of oxidative stress-inducible genes in macrophages. *J Biol Chem*, 275, 16023–9.
13. Numazawa, S. and Yoshida, T. (2004) Nrf2-dependent gene expressions: a molecular toxicological aspect. *J Toxicol Sci*, 29, 81–9.
14. Itoh, K., Wakabayashi, N., Katoh, Y., Ishii, T., Igarashi, K., Engel, J.D. and Yamamoto, M. (1999) Keap1 represses nuclear activation of antioxidant responsive elements by Nrf2 through binding to the amino-terminal Neh2 domain. *Genes Dev*, 13, 76–86.
15. Kobayashi, A., Kang, M.I., Okawa, H., Ohtsuji, M., Zenke, Y., Chiba, T., Igarashi, K. and Yamamoto, M. (2004) Oxidative stress sensor Keap1 functions as an adaptor for Cul3-based E3 ligase to regulate proteasomal degradation of Nrf2. *Mol Cell Biol*, 24, 7130–9.
16. Cullinan, S.B., Gordan, J.D., Jin, J., Harper, J.W. and Diehl, J.A. (2004) The Keap1-BTB protein is an adaptor that bridges Nrf2 to a Cul3-based E3 ligase: oxidative stress sensing by a Cul3-Keap1 ligase. *Mol Cell Biol*, 24, 8477–86.
17. Furukawa, M. and Xiong, Y. (2005) BTB protein Keap1 targets antioxidant transcription factor Nrf2 for ubiquitination by the Cullin 3-Roc1 ligase. *Mol Cell Biol*, 25, 162–71.
18. Dinkova-Kostova, A.T., Holtzclaw, W.D., Cole, R.N., Itoh, K., Wakabayashi, N., Katoh, Y., Yamamoto, M. and Talalay, P. (2002) Direct evidence that sulfhydryl groups of Keap1 are the sensors regulating induction of phase 2 enzymes that protect against carcinogens and oxidants. *Proc Natl Acad Sci U S A*, 99, 11908–13.
19. Kobayashi, A., Kang, M.I., Watai, Y., Tong, K.I., Shibata, T., Uchida, K. and Yamamoto, M. (2006) Oxidative and

electrophilic stresses activate Nrf2 through inhibition of ubiquitination activity of Keap1. *Mol Cell Biol*, 26, 221–9.

20. Huang, H.C., Nguyen, T. and Pickett, C.B. (2000) Regulation of the antioxidant response element by protein kinase C-ediated phosphorylation of NF-E2-related factor 2. *Proc Natl Acad Sci U S A*, 97, 12475–80.
21. Huang, H.C., Nguyen, T. and Pickett, C.B. (2002) Phosphorylation of Nrf2 at Ser-40 by protein kinase C regulates antioxidant response element-mediated transcription. *J Biol Chem*, 277, 42769–74.
22. Numazawa, S., Ishikawa, M., Yoshida, A., Tanaka, S. and Yoshida, T. (2003) Atypical protein kinase C mediates activation of NF-E2-related factor 2 in response to oxidative stress. *Am J Physiol Cell Physiol*, 285, C334–42.
23. Bloom, D.A. and Jaiswal, A.K. (2003) Phosphorylation of Nrf2 at Ser40 by protein kinase C in response to antioxidants leads to the release of Nrf2 from INrf2, but is not required for Nrf2 stabilization/accumulation in the nucleus and transcriptional activation of antioxidant response element-mediated NAD(P)H:quinone oxidoreductase-1 gene expression. *J Biol Chem*, 278, 44675–82.
24. Kang, K.W., Lee, S.J., Park, J.W. and Kim, S.G. (2002) Phosphatidylinositol 3-kinase regulates nuclear translocation of NF-E2-related factor 2 through actin rearrangement in response to oxidative stress. *Mol Pharmacol*, 62, 1001–10.
25. Cullinan, S.B., Zhang, D., Hannink, M., Arvisais, E., Kaufman, R.J. and Diehl, J.A. (2003) Nrf2 is a direct PERK substrate and effector of PERK-dependent cell survival. *Mol Cell Biol*, 23, 7198–209.
26. Motohashi, H., Katsuoka, F., Engel, J.D. and Yamamoto, M. (2004) Small Maf proteins serve as transcriptional cofactors for keratinocyte differentiation in the Keap1-Nrf2 regulatory pathway. *Proc Natl Acad Sci U S A*, 101, 6379–84.
27. He, C.H., Gong, P., Hu, B., Stewart, D., Choi, M.E., Choi, A.M. and Alam, J. (2001) Identification of activating transcription factor 4 (ATF4) as an Nrf2-interacting protein. Implication for heme oxygenase-1 gene regulation. *J Biol Chem*, 276, 20858–65.
28. Venugopal, R. and Jaiswal, A.K. (1998) Nrf2 and Nrf1 in association with Jun proteins regulate antioxidant response element-mediated expression and coordinated induction of genes encoding detoxifying enzymes. *Oncogene*, 17, 3145–56.
29. Katoh, Y., Itoh, K., Yoshida, E., Miyagishi, M., Fukamizu, A. and Yamamoto, M. (2001) Two domains of Nrf2 cooperatively bind CBP, a CREB binding protein, and synergistically activate transcription. *Genes Cells*, 6, 857–68.
30. Zhu, M. and Fahl, W.E. (2001) Functional characterization of transcription regulators that interact with the electrophile response element. *Biochem Biophys Res Commun*, 289, 212–9.
31. Strachan, G.D., Kopp, A.S., Koike, M.A., Morgan, K.L. and Jordan-Sciutto, K.L. (2005) Chemokine- and neurotrophic factor-induced changes in E2F1 localization and phosphorylation of the retinoblastoma susceptibility gene product (pRb) occur by distinct mechanisms in murine cortical cultures. *Exp Neurol*, 193, 455–68.
32. Ramsey, C.P., Glass, C.A., Montgomery, M.B., Lindl, K.A., Ritson, G.P., Chia, L.A., Hamilton, R.L., Chu, C.T. and Jordan-Sciutto, K.L. (2007) Expression of Nrf2 in neurodegenerative diseases. *J Neuropathol Exp Neurol*, 66, 75–85.
33. Lindl, K.A., Akay, C., Wang, Y., White, M.G. and Jordan-Sciutto, K.L. (2007) Expression of the endoplasmic reticulum stress response marker, BiP, in the central nervous system of HIV-positive individuals. *Neuropathol Appl Neurobiol, 33,* 658–69.

Chapter 19

Determination of Oxidized and Reduced CoQ10 and CoQ9 in Human Plasma/Serum Using HPLC-ECD

Ian N. Acworth, Paul A. Ullucci, and Paul H. Gamache

Abstract

This chapter describes the use of reversed-phase HPLC with multichannel coulometric electrochemical detection for the routine, sensitive, and simultaneous measurement of oxidized and reduced CoQ10 and CoQ9 in human plasma and serum. Analytes are first resolved chromatographically prior to electrochemical detection using three serially placed flow-through coulometric sensors set for oxidation—reduction—re-oxidation. Such electrochemical manipulation of analytes not only improves selectivity and specificity (decreasing the likelihood of co-elution), but also leads to improved sensitivity and decreased noise. The method is completed in <18 min, shows excellent linearity, good intra-day (% RSD = 1.2–2.3) and inter-day (% RSD 2.2–3.9) precision, and has a limit of detection to low pg levels (on column). This approach was used to measure oxidized and reduced CoQ10 and CoQ9 in 30 human plasma samples, and oxidized and reduced CoQ10 in 10 human serum samples (NIST Micronutrients Measurement Quality Assurance Program for fat-soluble vitamins).

Key words: Ubiquinone, ubiquinol, Coenzyme Q10, Coenzyme Q9, CoQ10, CoQ9, CoQ10 total, CoQ10 speciated, HPLC, Electrochemical detection, Human plasma, Human serum.

1. Introduction

CoQ10 (Coenzyme Q10; Ubiquinone/ubiquinol; Ubiquinone 50; 2,3-dimethoxy-5-methyl-6-decaprenylbenzoquinone or "Vitamin Q") is an ubiquitous, lipid-soluble, redox active compound, found in most cell membranes and in the circulation, where it is carried by lipoproteins. Structurally, CoQ10 is just one member of a family of related compounds each containing a benzoquinone head group

From: *Methods in Molecular Biology, vol. 477: Advanced Protocols in Oxidative Stress I,*
Edited by: D. Armstrong, DOI: 10.1007/978-1-60327-517-0_19, © Humana Press, New York, NY

attached to a lipophilic side chain consisting of between 1 and 12 isoprenyl subunits. The side-chain serves to "anchor" the molecule to the membrane. In humans, the predominant form contains 10 subunits (CoQ10) whereas in rodents nine subunits (CoQ9) is typically the major form. Recent evidence suggests that CoQ9 may also play an important physiological role in humans *(1)*.

The biological activity of CoQ10 results from its ability to redox cycle–accept or donate electrons as the benzoquinone head group inter-converts between the oxidized (quinone – CoQ10 Ox) and reduced (quinol – CoQ10 Red) forms (**Fig. 19.1**). Consequently, CoQ10 acts as an electron carrier of the electron transport chain of the inner mitochondrial membrane *(2)*; it is a chain-breaking antioxidant either reacting directly with lipid peroxyl radicals *(3)*, or through regeneration of α-tocopherol *(3)*, and may be critical in the prevention of tocopherol-mediated peroxidation of LDL *(4)*. By transporting protons across the lysosomal membrane, CoQ10 is involved in maintaining optimal pH of this organelle *(5)*.

CoQ10 is synthesized in most human tissues through a complex biosynthetic pathway involving multiple enzymes and

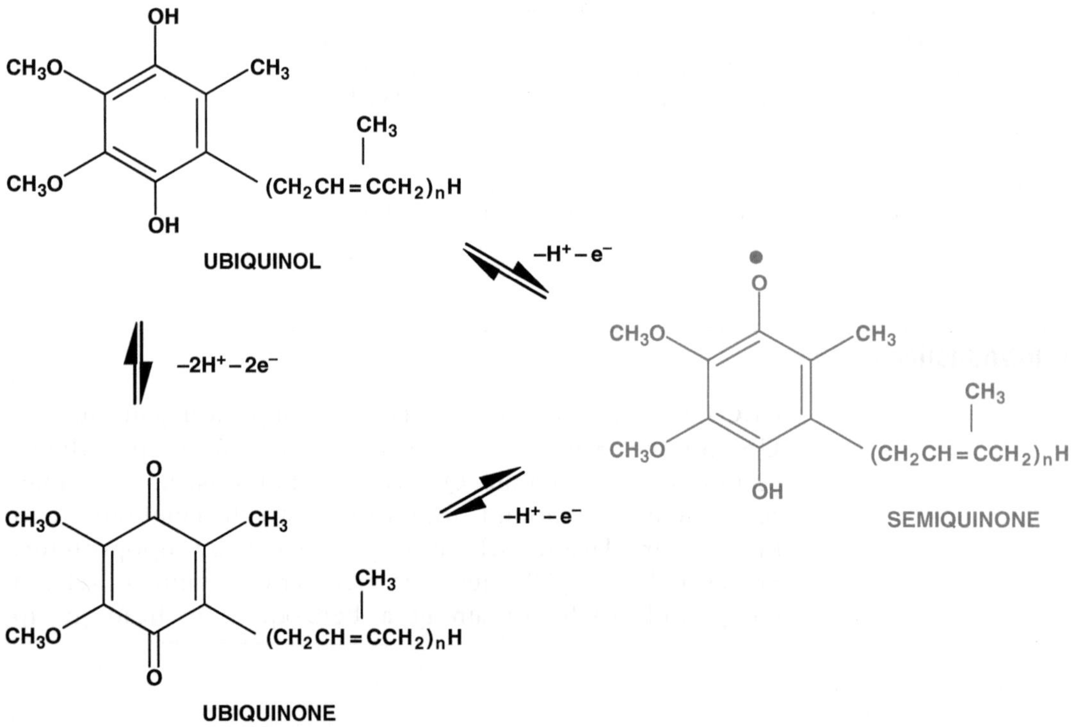

Fig. 19.1. Structures of the various redox forms of ubiquinone (n = 9 or 10). Electrochemical detection involves a 2-electron mechanism (presented to the left), whereas many biologically important mechanisms involve single electron reactions (right).

regulatory proteins *(6)*; only a small proportion is acquired through the diet (typically from meat and fish). To date, no CoQ10 deficiency symptoms have been reported in the general population, so it is assumed that both endogenous synthesis and diet provide sufficient CoQ10 for healthy individuals. CoQ10 levels can be decreased by aging, disease, and drugs. For example, primary human CoQ10 deficiency, a rare autosomal recessive condition, is associated with five different clinical presentations including encephalomyopathy, cerebellar ataxia, and Leigh syndrome *(6)*. Cholesterol lowering drugs such as statins work by inhibiting the key enzyme in cholesterol biosynthesis, hydroxymethylglutaryl (HMG)-CoA reductase. Unfortunately, this enzyme also plays a critical role in the regulation of CoQ10 biosynthesis. Recent research now suggests that statin-induced decreases in CoQ10 levels are implicated in cardiomyopathy, and that patients taking statins should also receive CoQ10 supplementation *(7, 8, 9)*.

The CoQ10 redox ratio [CoQ10 Red]/([CoQ10 Red] + [CoQ10 Ox]) is often used as an indicator of oxidative stress *(3, 10)*, and is reported to be decreased in a variety of diseases including Parkinson's, hepatitis, liver cirrhosis, carcinoma, hyperlipidemia, heart disease, and β-thalassemia (*(10)* and references therein).

Presented in this chapter is a modified method based on that of Tang *et al. (10, 11)*, that is capable of simultaneously and routinely measuring the oxidized and reduced forms of CoQ9 and CoQ10 in human plasma/serum extracts by high performance liquid chromatography with electrochemical detection (HPLC-ECD). A series of three coulometrically efficient working electrodes are used to electrochemically manipulate the analytes, thereby improving selectivity and sensitivity. This controlled-potential electrochemical approach also provides greater simplicity and reproducibility when compared to methods that involve chemical treatment of samples or on-line catalysis (e.g., Jones reactor) to generate the Red or Ox forms.

2. Materials

2.1. Analytical HPLC System

All from ESA Biosciences, Inc.

1. Pump – Model 584 – EC compatible.
2. Autosampler – Model 540, or 542 – biocompatible with tray cooling.
3. CoulArray® – Model 5600A – 4-channel system.
4. Analytical cell (dual coulometric electrodes) – Model 5011A.

5. Conditioning cell (single coulometric electrode) – Model 5021A.
6. Thermal organizer.
7. PEEK Pulse Damper.
8. Software – CoulArray®.
9. Column – Shiseido Capcell Pak C18 MG 100 (4.6 × 50 mm; 3μ).
10. Filters – post pump & pre-column, 0.5 μm PEEK.

2.2. Miscellaneous Equipment

1. Vortex-Genie 2 Mixer (or equivalent).
2. Microcentrifuge, Fisher accuSpin, Micro AR.
3. pH Meter, Fisher Accumet AB15 or equivalent and appropriate pH linearity calibration solutions (e.g., pH 1 and pH 5).
4. UV filtered lighting: Amber sleeves to enclose fluorescent light bulbs to block light from approximately 550 nm and below.
5. 5 and 10 mL Class A volumetric flasks.
6. Positive displacement microliter pipettes: 20, 50, 100, 200, 500 μL and replacement tips.
7. Microcentrifuge tubes, 1.5 mL.
8. Volumetric pipettes (Class "A") − 0.5, 1.0, 2.0 mL.
9. 12 × 32 mm, 750 μL Limited Volume autosampler vials, polypropylene, (e.g., Alltech 98050 or equivalent).
10. Glass Pasteur pipettes.
11. Zymark TurboVap evaporation unit.
12. Rack for 16 × 100 mm tubes.
13. 16 × 100 mm disposable glass tubes (e.g., Fisher (14-961-29) or equivalent.
14. Pre-purified nitrogen.
15. Nitrogen tank and regulator.
16. Centrifuge.
17. Vacutainer tubes containing Heparin as anticoagulant (Baxter Scientific).
18. Spectrophotometer, Shimadzu, UV-1601 or equivalent.

2.3. Reagents

1. 1-Propanol: HPLC Grade.
2. 2-Propanol: HPLC Grade.
3. Methanol: HPLC Grade.

4. Ammonium acetate: HPLC Grade.
5. Acetic acid, glacial.
6. De-ionized water: 18.2 MΩ-cm.
7. Coenzyme Q10 (Ubiquinone-50): Sigma C 9538, 100 mg.
8. Coenzyme Q9 (Ubiquinone-45): Fluka 27597, 1 mg ampoule or Sigma C9888, 2 mg ampoule.
9. Human Serum, Analyte Stripped, Delipidized: Biocell Laboratories, Inc. Rancho Dominguez, CA, Cat. No. 1131–00. Material must be free of both CoQ10 and CoQ9.
10. Ethanol, anhydrous reagent.
11. Hexane: HPLC Grade
12. Sodium borohydride, granular.

3. Methods

3.1. Samples (see Notes 1 and 2*)*

CoQ9 and CoQ10 are light sensitive. Protect samples from light. Laboratory should be equipped with UV-filtered lighting.

3.1.1. Plasma

Collect whole blood into a 7 mL Vacutainer-type tube containing heparin as anticoagulant. EDTA cannot be used as an anticoagulant for this method (*see* **Note 3**). Mix by inversion a few times, then immediately centrifuge the samples for 10 min, 10,000 rpm, at 4°C. Transfer plasma to an amber plastic vial and store at −80°C until analyzed. Do not subject samples to additional freeze-thaw cycles.

3.1.2. Serum

Collect whole blood into a 7 mL Vacutainer-type tube containing no anticoagulant. Let the specimen stand at room temperature for 15 min to allow clotting. Centrifuge, transfer serum, and store as in 3.1.1. Do not subject samples to additional freeze-thaw cycles.

3.2. Standards (see Notes 1 and 4*)*

Prepare standard diluent: 80/20 v/v 2-propanol/de-ionized water.

3.2.1. Stock Standards

1. CoQ10 Ox Stock Standard, nominally 100 μg/mL actual concentration must be determined, *see* **Section 3.2.7**. Weigh 10.0 mg of Coenzyme Q10 and dissolve in 100 mL of 2-propanol.
2. CoQ9 Ox Stock Standard, nominally 100 μg/mL actual concentration must be determined, *see* **Section 3.2.7**. Dissolve contents of 1 mg vial into 10 mL of 2-propanol.

3.2.2. Ox Sub Stock Standard Mixture

1. CoQ9/CoQ10 Ox Sub-Stock standard mix, 1 μg/mL. To a 10.0 mL volumetric, flask, add 100 μL of the CoQ9 Ox Stock and 100 μL of the CoQ10 Ox Stock. Dilute to volume with standard diluent.

3.2.3. Ox Working Calibrators

1. Dilute Ox Sub Stock Standard Mixture using standard diluent according to **Table 19.1** below. Use standard diluent for the zero calibrator.

3.2.4. Red Stock Standard

1. CoQ9/CoQ10 Sub-Stock, 1 μg/mL: freshly prepared as in 3.2.6.

3.2.5. Red Working Calibrators

1. Dilute Red Stock Standards using standard diluent according to **Table 19.2** below. Use standard diluent for the zero calibrator.

3.2.6. Preparation of CoQ9 Red and CoQ10 Red

Reduced Q9 and Q10 are prepared from the oxidized form by borohydride reduction *(11)* (*see* **Note 5**).

1. Into a 15 mL polypropylene tube add 100 μL of the 100 μg/mL Q9 Ox Stock Standard and 100 μL of the 100 μg/mL Q10 Ox Stock Standard prepared in **Section 3.2.1**.
2. Prepare 0.05 M sodium borohydride ($NaBH_4$) by dissolving 200 mg of $NaBH_4$ in 100 mL of deionized water. Must be prepared daily.
3. Add 0.80 mL of ethanol, and 100 μL of 0.05 M $NaBH_4$.
4. Vortex for 1–2 min.

Table 19.1
Working ox calibration standards

Volume (mL) of 1 μg/mL mixture[a]	Final volume (mL)[b]	Nominal concentration (ng/mL)[c]
4.00	10.0	400
2.00	10.0	200
1.00	10.0	100
0.50	10.0	50
0.20	10.0	20
0	10.0	0

[a]CoQ9/CoQ10 Ox Sub-Stock standard mix, 1 μg/mL, prepared in **Section 3.2.2**.
[b]Dilute with standard diluent, prepared in **Section 3.2**.
[c]Before correction based on molar absorbtivity, *see* **Section 3.2.7**.

Table 19.2
Working red calibration standards

Volume (mL) of 1 μg/mL mixture[a]	Final volume (mL)[b]	Nominal concentration (ng/mL)[c]
2.00	5.0	400
1.00	5.0	200
0.50	5.0	100
0.25	5.0	50
0.10	5.0	20
0	5.0	0

[a]CoQ9/CoQ10 Red Sub-Stock standard mix, 1 μg/mL, prepared in **Section 3.2.6**.
[b]Dilute with standard diluent, prepared in **Section 3.2**.
[c]Before correction based on molar absorbtivity, *see* **Section 3.2.7**.

5. Incubate at RT for 30 min in the dark.
6. Add 10 mL of hexane and mix well.
7. Add 3.0 mL of Milli-Q_{PLUS} Water. Mix well.
8. Centrifuge, carefully remove lower water layer with a Pasteur pipette.
9. Repeat steps 6–7, two more times.
10. Place two 5 mL aliquots of the remaining hexane into separate 16 × 100 mm evaporation tubes
11. Evaporate hexane at 40°C and N_2 80 psi in the TurboVap evaporator.
12. Extract the residue from each tube with three separate 0.5 mL volumes of standard diluent.
13. Combine extracts in a 10-mL volumetric flask and dilute to volume with standard diluent.
14. Result is a 1.0 μg/mL (nominal concentration) Standard of CoQ10 Red and CoQ9 Red.
15. Prepare daily.
16. Purity check: Analyze by HPLC, *see* **Section 3.5** using the 400 ng/mL CoQ10/CoQ9 Red Working Calibrator. It should contain <10% CoQ10/CoQ9 Ox.

3.2.7. Determination of CoQ9 Ox and CoQ10 Ox Stock Concentration

Due to the instability and variable purity of Coenzyme Q9 and Q10, the concentration of the stock standard must be determined spectrophotometrically using published molar absorptivity values *(11)*.

1. Volumetrically prepare individual 10-fold dilutions of the CoQ9 and CoQ10 Stock Standards to produce a nominal concentration of 10 μg/mL in ethanol.
2. Measure the absorbance of these solutions in the spectrophotometer at 275 nm using Quartz 1 cm matched cells. Zero the spectrophotometer with ethanol, measure and record the absorbance of the CoQ9 and CoQ10 solutions.
3. Calculations:
   - Q9 (MW = 795.2) has an absorptivity value of 0.0185 mL/μg cm (14,700 L/mole cm) in ethanol at 275 nm.
   - Q10 (MW = 863.4) has an absorptivity value of 0.0164 mL/μg cm (14,200 L/mole cm) in ethanol at 275 nm.

   Concentration is then calculated using the Beer-Lambert law:

   concentration (c) = (A) Absorbance/[absorbtivity (ε) x path length (l)]

Example: If the nominal 10 μg/mL Q9 solution (10-fold dilution of stock) being tested has a measured absorbance of 0.150 A, the concentration of the stock standard would be 81 μg/mL instead of 100 μg/mL.

$$c = (0.150/[0.0185 \times 1]) \times 10 \text{ (for dilution)} = 81\ \mu g/mL$$

The concentration of the calibrators for CoQ9 and CoQ10 (Ox and Red forms) must be adjusted accordingly in the instrument's calibration method, *see* **Table 19.3**.

4. The concentration of the standard should be checked weekly.

Table 19.3
Example calibration scheme for either red or ox species

Nominal concentration (ng/mL)[a]	Adjustment for 5-fold sample dilution	Value to enter in the instrument's calibration method (ng/mL)[b]
400	2000	1620
200	1000	810
100	500	405
50	250	203
20	100	81
0	0	0

[a]From **Table 19.1** or **19.2**.
[b]After correction based on molar absorbtivity, *see* **Section 3.2.7**. Example shown is based on a determined value of 81 μg/mL.

3.3. Sample Preparation

1. Hundred microlitres of plasma (or serum) or quality control (QC) sample or water (reagent blank) is placed in a 1.5 mL microcentrifuge tube.
2. Add 400 μL of 2-propanol.
3. Cap and Vortex for 1 min.
4. Centrifuge at 10,000 rpm for 5 min, 4°C.
5. Pour supernatant into 12 × 32 mm 750 μL autosampler vials.

3.4. QC Sample Preparation

Plasma QC samples can be prepared from commercial converted plasma (e.g. Biocell analyte stripped human serum, delipidized), which is free of CoQ10 and CoQ9. The plasma is spiked with CoQ10 and CoQ9 Ox at 500, 1000, and 2000 ng/mL and separately with CoQ10 and CoQ9 Red at 500, 1000, or 2000 ng/mL.

3.5. HPLC Analysis (*see* Note 6)

1. The system, in sequence, consists of: pump, pulse dampener, PEEK in-line filter, autosampler, column, conditioning cell, analytical cell.
2. Mobile Phase: Make 1 M ammonium acetate buffer, pH 4.4, stock solution by dissolving 77.08 g of ammonium acetate in approximately 400 mL of de-ionized water. Adjust pH to 4.4 with glacial acetic acid. Make to a final volume of 1 L with de-ionized water. Filter through 0.2 μm nylon filter. Mix 40 mL ammonium acetate buffer with 400 mL 1-propanol and 1,560 mL methanol. Degas prior to use. Pass fresh mobile phase through the system overnight, prior to analysis.
3. Flow rate: 1.0 mL/min.
4. Column oven temperature: 32°C.
5. Injection volume: 20 μL.
6. Autosampler tray temperature: 4°C.
7. Applied potentials (mV vs Pd reference): Conditioning cell = +700. Analytical cell, E1 = −700; E2 = +500.
8. Once the system is set-up and equilibrated, run the 400 ng/mL (nominal) Ox calibrator and record the peak height and area. For initial qualification of the instrument, these values must be within 50% of previously recorded values.
9. Calibration. Directly analyze each level of the Ox and Red working calibrators. **Table 19.3** provides an example of the values that should be used in the instrument's calibration method. These values take into account the 5-fold dilution of the sample (*see* **Section 3.3**) and a correction based on molar absorbtivity (*see* **Section 3.2.7**). This scheme is applied to both the Red and Ox forms. Sample concentration should be determined based on peak height using a quadratic fit.

4. Results

4.1. Method Performance

All analytes are well resolved in <18 min (**Fig. 19.2**). Voltammetric resolution greatly improved analytical selectivity and specificity. Analyte recovery (analyte spiked at 500, 1000, or 2000 ng/mL into Biocell analyte stripped human serum) averaged 93–110% (**Table 19.4**). Intra-day precision (five repeats) and inter-day precision (over 3 days), were 1.2–2.3, and 2.2–3.9% RSD, respectively (**Table 19.4**).

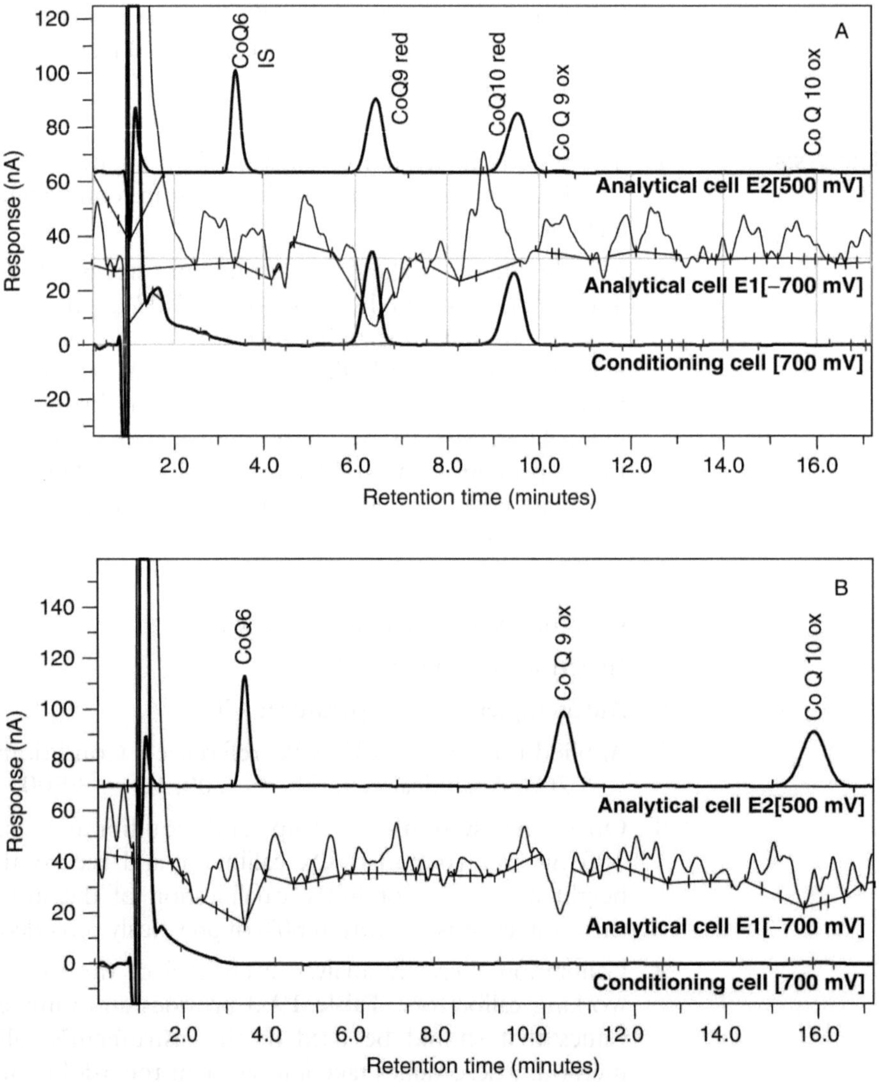

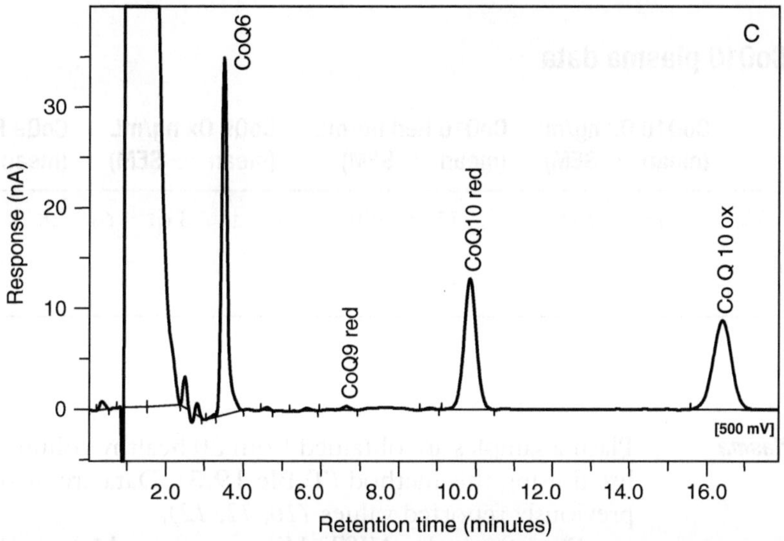

Fig. 19.2. (**A**) Analysis of CoQ9 Red and CoQ10 Red Calibrators (1000 ng/mL). The reduced analytes are initially oxidized on the conditioning cell, reduced on E1 of the analytical cell, and finally re-oxidized and measured on E2 (*see* **Notes 7** and **8**). (**B**) Analysis of CoQ9 Ox and CoQ10 Ox Calibrators (1000 ng/mL). The oxidized analytes do not react on the conditioning cell (they are already oxidized). Analytes are reduced on E1 and finally measured on E2 (*see* **Notes 7** and **8**). (**C**) Extracted human plasma. Only the analytical channel (E2) is shown, for clarity (*see* **Note 8**).

The stability of CoQ9 Red and CoQ10 Red during sample extraction was studied by spiking the reduced analytes (at different levels) into Biocell analyte stripped human serum. CoQ9 oxidation: 3.7% (500 ng/mL); 4.5% (1000 ng/mL); 5.1% (2000 ng/mL); CoQ10 oxidation: 4.2% (500 ng/mL); 4.6% (1000 ng/mL); 5.6% (2000 ng/mL).

Table 19.4
Recovery and precision data

	Average recovery (%)	Intra-day precision (% RSD)	Inter-day precision (% RSD)
CoQ10Ox	92.7	1.3	2.9
CoQ10Red	102.4	2.1	3.9
CoQ9Ox	98.3	1.2	2.2
CoQ9Red	109.7	2.3	3.5

Table 19.5
CoQ9 and CoQ10 plasma data

	CoQ10 Ox ng/mL (mean ± SEM)	CoQ10 Red ng/mL (mean ± SEM)	CoQ9 Ox ng/mL (mean ± SEM)	CoQ9 Red ng/mL (mean ± SEM)
Plasma (n = 30)	48 ± 23	775 ± 290	8.5 ± 3.3 (n = 6)	28 ± 6 (n = 6)
% Red/Total [Red + Ox]		94		78

4.2. Human Plasma Sample Data

Plasma samples are obtained from 30 healthy volunteers and analyzed using this method (**Table 19.5**). Data are in agreement to previously reported values *(10, 11, 12).*

Data from the NIST Micronutrients Measurement Quality Assurance Program for fat-soluble vitamins in human serum are presented in **Table 19.6**. Values (± 1 SD) are assigned by NIST for 10 samples based on data from six participating laboratories. Only the total CoQ10 levels were reported. Using this method, it is possible to measure both the Ox and Red forms. Data in **Table 19.6**

Table 19.6
CoQ10 NIST serum data (values in ng/mL)

Sample ID	314	315	316	317	318	324	325	326	327	328
TOTAL CoQ10	611	618	553	776	620	770	986	1190	740	350
NIST – assigned value (SD)	(54)	(103)	(89)	(135)	(90)	(77)	(99)	(119)	(74)	(35)
CoQ10 Red	150	370	540	585	600	51	522	1042	658	192
CoQ10 Ox	310	220	50	160	105	561	381	115	58	69
Total CoQ10 [RED + OX]	460	590	590	745	705	612	903	1157	716	261
% Red/ Total	32	63	92	79	85	8	58	90	92	74

show that the values obtained for total CoQ10 by this HPLC-ECD method are in agreement with the NIST assigned values. These data show significant variability between samples in the redox ratio (% Red/Total). The extent to which this variability is specifically related to actual differences in redox ratio between samples or to the unique nature and handling requirements of the NIST material remains to be determined. This does, however, highlight the need to exercise care during sample collection, processing, and storage to minimize ex vivo oxidation of the quinol forms.

5. Notes

1. All sample and standard preparation procedures must be performed under UV-filtered lighting conditions, as described.
2. CoQ9 Red and CoQ10 Red are susceptible to ex vivo oxidation. Care must be exercised to prevent oxidation and artifactual change of the ratio of Red to Ox forms. This includes care in sample collecting, processing, and storage, as well as sample preparation for the analysis. Plasma or serum must be prepared from whole blood as soon as possible, and immediately stored at −80°C.
3. DO NOT use Vacutainers containing EDTA as an anticoagulant. *(13)*
4. The actual concentration of the stock standards must be determined spectrophotometrically using published molar absorptivity values (*see* **Section 3.2.7**). Concentration values stated for the stock standards, and solutions derived from these standards, are therefore target (nominal) concentrations. Any difference between target and determined concentration should therefore be accounted for by entering the appropriate concentration value in the instrument's calibration method.
5. Although hexane (and nitrogen dry down) is used in the preparation of CoQ10 Red without issue, it is not recommended for preparation of samples. Considerable oxidation >50% of CoQ10 Red to CoQ10 Ox in samples has been observed in our lab and reported in the literature *(10)*.
6. Glass autosampler vials promote oxidation of CoQ10 Red and CoQ9 Red. Do not use.
7. (−700 mV). This channel should be regarded as a reactor channel, and due to noise, should not be used for quantitation.
8. These chromatograms include a peak for CoQ6, which was included as a potential internal standard. In our validation studies, variability was found in the peak shape for CoQ6

that may be attributable to its early retention time. Since the sample preparation does not involve an extraction procedure, the use of an internal standard was found to be unnecessary for this method.

References

1. Tang, P. H., and DeGrauw, T. (2004). Redox cycling of Coenzyme Q9 as a new measure of plasma reducing power. Clin. Chem., **50**, 1930–1932.
2. Ernster, L., and Dallner, G. (1995). Biochemical, physiological and medical aspects of ubiquinone function. Biochim. Biophys. Acta, **1271**, 195–204.
3. Acworth, I. N. (2003). The Handbook of Redox Biochemistry. CD-ROM. ESA Biosciences. Part number: 70–6090.
4. Stocker, R., Bowry, V. W., and Frei, B. (1991). Ubiquinol-10 protects human low density lipoprotein more efficiently against lipid peroxidation than does alpha-tocopherol. Proc. Natl. Acad. Sci. USA, **88**, 1646–1650.
5. Nohl, H., and Gille, L. (2001). The role of coenzyme Q in lysosomes. In: Kagan VEQ, P. J. (ed). Coenzyme Q: Molecular Mechanisms in Health and Disease. Boca Raton: CRC Press. Pp. 99–106.
6. Quinzii, C. M., DiMauro, S., and Hirano, M. (2007). Human coenzyme Q10 deficiency. Neurochem. Res., **32**, 723–727.
7. Silver, M. A., Langsjoen, P. H., Szabo, S., Patil, H., and Zelinger, A. (2003). Statin cardiomyopathy? A potential role for CoQ10 therapy for statin-induced changes in diastolic LV performance: description of a clinical protocol. Biofactors, **18**, 125–127.
8. Langsjoen, P. H., and Langsjoen, A. M. (2003). The clinical use of HMG CoA-reductase inhibitors and the associated depletion of coenzyme Q10. A review of animal and human publications. Biofactors, **18**, 101–111.
9. Folkers, K., Langsjoen, P., Willis, R., Richardson, P., Xia, L.-J., Ye, C.-Q., and Tamagawa, H. (1990). Lovastatin decreases coenzyme Q levels in humans. Proc. Natl. Acad. Sci. USA, **87**, 8931–8934.
10. Tang, P. H., Miles, M. V., DeGrauw, A., Hershey, A., and Pesce A. (2001). HPLC analysis of reduced and oxidized coenzyme Q10 in human plasma. Clin. Chem., **47**, 256–265.
11. Tang, P. H., Miles, M. V., Miles, L., Quinlan, J., Wong, B., Wenisch, A., and Bove, K. (2004). Measurement of reduced and oxidized coenzyme Q9 and coenzyme Q10 levels in mouse tissues by HPLC with coulometric detection. Clin. Chim. Acta, **341**, 173–184.
12. Molyneux, S., Lever, M., Florkowski, C., and George, P. (2007). Plasma total CoQ9 in the New Zealand population: reference interval and biological variation. Clin. Chem., **53**, 802–803.
13. Tang, P. H., Miles, M. V., Steele, P., DeGrauq, A., Chuck, G., Schroer, L., and Pesce, A. (2002). Anticoagulant effects on plasma coenzyme Q10 estimated by HPLC with coulometric detection. Clin. Chim. Acta, **318**, 127–131.

Chapter 20

Paraoxonases (PON1, PON2, PON3) Analyses *In Vitro* and *In Vivo* in Relation to Cardiovascular Diseases

Michael Aviram and Mira Rosenblat

Abstract

Mammalian paraoxonases (PON1, PON2, PON3) are a unique family of calcium-dependent hydrolases, with enzymatic activities toward a broad range of substrates (lactones, thiolactones, carbonates, esters, phosphotriesters). Although PONs physiological substrates were not yet identified, some studies suggest that they could be some lactones, or some specific oxidized phospholipids, or products of both enzymatic and nonenzymatic oxidation of arachidonic and docosahexaenoic acid, as well as N-acyl-homoserine lactones (which are quorum-sensing signals of pathogenic bacteria). Since no endogenous substrates for PONs activity determination are available yet, synthetic substrates such as paraoxon, phenyl acetate, and several lactones are used for PONs activity assays. All three members of the PON family (PON 1/2/3) were shown to protect from atherosclerosis development. Their anti-atherogenic biological activities were studied in vitro using serum or cell cultures, and also in vivo, using PON 1/2/3 knockout or transgenic mice, as well as humans – healthy volunteers and atherosclerotic patients (diabetics, hypercholesterolemics, and hypertensives).

Key words: Paraoxonase, Arylesterase, Lactonase, HDL, LDL, Oxidative stress, Macrophages, Mice, Atherosclerosis.

1. Introduction

PONs exhibit a range of important activities, including drug metabolism and detoxification of organophosphates such as nerve agents, and they were also shown to be involved in the protection against atherosclerosis *(1)*. Enzymatic characterization of the purified PONs revealed them to be lactonases/ lactonizing enzymes, with some overlapping substrates (hydroxyl acid derivatives of polyunsaturated fatty acids, estrogen esters).

From: *Methods in Molecular Biology, vol. 477: Advanced Protocols in Oxidative Stress I,*
Edited by: D. Armstrong, DOI: 10.1007/978-1-60327-517-0_20, © Humana Press, New York, NY

Organophosphates are hydrolyzed almost exclusively by PON1. Drug substrates (lovastatin, spironolactone) are hydrolyzed only by PON3, whereas PON2 preferentially hydrolyzes N-acyl-homoserine lactones.

1.1. PON1

Most of serum PON1 is HDL-associated, but low PON1 levels are also associated with chylomicrons and VLDL, but not LDL *(2)*. The native activity of PON1 is likely that of lactonase, whereas its arylesterase and phosphotriesterase activities are promiscuous *(3)*. Apolipoprotein A-I in HDL was shown to stabilize PON1 and to significantly stimulate its lactonase activity *(4)*.

The crystal structure of a variant of PON1 obtained by directed evolution was shown to consist of a six-bladed β-propeller with a unique active site and a catabolic mechanism based on its His-His dyad *(5)*.

PON1 have two common polymorphisms in the coding region: leucine (L)/methionine (M) at position 55 and glutamine (Q)/arginine(R) at position 192.

The role of PON1 in atherosclerosis development was demonstrated in studies using mice lacking PON1 *(6)*, or over expressing PON1 *(7)*. Attenuation of atherosclerosis by PON1 can result from its ability to hydrolyze specific oxidized lipids in lipoproteins *(8)*, in arterial wall cells (including macrophages), and in atherosclerotic lesions *(9)*. Macrophage cholesterol accumulation and foam cell formation is the hallmark of early atherosclerosis *(1)* and PON1 was shown to inhibit cholesterol influx, by reducing the formation of oxidized LDL (Ox-LDL), and its uptake by macrophages. Furthermore, PON1 inhibits cholesterol biosynthesis in macrophages *(10)*, and stimulates HDL-mediated cholesterol efflux from the cells *(11, 12)*. Recently, it was shown that the 192R/Q polymorphs of PON1 differ in HDL binding, in the extent of stimulating PON1 lipo-lactonase activity, as well as in HDL-mediated macrophage cholesterol efflux *(13)*.

1.2. PON2

Unlike PON1, PON2 could not be detectable in serum. Whereas PON1 is expressed mainly in the liver, but not in arterial macrophages, PON2 is expressed in most tissues, including human and mouse macrophages *(14, 15)*. PON2, like PON1, was shown to have a protective role against atherosclerosis development *(16, 17)*. PON2 abilities to inhibit LDL oxidation by the cells and Ox-LDL-induced monocyte chemotaxis and reactive oxygen species (ROS) formation in HeLa and in vascular cells were recently shown *(15, 18)*. We have demonstrated that cellular oxidative stress affects PON2 expression and enzymatic activities in a biphasic U-shape *(19)*. Finally, PON2 expression is upregulated via an NADPH-oxidase–dependent mechanism during monocytes to macrophages differentiation *(20)*, and also by pomegranate juice phenolic antioxidants *(21)*.

1.3. PON3

PON3 is present in tissues and to a minor extent also in serum where it is HDL-associated *(22, 23)*. PON3 mRNA is expressed in the liver, kidney, gastrointestinal tract, and also in mouse (but not human) macrophages. Recently it was shown that in human PON3 transgenic mice *(24)*, as well as in apolipoprotein E-deficient mice that were treated with adenovirus expressing human PON3 *(25)*, there was a significant attenuation of atherosclerotic lesion development, indicating that PON3, like PON1 and PON2, has also a role in atherosclerosis protection and it also protects LDL against oxidation *(22, 23)*.

The chapter objective is to describe all methods to be used in order to analyze PON1/2/3 roles in atherosclerosis.

2. Materials

2.1. Equipment

1. Elisa Reader.
2. Pipettors.
3. Multichannel pipettors.
4. Water bath.
5. Warm plate up to 100°C.
6. Centrifuge and microcentrifuge.
7. HPLC (Varian).
8. Spectrophotometer.
9. Real time PCR instrument.
10. Mini-protein 3 well for SDS-PAGE (Bio Rad).
11. Transfer – Blot Electrophoresis (Bio Rad).
12. Power Supply.
13. β-Counter.
14. Sorval ultracentrifuge.
15. Rotor SW-41 and swinging buckets.
16. FPLC recorder.
17. C18 column, 250 × 4.6 × 5 mm.

2.2. Reagents and Buffers

2.2.1. Purified PON1 Preparation

1. Purification of PON1 from serum (see reference 26).
2. Recombinant PON1 by directed evolution (see reference 27).

2.2.2. Arylesterase Activity

1. Phenyl acetate.
2. Methanol.
3. Reaction Tris buffer: 50 mM Tris HCL, pH 8.0, 1 mM $CaCl_2$.

2.2.3. Paraoxonase Activity

1. Purified paraoxon cat no. PS-610 from Chem Service (Supleco, Sigma).
2. Methanol.
3. Reaction glycine buffer: 50 mM glycine/NaOH, pH 10.5, 1 mM $CaCl_2$.

2.2.4. Lactonase Activity Toward Dihydrocoumarin

1. Dihydrocoumarin.
2. Dimethyl sulfoximine (DMSO).
3. Reaction Tris buffer: 50 mM Tris HCl, pH 7.5, 1 mM $CaCl_2$.

2.2.5. Lactonase Activity Toward Thiobutyl Butyrolactone (TBBL)

1. TBBL (synthesized, as previously described, reference 28).
2. Acetonitrile.
3. 5,5′-Dithiobis-2-nitrobenzoicacid (DTNB).
4. DMSO.
5. Reaction Tris buffer: 50 mM Tris HCl, pH 8.0, 1 mM $CaCl_2$.

2.2.6. Serum PON1 Phenotyping

1. Same materials as in **Section 2.2.2**.
2. Same materials as in **Section 2.2.3** except for the use of glycine/NaOH buffer: 1 M NaCl.

2.2.7. PON1 Stability: Inactivation Assay

1. β-Mercaptoethanol.
2. Nitrilotriacetic acid (NTA).
3. Tris buffer: 10 mM Tris-HCL, 150 mM NaCl, pH 8.0.
4. Reaction buffer: Tris buffer (from 3), 1 mM $CaCl_2$.
5. PCR strips.

2.2.8. Serum PON1 Concentration

1. PONI concentration determination by ELISA (see reference 29)

2.2.9. Biological Activities of Serum PON1

2.2.9.1. Serum Lipid Peroxidation

1. 2-Hydroxyquinoline.
2. Phosphate buffered saline (PBS).
3. 2, 2' Azobis-amidinopropane (AAPH).
4. TBARS reagent (0.375 g thiobarbituric acid, 15 ml of 100% trichloroacetic acid, 2.5 ml concentrated HCL, in 100 ml).
5. Tetramethoxypropane for the TBARS standard curve.

2.2.9.2. Serum-mediated Cholesterol Efflux

1. [^{3}H]-Cholesterol (Amersham).
2. 2-Hydroxyquinolin.
3. Quicksafe A scintillation fluid.

2.2.10. HDL-bound PON1

2.2.10.1. HDL Separation by Ultracentrifugation

1. Potassium bromide (1.59 g for 4 ml of serum).
2. NaCl solution (*d* = 1.084), 2 mM $CaCl_2$, 100 μM diethylenetriaminepeta-acetic acid (DTPA), pH 8.0.
3. NaCl solution (*d* = 1.006), 2 mM $CaCl_2$, 100 μM DTPA, pH 8.0

2.2.10.2. HDL Separation by Fast-Protein Liquid Chromatography (FPLC)

1. Filtered and degassed PBS.

2.2.11. Biological Activities of HDL-associated PON1

Stimulation of HDL-mediated Macrophage Cholesterol Efflux

(same materials as in **Section 2.2.9.2**).

Inhibition of Macrophage-mediated LDL Oxidation

1. LDL.
2. $CuSO_4$.
3. Isolated HDL.
4. RPMI medium without phenol red.
5. TBARS reagent and standard as in **Section 2.2.9.1**.

2.2.12. Serum PON3 Statinase Activity

1. Acetic acid.
2. Acetonitrile.
3. Methanol.
4. Lovastatin.
5. Reaction Tris buffer: 25 mM Tris HCl, pH 7.6, 1 mM $CaCl_2$.

2.2.13. Cell Cultures

1. RPMI medium with phenol red.
2. Dulbeccos Modified Eagels Medium (DMEM).
3. Fetal Calf serum (FCS).
4. Glutamine.
5. Pyruvate.
6. Pen-strep antibiotic.
7. Bovine serum albumin (BSA).

2.2.14. Macrophage PON2 Lactonase Activity

1. With dihydrocoumarin substrate and reaction buffer as in **Section 2.2.4**.
2. With TBBL. Same materials as in **Section 2.2.5**.

2.2.15. Macrophage PON2 mRNA by Real Time PCR

1. UltraPure water PCR grade (Fisher).
2. cDNA template.
3. Lyophilized primer and probe set for PON2 selected by the user.

2.2.16. Macrophage PON2 Protein by Western Blot Analysis

1. Lysis buffer.
2. Molecular weight standard.
3. Monoclonal antibody toward human or mouse PON2.
4. Secondary antibody coupled to peroxidase.

2.2.17. Macrophage PON3 Statinase Activity

1. Same materials as in **Section 2.2.12**.

2.2.18. Macrophage PON3 Protein by Western Blot Analysis

Same materials as in **Section 2.2.16**, except the use of primary monoclonal antibody for mouse PON3.

2.3. Supplies

1. 96-Well plates not sterile (UV and visible).
2. 12-Well plates for tissue culture.
3. 6-Wells plates for tissue culture.
4. 24-Well plates for tissue culture.
5. Eppendorf tubes.
6. UV cuvettes (disposable, semi micro, 1.5 ml, 12.5 × 12.5 × 15 mm).
7. Thin-walled reaction tubes.
8. Disposable sterile tips with filter, RNAse and DNAse free.
9. Nitrocellulose membrane.
10. Filter paper.
11. Films.
12. Scintilation vials.
13. Ultra-Clear centrifuge tubes (14 × 89 mm) from Beckman.
14. Dialyzing tubing width 10 mm, MWCO 12,000–14,000.
15. Syringes, 10 ml with needles.
16. Pasture pipettes.
17. Superose 6 column (1 × 30 cm, Pharmacia).
18. ECL kit (Biological Industries, Israel).
19. Master Pure RNA purification kit (Cat no. MCR-85102) – from EPICENTER, 726 Post Road, Madison, WI, 53713 USA.
20. Reverse-iT™ MAX 1st Strand Synthesis Kit –from ABgene, Abgene House, Blenheim Road, Epson, Surrey, KT19 9AP, UK
21. PrimerDesign 2× qPCR Mastermix (ABgene).

3. Methods

3.1. Serum, HDL or Isolated PON1 Activities (see Notes 1 and 2*)*

The serum samples are collected from healthy subjects, from atherosclerotic patients (hypertensive, diabetic, hypercholesterolemic), or from mice (C57BL/6 mice, the atherosclerotic apolipoprotein E-deficient mice, PON1, PON2, PON3 knockout mice, human PON1 transgenic mice).

3.1.1. Arylesterase Activity (see Note 3*)*

1. Prepare stock solution of 500 mM phenyl acetate in methanol (keep for up to 2 weeks at 4°C).
2. Dilute the stock to 2 mM with reaction buffer (freshly prepared), and use final concentration of 1 mM.
3. Before performance of the assay, dilute the serum 1:10, and the purified PON1 1:200 with reaction buffer (dilution factor depends on PON1 activity). Use undiluted HDL.
4. Add to the wells of a UV microplate 5 μl of the diluted serum or purified PON1, or of human HDL. For mouse HDL, use 10–15 μl. Complete to 100 μl with reaction buffer.
5. Add 100 μl of 2 mM phenyl acetate.
6. Read absorbance at 270 nm for 2 minutes every 15 seconds. Get the Mean V mOD/minute values.
7. 1 unit = 1 μmol of phenyl acetate hydrolyzed per minute per 1 ml.
8. The extinction coefficient is: $E_{270} = 1310/M/cm$
9. Calculation − (Mean V × 5 × dilution factor)/700 = units/ml. Serum arylesterase activity in healthy subjects is in the range of 100–200 units/ml.

3.1.2. Paraoxonase Activity

1. Prepare stock solution of 400 mM paraoxon in methanol (keep for up to 2 weeks at 4°C).
2. Dilute the stock to 4 mM with water (freshly prepared), and use final concentration of 2 mM.
3. Use undiluted serum or HDL. Dilute PON1 1:20 (Dilution factor depends on the activity of PON1).
4. Add to the wells of a visible microplate 5 μl of the serum or purified PON1, or 50 μl of human HDL. For mouse HDL use 100 μl. Complete to 100 μl with reaction buffer.
5. Measure also self hydrolysis of paraoxon.
6. Add 100 μl of 4 mM paraoxon.
7. Read absorbance at 412 nm for 3 minutes every 15 seconds. A yellow color is produced. Get the Mean V mOD/minute values.

8. 1 unit = 1 nmol of paraoxon hydrolyzed per minute per 1 ml.
9. Calculation – [(Mean V of sample – Mean V of self hydrolysis) × 5 × dilution factor]/11.725 = units.
   Serum paraoxonase activity in healthy subjects is in the range of 1000–5000 units/ml.

3.1.3. Lactonase Activity Toward Dihydrocumarin

1. Prepare stock solution of 100 mM dihydrocumarin in DMSO (keep for up to 2 weeks at 4°C).
2. Dilute the stock to 2 mM with reaction buffer (freshly prepared), and use final concentration of 1 mM.
3. Dilute serum 1:10 or purified PON1 1:200 with reaction buffer (dilution factor depends the activity of PON1). Use nondiluted HDL.
4. Add to the wells of a UV microplate, 3 μl of the diluted serum or purified PON1, or of human HDL. For mouse HDL, use 10–15 μl. Complete to 100 μl with reaction buffer.
5. Measure also self hydrolysis of dihydrocoumarin.
6. Add 100 μl of 2 mM dihydrocumarin.
7. Read absorbance at 270 nm for 10 minutes every 15 seconds. Get the Mean V mOD/minute values.
8. 1 unit = 1 μmol of dihydrocumarin hydrolyzed per minute per 1 ml.
9. The extinction coefficient is: E_{270} = 1295/M/cm
10. Calculation: – [(Mean V of sample – Mean V of self hydrolysis) × 5 × dilution factor]/700 = units/ml.
    Seum lactonase activity toward dihydrocoumarin in healthy subjects is in the range of 13–20 units/ml.

3.1.4. Lactonase Activity Toward TBBL (*see* Note 4)

1. Prepare stock solution of 240 mM TBBL in acetonitrile (keep for several months at 4°C).
2. Dilute the stock to 1 mM with reaction buffer (freshly prepared), and use final concentration of 0.25 mM.
3. Prepare a stock solution of 100 mM DTNB in DMSO. Use final concentration of 0.5 mM (1 μl per 200 μl final volume).
4. Dilute serum 1:10 or purified PON1 1:200 with reaction buffer. Use nondiluted HDL.
5. Calculate the total number of samples and prepare a stock of DTNB in reaction buffer (1 μl DTNB + 144 μl reaction buffer per one sample).
6. Add to the wells of a visible microplate, 5 μl of the diluted serum or purified PON1, or human HDL. For mouse HDL, use 10–15 μl. Then complete the volume to 150 μl with reaction buffer + DNTB.
7. Measure also self hydrolysis of TBBL.

8. Add 50 μl of 1 mM TBBL.
9. Read absorbance at 412 nm for 10 minutes every 15 seconds. A yellow color is produced. Get the Mean V mOD/minute values.
10. 1 unit = 1 μmol of TBBL hydrolyzed per minute per 1 ml.
11. Calculation – [(Mean V in sample − Mean V of self hydrolysis) × 5 × dilution factor)]/7000 = units/ml. Serum lactonase activity toward TBBL in healthy subjects is in the range of 10–40 units/ml.

3.2. Serum PON1 Phenotyping

1. Measure serum arylesterase activity as described in **Section 3.1.1**.
2. Measure serum paraoxonase activity as described in **Section 3.1.2** except for the use of glycine/NaOH buffer with 1 M NaCl.
3. Calculate the ratio paraoxonase activity/arylesterase activity.
4. The ratio <2 is for QQ, between 2 and 8 is for QR, and >8 is for RR (see reference 30).

3.3. Serum PON1 Stability: Inactivation Assay (see Note 5)

1. Prepare 50 mM β-mercaptoethanol in Tris buffer (keep at RT).
2. Prepare 100 mM NTA solution (keep at RT).
3. Prepare 500 mM phenyl acetate in methanol.
4. Prepare inactivation solution (0.5 mM NTA, 2 mM β-mercaptoethanol in Tris buffer).
5. Divide the serum samples to aliquots of (20–50 μl), and keep at −20°C.
6. A day before performing the assay, thaw the amount of serum needed, and add β-mercaptoethanol (final concentration of 5 mM).
7. Dilute the serum 1:10 with Tris buffer (final volume of 50 μl), in a strip of PCR small tubes.
8. Add 50 μl inactivation solution to 50 μl of the diluted serum. Incubate at 25°C.
9. Check arylesterase activity in microplate as described in **Section 3.1.1** at 0 time, after 30 minutes, 1, 2, 4, 6, 9, 15, and 24 hours. Add 100 μl of 4 mM phenyl acetate.
10. Calculations-The activity at 0 time is 100%. Calculate at each time point, the percentage of residual activity.

3.4. Biological Activities of Serum PON1

3.4.1. Serum Lipids Peroxidation

1. Prepare freshly 20 mM hydroxyquinoline in DMSO.
2. Incubate 250 μl of serum with 500 μM of hydroxyquinoline (PON1 specific inhibitor) or with a similar volume of DMSO at 25°C for 2 hours. Then add 750 μl PBS. Divide each sample into two eppendorf tubes.

3. Prepare AAPH solution in water (0.2 g in 1 ml), and add 50 μl to one set of tubes.
4. To the other set, add only 50 μl water.
5. Incubate the tubes in a water bath at 37°C for 2 hours.
6. Measure the extent of serum lipids peroxidation by the TBARS assay (see reference 31) and by the lipid peroxides assay (see reference 32), and calculate their levels as nmol/ml. In each system, subtract the values observed with no AAPH. In each system, compare the values with no inhibitor to the values observed with the addition of the inhibitor to indicate PON1 contribution to AAPH-induced serum lipids peroxidation.

The basal levels of TBARS and lipid peroxides in healthy subjects are: 2–4 and 4–10 nmol/ml, respectively. The AAPH-induced TBARS and lipid peroxides values in healthy subjects are: 10–14 and 400–500 nmol/ml, respectively.

3.4.2. Serum-mediated Cholesterol Efflux from Macrophages

1. Plate J774 A.1 macrophages, or mouse peritoneal macrophages (MPM) from the different mice types, or human monocytes-derived macrophages (HMDM) from healthy subjects or atherosclerotic patients in a 24 well plate.
2. Incubate 100 μl of serum with 500 μM hydroxyquinoline, or with a similar volume of DMSO at 25°C for 2 hours.
3. Prepare $[^3H]$-cholesterol in medium. Take 0.5 μl for each well from the stock (250 μCi in toluene), evaporate the toluene under air. Add absolute ethanol (1 μl/ml medium), and vortex. Then add the volume of medium needed (1 ml per well), and 20 μl BSA per 1 ml medium.
4. Wash the cells and add 1 ml of the $[^3H]$-cholesterol solution per well, and incubate the cells for 1 hour at 37°C.
5. Aspirate the medium and wash the cells ×3 with PBS. Add 10 μl serum in medium to each well (perform each serum sample in triplicates). One system of the cells will receive only medium with no serum.
6. After 3 hours incubation at 37°C, take 500 μl of the medium into scintillation vials. Wash the cells ×3 with PBS, and add 0.1 N NaOH (incubate over night at RT).
7. Take 500 μl from the cells into scintillation vials.
8. Read radioactivitiy in the medium and cells samples.
9. Calculate % efflux: (cpm in medium)/(cpm in cells + cpm in medium).
10. Subtract the value obtained in cells with no serum from those obtained with the addition of serum.
11. For each serum sample, compare the values with no PON1 inhibitor to those with the inhibitor in order to demonstrate

PON1 contribution to serum-mediated cholesterol efflux from the macrophages.

Cholesterol efflux rate with no addition is about 4%, and with the addition of healthy subject serum in the range of 20–30%.

3.5. HDL-bound PON1

3.5.1. HDL Separation by Ultracentrifugation

1. Human HDL is separated by discontinuous density gradient ultracentrifugation (please see reference 33), by using 4 ml of serum and potassium bromide to increase the serum density to 1.25 g/ml, and NaCl solutions of 1.084 g/ml and 1.006 g/ml (as indicated in the materials **Section 2.2.10(1)**).
2. For mouse HDL separation, collect blood from 10 mice to get 2 ml serum. Separate the HDL by continuous density gradient ultracentrifugation by three steps, each step 24 hours run at 35,000 rpm using SW 41 rotor: first step-VLDL separation, second step - LDL separation by increasing the density to 1.063 g/ml with potassium bromide, final step-HDL separation by increasing the density to 1.25 g/ml.
3. The isolated HDL samples are dialyzed against Tris buffer, 1 mM $CaCl_2$. The HDL protein concentration is determined by the Lowry method (see reference 34).

3.5.2. HDL Separation by FPLC

1. Two hundred microliters of serum were diluted ×2 with PBS and filtered.
2. These diluted serum samples were fractionated through Superose 6 column (1 × 30 cm, Pharmacia) using prefiltered and degassed PBS pH 7.5. The flow rate was 0.5 ml/minute, and sixty 0.5 ml fractions were collected and immediately analyzed for PON1 arylesterase and paraoxonase activities.
3. VLDL was eluted between fractions 15 and 17, LDL between 23 and 26, and HDL between 27 and 36.

3.6. Biological Activities of HDL-associated PON1

3.6.1. Stimulation of HDL-mediated Macrophage Cholesterol Efflux

Same procedure as in **Section 3.4.2**, but use HDL (100 μg protein/ml) that was preincubated with 500 μM of the PON1 inhibitor 2-hydroxyquinoline, or with similar volume of DMSO (control HDL). Compare in each system the extent of cholesterol efflux by control HDL to that observed with HDL and the PON1 inhibitor, to demonstrate the contribution of HDL-associated PON1.

Cholesterol efflux rates with healthy subject HDL are in the range of 25–35%.

3.6.2. Inhibition of Macrophage-Mediated LDL Oxidation (see Note 8)

1. Plate J774 A.1 macrophages in a 12 wells plate (0.5 × 10^6 per well).
2. Dialyze ×3 the LDL (1mg protein/ml) at 4°C against PBS: first and second dialysis for 1 hour and the third for O.N.

3. The cells are incubated at 37°C for 3–6 hours (depend on the LDL) with 1 ml of RPMI medium without phenol red, which includes LDL (50 μg protein/ml), or LDL + HDL (50 μg protein/ml). Use HDL samples that were preincubated with 500 μM of hydroxyquinoline, or with a similar volume of DMSO.
4. Incubate the above samples also in a cell-free system under the same conditions.
5. Add 5 μM $CuSO_4$ to all the plate wells (cells and cell-free).
6. Measure the extent of LDL oxidation by the TBARS assay.
7. Compare in each system the extent of inhibition of LDL oxidation by the macrophages in the presence of control HDL to that observed with HDL + PON1 inhibitor, to demonstrate the contribution of HDL-associated PON1.

The TBARS values for LDL from healthy subjects in cell-free system are 5–10 nmol/mg LDL protein; for LDL incubated with J774 A.1 macrophages, 20–30 nmol/mg LDL protein; and with peritoneal macrophages, 30–50 nmol/mg LDL protein. HDL inhibits LDL oxidation by the cells by 30%.

3.7. Serum PON3 Statinase Activity (*see* Note 6)

1. For *mouse serum* statinase activity measurement, take 5 μl serum and put into an eppendorf tube.
2. Add 490 μl reaction buffer and mix.
3. Add 5 μl of lovastatin (final concentration 5 μg/ml), mix and incubate for 1 hour at room temperature.
4. Prepare one additional eppendorf tube with buffer and lovastatin alone, in order to measure self hydrolysis of lovastatin.
5. For *human serum* statinase activity measurement, take 100 μl serum and add 390 μl reaction buffer and 5 μl of lovastatin. Mix and incubate for 4 hours at RT.
6. At the end of the incubation period, put the tubes on ice. Transfer an aliquot of 100 μl to another eppendorf tube that contains 100 μl of cold acetonitrile (to stop the reaction); mix.
7. Centrifuge for 1 minute in a cold microcentrifuge at 10,000 g.
8. Inject the samples (30 μl) to the HPLC, wavelength 238 nm, flow 1 ml/minute, run time 7 minutes, mobile phase consisting of the following:

A = acetic acid/acetonitrile/water (2:249:249, v/v/v) and B = acetonitrile, at the ratio A/B 35/65. Retention time for the acid 3.1 minutes and for the lovastatin 4.7 minutes.

9. Calculations:
   a. Peak area acid/(peak area acid + peak area lovastatin) = % lovastatin hydrolyzed.
   b. Subtract from % lovastatin hydrolyzed in the presence of serum the % self hydrolysis of lovastatin.

c. 1 unit = 1 pmol lovastatin hydrolyzed per minute per ml.
d. (12,400 × % lovastatin hydrolyzed × 100)/6000 = units/ml.

3.8. Cell Cultures

1. The cell lines J774 A.1 macrophages (murine) or THP-1 (human) were purchased from the American Tissue Culture Collection (ATCC, Rockville, MD). The cells were grown in DMEM + 5% FCS, or in RPMI medium + 5% FCS, respectively. In order to induce differention of THP-1 monocytes to macrophage, 10 ng/ml PMA was added to the medium and the cells were incubated for 3 days. Prior to the experiments, the PMA was aspirated, and the cells were incubated for 1 day with medium and FCS.
2. MPM were harvested 3 days after intraperitoneal injection of thioglycollate (40 g/l). MPMs were collected from C57BL/6 mice, the atherosclerotic apolipoprotein E-deficient mice, PON1, PON2, PON3 knockout mice, or human PON1 transgenic mice.
3. HMDM were isolated from whole blood using Ficoll gradient (please see reference 35). HMDMs were isolated from healthy subjects, or from atherosclerotic patients (hypertensive, hypercholesterolemic, and diabetic patients) (*see* **Note 7**).

3.9. Macrophage PON2 Lactonase Activity

3.9.1. Use Dihydrocumarin

1. Seed the cells (murine or human macrophages) in a 12 well plate, 1 × 10^6 per well.
2. Prepare freshly 100 mM dihdrocumarin in methanol.
3. Wash the cells with reaction buffer.
4. Incubate the cells with 1 ml of reaction buffer containing 10 mM dihdrocumarin; add a cell-free sample. Mix gently on a shaking plate for 10 minutes at RT.
5. Stop the reaction by placing the plate on ice.
6. Aspirate 1 ml from each well into UV cuvettes. Read absorbance at 270 nm against reaction buffer.
7. Calculations:
   a. One unit is equal to 1 μmol of dihydrocumarin hydrolyzed per minute per ml.
   b. Calculate the slope (o.d. 10 minutes in the wells with cells − o.d. 10 minutes dihydrocumarin self hydrolysis)/10 minute.
   c. The extinction coefficient is E_{270} = 1295/M/cm.
   d. Slope × 0.774 = units/ml. Express the results as (units/ml)/(mg cell protein).

Lactonase activity toward dihydrocoumarin in control mice peritoneal macrophages is 0.25 ± 0.05 units/mg cell protein.

3.9.2. Use TBBL

1. Seed the cells in a 96 well dish (the cells should be at about 80% confluency before performing the assay).
2. Wash the cells with reaction buffer.
3. The assay is then performed as described in **Section 3.1.4**.
4. Calculate the activity as units/ml as described in **Section 3.1.4** (no dilution factor is needed). Divide units per mg cell protein.

Lactonase activity toward TBBL in control mice peritoneal macrophages is 0.0062 ± 0.0018 units/mg cell protein.

3.10. Macrophage PON2 mRNA by Real Time PCR

1. Extract RNA from 2–4 × 10^6 cells according to the instructions provided with the RNA Purification Kit (Epicentre).
2. Prepare cDNA from 1 μg RNA using oligo-dT primer according to instructions provided with the Reverse-iTTM Max 1st Strand Synthesis Kit instructions (ABgene).
3. Reaction mix for real time PCR contains: 10 μl of 2xqPCR Mastermix (ABgene), 1 μl of reconstituted primer and probe mix (designed by "PrimerDesign"), 4 μl PCR-Grade water (Fisher), and 5 μl of diluted cDNA (1:20, v:v).
4. Run samples on the Real Time PCR "Rotor-Gene 6000" (Corbett).
5. Program: Enzyme activation (15 minute, 95°C); Cycling: Denaturation (15 seconds, 95°C); Data collection (30 seconds, 60°C); Extension (15 seconds, 72°C), for 45 cycles.

3.11. Macrophage PON2 Protein by Western Blot Analysis (see* Note 9*)

1. Extract total cell protein using 80–150 μl lysis buffer (50 mM Hepes pH = 7.9, 150 mM NaCl, 0.5% triton-X100). Vortex and centrifuge, 10 minute, 13,000 g, 4°C. Measure protein concentration in the supernatant by the Bradford assay. In this assay, in contrast to the Lowry assay, there is less interference of detergents to the assay procedure.
3. Upload samples (40–60 μg protein) on 12.5% polyacrylamide gel. Use purified human or mouse PON2 as positive control and the appropriate molecular weight standard is PON2 is 43 KD.
4. Transfer the proteins to a nitrocellulose membrane, block the membrane with 2% BSA in TBST (1% tween 20), 2 hours at RT.
5. Incubate membranes with rabbit antihuman PON2 antibody diluted 1:5000 in 1% BSA in TBST (1% tween 20), over night at 4°C.
6. Wash the membranes five times for 5 minutes in TBST (1% tween 20).

7 Incubate membranes with secondary anti-rabbit antibody conjugated to HRP, diluted 1:5000 (Chemicon) in TBST (1% tween 20), 1 hour at RT.

8. Wash the membranes five times for 5 minutes in TBST (1% tween 20).

9. Develop the membrane with ECL according to the instructions provided with the ECL kit.

3.12. Macrophage PON3 Statinase Activity

1. Seed the cells in a 12 well plate, 2 × 10^6 cells/ml.
2. Wash the cells with Tris buffer.
3. Calculate the number of wells with cells. Add one additional well for determination of lovastatin self hydrolysis.
4. Prepare the reaction solution for all wells: 990 μl reaction buffer per well + 10 μl lovastatin (final concentration 5 μg/ml). Mix and add 1 ml per well.
5. Incubate the cells, 4 hours in the incubator.
6. At the end of the incubation period, put the dishes on ice. Transfer an aliquot of 100 μl to another eppendorf tube containing 100 μl of cold acetonitrille (to stop the reaction), mix.
7. Continue the assay as described in **Section 3.7**.
8. Calculations:
   a. Peak area acid/(peak area acid + peak area lovastatin) = % lovastatin hydrolyzed.
   b. % lovastatin hydrolyzed in the presence of macrophages − % self hydrolysis of lovastatin.
   c. 1 unit = 1 pmol lovastatin hydrolyzed per minute per ml.
   d. (12,400 × % lovastatin hydrolyzed × 5)/24,000 = units/ml. Express the results as units/mg cell protein.

Statinase activity in control mice peritoneal macrophages is 1251 ± 349 units/mg cell protein.

3.13. Macrophage PON3 Protein by Western Blot Analysis (*see* Note 9)

1. Use the same procedure as described in **Section 3.11** for PON2.

4. Notes

1. It is very important to perform all PON1 assays in serum and HDL at 25°C. Dilute serum HDL or purified PON1 with Tris buffer just before performance of the assay.

2. Organophosphates like phenyl acetate or paraoxon are dangerous. Thus, prepare the stock solutions in a chemical hood, wearing lab coat, gloves, and face mask. Collect the used tips, eppendorfs, and microplates in a specific container labeled with oraganophoaphate stickers.
3. No need to measure self hydrolysis of phenyl acetate.
4. Keep the TBBL stock solution on ice during performance of the lactonase assay, since its self hydrolysis is quite high.
5. Use (nitrilotriacetic acid) NTA also for HDL-associated PON1 inactivation.
6. For PON3 statinase activity, use only filtered water and acetonitrille and acetic acid which are HPLC grade.
7. Human monocyte derived macrophages (HMDM) can be used on day 7–8 after their incubation with 15% autologus serum to induce differentiation.
8. The time required for LDL oxidation by macrophages depends on the LDL used. Thus, it is recommended to check the susceptibility of the specific LDL to oxidation by macrophages before performance of the big experiment, in order to know the exact time to measure the extent of LDL oxidation by the TBARS assay.
9. For western blot analysis of PON2/3, use BSA specific for electrophoresis. The antibody for human PON2 must recognize also mouse PON2 and mouse PON3. Thus, if there is no monoclonal antibody specific for mouse PON3, use the antibody for human PON2. After incubation of the membrane with the primary antibody, collect the antibody, and keep at −20°C without the addition of sodium azide. It can be used several times. The results for PON2 or PON3 must be normalized according to the cellular levels of β-actin, using specific antibody for β-actin.

References

1. Aviram M, Rosenblat M (2004) Paraoxonases 1, 2 and 3, oxidative stress, and macrophage foam cell formation during atherosclerosis development. *Free. Radic. Biol. Chem.* 37:1304–1316.
2. Fuhrman B, Volkova N, Aviram M (2005) Paraoxonase 1 (PON1) is present in postprandial chylomicrons. *Atherosclerosis* **180**:55–61.
3. Khersonsky O, Tawfik DS (2005) Structure-reactivity studies of serum paraoxonase PON1 suggest that its native activity is lactonase. *Biochemistry* **44**:6371–6382.
4. Gaidukov L, Tawfik DS (2005) High affinity, stability, and lactonase activity of serum paraoxonase PON1 anchored on HDL with apoA-I. *Biochemistry* **44**:11843–11854.
5. Harel M, Aharoni A, Gaidukov L, Brumshtein B, Khersonsky O, Meged R, Dvir H, Ravelli RBG, McCarthy A, Toker L, Silman I, Sussman JL, Tawfic DS (2004) Structure and evolution of the serum paraoxonase family of detoxifying and anti-atherosclerotic enzymes. *Nat. Struc. Mol. Biol.* **11**:412–419.

6. Rozenberg O, Rosenblat M, Coleman R, Shih DM, Aviram M (2003) Paraoxonase (PON1)-deficiency is associated with increased macrophage oxidative stress: studies in PON1-knockout mice. *Free. Radic. Biol. Med.* **34**:774–784.
7. Rozenberg O, Shih DM, Aviram M (2005) Paraoxonase 1 (PON1) attenuates macrophage oxidative status: studies in PON1 transfected cells and in PON1 transgenic mice. *Atherosclerosis* **181**:9–18.
8. Aviram M, Rosenblat M, Bisgaier CL, Newton RS, Primo-Parmo SL, La Du BN (1998) Paraoxonase inhibits high-density lipoprotein oxidation and preserves its functions. A possible peroxidative role for paraoxonase. *J. Clin. Invest.* **101**:1581–1590.
9. Aviram M, Hardak E, Vaya J, Mahmood S, Milo S, Hoffman A, Billicke S, Daraganov D, Rosenblat M (2000) Human serum paraoxonases (PON1) Q and R selectively decrease lipid peroxides in human coronary and carotid atherosclerotic lesions: PON1 esterase and peroxidase-like activities. *Circulation* **101**:2510–2517.
10. Rozenberg O, Shih DM, Aviram M (2003) Human serum paraoxonase (PON1) decreases macrophage cholesterol biosynthesis: a possible role for its phodpholipase-A_2 activity and lysophosphatidylcholine formation. *Arterioscler. Thromb. Vasc. Biol.* **23**:461–467.
11. Rosenblat M, Vaya J, Shih DM, Aviram M (2005) Paraoxonase 1 (PON1) enhances HDL-mediated macrophage cholesterol efflux via the ABCA1 transporter in association with increased HDL binding to the cells: a possible role for lysophosphatidylcholine. *Atherosclerosis* **179**:69–77.
12. Rosenblat M, Gaidukov L, Khersonsky O, Vaya J, Oren R, Tawfik DS (2006) The catalytic histidine dyad of high density lipoprotein-associated serum paraoxonase-1 (PON1) is essential for PON1-mediated inhibition of low density lipoprotein oxidation and stimulation of macrophage cholesterol efflux. *J. Biol. Chem.* **281**:7657–7665.
13. Gaidukov L, Rosenblat M, Aviram M, Tawfik DS (2006) The 192R/Q polymorphs of serum paraoxonase PON1 differ in HDL binding, lipolactonase stimulation, and cholesterol efflux. *J. Lipid. Res.* **47**:2492–2502.
14. Rosenblat M, Draganov D, Watson CE, Bisgaier CL, La Du BN, Aviram M (2003) Mouse macrophage paraoxonase 2 (PON2) activity is increased whereas cellular PON3 activity is decreased under oxidative stress. *Arterioscler. Thromb. Vasc. Biol.* **23**:468–474.
15. Ng CJ, Wadleigh DJ, Gangopadhyay A, Hama S,, Grijalva VR, Navab M, Fogelman AM, Reddy ST (2001) Paraoxonase-2 is an ubiquitously expressed protein with antioxidant properties, and is capable of preventing cell-mediated oxidative modification of low-density lipoprotein. *J. Biol. Chem.* **276**:44444–44449.
16. Ng CL, Hama SY, Bourquard N, Navab M, Reddy ST (2006) Adenovirus mediated expression of human paraoxonase 2 protects against the development of atherosclerosis in apolipoprotein E-deficient mice. *Mol. Genet. Metab.* **89**:368–373.
17. Ng CJ, Bourquard N, Grijalva V, Hama S, Shih SM, Navab M, Fogelman AM, Lusis AJ (2006) Paraoxonase-2-deficiency aggravates atherosclerosis in mice despite lower apolipoprotein-B-containing lipoproteins: anti-atherogenic role for paraoxonase 2. *J. Biol. Chem.* **281**:29491–29500.
18. Horke S, Witte I, Wilgenbus P, Kruger M, Strand D, Forstermann U (2007) Paraoxonase-2 reduces oxidative stress in vascular cells and decreases endoplasmic reticulum stress-induced caspase activation. *Circulation* **115**:2055–2064.
19. Shiner M, Fuhrman B, Aviram M (2006) A biphasic U-shape effect of cellular oxidative stress on the macrophage anti-oxidant paraoxonase 2 (PON2) enzymatic activity. *Biochem. Biophys. Res. Commun.* **349**:1094–1099.
20. Shiner M, Fuhrman B, Aviram M (2004) Paraoxonase 2 (PON2) expression is upregulated via a reduced nicotinamide-adenine-dinucleotide-phosphate (NADPH)-oxidase-dependent mechanism during monocytes differentiation into macrophages. *Free. Radical. Biol. Med.* **37**:2052–2063.
21. Shiner M, Fuhrman B, Aviram M (2007) Macrophage paraoxonase 2 (PON2) expression is up-regulated by pomegranate juice phenolic anti-oxidants via PPARgamma and AP-1 pathway activation. *Atherosclerosis.* **195**:313–321.
22. Reddy ST, Wadleigh DJ, Grijalva V, Ng C, Hama S, Gangopadhyay A, Shih DM, Lusis AJ, Navab M, Fogelman AM (2001) Human paraoxonase-3 is an HDL-associated enzyme with biological activity similar to paraoxonase-1 protein but is not regulated by oxidized lipids. *Arterioscler. Thromb. Vasc. Biol.* **21**:542–547.
23. Draganov D, Stetson PL, Watson C, Billecke S, La Du BN (2000) Rabbit serum paraoxonase 3 (PON3) is a high density

lipoprotein-associated lactonase and protects low density lipoprotein against oxidation. *J. Biol. Chem.* **275**:33435–33442.

24. Shih DM, Xia YR, Wang XP, Wang SS, Bourquard N, Fogelman AM, Lusis AJ, Reddy ST (2007) Decreased obesity and atherosclerosis in human paraoxonase 3 transgenic mice. *Circ. Res.* **100**:1200–1207.
25. Ng. CJ, Bourquard N, Hama SY, Shih D, Grijalva VR, Navab M, Fogelman AM, Reddy ST (2007) Adenovirus-mediated expression of human paraoxonase 3 protects against the progression of atherosclerosis in apolipoprotein E-deficient mice. *Arterioscler. Thromb. Vasc. Biol.* **27**:1368–1374.
26. Gan KN, Smolen A, Eckerson HW, La Du BN (1991) Purification of human serum paraoxonase/arylesterase. Evidence for one esterase catalyzing both activities. *Drug. Metab. Dispos.* **19**:100–106.
27. Aharoni A, Gaidukov L, Yagur S, Toker L, Silman I, Tawfik DS (2004) Directed evolution of mammalian paraoxonases PON1 and PON3 for bacterial expression and catalytic specialization. *Proc. Natl. Acad. Sci. USA* 101:482–487.
28. Khersonsky O, Tawfik DS (2005) Chromogenic and fluorogenic assays for the lactonase activity of serum paraoxonases (PONs). *ChemBiochem* 7:49–53.
29. Garin MCB, Abbott C, Messmer S, Mackness M, Durrington P, Pometta D, James RW (1994) Quantification of human serum paraoxonase by enzyme-linked immunoassay: population differences in protein concentrations. *Biochem J.* **304**:549–554.
30. Eckerson HW, Wyte CM, La Du BN (1983) The human serum paraoxonase/arylesterase polymorphism. *Am. J. Hun. Genet.* **35**: 1126–1138.
31. Buege JA, Aust SD (1978) Microsomal lipid peroxidation. *Methods Enzymol.* **52**:302–310.
32. El-Saadani M, Esterbauer N, El-Sayed M, Goher M, Nassar AY, Jurgens G (1989) Spectrophotometric assay for lipid peroxides in serum lipoproteins using commercially available reagent. *J. Lipid Res.* **30**:627–630.
33. Aviram M (1983) Plasma lipoprotein separation by discontinuous density gradient ultracentrifugation in hyperlipoproteinemic patients. *Biochem. Med.* **30**:111–118.
34. Lowry OH, Rosebrough NJ, Farr AL, Randall RJ (1951) Protein measurement with the Folin phenol reagent. *J. Biol. Chem.* **193**:265–275.
35. Dobrian AD, Lazar V, Sinescu C, Mincu D, Simionescu M (2000) Diabetic state induces lipid loading and altered expression and secretion of lipoprotein lipase in human monocytes-derived macrophages. *Atherosclerosis* **153**:191–201.

Chapter 21

Preparation, Characterization, and Use of Antioxidant-Liposomes

Hongsong Yang, Victor Paromov, Milton Smith, and William L. Stone

Abstract

Antioxidant liposomes provide a unique means of delivering both water and/or lipid soluble antioxidants to tissues thereby affecting disease states or signal transduction pathways modulated by oxidative stress. Considerable evidence suggests that liposome-encapsulated antioxidants can be superior to the corresponding free antioxidants in this regard. This chapter will provide practical details on the preparation, characterization, and use of antioxidant liposomes. Methods will be described for the small-scale preparation (1 ml) and large-scale (100 ml/hour) preparation of antioxidant liposomes as well as the techniques for characterizing their size distribution and their physical and chemical stability. The use of antioxidant liposomes in an in vitro situation will also be detailed.

Key words: Antioxidants, Liposomes, Vitamin E, Oxidative stress, Cellular uptake, Glutathione, phospholipid.

1. Introduction

A liposome is a vesicle composed of a closed phospholipid bilayer (the lipid phase) encapsulating an aqueous phase: they are also the smallest artificial vesicles that can be produced from natural nontoxic phospholipids and cholesterol. Liposomes have often been used as a drug delivery system since they have a unique ability to simultaneously deliver lipid-soluble as well as water-soluble agents *(1, 2)*. Similarly, antioxidant liposomes provide a unique molecular tool for delivering both water- and lipid-soluble antioxidants to tissues or cells. The therapeutic use of antioxidant liposomes is particularly

From: *Methods in Molecular Biology, vol. 477: Advanced Protocols in Oxidative Stress I,*
Edited by: D. Armstrong, DOI: 10.1007/978-1-60327-517-0_21, © Humana Press, New York, NY

relevant for diseases in which oxidative stress plays a key role *(3, 4)*. Oxidative stress is a condition in which the production of damaging free radical species/reactive oxygen species exceeds the capacity of antioxidant protection mechanisms to prevent cellular damage. Free radicals are molecules with unpaired electrons, which are often highly reactive and damaging to biological systems, particularly the biological membranes of subcellular organelles.

Lipid-soluble antioxidants such as vitamin E or coenzyme Q10 (CoQ10) can be incorporated into the hydrophobic lipid phase of liposomal bilayers. Vitamin E is a term referring to at least eight different isoforms including four tocopherols (alpha-, beta- gamma-, and delta) and four corresponding tocotrienols (alpha-, beta- gamma-, and delta-). Alpha-tocopherol is the primary form of vitamin E found in human plasma since the other isoforms are rapidly metabolized by the liver before being secreted into plasma *(5)*. Moreover, the liver also contains an alpha-tocopherol transfer protein, which selectively promotes the secretion of alpha-tocopherol out of the liver into very low density lipoprotein *(6)*. Oral administration with forms of vitamin E other than alpha-tocopherol is not, therefore, effective at increasing their plasma levels.

Increasing evidence shows that the different vitamin E isoforms have distinct chemical and biological properties *(7, 8, 9)*. Moreover, many of the biological properties of vitamin E isoforms appear unrelated to their role as antioxidants *(10)*. Evidence suggest, for example that gamma-tocopherol *(11, 12, 13)* as well as tocotrienols have the potential to kill many types of cancer cells by activating apoptosis *(14, 15, 16, 17)*. Intravenous injection with liposomes containing non-alpha-tocopherol isoforms would by-pass the selectivity imposed by the liver, allowing the unique properties of each vitamin E isoform to be therapeutically exploited.

Water-soluble antioxidants such as urate *(18)*, ascorbate, N-acetyl-L-cysteine (NAC), or glutathione (GSH) can be encapsulated into the interior aqueous domain of antioxidant-liposomes. NAC is the acetylated from of L-cysteine and it is an excellent source of sulfhydryl (SH) groups. NAC is converted in the body into metabolites capable of stimulating GSH synthesis, promoting detoxification, and it also acts directly as a free radical scavenger. Fan et al. *(19)* have found that liposomal encapsulated NAC provides a longer lasting protection against acute respiratory distress syndrome than does free unencapsulated NAC. Water-soluble antioxidant enzymes such as superoxide dismutase or glutathione peroxidase can also be incorporated into the aqueous domain of antioxidant liposomes *(4)*.

Although general guidelines for the preparation and size characterization of liposomes have been described *(20)*, this chapter will focus on the specific preparation and characterization of antioxidant liposomes.

2. Materials

2.1. Equipment

1. Mini-Extruder set including extruder, stand/heating block, filter support, gas-tight syringe, and polycarbonate membranes, Avanti Polar Lipids Inc. (Alabaster, AL).
2. Model M-100L Microfluidizer (a bench top instrument) from Microfluidics (Newton, MA).
3. Nicomp 380 DLS from Particle Sizing Systems (Port Richey, FL).
4. Spectra Max Plus 384 microplate reader (Molecular Devices, Sunnyvale, CA).
5. Coulochem II Electrochemical Detector, ESA Model 580 Solvent Delivery, Model 5011 Analytical Cell, a Model 5020 Guard Cell (Chelmsford, MA).

2.2. Reagents and Buffers

2.2.1. Lipids for Liposome Formulation

1. Cholesterol, C8667 (Sigma-Aldrich.com, St. Louis, MD).
2. Phospholipon 85G (PL8G) and phospholipon 90H (PL90H)(Lipoid LLC, Newark, NJ).
3. L-α-phosphatidic acid, Avanti Product Number 109774 (Avanti Polar Lipids Inc., Alabaster, AL).

2.2.2. Organic Solvents for Liposome Formulation

1. Methylene chloride (dichloromethane), D151 (Fisher Scientific).
2. Ethanol, 042104 (Aaper Alcohol and Chemical Co., Shelbyville, KY).
3. Chloroform, C606 (Fisher Scientific).

2.2.3. Lipid-soluble Antioxidants

1. RRR-α-tocopherol (AT), R951027A1 and RRR-γ-tocopherol (DT), R951027A5 (Cognis Corporation (LaGrange, IL).
2. δ-Tocotrienol (DT3), Cayman Product 10008513 (Cayman Chemicals, Ann Arbor, Michigan).

2.2.4. Water-soluble Antioxidants

1. N-Acetyl-L-cysteine (NAC), A8199 (Sigma).
2. L-Glutathione reduced (GSH), G4251 (Sigma).

2.2.5. GSH/GSSG Assay

1. The assay buffer is Dulbecco's Phosphate-Buffered Saline (PBS) (Mediatech Inc., Herdon, VA) supplemented with 2 mM ethylenediaminetetraacetic acid (EDTA) Sigma-Aldrich, St. Louis, MO.
2. A 0.1 mM GSH solution (30.7 mg/l) is prepared in PBS and stored at 4°C (for up to 2 weeks). The solution is used to prepare GSH standards.
3. A 108 μl aliquot of 2-vinylpyridine (2-VP) (Sigma) is added to 1.00 ml of ethanol (1.00 M) and prepared immediately prior to the assay.
4. A 1.00 mM Ellman's reagent is prepared by dissolving 40 mg of 5,5′-dithio-bis(2-nitrobenzoic acid), (DTNB) (Invitrogen, Eugene, OR) in 100 ml of PBS immediately prior to the assay.
5. A 2 mM stock solution of reduced β-nicotinamide adenine dinucleotide 2′-phosphate tetrasodium salt (NADPH) (Sigma-Aldrich, St. Louis, MO) is prepared by dissolving 1.7 mg of NADPH in 1.00 ml of PBS immediately prior to the assay (stored at 4°C).
6. Glutathione Reductase (GR) (Roche Diagnostics, Indianapolis, IN).
7. Fisherbrand Flat-bottom 96-well clear plastic plate (Fisher Scientific, Hampton, NH).

2.2.6. Vitamin E Assays

1. A lysis buffer is prepared by mixing a 40.0 ml aliquot of 50 mM HEPES/NaOH buffer (pH = 7.50) with 1.0 ml of Triton X-100, 5.0 ml of 2 mM EDTA, and 54.0 ml of distilled water and stored at 4°C.
2. Stock solutions for the mobile phase are: (a) 1 M ammonium acetate/acetic acid (NH_4OAc/HOAc) Buffer (pH = 4.40) prepared by dissolving 77.10 g of NH_4OAc in 1.0 l of polished water and adjusting the pH to 4.40 with HOAc (J.T. Baker Chemical Co., Phillipsburg, NJ); (b) 3.0 mM of citric acid is prepared by dissolving 0.1441 g of citric acid in polished water to a final volume of 250 ml. All solutions in the mobile phase are prepared from "polished water" which is distilled water filtered through a Sep-Pak cartridge (Water, Atlanta, GA).
3. A 10 μg/ml BHT solution in hexane is prepared by dissolving 10.0 mg of 2,6-ditertiary-butyl-4-methyl phenol (BHT) from Aldrich (Milwaukee, WIS) in 1.0 l hexane.
4. Mobile Phase is prepared by mixing a 80.0 ml aliquot of 1 M NH_4OAc/HOAc buffer (pH = 4.40) with 4.0 ml of 3.0 mM citric acid, adding polished water to give a final volume of 400 ml followed by filtration through a 0.22 μm filter

Millipore (Millipore Corp., Bedford, MA). This solution is then mixed with 3,600 ml of HPLC grade methanol (MeOH).

5. rac-5, 7-dimethyl tocol (Matreya, Pleasant Gap, PA).

2.2.7. Cell Culture with Antioxidant Liposomes

1. RAW264.7 murine macrophage-like cells are obtained from the American Type Culture Collection, Rockville, MD.

3. Methods

Antioxidant liposomes can be formulated with a wide variety of phospholipids, stabilizers, lipid-soluble antioxidants, and water-soluble antioxidants. It is critical to have an overall experimental design in mind for the use of a given liposomal formulation since this will dictate compositional issues. For example, for an in vitro tissue culture experiment, the stock liposomes will often be diluted to provide physiological levels of antioxidants.

3.1. Concentration Issues

Liposomes have two phases, an interior aqueous phase and a hydrophobic lipid bilayer phase. For the bilayer phase, the most useful units of concentration are mole fraction but for the interior aqueous phase it is mM or μM. The mole fraction of component i, x_i, in the lipid bilayer is given by $x_i = n_i/n$ where n_i is moles of component i in the bilayer and n is total number of moles of all components in the bilayer. When a liposomal formulation is diluted into a buffer/culture medium, the mole fraction of any component (like a lipid-soluble antioxidant) in the lipid bilayer will not change. It is also useful, however, to express the concentration of a lipid-soluble component in molar units so one can evaluate if a "physiological" level of a lipid component is being utilized. In the case of vitamin E, the molar concentration is given as moles of liposomal vitamin E added to the buffer divided by the volume of the buffer; a physiological concentration is about 30 μM for α-tocopherol. It should be kept in mind, however, that all the vitamin E is in the liposomal lipid phase, not "free" in solution.

A water-soluble antioxidant can also be incorporated into the interior aqueous phase of liposomes and its initial concentration in the liposomal aqueous phase (assuming no in vitro loss) would be the same as its concentration in the aqueous phase of the "formulation buffer". If the water-soluble antioxidant were an ionic, hydrophilic compound (such as ascorbate or GSH) it would not diffuse through the lipid bilayer of the liposome and its concentration in the interior liposomal aqueous phase would stay constant even after subsequent dilution of the liposomes or removal of the exterior formulation buffer. It is still, however, useful to calculate the molar concentration of the liposomal water-soluble antioxidant in the buffer or medium into which it is diluted, i.e., the moles of water-soluble antioxidant per liter of buffer/medium: again keeping in mind, that this is not the "true" free concentration of antioxidant.

3.2. Preparation of Antioxidant Liposomes by Extrusion (Small Scale)

Antioxidant liposomes can be made with a variety of techniques and compositional variations. Some liposomal components such as DPPC and tocotrienols can be very expensive and it might be necessary to make small-scale preparations for use in tissue culture experiments. In this case, the "mini" extrusion technique is a very useful apparatus and is available from Avanti Polar Lipids (see http://www.avantilipids.com/extruder.html) where its operation is also described in some detail.

As is the case for all liposomal preparations, the desired lipid components are first added together to form a homogenous solution using an organic solvent (*see* **Note 1**). The solvent is evaporated in a glass container leaving a thin lipid film, which is then hydrated and detached by vigorous mixing with the formulation buffer (which could contain water-soluble antioxidants) to form large multilamellar vesicles (LMVs), which are then reduced in size by energy input. For the extrusion technique, this energy input is accomplished by mechanically extruding the LMVs through a polycarbonate filter with a fixed pore size. This will yield large unilamellar vesicles (LUVs) with a diameter similar to the pore size of the polycarbonate filter.

3.2.1. Preparation of δ-Tocotrienol-Liposomes with NAC (NAC-DT3-Liposomes) by Mini-extrusion

Below, we provide a detailed procedure for the preparation of liposomes containing δ-tocotrienol (DT3) as the lipid-soluble antioxidant and NAC as the water-soluble antioxidant. In this example, phosopholipon 85G (PL85G) is used as the phospholipid source (soybean lecithin) and this particular phospholipid forms "liquid" bilayers at room temperature. In this example, the mole fractions (in the lipid bilayer) of PL85G to cholesterol to delta-tocotrienol are 0.666, 0.266, and 0.066, respectively (*see* **Note 2**). Cholesterol and vitamin E both make liposomes less permeable to aqueous dyes and more resistant to protein-induced disruption *(21)*. In this example, there are a total of 50 μmol of lipid per ml of formulation buffer and the concentration of NAC and DT3 are 75 and 3.33 mM (in the formulation buffer), respectively.

1. Stock solutions of PL85G (30 mg/ml), cholesterol (20 mg/ml), and DT3 (24 mg/ml) are made in methylene chloride (*see* **Note 3**). The formulation buffer contains 61.2 mg of NAC in 5.00 ml of PBS with the pH adjusted to 7.4 (*see* **Note 4**). The NAC solution is passed through a sterile 0.2 μm filter into a sterile 10 ml test tube.
2. The stock solutions (4.5 ml of PL85G, 1.3 ml of cholesterol, and 275 μl of delta-T3) are added to a round bottom

test tube (55 ml, Pyrex, No. 9826, 25 mm OD × 150 mm) with a Teflon-lined screw cap. The dichloromethane solvent is evaporated using a stream of nitrogen in a fume hood.

3. After the solvent is completely removed, the lipid film is hydrated in 5 ml of sterile 75 mM NAC in PBS with vigorous vortexing for at least 2 min (*see* **Note 5**).
4. A 1 ml aliquot of the fully hydrated lipid sample is loaded into one of the gas-tight syringes and placed into one end of the mini-extruder. A second, empty, gas-tight syringe is placed into the other end of the mini-extruder and the fully assembled apparatus placed in the extruder stand. The syringe plunger containing the sample is transferred through the polycarbonate filter to the receptor syringe and this process is repeated at least five times (or 10 passes). Polycarbonate filters are available from 0.2 to 1.0 μm.
5. If liposomes are made with a 0.2 mm polycarbonate filter, they can be passed through a 0.4 μm filter for sterilization.

3.3. Preparation of Antioxidant Liposomes by High Pressure Homogenization (Large Scale)

For in vivo animal experiments, it is often necessary to have liposomal volumes much greater than the 1 ml produced by the mini-extruder. In addition, it may be required that the liposomes used for an in vivo animal experiment be identical to those used in a potential large-scale clinical experiment. In this case, the ability to scale-up a bench size production to a commercial scale production, without loss of liposomal characteristics, would be a very important quality control issue. High pressure homogenization (HPH), as performed by the Model M-100L Microfluidizer (a bench top instrument) with a throughput of up to 270 ml/minute can be scaled-up without loss of liposomal physiochemical properties. We routinely use this apparatus to produce 100 ml aliquots of antioxidant liposomes for use in animal models. Avestin Inc. (Ottawa, Canada) produces an apparatus that produces liposomes by a combination of homogenization and extrusion. HPH works by using air pressure to push a large piston which, in turn, pushes a smaller plunger thereby transferring the intensified pressure to the product stream (LMVs in our case) contained within an interaction chamber with microchannels. This process causes the LMVs to undergo shear, impact, and cavitation forces resulting in unilamellar vesicles.

Talsma et al. *(22)* have compared the HPH technique for preparing liposomes with the extrusion method and the ultrasonication method. In agreement with the data presented by Talsma et al. *(22)*, we have found that about three to five cycles of homogenization, with chamber pressure of 17,000 psi, are sufficient to produce liposomes with a minimum mean particle size. In general, ultrasonication produces liposomes with a mean particle size similar to that achieved by HPH *(22)*. We have found, however, that sonication is not suitable for the preparation of antioxidant liposomes since it promotes the oxidative loss of any added antioxidants.

The total amount of lipid per ml of formulation buffer is also a factor influencing liposomal characteristics. In general, we have found that concentrations of 50 μmol of lipids per ml of formulation buffer produce liposomes with a narrow size distribution but higher levels increase the solution viscosity and result in liposomes with an increased size distribution as well as an increased mean particle size.

In the example below, we describe the details to produce 50 ml of antioxidant liposomes containing both alpha-tocopherol and gamma-tocopherol as lipid-soluble antioxidants and GSH as a water-soluble antioxidant. Phospholipon 90 H (PL90H) is the primary phospholipid used but phosphatidic acid (PA) was also used to impart a negative charge to the liposomes to help maintain a fully dispersed particulate state. The mole ratios of PL90H: cholesterol: PA: AT: GT in this preparation are 10:4:0.1:1:1, respectively.

1. A stock solution of PL90H (*see* **Note 6**) is made in chloroform (20.0 g in 100.0 ml of $CHCl_3$). Stock solution of cholesterol (2.0 g/100.0 ml), phosphatidic acid (2.5 g/100 ml), α-tocopherol (0.50 g AT/10.0 ml), and γ-tocopherol (0.50 g GT/10.0 ml) are prepared in methylene chloride.
2. The stock solutions (12.3 ml each of PL90H, 0.86 ml of PA, 24 ml of cholesterol, 2.7 ml of AT, and 2.6 ml of GT) are added to a 500 ml Pyrex round bottom flask with a 24/40 ground glass joint. The solvent is removed by rotary evaporation producing a thin film of lipid on the inside of the round bottom flask. The solvent should be completely removed by application of a vacuum with mild heating (37°C).
3. After the solvent is completely removed, the lipid film is hydrated in 100 ml of sterile 75 mM GSH in PBS with vigorous agitation until all the lipid is removed from the inner glass surface of the round bottom flask (*see* **Note 5**). The temperature of the PBS solution should be above the gel-liquid transition temperature for PL90H, which is 51°C.
4. The HPH pump is primed with ethanol, then with distilled water followed by sterile PBS. The 100 ml of fully hydrated lipid sample is then placed in the product inlet reservoir and pumped through the Microfluidizer using a pressure of 17,000 psi until the reservoir is almost drained. The product is collected and passed through pump for an additional four times (resulting in a total of five passes).
5. It is sometimes desirable to remove the formulation buffer containing the external water-soluble antioxidant(s), leaving only the liposomal encapsulated water-soluble antioxidant(s).

This can be done in large scale by sedimenting the liposomes (three times) by centrifugation (15,000 rpm in an SS-34 rotor in a Sorvall RC-5B centrifuge for 15 min) followed by their resuspension in PBS buffer. Alternatively, the liposome sample can be placed inside dialysis tubing and dialyzed against multiple changes of PBS buffer.

3.4. Size Measurements by Dynamic Light Scattering

Particle size is an important factor modulating the uptake of liposomes *(23, 24)*. Chono et al. *(23)* have found, for example, that the uptake of liposomes by rat alveolar macrophages increases with particle size over a range from 100 to 1000 nm. It is important, therefore, to characterize antioxidant liposomes by measuring their particle size distribution. This is most readily accomplished by the use of dynamic light scattering (DLS), which has a very large size range from 0.005 to 2 microns (1 micron is 10^{-6} m or 1000 nm). For DLS, light from a laser is focused into a glass tube containing a diluted sample of the liposome suspension, which scatters some of the light in all directions. For a laser light beam, it is possible to observe time-dependent fluctuations in the scattered intensity due to the Brownian motion of the liposomes. Analysis of the time dependency of the intensity fluctuations yields the diffusion coefficient of the liposomes. Knowing the viscosity of the medium, the hydrodynamic diameter of the liposomes can be calculated from the Stokes Einstein equation. In the example provided below, a Nicomp 380 DLS from Particle Sizing Systems (Port Richey, FL) is utilized.

1. Dilute 10 μl the liposome sample (assuming a concentration of about 50 μmol of total lipids/ml) into 1.00 ml of PBS buffer, which should give a scattering intensity of 300 kHz.
2. The particle size distribution is analyzed using the software (CW388) accompanying the Nicomp instrument. The simplest particle distribution for liposomes is a smooth Gaussian unimodal distribution with a well-defined mean diameter and half width. If, however, this distribution does not fit the raw data very well (*see* **Note 7**), it is then necessary to attempt a fit assuming a discrete distribution of particle sizes, i.e., a distribution analysis.

3.5. Chemical Stability of Antioxidant Liposomes

The chemical stability of the antioxidants incorporated into antioxidant liposomes is an important quality control issue. Below, we describe the analytical procedures to measure GSH (a water soluble) or vitamin E (lipid soluble) isomers in antioxidant liposomes.

3.5.1. GSH Assay Using the Enzymatic Recycling Method

The assay detailed below provides specific details for measuring both reduced GSH and oxidized GSH (GSSG) in liposomes using the Tietze's enzymatic recycling method *(25)*, but this method can easily be adopted to measuring cellular level as well.

This assay uses the Tietze's enzymatic recycling method *(25)* in which GSH reacts with DTNB (5,5′-dithio-bis-2-nitrobenzoic acid) yielding 5-thio-2-nitrobenzoic acid (TNB) as well as a mixed disulfide between GSH and TNB that is subsequently reduced by glutathione reductase, recycling GSH and producing more TNB. The rate of TNB production is directly proportional to the concentration of total GSH in the sample. Since this assay utilizes glutathione reductase, it measures the concentration of both GSH and GSSG (total GSH). In order to measure just GSSG, 2-vinylpyridine is first used to derivatize GSH alone *(26)* leaving just GSSG. The concentration of GSH in a sample can be calculated as the difference between total GSH and GSSG. We have found it convenient to utilize a GSH assay kit from World Precision Instruments, (Sarasota, FL) but other sources are also available (see 703002 Glutathione Assay Kit from Cayman Chemical Company, Ann Arbor, MI).

1. Liposome samples containing GSH are diluted in PBS to give a final GSH concentration in the range of 5–30 μM. A 100 μl of each diluted sample is pipetted into two plastic microcentrifuge tubes, one for total GSH and the other for GSSG measurements.
2. GSH standards (6, 12, 18, 24, and 30 μM) are prepared from the 0.1 mM stock solution by dilutions in PBS and assayed in parallel with the samples. PBS is used as a blank reference (0 μM GSH).
3. A 2-μl aliquot of 1.00 M 2-VP solution in ethanol is added to each GSSG sample tube (*see* **Note 8**).
4. After vortexing for 30 seconds, the sample and standard tubes are incubated at room temperature for 60 minutes. A 25-μl aliquot of sample/standard is placed in each well of a clear plastic 96-well plate and all samples/standards assayed in triplicate.
5. The assay cocktail is prepared by mixing 1 ml of DTNB stock, 1 ml of NADPH stock, 0.2 ml of glutathione reductase stock, and 7.8 ml assay buffer. This cocktail should be used within 20 min after preparation.
6. A 100 μl of the assay cocktail is added to each sample/standard followed by careful shaking or 2 seconds and then inserted into the Spectra Max Plus 384 microplate reader. The absorbance at 405 nm is measured for 10 min at 2 min intervals at room temperature.
7. For each standard/sample, the average OD (of the triplicates) is plotted as function of time, and slopes for each curve (δOD) calculated (only in the linear part of the curves) by **δOD = OD final − OD initial**. The δOD values for the standards are plotted as a function of the GSH concentration

and the resulting graph should be linear, i.e., δOD = ***a*** · **GSH concentration** + ***b***. The ***a*** and ***b*** parameters of the formula for the linear trend line of the plot are determined using Microsoft Excel (graph function "Add Trend line"). The total GSH or GSSG concentrations can be calculated by:

$$\text{Total GSH or GSSG} = (\delta\ \text{OD-}b)/a$$

The GSH concentrations in the diluted samples are calculated as the differences in total GSH and GSSG. Concentrations that are out of the 5–30 μM range are disregarded. Finally, GSH and GSSH concentrations in the liposomal samples are calculated using the appropriate dilution ratios.

Figure 21.1 shows ratio of GSSG (in GSH equivalents) to GSH assayed as a function of storage time for GSH-liposomes and GSH-α-tocopherol liposomes stored at room temperature (RT) or 4°C. These data show that GSH is oxidized fairly rapidly at room temperature compared to 4°C. Moreover, the presence of α-tocopherol does not retard GSH oxidation. These data suggest that GSH-liposomes would have to be used relatively soon after being formulated. A chelating agent such as ethylenediaminetetraacetate (EDTA) or urate *(18)* could be used as a preservative for both lipid- and water-soluble antioxidants.

3.5.2. Vitamin E Assay Using HPLC with Electrochemical Detection (HPLC-ECD)

The HPLC method described below utilizes a highly sensitive coulometric electrochemical detection technique suitable for measuring vitamin E isomers in 10^6 cells. Although we describe an assay for measuring vitamin E in adherent cultured cells, this technique

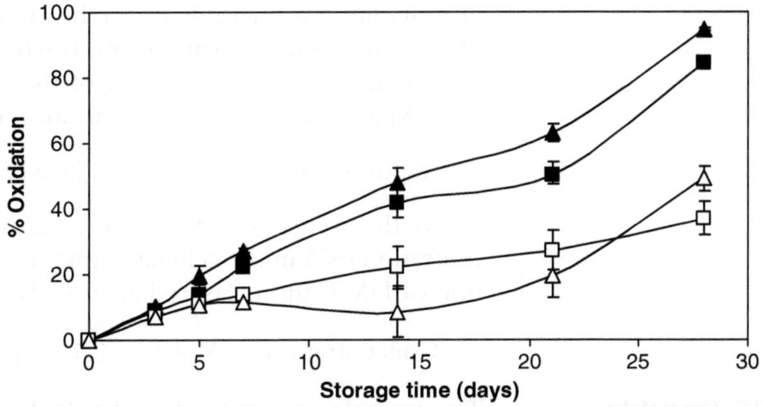

Fig. 21.1. GSH oxidation to GSSG in liposomes during storage. GSH-liposomes (square symbols) and GSH-α-tocopherol liposomes (triangular symbols) were stored at room temperature (solid symbols) or at 4°C (open symbols). Total GSH (GSH + GSSG) and GSSG concentrations within the liposome samples were measured after 3, 5, 7, 14, 21, and 28 days of storage using the Tietze method. Oxidation of GSH encapsulated in the liposomes was expressed as a percent of GSH converted to GSSG relatively to the total GSH content of each sample.

can be readily adopted for measuring vitamin E levels in liposomes or tissue samples. The particular HPLC system we use consists of a Coulochem II Electrochemical Detector, an ESA Model 580 Solvent Delivery Module, a HR-80 column (C 18, 3 μm, 8 cm), a Model 5011Analytical Cell, a Model 5020 Guard Cell.

1. The culture medium is removed from 2×10^6 cells in a microplate by washing three times with cold PBS buffer. The cells are then overlaid with 400 μl of lysis buffer, scraped, and collected into a glass vial.
2. The lysed cell sample is mixed with 1.0 ml ethanol containing rac-5, 7-dimethyl tocol as an internal standard and 1.0 ml hexane containing 10 μg/ml BHT.
3. After vortexing for 4 min, the mixture is centrifuged at $1020 \times g$ for 5 min, and a 400-μl aliquot of the supernatant fluid is removed, dried under nitrogen, and taken up in 100 μl of mobile phase and filtered using 0.45 μm syringe filter.
4. The flow rate is adjusted to 1.2 ml/min and the second analytical cell potential set to 400 mV and the potential of the guard cell electrode, used to eliminate interference by electro-active impurities in the mobile phase, set to 300 mV.

It is important to measure the response factor for each vitamin E isomer with respect to the internal standard (they are all different). This is accomplished by preparing standard solutions with known concentrations of vitamin E isomer (the extinction coefficients are listed in **Note 9**). The standard solutions are mixed with internal standard and the areas of peaks measured using the HPLC-ECD method. The response factors of a particular vitamin E isomer relative to the internal standard (DMT) are calculated by the following formula where $\text{area}_{(DMT)}$ and [DMT] are the area and concentration of the internal standard; $\text{area}_{(A)}$ and [A] are the area and concentration of the vitamin E isomer:

$$\text{Response Factor of A} = [\,A]/[DMT] \times \text{area}_{(DMT)}/\text{area}_{(A)}$$

With the response factor (F), we can easily obtain the concentration of A in the cellular extract from the known concentration of DMT and measured areas of DMT and A by:

$$\text{Concentration of A} = F \times [DMT] \times \text{Area}_{(A)}/\text{Area}_{(DMT)}$$

3.6. The Uptake of liposomes Containing Both α-Tocopherol and γ-Tocopherols by RAW 264.7 Murine Macrophages

The example provided below details the procedure for measuring the cellular uptake of vitamin E by RAW 264.7 murine macrophages incubated with liposomes containing both α-tocopherol and γ-tocopherol (AT,GT-liposomes). The liposomes used here were prepared by the HPH technique using the protocol described above except no NAC was present. Blank liposomes were also prepared in this experiment and they contain neither lipid- nor water-soluble antioxidants.

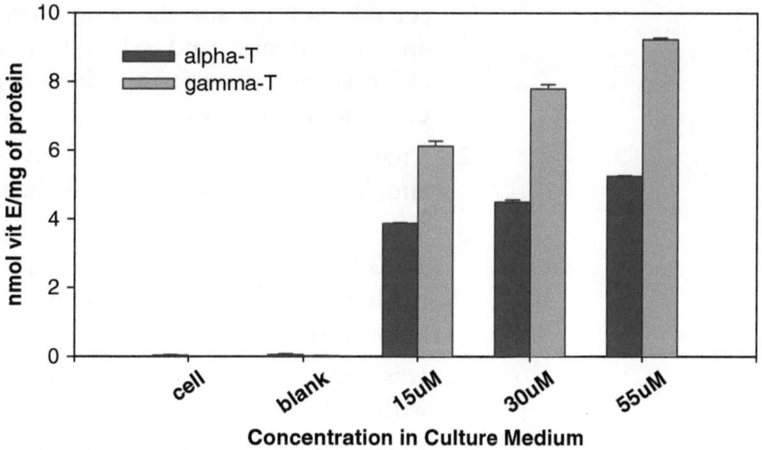

Fig. 21.2. RAW264.7 macrophage cellular uptake of tocopherols after incubation with α/γ-tocopherol antioxidant liposomes for 24 hours. The vitamin E content of RAW264.7 macrophages dramatically increases by incubation (for 24 hours) with liposomes containing both alpha- and gamma-tocopherol. The levels of liposomal vitamin E present in the medium varied from 0 to 55 μM. Physiological levels of vitamin E are in the 10–30 μM range.

1. The stock liposomes contained 3.2 mM of α-tocopherol (AT) and 3.2 mM of γ-tocopherol (GT).
2. A 2.5 ml aliquot of 5×10^5 cells/ml is placed in each well of a 12-well plate and the cells incubated overnight to about 90% confluence.
3. The medium is removed from each well and fresh medium added with different concentrations of AT,GT-liposomes, black liposome or PBS into wells. After 24 hours of incubation, the medium from each well is removed, the cells lysed, and assayed for tocopherols as detailed in **Section 3.5.2**.
4. The total proteins in the lysed cell samples (*see* **Note 10**) are measured using BCA assay and the final data presented as nmol of tocopherols/mg of protein.

Figure 21.2 shows the cellular uptake of tocopherols incubated with diluted samples of the AT,GT-liposomes for 24 hours. It is clear that the cells alone or cells incubated with the blank liposomes have very low endogenous tocopherol levels. In marked contrast, the cells incubated with the AT,GT-liposomes show a marked increase in the intracellular levels of both α-tocopherol and γ-tocopherol.

4. Notes

1. Vitamin E rapidly decays in halogenated solvents like chloroform or methylene chloride and therefore cannot be stored in these solvents for any length of time. Ethanol is a

suitable solvent for the long-term storage of tocopherols and tocotrienols and ethanol is miscible with methylene chloride or chloroform. Moreover, the molar extinction coefficients for most vitamin E isomers are given in ethanol.

2. Vitamin E can be added to liposomes at a level up to 33 mol% and, in general, has an effect similar to cholesterol *(21)*.
3. The total lipid concentration is typically in the range of 10–30 mg of lipid per ml of organic solvent. Also note that methylene chloride is generally considered less toxic than chloroform.
4. It is important to adjust the pH of the NAC solution to 7.4 with NaOH or it will be acidic and potentially toxic.
5. When a water-soluble antioxidant is added to the PBS or culture medium, it is possible to increase the efficiency of entrapment into the liposomal aqueous phase by 3–5 freeze/thaw cycles.
6. Phospholipon 90H (PL90H) is not very soluble in methylene chloride and it is, therefore, necessary to use chloroform as the solvent.
7. The Nicomp software provides a "Chi Squared" parameter to help evaluate the type of particle distribution in a sample. A value less than 2 indicates a unimodal distribution and a value greater than 3 suggests a bimodal (or greater) distribution.
8. All work with 2-VP should be kept under the hood as the chemical is both toxic and volatile.
9. The molar extinctions coefficients for tocopherols (Ts) and tocotrienols (T3) are provided below.

Substance	M.W.	λ_{Max} (nm) in ethanol	ε (per cm per M) in ethanol
α-T	430.7	292	3270
β-T	416.7	296	3730
γ-T	416.7	298	3810
δ-T	402.7	298	3520
α-T3	424.7	292	3870
β-T3	410.7	295	3600
γ-T3	410.7	298	4230
δ-T3	396.7	292	3300
rac-5,7-dimethyl tocol	416.7	292	3460

10. It is important to make the BSA protein standard in the exact solution that the protein to be assayed is in, i.e., the lysis buffer.

Acknowledgments

This work was support by two United States Army Medical Research Command Grants: "The Influence of Antioxidant Liposomes on Macrophages Treated with Mustard Gas Analogues", USAMRMC Grant No. 98164001, and "Topical Application of Liposomal Antioxidants for Protection against CEES Induced Skin Damage", USAMRMC Grant No. W81XWH-05-2-0034.

References

1. Ciobanu, M., Heurtault, B., Schultz, P., Ruhlmann, C., Muller, C. D., and Frisch, B. (2007) Layersome: development and optimization of stable liposomes as drug delivery system. *Int J Pharm.* ***344***, 154–157.
2. Zaru, M., Mourtas, S., Klepetsanis, P., Fadda, A. M., and Antimisiaris, S. G. (2007) Liposomes for drug delivery to the lungs by nebulization. *Eur J Pharm Biopharm.* *67*, 655–666.
3. Stone, W. L., Mukherjee, S., Smith, M., and Das, S. K. (2002) Therapeutic uses of antioxidant liposomes. *Methods Mol. Biol.* ***199***, 145–161.
4. Stone, W. L., and Smith, M. (2004) Therapeutic uses of antioxidant liposomes. *Mol. Biotechnol.* *27*, 217–230.
5. Sontag, T. J., and Parker, R. S. (2007) Influence of major structural features of tocopherols and tocotrienols on their omega-oxidation by tocopherol-omega-hydroxylase. *J. Lipid Res.* ***48***, 1090–1098.
6. Mustacich, D. J., Bruno, R. S., and Traber, M. G. (2007) Vitamin E. *Vitam. Horm.* *76*, 1–21.
7. Krishnan, K., Campbell, S., Abdel-Rahman, F., Whaley, S., and Stone, W. L. (2003) Cancer chemoprevention drug targets. *Curr. Drug Targets* *4*, 45–54.
8. Stone, W. L., Krishnan, K., Campbell, S. E., Qui, M., Whaley, S. G., and Yang, H. (2004) Tocopherols and the treatment of colon cancer. *Ann. N Y Acad. Sci.* ***1031***, 223–233.
9. Stone, W. L., and Papas, A. M. (1997) Tocopherols and the etiology of colon cancer. *J. Natl. Cancer Inst.* ***89***, 1006–1014.
10. Azzi, A. (2007) Molecular mechanism of alpha-tocopherol action. *Free Radic. Biol. Med.* ***43***, 16–21.
11. Campbell, S. E., Stone, W. L., Lee, S., Whaley, S., Yang, H., Qui, M., Goforth, P., Sherman, D., McHaffie, D., and Krishnan, K. (2006) Comparative effects of RRR-alpha- and RRR-gamma-tocopherol on proliferation and apoptosis in human colon cancer cell lines. *BMC Cancer.* *6*, 13.
12. Jiang, Q., Wong, J., and Ames, B. N. (2004) Gamma-tocopherol induces apoptosis in androgen-responsive LNCaP prostate cancer cells via caspase-dependent and independent mechanisms. *Ann. N Y Acad. Sci.* ***1031***, 399–400.
13. Jiang, Q., Wong, J., Fyrst, H., Saba, J. D., and Ames, B. N. (2004) Gamma-tocopherol or combinations of vitamin E forms induce cell death in human prostate cancer cells by interrupting sphingolipid synthesis. *Proc. Natl. Acad. Sci. U S A.* ***101***, 17825–17830.
14. Har, C. H., and Keong, C. K. (2005) Effects of tocotrienols on cell viability and apoptosis in normal murine liver cells (BNL CL.2) and liver cancer cells (BNL 1ME A.7R.1), in vitro. *Asia Pac. J. Clin. Nutr.* ***14***, 374–380.
15. McIntyre, B. S., Briski, K. P., Gapor, A., and Sylvester, P. W. (2000) Antiproliferative and apoptotic effects of tocopherols and tocotrienols on preneoplastic and neoplastic mouse mammary epithelial cells. *Proc. Soc. Exp. Biol. Med.* ***224***, 292–301.
16. Srivastava, J. K., and Gupta, S. (2006) Tocotrienol-rich fraction of palm oil induces

cell cycle arrest and apoptosis selectively in human prostate cancer cells. *Biochem. Biophys. Res. Commun.* **346**, 447–453.

17. Sylvester, P. W. (2007) Vitamin E and apoptosis. *Vitam. Horm.* **76**, 329–356.
18. Ma, Y. S., Stone, W. L., and LeClair, I. O. (1994) The effects of vitamin C and urate on the oxidation kinetics of human low-density lipoprotein. *Proc. Soc. Exp. Biol. Med.* **206**, 53–59.
19. Fan, J., Shek, P. N., Suntres, Z. E., Li, Y. H., Oreopoulos, G. D., and Rotstein, O. D. (2000) Liposomal antioxidants provide prolonged protection against acute respiratory distress syndrome. *Surgery* **128**, 332–338.
20. Woodle, M. C., and Papahadjopoulos, D. (1989) Liposome preparation and size characterization. *Methods Enzymol.* **171**, 193–217.
21. Halks-Miller, M., Guo, L. S., and Hamilton, R. L., Jr. (1985) Tocopherol-phospholipid liposomes: maximum content and stability to serum proteins. *Lipids* **20**, 195–200.
22. Talsma, H., Ozer, A. Y., van Bloois, L., and Crommelin, D. J. A. (1989) The size reduction of lipsomes with a high pressure homogenizer (microfluidicsTM). Characterization of a prepared dispersions and comparison with conventional methods. *Drug Dev. Ind. Pharm.* **15**, 197–207.
23. Chono, S., Tanino, T., Seki, T., and Morimoto, K. (2007) Uptake characteristics of liposomes by rat alveolar macrophages: influence of particle size and surface mannose modification. *J. Pharm. Pharmacol.* **59**, 75–80.
24. Chono, S., Tauchi, Y., and Morimoto, K. (2006) Influence of particle size on the distributions of liposomes to atherosclerotic lesions in mice. *Drug Dev. Ind. Pharm.* **32**, 125–135.
25. Tietze, F. (1969) Enzymic method for quantitative determination of nanogram amounts of total and oxidized glutathione: applications to mammalian blood and other tissues. *Anal. Biochem.* **27**, 502–522.
26. Griffith, O. W. (1980) Determination of glutathione and glutathione disulfide using glutathione reductase and 2-vinylpyridine. *Anal. Biochem.* **106**, 207–212.

Chapter 22

Antioxidant Activity of Biotransformed Sex Hormones Facilitated by *Bacillus stearothermophilus*

Mohammad Afzal and Sameera Al-Awadi, Sosamma Oommen

Abstract

Bacillus stearothermophilus, a thermophilic bacterium isolated from Kuwaiti desert, when incubated with exogenous progesterone for 10 days at 65°C produced two monohydroxylated, two dihydroxy isomers of progesterone and a B-Seco compound. These metabolites were purified by TLC and HPLC followed by their identification through ^{1}H, ^{13}C NMR and other spectroscopic data. Microbial hydroxylation of 17β-estradiol resulted in the production of estrone. The effect of some inducers resulted in the production of two metabolites from 17β-estradiol, one of which was identified as 3,6β,17β-trihydroxyestra-1,3,5,14(10)-tetrene and the other metabolite remains unidentified. The transformation products were identified through their spectral data and comparison with reference compounds. Antioxidant activities of progesterone transformed mixture and purified metabolites of 17β-estradiol were studied by linoleic acid/β-carotene assay. An enhanced antioxidant activity for progesterone transformation products was observed when compared to progesterone. A comparison of antioxidant activity of progesterone and 17β-estradiol transformation products is reported.

Key words: Antioxidant activity, Steroids, Biotransformation, *Bacillus stearothermophilus.*

1. Introduction

Oxidative stress is a principal cause of aging and chronic diseases such as inflammation, infection, cancer, and cardiovascular disorders *(1, 2)*. Exogenous or endogenous sources of oxidative stress and weakened antioxidative defenses can damage macromolecules such as DNA, lipids, and proteins.

Estrogens are powerful antioxidants and through their receptor status, estrogen/progesterone play an important role,

From: *Methods in Molecular Biology, vol. 477: Advanced Protocols in Oxidative Stress I,*
Edited by: D. Armstrong, DOI: 10.1007/978-1-60327-517-0_22, © Humana Press, New York, NY

in development, growth, and the differentiation of both male and female secondary sex characteristics *(3)* and protection against neurodegeneration and oxidative stress *(4)*. In addition, estrogen has favorable effects on lipid profiles, endothelial cell function, vascular reactivity, and haemostatic factors. Thus estrogens, such as 17β-estradiol (E_2), have a protective effect on oxidative stress mediated by estrogen receptor α (ERα) *(5)* and may also protect from cardiovascular disease and hepatic fibrosis *(6)*. Exogenous estrogen is known to protect against atherosclerosis by modulating low-density lipoprotein oxidation, binding free radicals, and lowering plasma cholesterol *(7)*.

Steroids are used as progestational, anabolic, antitumor agents and oral contraceptives as well as sedatives. Hydroxylation of steroids is valuable due to its physiological and clinical bearing. Thus hydroxylated estrogens are important therapeutic agents used in the treatment of breast tumors, etc. However, chemical synthesis of these new estrogens is difficult to achieve. We have investigated a bacterial transformation of these hormones and have achieved a new hydroxylated estrogens identified through their spectral data. This paper reviews the newly biotransformed sex hormones by a thermophilic bacteria *B. stearothermophilus* isolated from oil polluted desert soil around Kuwait oil fields. We have studied biotransformation of progesterone, testosterone, and 17β-estradiol. Transformation of progesterone for 24 h resulted in three hydroxylated progesterone derivatives 20α-hydroxyprogesterone, 6β-hydroxyprogesterone, and the rare 6α-hydroxyprogesterone. A new biotransformed metabolite 9,10-seco-4-pregnene-3,9,20-trione was also purified and identified *(8)*. Prolonged incubation of progesterone resulted in dihydroxy and 5α-progesterone along with B-seco-progesterone *(9)*. Some of the products identified from testosterone biotransformation were 6α-hydroxyandrost-4-en-3,17-dione, 6β-hydroxyandrost-4-en-3,17-dione, 6α-hydroxytestosterone, 6β-hydroxytestosterone, and androst-4-en-3,17-dione *(10)*. The metabolism of 17β-estradiol resulted in the formation of estrone, 3,6β,17β-trihydroxyestra-1,3,5,14(10)-tetrene and an unidentified compound.

The steroidogenic capacity and oxidative stress-related parameters of the human corpus luteum (CL) at different stages of the luteal phase have been studied under basal and human chorionic gonadotropin (hCG)-stimulated conditions *(11)*. Oxidative stress is known to inhibit ovarian and testicular steroidogenesis, and progesterone receptor modulators with antioxidant effects have been investigated as possible treatments for endometriosis *(12)*.

2. Materials

2.1. Equipments

2.1.1. Analytical HPLC System (GBC Australia)

1. UV detector (GBC Model LC 1205 UV/ visible detector).
2. Winchrome chromatography software.
3. HPLC column (Waters C_{18} symmetry column 4.6 × 150 mm).

2.1.2. Spectroscopic Measurements

1. ^{1}H and ^{13}C NMR spectroscopy techniques (400 MHz NMR – Bruker AC 400).
2. Infrared data (Perkin Elmer System 2000 FTIR).
3. Mass spectra were determined using a V.G. Analytical Ltd, (Manchester, England) model 305 mass spectrometer-2025 attached with a library data system.

2.2. Reagents

1. Culture media were purchased from Fluka Riedel-deHaën (Germany) and Scharlau (Barcelona, Spain).
2. All chemicals were of analytical grade and were supplied by Merck (Darmstadt, Germany) and Scharlau.
3. Kieselgel-60 F_{254} fluorescent thin layer chromatographic plates (TLC) were obtained from Merck (Darmstadt, Germany).
4. Progesterone, 17β-estradiol, and their derivatives for the identification of metabolites were obtained from Sigma-Aldrich Co. UK.

3. Methods

3.1. Biotransformation of 17β-estradiol Using *Bacillus stearothermophilus*

Starter cultures (50 ml) of *B. stearothermophilus* were grown overnight on TYE medium at 65°C in a shaker incubator (*see* **Note 1**). The culture was transferred to 500 ml of fresh TYE media in a 1-l flask and was kept under the same conditions until the end log phase of the growth (4–4.5 h). Cells were collected by centrifugation and were washed with 0.05 M sodium phosphate buffer pH 7 and mixed with 10 ml of the same buffer and transferred to 500 ml of the phosphate buffer containing 10% Castenholtz mineral salt solution with 0.1% ascorbic acid (*see* **Notes 2** and **3**). A final concentration of 0.3 mg/ml of 17β-estradiol was added and the cultures were re-incubated at 65°C for 72 h (*see* **Note 4**). Inducers for the transformation of 17β-estradiol like chloramphenicol (50 μg/ml), cyclodextrin (2 moles/mole of substrate), and riboflavin (1 mg/50 ml) were individually studied. The metabolites were extracted with ethyl acetate and the solvent was evaporated under vacuum. The residue, redissolved in methanol, was resolved on silica gel TLC plates (Merck 60 F_{254}) using benzene:dichloromethane:acetone:ethylacetate (22:22:3:3 v/v/v) as a mobile phase. The purified metabolites were identified from their respective spectral data including ^{1}H, ^{13}C NMR, mass spectroscopy, and FT infrared spectrophotometry.

3.2. Biotransformation of Progesterone by B. stearothermophilus

Cells of *B. stearothermophilus* were collected after incubation as described for 17β-estradiol. Cell suspension was transferred to phosphate buffer containing 10% Castenholtz mineral salt solution (*see* **Note 3**). An ethanolic solution of progesterone was added to this flask to achieve a final concentration of 50 μg/ml and was re-incubated for a further period of 10 days (*see* **Note 5**).

Progesterone metabolites were extracted by shaking the buffer with an equal volume of chloroform. Solvent from the dried extract was evaporated under reduced pressure and the residue was redissolved in methanol and chromatographed using 20 × 20 cm Kieselgel-60 F_{254} fluorescent TLC plates, developed in diethylether:toluene:ethylacetate (3:5:2.v/v/v) as a mobile phase. The metabolites were viewed under UV light, marked with a pencil and edge of the TLC chromatogram was stained with an anisaldehyde reagent, which produced a spectrum of different colors. An isocratic elution of these compounds using acetonitrile:water mixed with 0.1% acetic acid (31:69 v/v) as a mobile phase in HPLC separation, resulted in good resolution of the dihydroxy compounds from the corresponding monohydroxy-progesterone derivatives (**Fig. 22.1**).

3.3. Results

3.3.1. Estradiol

Metabolism of 17β-estradiol resulted in the formation of estrone, which is a commonly reported conversion product. Biotransformation was also studied by an addition of inducers such as cyclodextrin, riboflavin, and chloramphenicol, that resulted in the production of other two additional metabolites, one of which was identified as 3,6β,17β-trihydroxyestra-1,3,5,14(10)-tetrene and the other remains unidentified. The molecular structures of these metabolites are shown in **Fig. 22.2**.

3.3.2. Progesterone

This thermophilic bacillus was capable of site-selective hydroxylation of progesterone at position C-6, producing a rare microbial transformation product 6α-hydroxyprogesterone and its congener 6β-hydroxy progesterone. In addition, two other metabolites were also produced, one of which was identified as

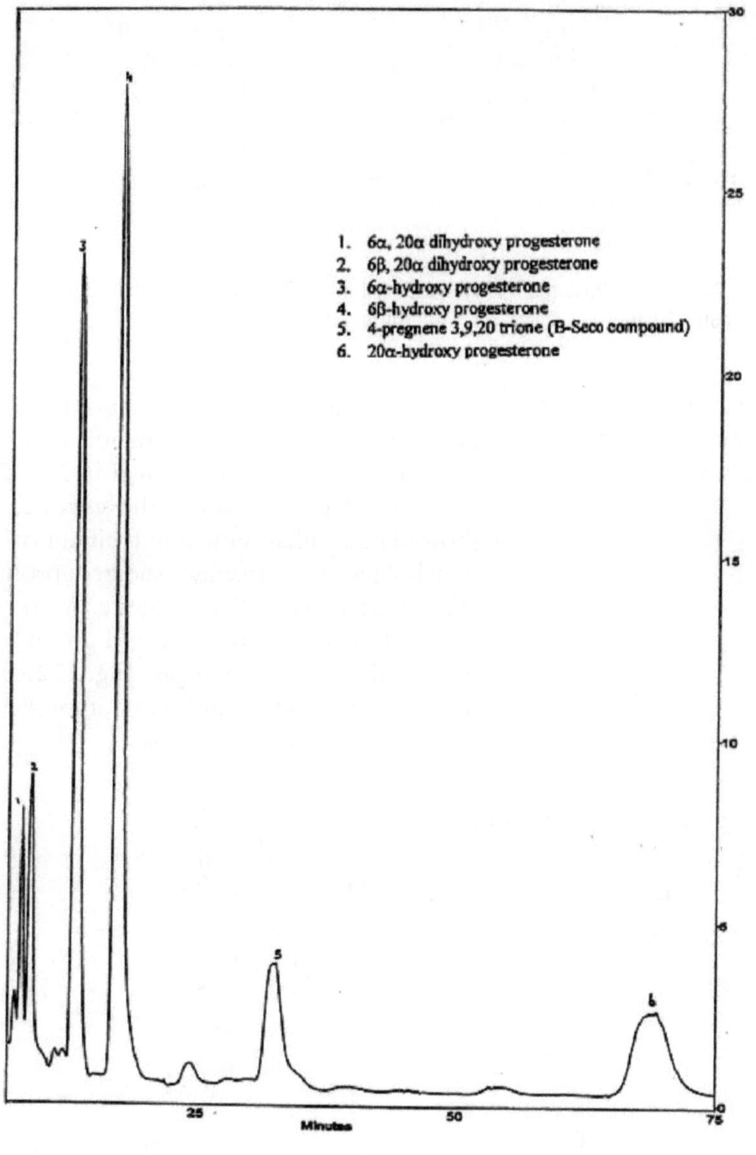

Fig. 22.1. HPLC chromatogram of 10 day progesterone transformation by *B. stearothermophilus*. From left to right, Peak 1, 6α,20α-dihydroxyprogesterone; Peak 2, 6β,20α-dihydroxyprogesterone; Peak 3, 6α-hydroxyprogesterone; Peak 4, 6β-hydroxyprogesterone; Peak 5, B-Seco compound; Peak 6, 20α-hydroxyprogesterone.

a common biotransformation isomer 20α-hydroxyprogesterone and the other 9,10-seco-4-pregnene-3,9,20-trione (B-seco compound), which arises from rare B ring cleavage of progesterone molecule. With prolonged incubation of up to 10 days, dihydroxylated products, 6α,20α-dihydroxyprogesterone, and 6β,20α-dihydroxyprogesterone were also formed. The structures of the metabolites and their formation are shown in **Fig. 22.3**.

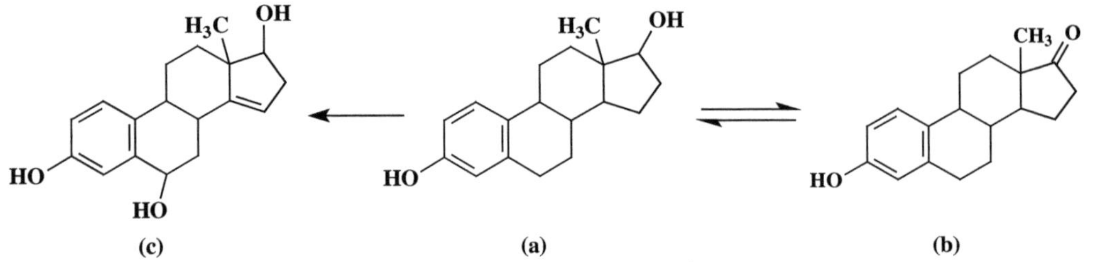

Fig. 22.2. The structures of 17β-estradiol and their transformation products. (**A**) 17β-estradiol, (**B**) Estrone, (**C**) 3,6β,17β-trihydroxyestra-1,3,5,14 (10)-tetrene.

3.3.3. Antioxidant Studies of the Transformed Products

Antioxidant activities of progesterone-transformed mixture and purified metabolites of 17β-estradiol were studied by linoleic acid/ β-carotene assay and are shown in **Fig. 22.4A, B**. Transformation products of 17β-estradiol, the estrone, and unidentified product showed little enhancement in their antioxidant properties since these metabolites were mainly oxidized products of the parent sterol (**Fig. 22.4A**). However, progesterone transformation products mixture with their multiple hydroxyl groups, showed enhanced antioxidant activity as shown in **Fig. 22.4B**. Thus, biotransformed progesterone metabolites carry improved antioxidant activity that may have clinical implications.

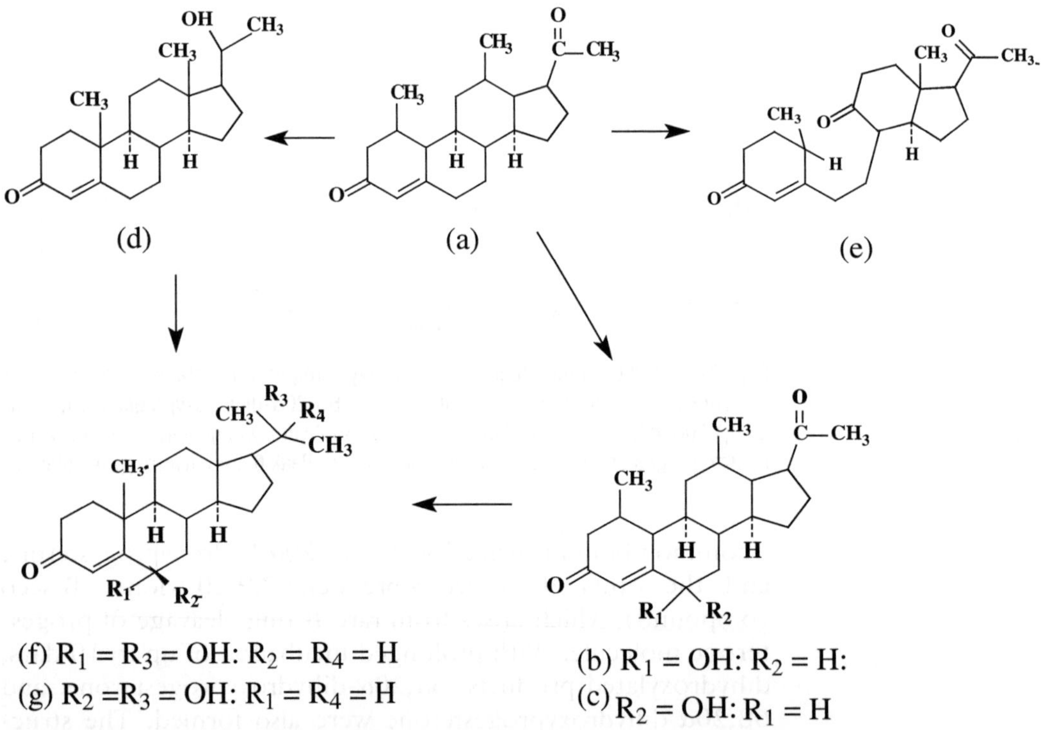

Fig. 22.3. Molecular structures of progesterone and its metabolites. (**A**) Progesterone, (**B**) 6α-hydroxyprogesterone, (**C**) 6β-hydroxyprogesterone, (**D**) 20α-hydroxyprogesterone, (**E**) 9,10-Seco-4-pregnene-3,9,20-trione, (**F**) 6α,20α-dihydroxyprogesterone, (**G**) 6β,20α-dihydroxyprogesterone.

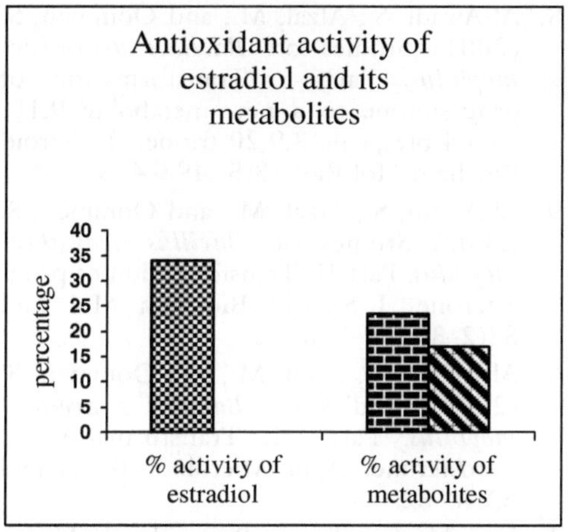

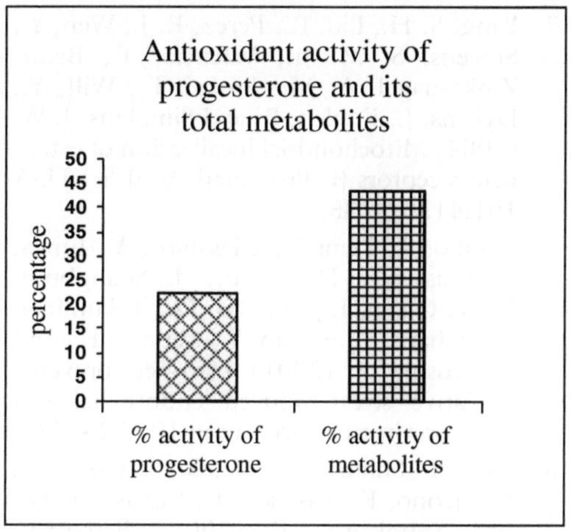

Fig. 22.4. (**A**) Antioxidant activities of 17β-estradiol, 3,6β,17β-trihydroxy-1,3,5,14 (10) tetrene and estrone, respectively. (**B**) Antioxidant activities of progesterone and its total metabolites as given in Fig. 22.3, respectively.

4. Notes

1. As we deal with thermophiles, it is preferred to autoclave media and buffer at higher temperature for a longer period of time.
2. Extreme care should be taken to avoid the contamination of the strains.
3. Transformation experiment in buffer with minerals is preferred rather than media to prevent the formation of bacterial products.
4. Stock solution of steroids should be freshly prepared to prevent decomposition while storing in solvents.
5. Increase in time of incubation can enhance the number and quantity of the metabolites produced.

Acknowledgement

This work was supported by a Kuwait University grant # SB040 and SL02/02 for which authors are thankful.

References

1. Yoshishige, U., Yoshito, I., Hiroaki, M., Shinji, G., Takehiko, K., Junji, Y., Satoshi, I., and Takahito, K. (2006) 17β-Estradiol protects against oxidative stress-induced cell death through the glutathione/glutaredoxin-dependent redox regulation of Akt in myocardiac H9c2 cells. J Biol Chem 218(19):13092–13102.
2. Finkel, T., and Holbrook, N. J. (2000) Oxidants, oxidative stress and the biology of aging. Nature 408:239–247.

3. Yang, S. H., Liu, R., Perez, E. J., Wen, Y., Stevens, S. M. Jr., Valencia, T., Brun-Zinkernagel, A. M., Prokai, L., Will, Y., Dykens, J., Koulen, P., and Simpkins, J. W. (2004) Mitochondrial localization of estrogen receptors β. Proc Natl Acad Sci USA 101:4130–4135.
4. Noriko, S., Kenji, T., Masahiro, A., Junya, A., Masayoshi, H., Katsuya, I., Seungbum, K., Yi-Qiang, L., Yumiko, O., Tokumitsu, W., Ieharu, Y., Masao, Y., Masato, E., and Yasuyoshi, O. (2001) Estrogen prevents oxidative stress–induced endothelial cell apoptosis in rats. Circulation 103:724–731.
5. Baba, T., Shimizu, T., Suzuki, Y., Ogawara, M., Isono, K., Koseki, H., Kurosawa, H., and Shirakawa, T. (2005) Estrogen, insulin, and dietary signals cooperatively regulate longevity signals to enhance resistance to oxidative stress in mice. J Biol Chem 280:16417–16426.
6. Yan, L., Ji, Z., Yan, L., and Xiao-Song, G. (2002) Protective effect of estradiol on hepatocytic oxidative damage. Gastroenterol. 8(2):363–366.
7. Kuhl, H. (1994) Cardiovascular effects and estrogen/gestagen substitution therapy. Ther Umsch 51:748–754.
8. Al-Awadi, S., Afzal, M., and Oommen, S. (2001) Studies on *Bacillus stearothermophilus.* Part 1. Transformation of progesterone to a new metabolite 9,10-seco-4-pregnene-3,9,20-trione. J Steroid Biochem Mol Biol 78(5):493–498.
9. Al-Awadi, S., Afzal, M., and Oommen, S. (2002) Studies on *Bacillus stearothermophilus.* Part II. Transformation of progesterone. J Steroid Biochem Mol Biol 82(2–3):251–256.
10. Al-Awadi, S., Afzal, M., and Oommen, S. (2003) Studies on *Bacillus stearothermophilus.* Part III. Transformation of testosterone. Appl Microbiol Biotechnol 62:48–52.
11. Vga, M., Castillo, T., Retamales, I., Heras, J. L., Devoto, L., and Videla, L. A. (1994). Steroidogenic capacity and oxidative stress-related parameters in human luteal cell regression. Free Rad Biol Med 17(6):493–499.
12. Gupta, S., Agarwal, A., Krajcir, N., Alvarez, J. G. (2006) Role of oxidative stress in endometriosis. Reprod Biomed 13(1):126–134.

Chapter 23

Biolistic Mediated Production of Transgenic Oil Palm

Ghulam Kadir Ahmad Parveez

Abstract

Physical and biological parameters affecting DNA delivery into oil palm embryogenic calli using the biolistic device are optimized. Five different promoters are also evaluated to identify the most suitable promoter for use in oil palm transformation. Finally, the effectiveness of kanamycin, geneticin (G418), neomycin, hygromycin, and herbicide Basta as selection agents to inhibit growth of oil palm embryogenic calli is evaluated. Combination of optimized parameters, best promoter and selection agent is later used to transform oil palm embryogenic calli for producing transgenic oil palm plants. Bombarded embryogenic calli are exposed to 50 mg/l of Basta after 3 weeks. Basta-resistant embryogenic calli started to emerge five to six months in medium containing Basta. The Basta-resistant embryogenic calli are proliferated until they reach a specific size, and the Basta-resistant calli are later individually isolated and regenerated to produce complete plantlets. The complete regenerated plantlets are evaluated for the presence of transgenes by PCR, Southern and thin layer chromatography analyses.

Key words: Biolistics, Transformation, Optimization, Transgenic oil palm, *Elaeis guineensis*, Oil crop.

1. Introduction

Palm oil has now become the largest source of edible oil in the world, overtaking soybean oil's position *(1)*. World palm oil is mainly produced from plantations in Malaysia and Indonesia. With the current premier position, it is essential for the oil palm industry to remain competitive by overcoming threats such as, shortage of labor and arable land though increasing yield per unit area as well as producing novel high-value products using approaches such as genetic engineering. From previous experience, it is estimated that only four to five

From: *Methods in Molecular Biology, vol. 477: Advanced Protocols in Oxidative Stress I,*
Edited by: D. Armstrong, DOI: 10.1007/978-1-60327-517-0_23, © Humana Press, New York, NY

years are required to produce transgenic oil palm plantlets from tissue culture explants *(2)*.

The main goal for genetic engineering of oil palm is to increase oleic acid content at the expense of palmitic acid *(3, 4, 5)*. Other transgenic products that have been targeted are stearic acid, palmitoleic acid, ricinoleic acid, and biodegradable plastics *(6)*. Currently, oil palm genetic engineering is mainly carried out for research purpose and transgenic plants are being planted in a fully contained biosafety greenhouse.

Transgenic plants are produced by a process called genetic transformation. It involves the uptake of naked DNA (in the form of a DNA fragment) by competent plant cells, followed by integration into the chromosome, and subsequently expression of the gene to synthesize the gene product *(7)*. The process starts with the penetration of DNA into a plant cell through the cell wall or protoplasts membrane and eventually into the nucleus. There are various methods available for plant gene transfer and are basically divided into two main groups, namely *Agrobacterium*-mediated gene transfer *(8)* and direct gene transfer such as protoplasts electroporation *(9)* and biolistic *(10)*. *Agrobacterium*-mediated is the first reported method for transferring foreign genes into plants and the most popular and widely used transformation system for plants. Initially, it is reported *Agrobacterium* only transfers genes into dicotyledonous plants, where monocotyledon species, particularly cereals, are resistant to *Agrobacterium* infection *(11)*. However, the scenario has now changed where many monocotyledonous plants have been transformed with *Agrobacterium (12, 13, 14, 15, 16, 17)*. The biolistic is the second most popular method of choice for plant. Biolistic method of transformation is simple, has no species dependency, and no cell type or tissue limitation. Most importantly, biolistics is reported to transform monocot plants and agronomically important crops *(11)*.

Biolistic, being the best method for monocot plants, still suffers from two main limitations: the efficiency of obtaining stable transformation has been very low *(11)* and integration of multiple copies of transgenes, which could result in transgene silencing *(18, 19, 20)*. It is estimated from various reports that the efficiency of obtaining stable transformation events over total number cells transiently expressing the transgene is around 0.1–2.0% *(21, 22, 23)*. Based on various reports on successful production of transgenic plants using biolistics, it could be concluded that optimization of physical and biological parameters affecting DNA delivery, identification of an efficient promoter, determination of a sensitive selection agent, and development of a reproducible regeneration protocol are essential.

Efforts to genetically engineer oil palm were not possible in early 1990s, as there was no transformation system established, which was considered a bottleneck *(24)*. At the initial stage of this research, when *Agrobacterium* is still not preferred method for monocot plants, biolistics is chosen for oil palm transformation. Optimization of physical and biological parameters affecting DNA delivery, identification of an efficient promoter, and determination of a sensitive selection agent are carried out and the optimized condition is used to transform oil palm. Transformed embryogenic calli are selected using the optimal concentration of the selection agent, Basta and followed by proliferation and regeneration of transgenic plants. Molecular and protein analyses are conducted to confirm the transgenic status of the plants. Problems faced during the research and approaches taken to overcome the problems are highlighted in the Notes section.

2. Materials

2.1. Equipment

1. Biolistics – PDS-1000/He Apparatus (Bio-Rad)
2. Autoclave (Tomy)
3. Incubator shaker (Innova)
4. Electrophoresis unit and power supplies (Bio-Rad)
5. PCR apparatus (MJ Research – PTC-100 and PTC-200)
6. Homogenization unit
7. TLC chamber
8. Centrifuge RC5C *Plus* (Sorvall)
9. Vacuum chamber
10. Hybridization oven (Techne Hybridiser HB-1D)
11. Vortex (Finemixer Sh2000, FinePCR)
12. pH meter (Mettler Toledo)
13. Hot plate and stirrer
14. Analytical balance
15. Plant tissue-culture incubator (Conviron TC-16)
16. Microcentrifuge (Hettich Mikro120).

2.2. Supplies

1. Gold microcarriers (Bio-Rad)
2. Macrocarrier (Bio-Rad)
3. Stopping screen (Bio-Rad)
4. Rupture disc (Bio-Rad)

5. QIAGEN Plasmid Maxi kit
6. Treff microtube (Switzerland)
7. Petri dish
8. Nylon membrane (Hybond-N+, Amersham)
9. 3 MM paper (Whatman)
10. X-ray film (Kodak)
11. TLC plate (Whatman, Cat: 4865–82)

2.3. Reagent for DNA Delivery in Oil Palm Cultures

2.3.1. Large Scale Plasmid Isolation

1. *E. coli* containing the desired plasmid are grown in LB medium (500 ml medium contains – 5 g NaCl {Sigma}, 5 g tryptone {Pronadisa}, and 2.5 g yeast extract {Pronadisa}, pH 7.0) in the presence of 50 μg/ml ampicillin {Sigma}. Ampicillin is added just before used.
2. DNA isolation is carried out using the QIAGEN Plasmid Maxi kit. The plasmid DNA obtained is dissolved in 1 ml of TE buffer (10 mM Tris {J.T. Baker – Harmful if swallowed}, 1 mM EDTA {Sigma, irritant}, pH 8.0).
3. The DNA is analyzed by electrophoresis on 1% agarose gels {Sigma} in TE buffer.

2.3.2. DNA-Microcarrier Preparation and Bombardment

1. Gold microcarriers are sterilized with absolute ethanol {Merck, Flammable} and finally dissolved in sterile distilled water. DNA is attached to gold microcarriers by the sequential addition of 50 μl of $CaCl_2$ (2.5 M) {Merck, irritant} and 20 μl spermidine (0.1 M, free base form) {Sigma}.
2. Transient expression is carried out using GUS assay buffer (0.1 M $NaPO_4$ buffer {pH 7.0} [mixture of NaH_2PO_4 and Na_2HPO_4 – Sigma], 0.5 mM K-Ferricyanide {Sigma, harmful, do not inhale and light sensitive}, 0.5 mM K-Ferrocyanide {Sigma, harmful, do not inhale}, 0.01 M EDTA {Sigma, irritant}, 1 mg/ml X-gluc (5-Bromo-4-Chloro-3-Indolyl-β-D-glucuronide) {Biosynth}, 1 μl/ml Triton-X {Sigma, harmful if swallow, do not breath vapor}, and 20% methanol (v/v){Merck}).

2.3.3. Optimization of Biological and Physical Parameters

For osmoticum treatment optimization, mannitol {Sigma, Plant Cell Culture Tested} and sorbitol {Sigma,} are used to adjust the osmotic pressure within the target cells.

2.3.4. Minimal Inhibitory Concentration of Selection Agents

Hygromycin {Sigma}, kanamycin {Sigma}, geneticin G418 {Sigma}, and neomycin {Sigma} are prepared as stock solutions of 50 mg/ml while herbicide Basta (Bayer CropScience) as a stock solution of 20 mg/ml. All selective agents are filter sterilized and stored at temperatures below 0°C.

2.4. Regeneration of Transgenic Oil Palm Plants

2.4.1. Callus Initiation from Oil Palm Leaflet and Roots

Oil palm leaf and roots explants are cultured on solid callus initiation medium [(MS salts *(25)* {Sigma, irritant and avoid contact} + Y3 vitamins *(26)*, (1 mg/ml Thiamine-HCl {Sigma, do not breath or contact}, 1 mg/ml Pyridoxine HCl {Sigma}, and 1 mg/ml Nicotinic Acid {Sigma, irritant} + 0.1 mg/ml myo-Inositol {Sigma, do not breath and avoid contact} + 0.1 mg/ml L-glutamine {Sigma} + 3% sucrose {Sigma} + 5×10^{-5} M, 2, 4-D {Sigma, toxic, can cause cancer} + 0.25% activated charcoal {Sigma, harmful, do not swallow and irritant} + 0.7% agar {Sigma}, pH 5.7] to induce callus.

2.4.2. Production of the Embryogenic Calli

Oil palm callus are cultured on agar solidified embryogenic media containing MS macro and micronutrients {Sigma, irritant and avoid contact} and Y_3 vitamins {Sigma} supplemented with 100 mg/l each of myo-inositol {Sigma, do not breath and avoid contact}, L-glutamine {Sigma}, L-asparagine {Sigma}, 10 μM 2,4-D {Sigma, toxic, can cause cancer}, 5 μM NAA {Sigma, harmful if swallow, irritant}, and 30 g/l sucrose {Sigma} until embryogenic calli are produced. The medium is adjusted to pH 5.7 with KOH {Sigma, corrosive, irritant} prior to autoclaving.

2.4.3. Transformation and Selection of Transformed Embryogenic Calli

Embryogenic calli are bombarded on a Petri dish containing solidified embryogenic media (*see* **Section 2.2**(**2**)) with 0.4 M mannitol {Sigma}. After 3 weeks, the cultures are transferred onto solidified embryogenic media (*see* **Section 2.2**(**2**)) containing 50 mg/l herbicide Basta (Bayer CropScience).

2.4.4. Production of Oil Palm Polyembryogenic Cultures

Basta-resistant embryogenic calli are transferred onto solidified polyembryogenic culture media containing MS macro and micronutrients {Sigma, irritant and avoid contact} and Y_3 vitamins {Sigma} supplemented with 100 mg/l each of myo-inositol {Sigma, do not breathe and avoid contact}, L-glutamine {Sigma}, L-arginine {Sigma} and L-asparagine {Sigma}, 5 μM IBA {Sigma}, 0.7% agar {Sigma}, and 30 gm/l sucrose {Sigma} to form polyembryogenic cultures. The medium is adjusted to pH 5.7 with KOH {Sigma, corrosive, irritant} prior to autoclaving.

2.4.5. Small Plantlets Production from Polyembryogenic Cultures

Shoots are produced from polyembryogenic cultures on media containing MS macro and micronutrients {Sigma, irritant and avoid contact} and Y_3 vitamins {Sigma} supplemented with 100 mg/l each of myo-inositol {Sigma, do not breathe and avoid contact}, L-glutamine {Sigma}, L-arginine {Sigma}, and L-asparagine {Sigma}, 0.1 μM NAA {Sigma, harmful if swallowed, irritant}, 0.4% agar {Sigma}, and 30 g/l sucrose {Sigma}. The medium is adjusted to pH 5.7 with KOH {Sigma, corrosive, irritant} prior to autoclaving.

2.4.6. Root Initiation from Oil Palm Cultures

Roots are induced from transformed shoots on liquid root inducing media containing MS macro and micronutrients {Sigma, irritant and avoid contact} and Y_3 vitamins {Sigma} supplemented with 300 mg/l L-glutamine, 100 mg/l myo-inositol {Sigma, do not breathe and avoid contact}, 10 μM 2,4-D {Sigma, toxic, can cause cancer}, 70 μM NAA {Sigma, harmful if swallowed, irritant}, 0.15% activated charcoal {Sigma, harmful, do not swallow and irritant}, and 60 g/l sucrose {Sigma}. The medium is adjusted to pH 5.7 with KOH {Sigma, corrosive, irritant} prior to autoclaving.

2.5. Reagents for Molecular and Protein Analyses of Transgenic Plants

2.5.1. Preparation of Total Plant DNA from Transformed Plants

1. Total DNA is extracted from grounded transformed and untransformed leaves in 10 ml CTAB Extraction buffer – 100 mM Tris-HCl {J.T. Baker – harmful if swallowed}, pH 8.0; 20 mM EDTA {Sigma, irritant}; 1.4 NaCl; 2% CTAB {Sigma, toxic and harmful, do not swallow, irritant}; 1% PVP {Sigma, avoid contact or inhalation}, and 0.2% 2-mercaptoethanol {Sigma, toxic, do not swallow or contact} *(27)*.
2. Debris and contaminating proteins are removed by chloroform: isoamyl alcohol (24:1) {System, harmful if swallowed: BDH, flammable and harmful if taken} extraction.
3. Total DNA is precipitated using adding 6 ml chilled isopropanol {Merck, irritant, flammable}. The pellet is washed with washing Buffer (76% ethanol {Merck, flammable}, 1 mM ammonium acetate {Hamburg Chemical GmbH}).
4. The pellet obtained is dried and dissolved in 500 μl TE Buffer (10 mM Tris-HCl, pH 7.5, (10 mM Tris {J.T. Baker – Harmful if swallowed}, 1 mM EDTA {Sigma, irritant}).

2.5.2. Polymerase Chain Reaction (PCR)

1. Amplification of transgenes is carried out using standard and touch-down PCR protocols *(28)*. PCR reactions are carried out in the following mixture: DNA (50 ng total DNA or 5 ng plasmid DNA) + reaction buffer (50 mM KCL and 10 mM Tris-Cl pH 8.3) {Promega} + dNTPs (200 μM each) {Promega} + 1.5 mM MgCl2 {Promega} + primers, and Taq DNA polymerase {Perkin-Elmer}.
2. The amplified DNA fragments are checked by running on a 1.4% agarose gel {Sigma} and electrophoresed in 0.5 X TBE buffer (45 mM Tris-Borate {J.T. Baker, harmful if swallowed – Sigma, avoid contact or inhaltion}, 1 mM EDTA {Sigma, irritant}, pH 8.0).

2.5.3. Southern Hybridization

1. Digested and undigested total DNA from transformed plants are separated on a 0.8% agarose gel {Sigma} in 1 X TBE agarose gel. The DNA from agarose gel is capillary transferred onto nylon membrane *(29)*.

Southern Transfer of DNA from Agarose Gel to Nylon Membrane

2. Agarose gel containing the digested (20 μg) and undigested (15 μg) DNA is soaked in depurinating buffer (0.2 M HCl)

{Merck, corrosive, no contact} for 10 min, denaturing buffer (1.5 M NaCl {Sigma} and 0.5 M NaOH {Sigma, corrosive, avoid contact, causes burn}) for 45 min, and transferred into neutralization buffer (1 M Tris {pH 8.0} {J.T. Baker – Harmful if swallowed}, and 1.5 M NaCl {Sigma}) for 1 h.

3. The gel is transferred onto 3 MM paper with its end soaked in 20 X SSC (175.3 g NaCl {Sigma} + 88.2 g sodium citrate {Sigma} in 1 l, pH 7.0).
4. A nylon membrane is placed on top of the gel followed by two sheets of 3 MM paper, paper towels, a glass plate, and a 500 g weight.
5. After transfer, the membrane is washed with 2 X SSC (17.53 g NaCl {Sigma} + 8.82 g sodium citrate {Sigma} in 1 l, pH 7.0) and baked at 80°C for 2–4 h.

Preparation of Probes

1. DNA fragments (~10 ng) to be used as probes are oligolabeled *(30)* by adding to 20 μl 5 X OLB (0.25 M Tris-HCL {pH 8.0}{J.T. Baker – harmful if swallowed}, 25 mM $MgCl_2$, {Sigma}, 0.36% (v/v) 2-mercaptoethanol {Sigma, toxic, do not swallow or contact}, 1 M hepes {Sigma} {pH 6.6}, 30% hexadeoxyribonucleotides {90 O.D. units/ml}) and boiled for 5 min, then chilled on ice.
2. The following are added to the above mixture: 2 μl 0.1 M dNTPs (except dCTP) {Amersham}, 5 μl $\propto^{32}$ p (dCTP) (370 KBq/μl) {Amersham}, 2 μl 10 mg/ml BSA {Amersham}, 1 μl klenow (6 U/μl) {Amersham}, and 14 μl distilled water.
3. The probe is denatured by the addition of 50 μl 1 M NaOH {Sigma, corrosive, avoid contact, causes burn} for 1 min, 50 μl 1 M HCl {Merck, corrosive, no contact} for 1 min, and 50 μl Tris-HCl (pH 7.5) {J.T. Baker – harmful if swallowed} for 1 min.

Hybridization

1. Pre-hybridization and hybridization are carried out using the pre-hybridization buffer (40% pipes/NaCl, pH 6.8, {1.5% pipes {Sigma}, 8.7% NaCl {Sigma} and 0.37% EDTA {Sigma, irritant}, [pH 8.0]}; 20% Denhardts 50X [1% BSA {Sigma}, 1% Ficoll {Sigma}, 1% PVP {Sigma, avoid contact or inhaltion} and 10% SDS] {Promega, harmful and avoid contact or swallowing}; 0.5% SS-DNA (10 mg/ml) and 39.5% distilled water).
2. After hybridization, the membrane is pre-washed once with 2 X SSC (17.53 g NaCl {Sigma}~8.82 g sodium citrate {Sigma} in 1 l, pH 7.0) and twice in 0.1 X SSC and 0.1% SDS {Promega, harmful and avoid contact or swallowing}.
3. The washed membrane is wrapped with saran wrap and exposed to X-ray film with an intensifying screen at −70°C.

2.5.4. Phosphinothricin Acetyltransferase (PAT) Analysis

1. Leaf tissue from transgenic plants and from nontransformed plants is homogenized in a PAT assay buffer composed of: 50 mM Tris-HCl, pH 7.5 {J.T. Baker – harmful if swallowed}; 2 mM EDTA {Sigma, irritant}; 0.1 mg/ml leupeptin {Sigma, do not breathe dust, avoid contact}; 0.3 mg/ml BSA {Sigma}; 0.3 mg/ml DTT {Sigma – harmful, avoid contact, irritant}; and 0.15 mg/ml PMSF {Sigma, toxic, avoid contact}. *(31)*.
2. After centrifugation, 12.5 μl of supernatant is incubated with 0.75 μl 1 mM PPT {Sigma} and 1.25 μl ^{14}C acetyl-CoenzymeA (57 mCi/mmol) {Amersham} at 37°C, for 30 min.
3. Six microlitres of the reaction is transferred to a thin-layer chromatography plate in a mixture of isopropanol: ammonium solution (3:2 parts by volume).
4. After chromatography is completed, the plate is air dried and wrapped with a saran wrap and exposed for 1 h or more to X-ray film.

2.5.5. Analysis of Herbicide Basta Resistance

Transformed and untransformed leaves are painted with 0.03% v/v Basta {Bayer CropScience} solutions containing 0.1% v/v Tween 20 {Promega} and 0.1% v/v Triton X-100 {Sigma – harmful if swallowed or breathed vapor} *(32)* to approximately two-thirds of each leaf.

3. Methods

3.1. Equipment

See **Section 2.1.**

3.2. Supplies

See **Section 2.2.**

3.3. DNA Delivery in Oil Palm Cultures

3.3.1. Large-scale Plasmid Isolation

1. *E. coli* culture containing the desired plasmid are inoculated from glycerol stock and streaked on agar medium containing antibiotic and incubated overnight at 37°C. Single colony is picked and inoculated into 10 ml LB containing antibiotic and incubated by shaking overnight at 37°C (*see* **Note 1**). One milliliter of the culture is later inoculated into 500 ml of LB containing antibiotic and incubated by shaking overnight at 37°C.
2. The overnight culture is transferred into 250 ml centrifuge bottles and cells are pelleted by centrifugation (4000 rpm, 10 min, and 4°C).
3. DNA isolation is carried out using the QIAGEN method (*see* **Note 2**). DNA solution is later stored at a concentration of 1 μg/μl and stored at −20°C until needed.

3.3.2. DNA-Microcarrier Preparation and Bombardment

1. DNA precipitation onto gold microcarriers is carried out according to manufacturer's instructions (*see* **Note 3**).
2. Sixty milligrams of gold microcarriers are sterilized three times with 1 ml absolute ethanol and the final pellet is resuspended in 1 ml of sterile distilled water, and 50 μl aliquoted into a microtube. The aliquot could be kept at 4°C (*see* **Note 4**).
3. Five microlitres of DNA solution (1μg/μl), $CaCl_2$ and spermidine are added sequentially into the 50 μl gold suspension. The mixture is vortexed for 3 min, spun for 10 s at 10,000 rpm and the supernatant discarded. The pellet is washed with 250 μl of absolute ethanol and finally resuspended in 60 μl of absolute ethanol.
4. Six microlitres of the solution is loaded onto the centre of the macrocarrier and is air dried (*see* **Note 5**). Bombardments are carried out at: 1100 psi rupture disc pressure; 6 mm rupture disc to macrocarrier distance; 11 mm macrocarrier to stopping plate distance, 75 mm stopping plate to target tissue distance and 67.5 mmHg vacuum pressure.
5. Two controls are incorporated i.e. tissues without bombardment and bombardment of microcarrier without DNA (*see* **Note 6**).
6. The bombarded tissues are then incubated for 48 h at 28°C in the dark prior to transient GUS assay overnight (20 h) at 37°C and blue spots are scored optically.
7. GUS (β-Glucuronidase) *(33)* assay is an assay for detecting the transient expression of the *E. coli gusA* gene histochemically (using X-gluc as a substrate). Each blue spot arising from the histochemical localization of GUS activity, whether in a single cell or a group of cells, was considered as one expression unit. This blue spot will be the indicator of gene transferred into plant cells.

3.3.3. Optimization of Biological and Physical Parameters

1. Optimization is not based on transient expression alone, but also on shot-to-shot variability and the physical impact on the target tissue. The physical parameters are used to determine the set of conditions where the best penetration of DNA will occur.
2. The biological parameters will determine the best biological material and conditions to be used for a higher chance of obtaining stable transformation.
3. Optimization is carried out as follows: A parameter such as helium pressure is optimized by bombarding the embryogenic callus using the same conditions except for varying the helium pressure. Five replicates are used to make the comparison statistically valid. After bombardment, the callus is kept for 48 h at 28°C in the dark and is later subjected to GUS staining.

4. The pressure that gave the highest number of blue spots, the least variability between shots, and the minimum tissue dislocation upon impact is selected (*see* **Note 7**). All the physical and biological parameters optimized are as in **Table 23.1**.

3.3.4. Minimal Inhibitory Concentration of Selection Agents

1. Embryogenic medium is autoclaved and cooled to 50°C followed by addition of selection agent at concentrations: 1, 5, 10, 50, 100, 250, 500, 1000, and 2000 mg/l. For control and each treatment, three replicates are used and are carried out twice.
2. In each plate, 300 mg of embryogenic callus is placed onto the media (incubated 28°C; dark). After 30 days, the percentage proliferation is measured using the formula below.

Table 23.1
Summary of optimization of physical and biological parameters for biolistic transformation of oil palm

Parameters	Optimized setting
Helium pressures	1100 psi
Distance from stopping plate to target tissue	7.5 cm
Distance from rupture disc to macrocarrier	6 mm
Distance from macrocarrier to stop plate	11 mm
Vacuum pressures	70 mmHg (NS)
Number of bombardments	Once (NS)
Particle types and sizes	G1.0 micron
Effect of $CaCl_2$ and spermidine on particle-DNA binding	WCS
Explant types	Leaf
Duration of calli in fresh medium prior to bombardment	1 day
Duration between bombardment and GUS staining	2 days
Genotype	Any genotype (NS)
Immature embryo preculture duration	7 days
DNA concentration (μg/bombardment)	1.5
Osmoticum type and concentration	0.4 M mannitol
Osmoticum treatment duration before and after bombardment	2–4 h

NS, = not significant; G, gold particle; WCS, with calcium chloride and spermidine

3. In this experiment, the growth of the control is considered as 100% proliferation assuming that there is no stress to reduce the proliferation rate. The increase of calli weight on the selected media is divided by the increased weight of control to give the percentage of proliferation relative to the control.

$$\%\ of\ proliferation = \frac{(Final\ weight - initial\ weight)\ of\ selected\ tissues}{(Final\ weight - initial\ weight)\ of\ control\ tissues} \times 100$$

3.3.5. Selecting the Best Constitutive Promoter for Oil Palm

Selection of a strong promoter is important to ensure survival and constitutive expression of the foreign gene in a transgenic plant. This will ensure that high levels of transient expression can be achieved. Selection is carried out by bombarding five well-documented plant promoters into oil palm using the optimized biological and physical parameters. The evaluation is not only based on the highest number of blue spots, but also on how fast the expression occurred (**Fig. 23.1**).

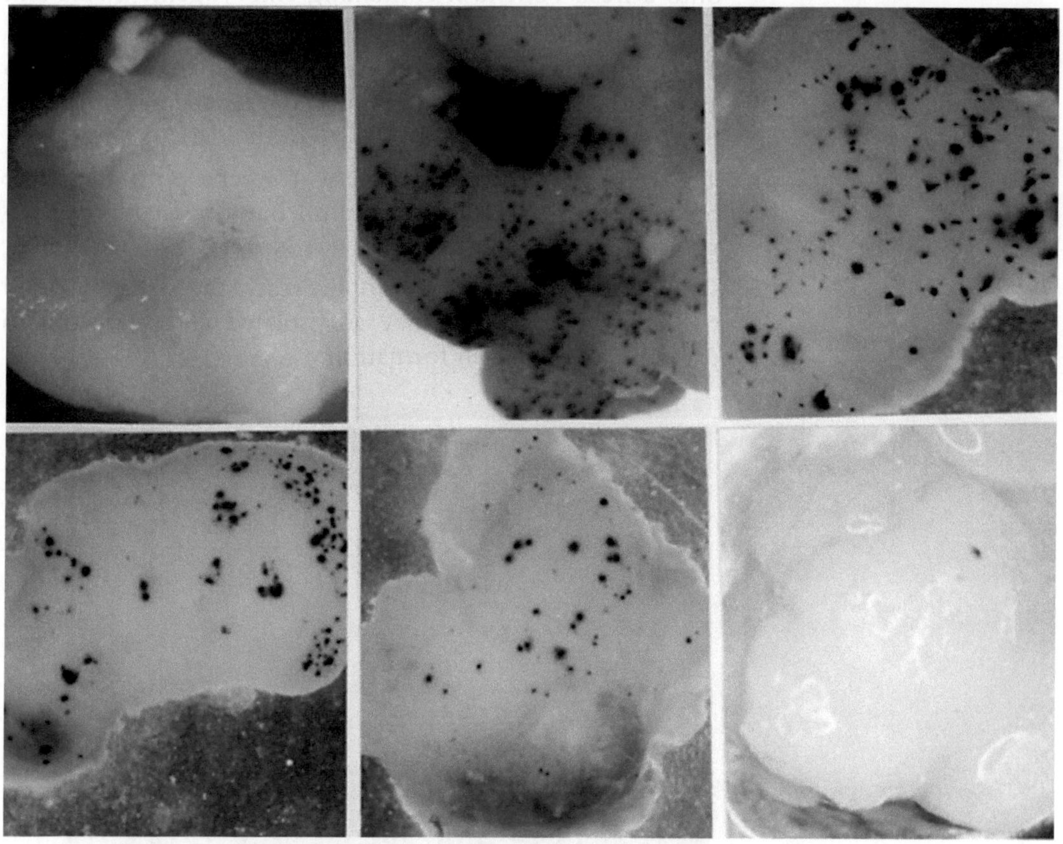

Fig. 23.1. Comparison of transient *gusA* gene expression in oil palm embryogenic calli after bombardment with plasmids carrying different promoters. (**A**) none (bombarded without plasmid DNA), (**B**) pEmuGN (Emu), (**C**) pAHC25 (Ubi1), (**D**) pAct1-F4 (Act1), (**E**) pGH24 (CaMV 35S) and (**F**) pBARGUS (Adh1).

3.4. Regeneration of Transgenic Oil Palm Plants

3.4.1. Callus Initiation from Oil Palm Leaflet and Roots

Leaflets from unopened (-6) frond are aseptically transferred onto solid callus initiation medium and incubated at 28°C in the dark. The explants are subcultured every 4 weeks onto the same medium until calluses are formed (*see* **Note 8**).

3.4.2. Production of the Embryogenic Calli

The calli obtained from **Section 3.2**(**1**) are transferred onto solidified embryogenic media and are incubated at 28°C in the dark. The calli are subcultured every 4 weeks onto the same medium and maintained until embryogenic calli are produced (*see* **Note 9**).

3.4.3. Transformation and Selection of Transformed Embryogenic Calli

1. Embryogenic calli are arranged as a circle in the middle of a Petri dish containing embryogenic media with osmoticum in a 2 cm diameter.
2. After bombardment, the embryogenic calli are transferred onto a fresh Petri dish containing embryogenic media and incubated at 28°C in the dark for 3 weeks.
3. The bombarded embryogenic calli are later transferred onto a fresh Petri dish containing embryogenic media with selection agent and incubated at 28°C in the dark.
4. The calli are subcultured every 4 weeks onto the same medium with selection agent until Basta-resistant embryogenic calli colonies are formed (*see* **Note 10**).
5. Once the Basta-resistant colonies reach the size of around 6–8 mm (**Fig. 23.2**), they are isolated and subjected to polyembryogenic calli formation.

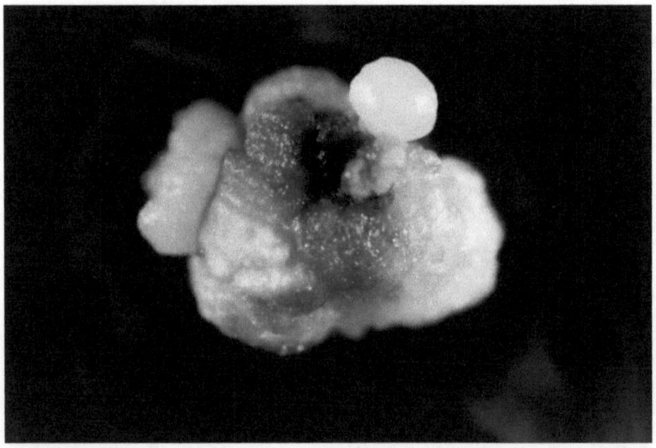

Fig. 23.2. Basta-resistant colonies that are ready for isolation and regeneration to polyembryogenic calli.

3.4.4. Production of Oil Palm Polyembryogenic Cultures

Basta-resistant embryogenic calli colonies are isolated and transferred onto a fresh Petri dish containing solidified polyembryogenic media and incubated at 28°C in the presence of light. The embryogenic calli colonies are subcultured every 4 weeks onto the same medium with selection agent until polyembryogenic cultures are obtained (*see* **Note 11**).

3.4.5. Small Plantlets Production from Polyembryogenic Cultures

Polyembryogenic calli are transferred to shoot induction medium (28°C: in light). The calli are subcultured every 4 weeks onto the same medium with selection agent until shoots are produced. Once the shoot is long enough, it will be subjected to root induction.

3.4.6. Root Initiation from Oil Palm Cultures

Transformed shoots are transferred to liquid roots induction medium in test tubes (28°C: in light). Small plantlets produced are later transferred onto soil in a biosafety screenhouse (**Fig. 23.3**).

3.5. Molecular and Protein Analyses of Transgenic Plants

3.5.1. Preparation of Total Plant DNA from Transformed Plants

1. One to ten grams of leaves from transformed and untransformed plants are chilled and grounded in liquid nitrogen prior to addition of 10 ml CTAB extraction buffer and incubated at 65°C for 1 h *(27)*. Protein and debris are later removed by 5 ml chloroform:isoamyl alcohol (24:1) extraction (4°C, 13,000 rpm for 5 min). The aqueous phase is transferred into new tube. This is repeated twice.
2. DNA is precipitated by adding 6 ml chilled isopropanol (20 min; room temperature) and centrifuged at 13,000 rpm for 5 min. Supernatant is discarded and pellet is dried at room temperature.

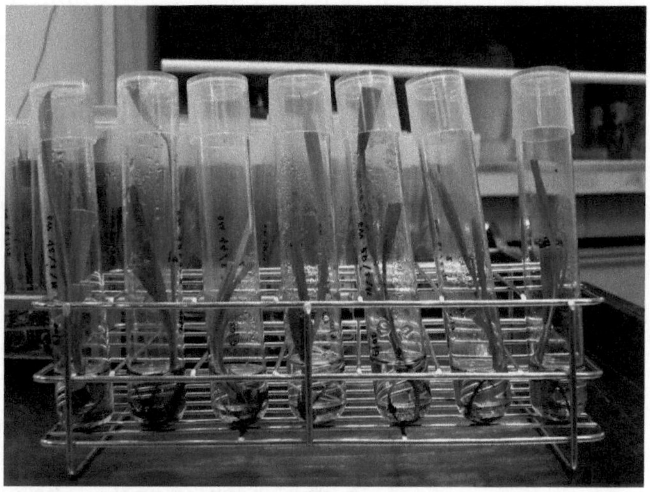

Fig. 23.3. Normal-looking transgenic oil palm plants in test tubes.

3. Pellet is washed (Washing Buffer) at room temperature for 20 min before centrifuged at 13,000 rpm for 5 min. Final pellet is dried and dissolved in TE Buffer to a concentration of 1 µg/µl and stored at −20°C until needed.

3.5.2. Polymerase Chain Reaction (PCR)

Standard PCR protocol involves 30 cycles: hot start 95°C (300 s), denature 92°C (50 s), annealing 60°C (50 s), elongation 72°C (60 s), and finally 120 s at 72°C *(28)*. The touch down procedure is as follows: hot start 95°C (300 s), denature at 94°C (45 s), annealing at 70°C (45 s; −0.5°C per cycle for 10 cycles), and elongation at 72°C (60 s) followed by 24 cycle of 94°C (45s), 65°C (45 s), 72°C (60 s) and finally 72°C for 120 s (*see* **Note 12**). The amplified fragments are checked using agarose gel electrophoresis.

3.5.3. Southern Hybridization

Transfer of DNA from Agarose Gel to Nylon Membrane

1. Agarose gel containing the digested DNA is depurinated for 10 min, denatured for 45 min, and neutralized for 1 h *(29)*.
2. The gel is later transferred onto 3 MM paper with the end of the paper soaked into 20 X SSC buffer. A nylon membrane is placed on top of the gel followed by two sheets of 3 MM paper, paper towels, a glass plate, and a 500 g weight.
3. The set-up is left for 16–20 h to allow the entire DNA transfer onto the membrane. The membrane is washed with 2 X SSC and baked at 80°C for 2–4 h.

Preparation of Probes

DNA fragments (prepared by PCR) for use as probes are oligolabeled in oligolabeling buffer (5 X OLB), boiled for 5 min, then chilled on ice *(30)*. After adding all the enzymes, labeled nucleotide and buffer, the reaction is incubated at 37°C for 30 min. The probe is later denatured and stored on ice until use.

Hybridization

1. Pre-hybridization and hybridization are carried out using the same buffer.
2. The membrane is pre-hybridized with pre-hybridization buffer for 90 min at 65°C.
3. Denatured DNA probe is added and hybridized for 20 h at 65°C.
4. After hybridization, the membrane is pre-washed for 1 min and washed twice: first for 30 min and the second for 45 min, both at 65°C. The washed membrane is wrapped

with saran wrap and exposed to X-ray film with an intensifying screen at −70°C.

3.5.4. Phosphinothricin Acetyltransferase (PAT) Analysis

1. About 100 mg of leaf tissue from transformed and control plants are homogenized in 50 μl PAT assay buffer and centrifuged at 12,000 rpm, 5 min at room temperature *(31)*.
2. After centrifugation, 12.5 μl of clear supernatant is incubated with PPT and radiolabeled acetyl-CoenzymeA at 37°C, for 30 min.
3. Six microlitres of the reaction is later transferred onto a thin-layer chromatography plate in a mixture of isopropanol: ammonium solution. Chromatography is carried out until the solvent had covered 80% of the plate height. The plate is air dried and saran wrapped.
4. Separated PPT and acetyl-PPT are detected by exposing the plate for 1 h or more to X-ray film (**Fig. 23.4**).

3.5.5. Analysis of Herbicide Basta Resistance

1. Leaves from transformed and untransformed oil palm plants are painted with full strength of Basta solution *(32)* to approximately two-thirds of each leaf.
2. Symptoms are evaluated 2 weeks after application (*see* **Note 13**).

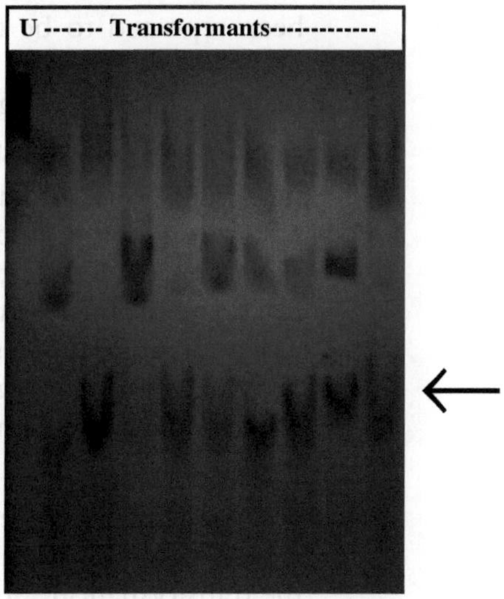

Fig. 23.4. Thin layer chromatography of PAT activity in Basta-resistant transgenic oil palm plants. Lane U is protein from untransformed control and lanes transformants (9 lanes) are proteins from transformed plants. Arrow indicates acetylated protein band.

4. Notes

1. The use of fresh bacterial colony from glycerol stock has been proven to result in the highest plasmid DNA yield when using QIAGEN Plasmid Maxi kit. Even using bacterial colony after 1 day storage at 4°C could result in up to 70% yield reduction.
2. Through experience, the QIAGEN Plasmid Maxi kit tubes could be reused for the same plasmid isolation up to four times without reduction in DNA yield. The tube should be rewashed and used as a new tube. Keeping the used tubes at 4°C for future use (maximum 1 week) is also recommended. However, reusing the tube for another plasmid is not recommended as contamination may occur. This is economical because the buffer used could be prepared in the lab as the formula is given in the brochure.
3. The use of gold microcarrier is better than tungsten because it is rounder and more uniform in size, biologically inert, non-toxic, and does not degrade DNA bonds. Tungsten, on the other hand, is highly heterogeneous in both size and shape, potentially toxic to some cell types, and subjected to surface oxidation. It also has been reported to acidify and degrade DNA bonds *(34)* and in comparison to gold, to have a greater tendency to aggregate during precipitation *(35)*.
4. A number of different brands of microfuge tubes were evaluated for biolistic transformation and Treff microtube has been reported to be the best tube for biolistic transformation *(34)*. Therefore, in all oil palm transformation experiments, Treff microfuge tubes are used.
5. Only 6 μl of DNA-gold solution is loaded onto the center of the macrocarrier, additional loading will be wasted as it will not pass through the stopping screen area. Furthermore, when aliquoting the DNA-gold solution from the microfuge tube, make sure the tube is immediately closed after use as the ethanol will evaporate. Thus, less number of bombardments can be carried out as well as a higher amount of DNA-gold be delivered, which could cause damage to the receiving target tissue.
6. It is important to have at least two controls for bombardment to ensure final interpretation of results not to be questionable. Tissue without bombardment is used as a negative control for the transformation. Bombardment without DNA is to ensure all the materials used, gold microcarrier, spermidine, $CaCl_2$ and water are not contaminated with any DNA source.
7. The advantage of this procedure is that the highest expression will result in a higher chance of getting stable transformation as there will be a lower number of stable transformants over

transients. Low variability will result in a consistent performance of bombardment. It is also expected that bombardment which resulted in massive tissue dislocation may result in high injury to the cells and ultimately reduce the chances of regenerating stable transgenic plants. Osmoticum agents (sorbitol and mannitol) used as high-velocity bombardments are required for the penetration of the microcarrier into cells to deliver DNA. This penetration can disturb the intracellular lipid membrane structure causing cell destruction and ethylene accumulation *(36)*. The use of an osmoticum can facilitate stabilization of cell membranes for faster healing of the lesion caused by microcarrier penetration and reduce turgor pressure of cells to prevent leakage and cell rupture.

8. Explants are handled in laminar flow to ensure its sterility. Oil palm callus normally appeared after a minimum period of 3 months. Sometimes, it may take a longer time, i.e., up to 12 months. After 12 months, the explants that failed to produce callus are discarded. It must be highlighted here that oil palm tissue culture process is very genotype dependent. Not all the explants cultured will eventually produce callus.
9. Embryogenic calli are normally produced from callus around 3–6 months after culturing. Some cultures may take a shorter and some may take a longer time. Not all calli will become embryogenic. Normally less than 50% of the calli will eventually produce embryogenic calli. The chance of embryogenic calli to produce polyembryogenic calli is high.
10. Selections are normally carried out 3 weeks post-bombardment to ensure that the bombarded embryogenic calli have recovered from stress and have doubled at least one time to be ready for exposure to selection agent. Selection is a slow process and quite discouraging, especially when the cultures turn dark brown in medium containing herbicide Basta. Normally, Basta-resistant embryogenic callus colonies will appear at least 5–6 months after exposure to the selection agent.
11. Once the Basta-resistant embryogenic callus colony reaches the size of around 6–8 mm, it will be individually isolated by removing the surrounding dead cultures and transferred onto polyembryogenic inducing medium. The size is important to ensure that the cultures are capable of proliferating and regenerating by themselves. Using smaller sizes of resistant embryogenic calli will cause most of the calli to lose their regeneration capability and will eventually die. This part of the selection process is very critical and important.
12. Initially, PCR is carried out using internal control primers for oil palm, which will specifically amplify a 1 Kb fragment oil palm genome. All samples (including untransformed

control) are subjected to amplification using the internal control primers. In the event, the control band is not observed, the DNA sample is further purified until the band is produced. The internal control is important to ensure that samples that do not amplify the transgenes are actually untransformed and not due to un-pure DNA or the amount of DNA used. Touch down PCR procedure is used for the Basta-resistant gene because the gene has a high GC content (more than 60%). The use of normal PCR protocols normally resulted in the amplification of nonspecific and false positive bands. The use of touch down is very successful in specifically amplifying the Basta-resistant gene.

13. One of the effective and easiest approaches to test for Basta resistance in transformed oil palm plants is by direct painting of leaves with a herbicide solution. The leaves from control palms became brown and wilted after 2 weeks, compared to the transformed plants, which remained healthy and green. These results showed that the transformed plants are resistant to Basta and that the transgene is stably expressed.

Acknowledgement

The author thanks Director-General of MPOB for permission to publish this chapter. Thanks are also due to Mdm. Fatimah Tahir of MPOB for her technical assistance and Mr. Omar Rasid for comments on the manuscript.

References

1. Anon (2007). Review of the Malaysian Oil Palm Industry 2006 – MPOB Empowering Change, Economics and Industry Development Division, Malaysian Palm Oil Board, Kelana Jaya, Selangor, Malaysia, pp.95.
2. Parveez, G. K. A. (2000). Production of transgenic oil palm (*Elaeis guineensis* Jacq.) using biolistic techniques, in *Molecular Biology of Woody Plants* (Jain S. M. and Minocha. S. C., eds.), Kluwer Academic Publishers, Netherlands, Vol. 2, pp. 327–350.
3. Parveez, G. K. A., Chowdhury, M. K. U. and Saleh, N. M. (1994). Current status of genetic engineering of oil bearing crops. *Asia Pac. J. Mol. Biol. Biotechnol.* **3**, 174–192.
4. Cheah, S. C., Sambanthamurthi, R., Siti Nor Akmar, A., Abrizah, O., Manaf, M.A.A., Umi Salamah, R. and Parveez, G. K. A. (1995). Towards genetic engineering of oil palm, in *Plant Lipid Metabolism* (Kader, J. C. and Mazliak, P., eds.), Kluwer Academic Publishers, Netherlands, pp. 570–572.
5. Sambanthamurthi, R., Siti Nor Akmar, A. and Parveez, G. K. A. (2002). Genetic manipulation of the oil palm – Challenges and prospects. *The Planter* **78**(919), 547–562.
6. Parveez, G. K. A. (2003). Novel products from transgenic oil palm. *AgBiotechNet* **5** (ABN113), 1–8.
7. Gruber, M. Y. and Crosby, W. L. (1993). Vectors for plant transformation, in *Methods in Plant Molecular Biology and Biotechnology*

(Glick B. R. and Thompson J. E., eds.) CRC Press, Boca Raton, pp. 89–119.

8. Braun, A. C. (1952). Conditioning of the host cell as a factor in the transformation process in crown gall. *Growth* **16**, 65–74.
9. Fromm, M. E., Taylor, L. P. and Walbot, V. (1986). Stable transformation of maize after electroporation. *Nature* **319**, 791–793.
10. Sanford, J. C., Klein, T. M., Wolf, E. D. and Allen, N. (1987). Delivery of substances into cells and tissues using a particle bombardment process. *J. Part. Sci. Tech.* **5**, 27–37.
11. Potrykus, I. (1991). Gene transfer to plants – assessment of published approaches and results. *Ann. Rev. Plant Physiol. Plant Mol. Biol.* **42**, 205–225.
12. Gould, J., Devey, M., Hasegawa, O., Ulian, E. C., Peterson, G. and Smith, R. H. (1991). Transformation of *Zea mays* L. using *Agrobacterium tumefaciens* and the shoot apex. *Plant Physiol.* **95**, 426–434.
13. Chen, M.T., Chang, H. H., Ho, S. L., Tong, W. F. and Yu, S. M. (1993). *Agrobacterium*-mediated production of transgenic rice plants expressing a chimearic α-amylase promoter/β-glucuronidase gene. *Plant Mol. Biol.* **22**, 491–506.
14. Hiei, Y., Otho, S., Komari, T. and Kumasho, T. (1994). Efficient transformation of rice *(Oryza sativa* L.) mediated by *Agrobacterium* and sequence analysis of the boundaries of the T-DNA. *The Plant J.* **6**, 271–282.
15. Rashid, H., Yokoi, S., Toriyama, K. and Hinata, K. (1996). Transgenic plant production mediated by *Agrobacterium* in *Indica* rice. *Plant Cell Rep.* **15**, 727–730.
16. Ishida, Y., Saito, H., Otho, S., Hiei, Y., Komari, T. and Kumashiro, T. (1996). High efficiency transformation of maize (*Zea mays* L.) Mediated by *Agrobacterium tumefaciens. Nature Biotechnol.* **14**, 745–750.
17. Tingay, S., McElroy, D., Kalla, R., Fieg, S., Wang, M., Thornton, S. and Brittell, R. (1997). *Agrobacterium tumefaciens*-mediated barley transformation. *The Plant J.* **11**, 1369–1376.
18. Spencer, T. M., O'Brien, J. V., Start, W. G. and Adams, T. R. (1992). Segregation of transgenes in maize. *Plant Mol. Biol.* **18**, 201–210.
19. Srivastava, V., Vasil, V. and Vasil, I. K. (1996). Molecular characterization of the fate of transgenes in transformed wheat (*Triticum aestivum* L.). *Theor. Appl. Genet.* **92**, 1031–1037.
20. Zhang, S., Warkentin, D., Sun, B., Zhong, H. and Sticklen, M. (1996). Variation in the inheritance of expression among subclones for unselected (*uid*A) and selected (*bar*) transgenes in maize (*Zea mays* L.). *Theor. Appl. Genet.* **92**, 752–761.
21. Gordon–Kamm, W. J., Spencer, T. M., Mangano, M. L., Adams, T. R., Daines, R. J., Start, W. G., O'Brien, J. V., Chambers, S. A., Adams, W. R., Willets, N. G., Rice, T. B., Mackey, C. J., Krueger, R. W., Kausch, A. P. and Lemaux, P. G. (1990). Transformation of maize cells and regeneration of fertile transgenic plants. *The Plant Cell* **2**, 603–618.
22. Vasil, V., Brown, S. M., Re, D., Fromm, E. M. and Vasil, I. K. (1991). Transformed callus lines from microprojectile bombardment of cell suspension cultures of wheat, *Biotechnology* **9**, 743–747.
23. Bower, R. and Birch. R. G. (1992). Transgenic sugarcane plants via microprojectile bombardment, *The Plant J.* **2**, 409–416.
24. Parveez, G. K. A., Chowdhury, M. K. U. and Saleh, N. M. (1994). Current status of genetic engineering in oil bearing crops. *AsPac. J. Mol. Biol. Biotechnol.* **2**, 174–192.
25. Murashige, T. and Skoog, T. (1962). A revised method for rapid growth and bioassays with tobacco tissue cultures. *Physiol. Plant* **15**, 473–497.
26. Eeuwens, C. J. (1976). Mineral requirement for growth and callus initiation of tissue explants from mature coconut palms (*Cocos nucufera*) cultured *in vitro. Physiol. Plant* **36**, 23–28.
27. Doyle, J. J. and Doyle, J. L. (1990). Isolation of plant DNA from fresh tissue. *Focus*, **12**(1), 13–15.
28. Sambrook, J., Fritsch E. F. and Maniatis T. (eds.) (1989). *Molecular Cloning – A Laboratory Manual.* Cold Spring Harbor, New York.
29. Southern, E. M. (1975). Detection of specific sequences among DNA fragments separated by gel electrophoresis. *J. Mol. Biol.* **98**, 503–517.
30. Feinberg, A. P. and Vogelstein, B. (1984). A technique for radiolabelling DNA restriction endonuclease fragments to high specific activity. *Anal. Biochem.* **137**, 266–267.
31. DeBlock, M., Botterman, J., Vanderwiele, M., Montagu, M. and Leemans, J. (1987). Engineering herbicide resistance in plants by expression of a detoxifying enzyme. *EMBO J.* **6**, 2513–2518.

32. Castillo, A. M., Vasil, V. and Vasil, I. K. (1994). Rapid production of fertile transgenic plants of rye (*Secale cereale* L.). *Biotechnology* **12**, 1366–1371.

33. Jefferson, R. A., Kavanagh, T. A. and Beven, M. W. (1987). GUS fusions: β-glucuronidase as a sensitive and versatile gene fusion marker in plants. *EMBO J.* **6**, 3901–3907.

34. Sanford, J., Smith, F. D. and Russel, J. A. (1993). Optimizing the biolistic process for different biological applications. *Meth. Enzymol.* **217**, 483–509.

35. Christou, P., McCabe, D. E., Martinell, B. J. and Swain W. F. (1990). Soybean genetic engineering – commercial production of transgenic plants. *Trends In Biotechnol.* **8**, 145–151.

36. Imaseki, H. (1986). Ethylene, in *Chemistry of Plant Hormones* (Takahashi, N. ed.), CRC Press, Boca Raton, pp. 249–264.

Chapter 24

Preparation of a Multi-antioxidant Formulation

John A. Mulnix and Brook E. Stoddard

Abstract

This chapter describes the need for, as well as the process of, development, manufacture, and the clinical evaluation of a multi-antioxidant formulation. The formulation is in a chewable tablet form, for use in dogs showing clinical signs of cognitive dysfunction syndrome.

Key words: Antioxidant, Cognitive dysfunction syndrome (CDS), Oxidative damage, Dogs, Clinical evaluation, Nutraceutical, Alzheimer's disease, Novel ingredients.

1. Introduction

The approach to the development of an antioxidant (AOX) formulation for management of cognitive dysfunction syndrome (CDS) in the dog is presented. Current treatment strategies in the United States for CDS in the dog are: nutritional and dietary (antioxidant-enriched diets), environmental enrichment and previous cognitive experience, drug therapy (Selegiline), anti-inflammatory drugs (particularly the non-steroidal anti-inflammatory drugs), hormone replacement (estrogen may have anti-inflammatory and antioxidant effects), weight control, and naturopathic supplements, nutraceuticals, and homeopathic remedies *(1)*.

The preparation of a multi-antioxidant formulation is the primary focus of this chapter. Combinations of AOX are synergistic and support cellular functions, effects that are often not apparent with individual agents *(2)*. Many recent therapeutic options for both pets and humans have focused on a synergistic combination of ingredients rather than monotherapy *(3)*.

From: *Methods in Molecular Biology, vol. 477: Advanced Protocols in Oxidative Stress I,*
Edited by: D. Armstrong, DOI: 10.1007/978-1-60327-517-0_24, © Humana Press, New York, NY

Combinations of AOX, certain fatty acids, essential minerals, vitamins, metabolic cofactors, and trophic nutrients may be effective in contributing to brain cell health and memory preservation *(4, 5, 6)*. Because oxidative damage is considered to be one of the main factors in age-related decline of dogs, this multimodal, synergistic approach has become a focus of therapeutic intervention for the treatment of cognitive dysfunction in dogs, and many of the ingredients studied have been selected for their ability to scavenge or prevent the production of oxygen-free radicals *(7, 8, 9)*.

The central nervous system has been the subject of considerable research in dietary management of certain diseases. CDS in dogs has received considerable attention due to the debilitating nature of the disease, its similarity to Alzheimer's disease in man, and the active role of reactive oxygen species (ROS) in the central nervous system. Studies have shown that dietary fortification with AOX and a program of behavioral enrichment can slow age-dependent cognitive decline in both pets and people *(10)*. The role of AOX in veterinary medicine continues to expand and evolve into an important area of therapeutic consideration.

From 1990 to 1997, the use of herbs and other ingredients in human medicine grew close to 400%, with current growth approximating 15% per year *(11)*. It appears that the use of novel ingredients in veterinary medicine has paralleled that in human medicine. The actual use of novel ingredients in animals is unknown with estimates of up to 30% *(11)*. The use of novel ingredients, such as AOX, is complicated by issues regarding their safety, efficacy, and manufacturing and they are often used without doctor supervision *(11)*. Combinations of AOX are best suited for clinical application in modulating disease and reducing premature aging when caused by excessive free radical accumulation *(2)*. Providing AOX in supplemental form improves the ability to control the dose of active ingredients, which is extremely difficult to accomplish in dietary form *(12)*. Secondly, it is possible to provide a therapeutic effect from ingredients, which may not normally occur in the diet of a dog, cat, or research animal.

A manufacturer of an AOX compound can pursue marketing through Food and Drug administration (FDA) approval, which is extremely expensive and time consuming, or through the American Association of Feed Control Officials (AAFCO) as a group of defined ingredients. Evidence of efficacy for drugs generally is supported by pharmacokinetic and pharmacodynamic information *(13, 14)*. Collection of pharmacodynamic information that describes clinical response may be difficult for novel ingredients, because the diseases being studied often are complex and characterized by outcome measures that are neither easily measured nor discrete *(11)*. We describe here steps in the process of preparing a multi-antioxidant formulation in the form of a chewable tablet for testing in the laboratory and in clinical trials; however, alternate formats may be chosen, such as capsules, liquids, powders.

2. Materials

2.1. Equipment (see Note 1)

1. Scale
2. Ribbon blender
3. Tablet press
4. Tablet counter
5. Bottle capper
6. In-line labeler
7. Bottle neck bander
8. Heat tunnel/induction sealer
9. Advia 120, Hitachi Corp., San Jose, CA – Complete Blood Count (CBC)
10. Hitachi 747, Hitachi Corp., San Jose, CA – General Chemistry Health Screen
11. Urinalysis (Bench method)
12. Agilent Technologies HPLC and ICP-MS, Santa Clara, CA

2.2. Ingredients

1. Choose ingredients and determine amounts required to achieve post-production dose and guaranteed analysis. Examples: vitamin E, alpha-lipoic acid. (*see* **Note 2**)
2. Binders and fillers, if necessary, such as microcrystalline cellulose.
3. Flavoring agents, if desired, such as garlic or liver.

2.3. ORAC and Antioxidant Potency Assays

1. TBARS Assay (thiobarbituric acid reacting substances) – ZeptoMetrix Corp., Buffalo, NY
2. TEAC (trolox equivalent antioxidant capacity) – Randox Laboratories, Oceanside, CA

2.4. Evaluation Forms

1. Research agreement
2. Owner consent, if applicable
3. Physical exam
4. Behavior evaluation

5. Urinalysis results
6. CBC results
7. Chemistry screen results
8. Palatability, if applicable

3. Methods

3.1. Determine the Target Body System

Review the current methods of treatment available for that particular disease or the body system to be evaluated. It is important to establish why AOX would be as good as or better than the current treatment methods, or if the AOX are to be used as adjunctive therapy. Do AOX gain easy access to the tissue to be treated and if so, which type of AOX would be most advantageous? Establish whether or not AOX combinations, or single AOX are indicated, and which ones will be most effective and complementary.

Research AOX and other complementary nutrients that directly benefit the disease in question. Establish why concentrated AOX extracts are more therapeutic than eating whole foods.

3.2. Research Ingredients

The American Association of Feed Control Officials (AAFCO) publishes a list of all substances that can be consumed orally by animals. They identify which of these foods are considered as safe for use in animals and in what amount. Each item is identified relative to the diet and the species of animal in which it is used.

Consult AAFCO, published research, and validated supplier data to narrow the list of potential ingredients further by considering the following.

1. Bioavailability
2. Where the ingredients are metabolized
3. Solubility and digestibility
4. Indications, contraindications, and toxicity of the ingredients
5. Side effects and adverse effects of the active ingredients
6. ORAC or AOX potency of the individual ingredients
7. Feasibility of ingredient combinations
8. Compatibility with chosen delivery format

3.3. Produce Pilot Batch

Once the ingredients have been determined, the raw materials are obtained along with a certificate of analysis. Each ingredient is weighed on a scale and added to a ribbon blender. After the mixture is blended sufficiently, it is put through the tablet press. Adjustments are made in pressure and binding agents, as needed.

If binding agents are increased, the mixture must be blended again and passed through the tablet press until a good tablet is produced.

When the pilot tablets are completed, they are analyzed by an independent laboratory for microbial contamination (microscopically) and label claim primarily by high pressure liquid chromatography (HPLC) and inductively coupled plasma-mass spectrometry (ICP-MS). The pilot may be used to test palatability before a larger production run is manufactured for testing safety and efficacy in a clinical setting.

3.4. Verify Product Safety

Product safety is determined by clinical evaluation in healthy young animals. All animals are evaluated before, during, and after the trial period. The test dose is determined and administered first as a single dose, and then it is doubled and finally, tripled. All animals are tested by using clinical and clinical laboratory methods of evaluation, including a complete blood count (CBC), chemistry screen, and urinalysis.

3.5. Clinical Evaluation of the Product

Animals that are showing disease of the system in question are then evaluated using generally accepted methods of clinical testing. Consult the VICH Good Clinical Practice guidelines.

1. Identify the necessary length of time to complete the study. (*see* **Note 3**)
2. Obtain signed agreement of the protocol.
3. Present explanatory written information if the patient is a client owned animal.
4. Obtain signed consent form to utilize an experimental supplement or product.
5. Double blind all test subjects including the supervising veterinarian and the owner.
6. Identify the statistical data needed to soundly evaluate the product.
7. Identify the expected results and prepare the proper forms for the client and supervising veterinarian to complete at the beginning, during, and at the end of the study period.
8. Establish post-study options for owners of animals that are given the placebo initially, so they can receive the active ingredients afterward.
9. Utilize ORAC or AOX potency values to compare the test group and control group and correlate the values to study results.

3.6. Protocol for Clinical Evaluation

3.6.1. Safety Evaluation

The study protocol to evaluate the safety of a multi-antioxidant formulation in healthy adult dogs follows.

1. Identify a group of healthy test subjects (dogs not showing signs compatible with CDS). Minimum of 25 dogs. (*see* **Note 4**)
   a. Standardize the safety group subjects by age selection (only dogs less than 7 years of age, of either sex, and of any breed).
2. Perform a full physical examination at the start of the study. Repeat at each interval of the study thereafter (e.g. 14, 28, 42 days)
3. Obtain suitable biochemical and other laboratory data, such as complete blood count, chemistry screen and urinalysis, to rule out co-existing disease(s) that could alter the results of the study. Repeat these tests at each interval of the study thereafter.
4. Each dog is dosed at the single dose for the first interval, double dose for the second interval, and triple dose for the third interval.
5. Do statistical analysis of the collected data.
6. Identify Type I and Type II error values in both study groups. A type I error allows for identification of a significant difference when one does not truly exist, generally 5% or $P \leq 0.05$. A Type II error, power, allows for failing to detect a significant difference when one truly does exist, usually 80% *(11)*.

3.6.2. Efficacy Evaluation

The study protocol to evaluate a multi-antioxidant formulation for use in dogs showing signs of CDS follows.

1. Prepare a suitable placebo that is indistinguishable from the antioxidant formula to be tested.
2. Randomize the distribution of placebo and test AOX formula. Avoid the use of placebo or AOX formula in patients receiving any other medications.
3. Identify the test subjects (dogs showing clinical signs that are considered to be compatible with CDS). Minimum of 25 dogs (*see* **Note 4**).
   a. Standardize the treatment group by age selection (only dogs over 9 years of age, of either sex, and of any breed).
4. Perform a full physical examination at the start of the study. Repeat at each interval of the study thereafter (e.g. 14, 28, 42 days)
5. Obtain suitable biochemical and other laboratory data, such as complete blood count, chemistry screen and urinalysis, to rule out co-existing disease(s) that could alter the results of the study. Repeat these tests at each interval of the study thereafter.

6. Each dog in the treatment group must be evaluated for behavioral changes associated with brain aging by the attending veterinarian with assistance from the owner at the beginning of the study and at each interval thereafter.
7. Do statistical analysis of the collected data.
8. Identify Type I and Type II error values in both study groups. A type I error allows for identification of a significant difference when one does not truly exist, generally 5% or $P \leq 0.05$. A Type II error, power, allows for failing to detect a significant difference when one truly does exist, usually 80% *(11)*.

4. Notes

1. Similar equipment can be used for laboratory analyses of product and test animal blood/urine.
2. The AOX compound described is for use in canines. The same safety protocol should be followed for animals other than dogs.
3. In a multi-center study, be prepared for the duration of the completed study to be longer than projected. Some participants will take longer than others to find qualifying participants.
4. It is wise to collect data from more than the minimum required as some data may require disqualification and some animals may be removed for reasons unrelated to the test product. In particular, a study involving older animals may result in natural death or euthanasia of patient(s) before completion. It is important to document such unanticipated events and as they may affect the quality and integrity of the study.

References

1. Landsberg, G. and Araujo, J.A. (2005) Behavior problems in geriatric pets, in *Veterinary Clinics of North America* (Fortney, W.D., ed.), WB Saunders, Philadelphia, PA, pp.675–698.
2. Mandelker, L. (2004) The natural activity of cells, the role of reactive oxygen species, and their relation to antioxidants, nutraceuticals, botanicals, and other biologic therapies, in *Veterinary Clinics of North America* (Mandelker, L., ed.), WB Saunders, Philadelphia, PA, pp.39–66.
3. Landsberg, G. (2006) Therapeutic options for cognitive decline in senior pets. *J. Am. Anim. Hosp. Assoc.* **42**, 407–413.
4. Zandi, P.P., Anthony, J.C., Khachaturian, A.S., et al. (2004) Reduced risk of Alzheimer disease in users of antioxidant vitamin supplements – The Cache County Study. *Arch. Neurol.* **61**, 82–88.
5. Standridge, J.B. (2005) The pharmacological prevention of Alzheimer's disease. *Alzheimers Dement.* **1**, 64.

6. Kidd, P.M. (2005) Neurodegeneration from mitochondrial insufficiency; nutrients, stem cells, growth factors, and prospects for brain rebuilding through integrative management. *Altern. Med. Rev.* **10**, 268–293.
7. Head, E., Liu, J., Hagen, T.M., et al. (2002) Oxidative damage increases with age in a canine model of human brain aging. *J. Neurochem.* **82**, 375–381.
8. Rofina, J.E., Singh, K., Skolumalova-Vesela, et al. (2004) Histochemical accumulation of oxidative damage products is associated with Alzheimer-like pathology in the canine. *Amyloid* **11**, 90–100.
9. Skoumalova, A., Rofina, J., Schwippelova, Z., et al. (2003) The role of free radicals in canine counterpart of senile dementia of the Alzheimer type. *Exp. Gerentol.* **38**, 711–719.
10. Milgram, N.W. and Head, E. (2005) Learning ability in aged beagle dogs is preserved by behavioral enrichment and dietary fortification: a two-year longitudinal study. *Neurobiol. Aging.* **26**, 77–90.
11. Boothe, D.M. (2004) Balancing fact and fiction of novel ingredients: definitions, regulations and evaluation, in *Veterinary Clinics of North America* (Mandelker, L., ed.), WB Saunders, Philadelphia, PA, pp.7–38.
12. Zicker, S.C., Wedekind, K.J., and Jewell, D.E. (2006) Antioxidants in veterinary nutrition, in *Veterinary Clinics of North America* (Kirk, C.A. and Bartges, J.W., ed.), WB Saunders, Philadelphia, PA, pp.1183–1198.
13. Booth, D.M. (2001) Principals of drug therapy, in *Small Animal Clinical Pharmacology and Therapeutics* (Boothe, D.M., ed.), WB Saunders, Philadelphia, PA, pp.3–17.
14. Mathieu, M.P. (1994) *New Drug Development: A Regulatory Overview.* Parexel International Corporation, Waltham, MA.

Part III
Gene Expression

Chapter 25

A Functional Genomics Approach to Identify and Characterize Oxidation Resistance Genes

Michael R. Volkert, Jen-Yeu Wang, and Nathan A. Elliott

Abstract

In order to develop a more complete understanding of the genes required for resistance to oxidative DNA damage, we devised methods to identify genes that can prevent or repair oxidative DNA damage. These methods use the oxidative mutator phenotype of a repair deficient *E. coli* strain to measure the antimutator effect resulting from the expression of human cDNAs. The method can be adapted to characterize the function, and to determine the active site domains, of putative antimutator genes. Since bacteria do not contain subcellular compartments, genes that function in mitochondria, the cytoplasm, or the nucleus can be identified. Methods to determine the localization of genes in their normal host organism are also described.

Key words: Oxidative DNA damage, DNA repair, Mutagenesis, Mitochondrial localization, Fluorescence microscopy, Functional gene cloning, 8-oxoguanine.

1. Introduction

Oxidative damage to DNA is a constant threat to genomic integrity that results when DNA is attacked by reactive oxygen species (ROS), which are by-products of normal aerobic metabolism. Oxidative DNA damage can result in mutagenesis and cell death. Exposure to oxidative agents can further increase the level of oxidative damage, leading to increased levels of mutagenesis and cell death. To mitigate oxidative DNA damage, cells have two lines of defense. The first line of defense is provided by the enzymes that detoxify ROS. The second line of defense is provided by the DNA repair enzymes that repair oxidative lesions in DNA. In a wild type, repair

From: *Methods in Molecular Biology, vol. 477: Advanced Protocols in Oxidative Stress I*,
Edited by: D. Armstrong, DOI: 10.1007/978-1-60327-517-0_25, © Humana Press, New York, NY

proficient cell, increasing ROS production, decreasing ROS detoxification, or decreasing repair of oxidative DNA damage can shunt more DNA damage down the pathway leading to mutagenesis (**Fig. 25.1**).

We developed methods to identify and characterize gene products that either prevent or repair oxidative DNA damage. These methods use a specific *lacZ* allele as a reporter of oxidative mutagenesis in a bacterial strain that is genetically modified to be deficient in repair of specific oxidative DNA lesions. The mutant *lacZ* allele results in colonies that are white, but contain dark blue *lacZ*$^+$ microcolonies, or papillae that arise from cells that acquire *lacZ*$^+$ reversion mutations during growth (**Fig. 25.2**). The specific *lacZ* allele present in the mutator strain we use has a mutation that converts its active site glutamic acid codon (GAG) to GCG, which reverts to GAG only by a GC → TA transversion at the central GC base pair. This transversion mutation results from the oxidation of G to 8-oxoG by endogenously produced ROS and the number of mutant papillae that result from this mutagenic event are very reproducible from colony to colony. This phenotype can be incorporated into a genetic screen by transforming cells with a cDNA library cloned into a bacterial expression library, then screening individual transformed colonies for ones that exhibit a reduced level of papillation. Genes that cause a reduction in mutagenesis are generally ones that either reduce or repair oxidative DNA damage.

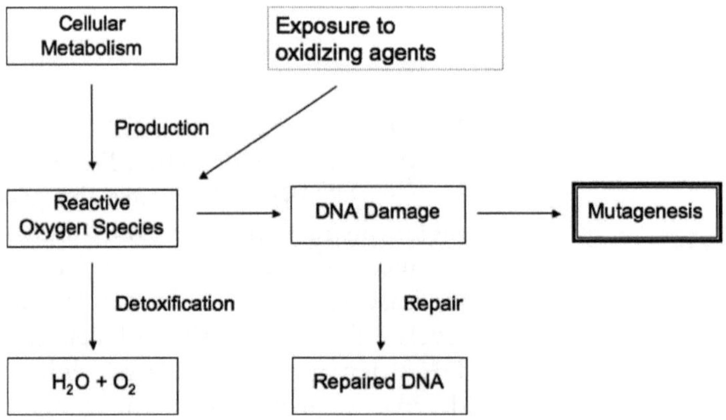

Fig. 25.1. A model of factors affecting ROS mutagenesis. In wild type cells, oxidative mutagenesis remains low because the constant production of ROS is in balance with detoxification and repair mechanisms. Factors increasing ROS production, or decreasing detoxification will cause more damage to be produced, causing more mutations. Factors decreasing repair will cause more damage to be processed into mutations. The effect of eliminating repair of 8-oxoG and its 8-oxoG:A mispaired intermediate by *mutM* and *mutY* mutations can be readily seen by comparing panels **A** and **B** in **Fig. 25.2**.

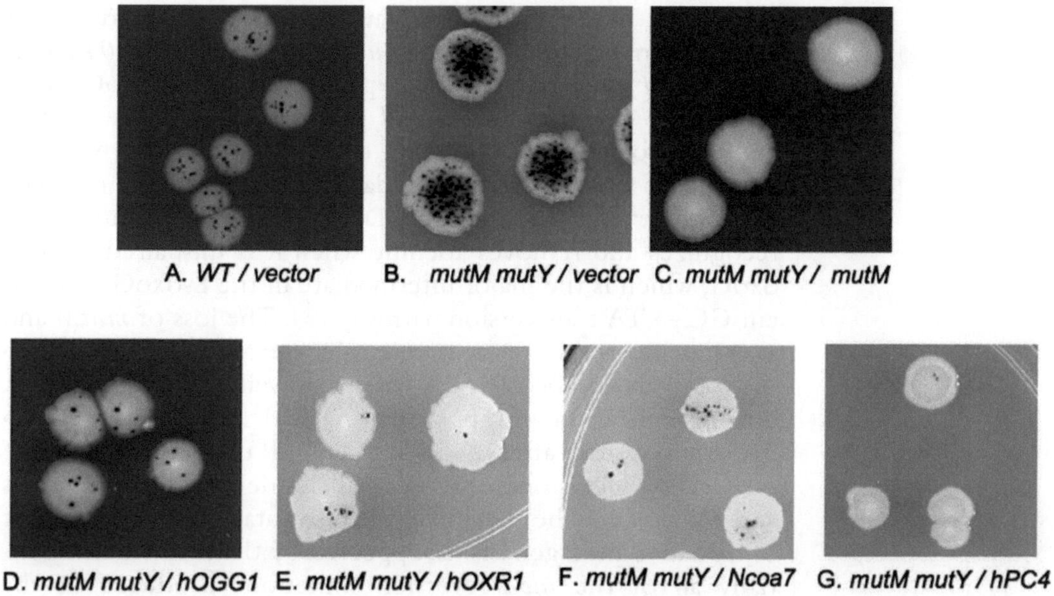

Fig. 25.2. Papillation assay. Light colored *lacZ*⁻ colonies containing dark microcolonies, or papillae due to reversion to *lacZ*⁺ are shown. (**A**) wild type *E. coli* carrying the vector plasmid; (**B**) *mutM mutY* carrying the vector plasmid; (**C**) *mutM mutY* strain expressing the plasmid borne bacterial *mutM* gene; (**D**) *mutM mutY* strain expressing the human OGG1 (8-oxoG DNA glycosylase) gene; Panels **E**, **F**, and **G** show colonies that exhibit reduced mutagenesis due to the expression of human cDNAs discovered in our screens: OXR1 (**E**) and Ncoa7 (**F**) are two different genes from humans that produce highly similar proteins that appear to affect ROS detoxification (5, 6, 8, 9, 10). PC4 (**G**) appears to function in repair *(11)*. All panels are labeled to indicate the genotype of the bacterial strain followed by a slash mark (/) and the gene present in the bacterial expression plasmid. Panels **A**, **B**, **C**, **D** and **G** are reprinted from *(11)* and used by permission from ASM Press; panels **E** and **F** are reprinted from *(10)* and used by permission of the authors.

To understand how this works and to fully exploit the system, it is useful to describe several features of the basic method in detail. In this assay, no treatment of the cells with oxidative agents is required, since the loss of DNA repair increases the spontaneous ROS-dependent mutagenesis to levels that are sufficiently high to allow identification of antimutator genes. In normal cells, a low level of GC → TA reversion can be detected (**Fig. 25.2**). This low level of spontaneous GC → TA transversion mutagenesis is due to oxidative damage, since it can be reduced, either by growing cells anaerobically, or by overexpressing the *mutM* gene, which repairs 8-oxoG *(1)*. *E. coli* strains that are deficient in repair of oxidative lesions in DNA exhibit high levels of spontaneous oxidative mutagenesis (**Fig. 25.2B**) that can be reduced upon introduction and expression of gene products that decrease ROS production, increase ROS detoxification, or increase repair. In either case, less DNA damage is shunted down the pathway to mutagenesis (**Fig. 25.1**), reducing papillation. Examples of the antimutator effects of bacterial and human genes are shown in **Fig. 25.2C–F**.

The *E. coli* strain used in the assay described here carries three key mutations: *mutM, mutY*, and the *lacZcc104* allele. The *mutM* gene encodes the Fpg protein, an 8-oxoG DNA glycosylase that recognizes and removes from DNA 8-oxoG and the formamidopyrmidine (FaPy) lesion, an oxidation product of G in which the imidazole ring is broken *(2)*. The *mutY* gene encodes a second DNA glycosylase (MutY) that recognizes and removes adenine when it is mispaired with 8-oxoG, which is the major intermediate in the 8-oxoG dependent GC → TA transversion pathway *(3)*. The loss of *mutM* and *mutY* genes results in a dramatically elevated level of GC → TA transversions because 8-oxoG pairs well with either C or A during replication and its mispairing with A leads to a GC → TA transversion after a second round of replication. The *lacZcc104* allele is one of six *lacZ* alleles developed by Cupples and Miller *(4)*. The specificity of this mutation as an indicator of oxidative mutagenesis is supported by the result that essentially all of the *lacZ* reversion mutations produced in the *mutM mutY* double mutant strain can be eliminated simply by over-expressing only the *mutM* encoded 8-oxoG DNA glycosylase (**Fig. 25.2**). Thus, essentially all of the lesions that give rise to *lacZ* reversion mutations in this strain are repairable by *mutM* and spontaneous lesions other than *mutM* repairable lesions do not contribute to any measurable extent to GC → TA transversion.

The six *lacZ* alleles produced by Cupples and Miller carry different base pair changes in the active site glutamic acid codon and each one reverts to wild type by a different base pair change. Together they can detect each of the six possible base pair substitution mutations *(4)*. It is noteworthy that other oxidative DNA damage repair deficiencies coupled with different alleles of this group *lacZ* alleles can be used as well. We have constructed a large set of strains carrying mutations in the *mutM, mutY, nei*, and *nth* oxidative damage repair glycosylases, either singly or in combination. Strains carrying these repair genes singly and in most of the possible combinations have been constructed in each of the six *lacZcc* reporter strains and can be used to search for additional classes of cDNAs.

This method greatly simplifies procedures used earlier *(5)*, because the initial cDNA screening can be performed simply by plating the primary transformants on indicator medium, incubating the cells, then screening colonies directly. Previous methods required purification, growth, and testing of individual transformants.

Since the assay described above can detect genes whose products function in any subcellular compartment of a eukaryotic cell, an important step in the characterization of a new protein is determining in which subcellular compartment the protein normally functions. In the case of the human OXR1 protein, both bioinformatics and biological data suggested the protein might be localized to the mitochondria. The first exon of the hOXR1 cDNA encodes a putative N-terminal mitochondrial targeting signal (MTS). Additionally, the yeast homolog of OXR1, scOxr1p, localizes to the yeast mitochondria *(6)*. We, therefore, chose to determine the localization of hOXR1 in human cells using immunofluorescence microscopy (**Fig. 25.3**). In order to definitively show co-localization of the protein with the mitochondrial compartment, a mitochondrial-specific marker is needed. We use the MitoTracker Red CMX-Ros probe (Invitrogen), a cell-permeant dye that is retained in the mitochondria following an oxidation and thiol-conjugation step. The thiol-conjugation and mitochondrial retention of this probe makes it desirable for use in immunostaining protocols in which cell permeabilization is required following fixation, as the fluorescent signal is not easily washed out of the fixed and permeabilized cells.

In addition to subcellular localization, immunofluorescence microscopy can detect qualitative changes in protein abundance. Since the yeast OXR1 gene has been shown to be stress inducible, we wanted to determine if human OXR1 protein levels were increased following oxidative stress by comparing the fluorescence signal from untreated cells and those treated with hydrogen peroxide. **Figure 25.3** shows this to be the case.

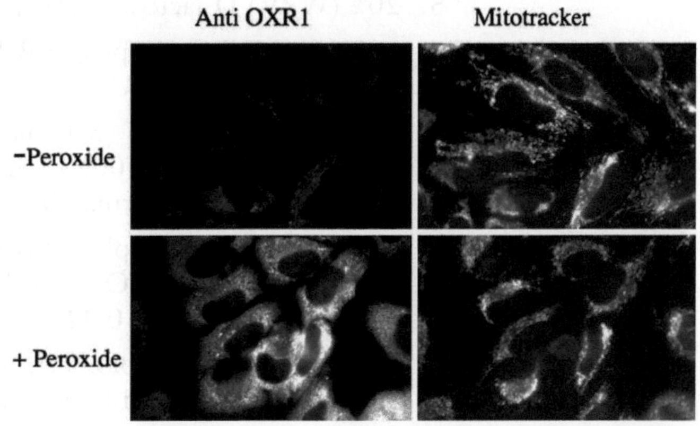

Fig. 25.3. Mitochondrial localization and oxidative stress induction of human OXR1. Antibody to OXR1 and the MitoTracker Red CMX-Ros dye were used in HeLa cells either exposed or not to hydrogen peroxide (for details see: *[6]*). The left panels show the mitochondrial localization of OXR1. By comparing the untreated and peroxide treated cells (upper and lower left panels respectively), relative OXR1 levels can be compared. The right panels using MitoTracker Red CMX-Ros serve as controls to identify the mitochondria and as an exposure control for the two OXR1 panels. **Figure 25.3** is reprinted from *(6)* and used by permission from ASM press.

2. Materials

2.1. Equipment

1. Nikon Eclipse E800 fluorescence microscope, Nikon Instruments Group, NY.
2. Canon. EOS 30 D digital camera, Canon, Inc, NY.

2.2. Bacterial Strains and Storage

All *E. coli* strains are derivatives of the *lacZ* mutant *E. coli* strain cc104 from the Cupples and Miller *(4)* collection of *lacZcc* strains, wherein a specific mutant *lacZ* allele is harbored on an F' episome. MV4704 and MV4705 are *ΔmutM::Tn10 (ΔmutM::Tn10)* and resistant to 10 μg/ml tetracycline. MV4706 and MV4707 are *ΔmutY::cat* and resistant to 10 μg/ml chloramphenicol. MV4708 and MV4709 are *ΔmutM::Tn10 ΔmutY::cat*. These strains are grown on 0.2% D-glucose minimal 1X A-salts plates to select for retention of the episome. For long-term storage, they are grown in liquid minimal glucose medium and stored as −80°C DMSO stocks [dimethyl sulfoxide; added to 10% (V/V) final concentration]. Glycerol (15% final conc.) can be substituted for 10% DMSO for storage at −80°C at any stage.

2.3. Reagents

1. 10X A-salts (per liter: 105 g K_2HPO_4; 45 g KH_2PO_4; 10 g $(NH_4)_2SO_4$; 5 g sodium citrate · $2H_2O$).
2. 1 M $MgSO_4$ (1000X).
3. 10 mg/ml thiamine-HCl (2000X).
4. 20% (W/V) D-glucose (100X).
5. 20% (W/V) D-lactose (100X).
6. 100 mM IPTG (isopropyl-β-D-galactopyranoside; 200X) (Sigma, St. Louis, MO)
7. 20 mg/ml X-Gal (5-bromo-4-chloro-3-indolyl-β-D-galactopyranoside; 500X) (Sigma, St. Louis, MO); dissolved in dimethylformamide.
8. 100 mg/ml P-Gal (phenyl-β-D-galactopyranoside; 200X) (Sigma, St. Louis, MO); dissolved in ddH_2O and sterilized by passing through a 0.22 μm filter.
9. Carbenicillin (US Biological, Swampscott, MA) 50 mg/ml in water (1000X) and filter sterilized as above.

2.4. Papillation Assay Medium

Papillation assay medium contains 1X A-salts, 1 mM $MgSO_4$, 5 μg/ml thiamine-HCl, 0.2% D-glucose, 0.5 mM IPTG, 40 μg/ml X-Gal, 500 μg/ml P-Gal, 50 μg/ml carbenicillin, and 2% agar *(7)* (*see* **Note 1**). Twenty grams of Bacto-agar in 870 ml deionized H_2O are first autoclaved, and A-salts and $MgSO_4$ are added immediately after sterilization to avoid precipitation. All other components are added after the medium has cooled (~60°C).

2.5. Minimal Media for Quantitation of the Mutation Frequency

These media contain 1X A-salts, 1 mM $MgSO_4$, 5 μg/ml thiamine-HCl, 0.5 mM IPTG, 50 μg/ml carbenicillin, and 2% agar, and are supplemented with either D-glucose or D-lactose to a final concentration of 0.2%. Liquid medium are identical, except for the absence of agar.

2.6. HeLa Cell Culture

1. Dulbecco's Modified Eagle's Medium (DMEM) (InVitrogen, Carlsbad, CA) supplemented with 10% fetal bovine serum (FBS)
2. 24 well tissue culture plates (Fischer)
3. MitoTracker Red CMX-Ros probe (1 mM) dissolved in dimethyl sulfoxide (DMSO).
4. Hydrogen peroxide (30% H_2O_2, Sigma)
5. Trypsin/EDTA solution (InVitrogen, Carlsbad, CA).

2.7. OXR1 Immunofluorescence

1. Poly-D-Lysine coated cover slips (Becton Dickinson, Franklin Lakes, NJ).
2. Phosphate Buffered Saline (10X stock): 1.37 M NaCl, 30 mM KCl, 120 mM Na_2HPO_4, 14 mM KH_2PO_4. For working stock, dilute 1:10 in water and filter sterilize through a 0.2 μm filter.
3. Paraformaldehyde (2.5% W/V) in PBS.
4. Permeabilization solution: 0.1% Triton X-100 in PBS.
5. Primary antibody: rabbit anti-mouse C7C (0.78 mg/ml stock) (gift of E. Engvall, Burnham Institute, LaJolla, CA).
6. Secondary antibody: anti-rabbit AlexaFluor 488 (Molecular Probes/InVitrogen, Carlsbad, CA)
7. Mounting medium: 10 mM Tris-Cl, pH 8.0, 90% glycerol, 50 ng/ml DAPI stain.

3. Methods

3.1. Primary Screening for Antimutator cDNAs by the Papillation Assay

1. A stock of MV4709 cells is first transformed with a cDNA library cloned into a suitable bacterial expression vector by any standard method of *E. coli* transformation or electroporation. After transformed cells are grown for 1 h, they are frozen at −80°C after adding either DMSO (10% final conc.), or glycerol (15% final conc.) to prevent ice crystal formation.
2. Next, an aliquot of MV4709 transformants harboring the cDNA library is thawed, serial dilutions performed, diluted cultures spread on glucose minimal medium plates, and

incubated at 37°C. The goal is to determine the appropriate dilution that gives rise to ~100 colonies/plate.

3. Once the dilution factor is determined, aliquots of the remaining frozen transformants can be thawed, diluted, and plated on papillation assay medium as determined in step 2, and incubated at 37°C for 5–6 days, until the colonies reach a diameter of approximately 3 mm. The long incubation is required to allow the mutants to arise and grow into visible microcolonies within the parent colony.

4. The surfaces of the majority of colonies will be covered with hundreds of dark blue papillae. Colonies lacking papillae, or containing a reduced number of papillae are readily apparent. These colonies harbor potential candidate cDNA that provide an antimutator function, either by preventing ROS from acting on DNA, or repairing the damage it produces. Once candidate colonies are identified, they should be picked and grown as 2 ml cultures in liquid glucose minimal medium for at least 18 h. Then the plasmid DNAs are isolated using any of the commercial miniprep kits. Each potential clone is thus isolated, numbered, and stored at −20°C. It is not advisable to store −80°C bacterial stocks of the cDNAs in strain MV4709 at this stage (*see* **Note 2**).

5. It is very important to confirm that the antimutator phenotype (i.e., colonies exhibiting either reduced numbers of papillae or their absence) is reproducible and due to the cDNA and not a false positive. False positives do occur because the test strain of bacteria has a high spontaneous mutation frequency, which can result in secondary mutations that prevent expression of the mutator phenotype. Most false positives are eliminated by simply re-transforming the MV4709 strain with each candidate cDNA, plating each at a density of ≤50 colonies/plate and examining to see if the majority shows the antimutator phenotype. This is usually achieved by plating 100 μl of the 10^{-3}, 10^{-4} and 10^{-5} dilutions on three plates, respectively. Some cDNAs prove to be less stable than others, suggesting a selection against their maintenance. This class of cDNAs tends to produce a higher frequency of colonies that lose the plasmid, or mutate the cloned cDNA. These types of cDNAs also tend to produce sectored colonies; ones in which pie shaped sectors within individual colonies revert to a high mutation frequency due to loss of the plasmid after a number of rounds of replication.

6. If the antimutator phenotype is reproducible, make a digital imagine using either a desktop scanner or a digital camera of the plate and keep it for future reference (*see* **Section 3.2**).

3.2. Qualitative Determination of Antimutator Strengths

Once a catalogue of images is made, it is possible to rank the antimutator strength of each candidate according to the numbers of papillae present on a single colony. This roughly reflects the mutation frequency and is a consistent characteristic for each clone. Candidate clones can be divided into three groups based on the strength of their antimutator phenotype: strong (≤5 papillae), medium (5–20 papillae), and weak (≥20 papillae) antimutators.

3.3. Quantitative Determination of Mutation Frequency

1. For a particular candidate cDNA, the purified plasmid DNA previously stored in −20°C is used to retransform strain MV4709. The transformants are plated on glucose minimal agar medium and incubated at 37°C until colonies grow to ≥1 mm in diameter (~24 h). Five to seven colonies are picked and used to inoculate 2-ml glucose minimal liquid medium and grown overnight at 37°C for ~20 h.
2. To determine *lac* reversion frequency, the 2-ml overnight cultures are washed once by centrifugation and resuspended in 1X A-salts solution, then centrifuged again and resuspended again in 2 ml of 1X A-salts solution. Total viable cells are measured by plating serial dilutions (10^{-4}, 10^{-5}) on glucose minimal medium and incubating at 37°C for ~24 h. The numbers of Lac^{+} revertants in each culture are determined by plating serial dilutions (undiluted to 10^{-3}) on lactose minimal medium at 37°C for ~3 days (*see* **Notes 3** and **4**). An average of the mutation frequencies (Lac^{+} revertants/total viable cells) is determined from at least three of the cultures after excluding the outliers.

3.4. Mitochondrial Localization and HeLa Cell Culture

1. The day before immunofluorescence staining, a sterile, poly-D-lysine coated cover slip is placed in the bottom of each well of a 24 well cell culture dish, then HeLa cells are seeded into wells.
2. A confluent culture of HeLa cells grown in a 25 cm flask is harvested by trypsinization with 1 ml of trypsin/EDTA for 5 min at room temperature (or until cells began to detach from the bottom of the flask)
3. Cells are washed off the bottom of the flask with 5 ml of fresh DMEM containing 10% fetal bovine serum (FBS), pre-heated to 37°C.
4. This cell suspension is diluted 1:3 in fresh DMEM/10% FBS and aliquoted into the wells containing the poly-D-lysine coated cover slips. A total volume of 2 ml is seeded into each well.
5. Cells were grown overnight in a 37°C/5% CO_2 incubator. Ideally, cells should be approximately 50% confluent after 24 h of growth.

3.5. H_2O_2 Treatment of Human Cells and MitoTracker Staining

1. If cells are to be treated with H_2O_2, remove medium and replace with fresh DMEM/10%FBS with 1 mM H_2O_2. For control cells, replace with medium lacking H_2O_2 (*see* **Note 5**).
2. Return plate to the incubator for desired time points. Increases in OXR1 protein level are usually observed by 4 h post-peroxide treatment.
3. Remove culture medium from wells and replace with fresh DMEM/10%FBS plus 100 nM MitoTracker Red CMX-Ros probe (*see* **Note 6**).
4. Return plate to the 37°C incubator for 15 min to allow mitochondrial labeling.

3.6. Immunofluorescence Staining and Mitochondrial Localization

1. Remove culture medium from each well by aspiration.
2. Add 1 ml filter-sterilized PBS to each well and gently rock for 5 min.
3. Remove PBS and repeat step 2 four more times.
4. Remove PBS and fix cells by the addition of 1 ml of freshly prepared 2.5% paraformaldehyde, and gently rock for 10 min at room temperature.
5. Wash cells one time in 1 ml PBS for 5 min.
6. Permeabilize cells by removing PBS, then adding 1 ml PBS/0.1% Triton X-100, and gently rock for 5 min at room temperature.
7. Dilute stock solution of rabbit anti-mouse C7C antibody 1:200 in 500 μl PBS. Add to cover slip, and incubate with gentle rocking for 1 h at room temperature.
8. Wash cover slip three times for 5 min with 1 ml PBS.
9. Dilute stock AlexaFluor 488 secondary antibody 1:200 in 500 μl PBS and add to cover slip. Incubate with gentle rocking for 1 h.
10. Wash cover slip three times for 5 min with 1 ml PBS.
11. Prepare a microscope slide by placing a small drop of mounting medium onto the middle of the slide.
12. Carefully remove cover slip from the well with fine-tipped forceps, invert, and place onto the drop of mounting medium on the microscope slide.
13. Seal the edges of the cover slip with clear nail polish, and allow it to dry completely.
14. Proceed to microscopy, or store the slides in the dark at −20°C.
15. Slides are viewed using fluorescence microscopy. DAPI excitation is 364 nm (blue emission), AlexaFluor 488 excitation

is 488 nm (green emission), and MitoTracker Red CMX-Ros excitation is 579 nm (red emission). Images are captured on a CCD camera, and processed with appropriate software. Three-color overlay images can be created in Photoshop.

4. Notes

1. Since it usually takes 5–6 days to allow blue papillae formation, it is important to pour at least 25 ml of medium in a 90-mm Petri dish to supply enough nutrients and moisture. Although more expensive, carbenicillin is much more stable than ampicillin and prevents the formation of satellite colonies; small, generally ampicillin sensitive colonies that grow around the primary colony after it destroys the ampicillin. We also found that A-salts based minimal medium gave better resolution of papillae than other minimal media.
2. Due to a very high intrinsic spontaneous mutation frequency of MV4709 (Δ*mutM* Δ*mutY*), it is not practical to maintain plasmids with a cloned cDNA in this strain. The amounts of initially isolated mini-prep plasmid DNAs stored at −20°C should be enough for phenotypic confirmation by retransformation of MV4709 in a papillation assay, for quantitative measurements of *lac* reversion frequencies and also for ~2 DNA sequencing reactions. For additional DNA sequencing and other functional analyses, we suggest transforming an ordinary, non-mutator cloning host for further propagation and maintenance of cDNAs, and freezing these cells for long term storage.
3. In the quantitative mutagenesis assay, tiny pinpoint-sized colonies begin to emerge after about 3 days incubation. It is prudent to count only the uniformly large colonies as Lac^+ revertants. Although the processes are a bit arbitrary, to obtain consistent and reliable measurements of mutation frequencies, it is best to conduct all steps identically, such as using only ≥1 mm colonies to set up 2-ml overnight cultures, keeping incubation times constant, as well as counting only large Lac^+ colonies on day 4 (~72 h).
4. For most of the strong and some medium antimutators, the mutation frequency is reduced to a level too low to measure. For this class of cDNAs, one can either grow the 2-ml culture to saturation (48 h), or grow the 2-ml overnight, then add 18 ml fresh medium and incubate for an additional day, then concentrate the culture 10-fold by centrifugation prior to washing and plating on lactose and glucose minimal media as above.

5. Since cell lines will vary in their sensitivity to H_2O_2, it is best to empirically determine the optimal concentration of peroxide to use in this step. The concentration should not cause a greater than 10% loss in viability of the cells.
6. HeLa cells appear to be sensitive to the DMSO used to dissolve MitoTracker. It is best to empirically determine the amount of dye used in this step. We see good results with 100 nM, though some cell lines may require more or less to achieve optimal mitochondrial staining.

Acknowledgments

This work was supported by NIH Grant RO1-CA100122.

References

1. Wyrzykowski, J., Volkert, M. R. (2003) The *Escherichia coli* methyl-directed mismatch repair system repairs base pairs containing oxidative lesions. *J. Bacteriol.* **185**, 1701–1704.
2. Tchou, J., Kasai, H., Shibutani, S., Chung, M., Laval, J., Grollman, A. P., et al. (1991) 8-oxoguanine (8-hydroxyguanine) DNA glyxosylase and its substrate specificity. *Proc. Natl. Acad. Sci. USA* **88**, 4690–4694.
3. Au, K. G., Cabrera, M., Miller, J. H., Modrich, P. (1988) *Escherichia coli mutY* gene product is required for specific AG → CG mismatch correction. *Proc. Natl. Acad. Sci. USA* **85**, 9163–9166.
4. Cupples, C. G., Miller, J. H. (1989) A set of *lacZ* mutations in *Escherichia coli* that allow rapid detection of each of the six base substitutions. *Proc. Natl. Acad. Sci. USA* **86**, 5345–5349.
5. Volkert, M. R., Elliott, N. A., Housman, D. E. (2000) Functional genomics reveals a family of eukaryotic oxidation protection genes. *Proc. Natl. Acad. Sci. USA* **97**, 14530–14535.
6. Elliott, N. A., Volkert, M. R. (2004) Stress induction and mitochondrial localization of Oxrl proteins in yeast and humans. *Molec. Cell Biol.* **24**, 3180–3187.
7. Nghiem, Y., Cabrera, M., Cupples, C. G., Miller, J. H. (1988) The *mutY* gene: A mutator locus in *Escherichia coli* that generates GC → TA transversions. *Proc. Natl. Acad. Sci. USA* **85**, 2709–2713.
8. Fischer, H., Zhang, X. U., O'Brien, K. P., Kylsten, P., Engvall, E. (2001) C7, a novel nucleolar protein, is the mouse homologue of the *Drosophila* late puff product L82 and an isoform of human *oxr1*. *Biochem. Biophys. Res. Commun.* **281**, 795–803.
9. Shao, W., Halachmi, S., Brown, M. (2002) Erap140, a conserved tissue-specific nuclear receptor coactivator. *Mol. Cell Biol.* **22**, 3358–3372.
10. Durand, M., Kolpak, A., Farrell, T., Elliott, N. A., Shao, W., Brown, M., et al. (2007) The Oxr domain defines a conserved family of eukaryotic oxidation resistance proteins. *BMC Cell Biol.* **8**, 13.
11. Wang, J.-Y., Sarker, A. H., Cooper, P. K., Volkert, M. R. (2004) The single-strand DNA binding activity of human PC4 functions to prevent mutagenesis and killing by oxidative DNA damage. *Mol. Cell Biol.* **24**, 6084–6093.

Chapter 26

Genome-Wide Overexpression Screen for Activators of Antioxidant Gene Transcription

Hendrik Luesch and Yanxia Liu

Abstract

Genome-wide screens have proven powerful in associating gene products with certain phenotypes or signal transduction pathways, and thus are valuable tools to ascribe gene function. These genomic screens can be extended to discover genes/proteins that attenuate oxidative stress-induced damage, which is implicated in aging, neurodegenerative disorders, and other diseases. One mechanism by which humans protect themselves from oxidative stress is through an endogenous stress response that leads to transcriptional activation of the antioxidant response element (ARE). The ARE is located in the 5′-flanking regions of many phase II detoxification and antioxidant enzymes such as NAD(P)H:quinone oxidoreductase 1 (NQO1) and regulates the expression of these genes. Increasing the levels of antioxidant enzymes without causing oxidative stress can potentially counteract degeneration and may be therapeutically useful. On a genomic scale, ARE activators can be identified by screening expression cDNA libraries in a high-throughput amenable reporter gene assay. Further validation of putative hits requires testing of cDNAs for their ability to upregulate the expression of endogenous ARE-regulated genes on the transcript and protein levels, and their ability to protect cells from oxidative insults. General screening procedure and subsequent hit validation are discussed in detail.

Key words: Oxidative stress, Genome-wide screen, cDNA library, Antioxidant response element, Reporter gene assay, Neuroprotection, Cytoprotection.

1. Introduction

Oxidative stress contributes to aging and age-related neurodegenerative disorders, such as Parkinson's disease, Alzheimer's disease, and Huntington's disease *(1, 2)*. The ARE, a *cis*-acting enhancer element found in the 5′-flanking regions of numerous

From: *Methods in Molecular Biology, vol. 477: Advanced Protocols in Oxidative Stress I,*
Edited by: D. Armstrong, DOI: 10.1007/978-1-60327-517-0_26, © Humana Press, New York, NY

genes including those encoding phase II detoxification enzymes, mediates the transcriptional activation of downstream genes in cells exposed to oxidative stress *(3)*. The activation of the ARE has neuroprotective effects; oxidative damage induced death of neuroblastoma cells, astrocytes, or neurons is attenuated or prevented *(4, 5, 6)*. Oxidative stress-independent ARE activation could provide a therapeutic approach to neurodegeneration by promoting the beneficial upregulation of neuroprotective enzymes without the detrimental effects of uncontrolled oxidative stress.

It is impossible to rapidly identify positive ARE regulators in traditional small-scale or one-gene-at-a-time studies. Here, we describe an approach to find genes involved in protection from oxidative stress by genome-wide high-throughput screening of thousands of expression cDNAs spatially arrayed in 384-well plates with an ARE reporter gene assay *(7)*. Putative hits need to be rigorously validated in secondary assays; examples are described in greater detail in this chapter as well.

2. Materials

2.1. Equipment

1. Dispense instrument (e.g., WellMate from Matrix Technologies Corp., Hudson, NH).
2. SpectraMax M5 or CLIPR luminescence plate reader (Molecular Devices, Sunnyvale, CA).
3. Real-time PCR system model 7300 (Applied Biosystems, Foster City, CA).
4. Gel Logic 2200 imaging system (Kodak, Rochester, NY).

2.2. Cell Culture and Transfection

1. IMR-32 neuroblastoma cells (ATCC, Manassas, VA).
2. Dulbecco's Modified Eagle Medium (DMEM) (1X) liquid (high glucose) (Invitrogen, Carlsbad, CA), supplemented with 10% fetal bovine serum (FBS, HyClone, Logan, UT) and 1% Antibiotic-Antimycotic (Invitrogen).
3. OPTI-MEM® I Reduced Serum Medium (Invitrogen).
4. FuGENE 6 (Roche, Indianapolis, IN).
5. siLentFect (Bio-Rad, Hercules, CA).
6. Versene 1:5000, liquid (Invitrogen).
7. CMV-*GFP* (green fluorescent protein) plasmid (Clontech, Palo Alto, CA).

2.3. Genome-wide Screening for ARE Activators

1. cDNA library can be obtained from Origene Technologies (Rockville, MD) or Mammalian Gene Collection (MGC; e.g., ATCC). Alternatively, pre-spotted (Assay-Ready) MGC cDNA screening sets are available from Open Biosystems (Huntsville, AL).
2. *NQO1*-ARE luciferase reporter construct *(8)*.
3. Expression cDNA encoding constitutively active phosphatidylinositol 3-kinase (PI3K), PI3K* *(9)* (*see* **Note 1**).
4. Bright-Glo™ luciferase assay system (Promega, Madison, WI).
5. Dispense instrument (WellMate from Matrix Technologies Corp., Hudson, NH).
6. Luminescence plate reader (SpectraMax M5 or CLIPR from Molecular Devices, Sunnyvale, CA).
7. 384-well solid white flat bottom tissue culture (TC)-treated microplates (Corning Incorporated, Corning, NY).

2.4. Hit Confirmation

1. β-Galactosidase plasmid (e.g., actin-*lacZ*).
2. *N*-acetyl cysteine (Sigma, St. Louis, MO).
3. Gal-Screen® assay system (Applied Biosystems, Foster City, CA).
4. Bright-Glo™ luciferase assay system (Promega).
5. 24-well TC-treated plates (Nunc, Rochester, NY).
6. 96-well solid white flat bottom TC-treated microplates (Corning Incorporated).
7. Vector control (e.g., pCMV6-XL4; OriGene Technologies).

2.5. Real-Time PCR

1. RNeasy Kit (Qiagen, Valencia, CA).
2. Oligo(dT)$_{12-18}$ Primer (Invitrogen).
3. 10 mM dNTP Mix, PCR Grade (Invitrogen).
4. RNaseOUT™ Ribonuclease Inhibitor (Invitrogen).
5. SuperScript™ II Reverse Transcriptase (Invitrogen).
6. Human *NQO1* and *GAPDH* TaqMan® Gene Expression Assays (Applied Biosystems).
7. TaqMan® Universal PCR Master Mix (Applied Biosystems).
8. DEPC-treated sterile water.
9. 7300 Real-Time PCR System (Applied Biosystems).
10. 6-well TC-treated plates (Nunc).

2.6. Immunoblot Analysis

Buffers (*see* **Note 2**)

1. Protein lysis buffer: PhosphoSafe extraction reagent (Novagen, Madison, WI).
2. Sample buffer: NuPAGE® LDS Sample Buffer (Invitrogen).

3. Running buffer: NuPAGE® MES SDS Running Buffer (Invitrogen).
4. Transfer Buffer: NuPAGE® Transfer Buffer (Invitrogen) with 10–20% Methanol (Fisher Scientific, Fair Lawn, NJ).
5. Wash buffer: Tris Buffered Saline, with Tween® 20 (TBS-T, pH 8.0, Sigma).
6. Blocking buffer: 5% (w/v) BSA (Sigma) or 5% nonfat dry milk in TBS-T.

PAGE Gels

1. NuPAGE® Novex® 4–12% Bis-Tris Mini Gels (Invitrogen).

Antibodies

1. Primary antibodies: anti-NQO1 (Abcam, Cambridge, MA), anti-β-actin (Cell Signaling, Beverly, MA).
2. Secondary antibodies: Anti-rabbit IgG, horseradish peroxidase (HRP)-linked antibody (Cell Signaling); Anti-goat IgG, HRP-linked antibody (Santa Cruz Biotechnology, Santa Cruz, CA).

Detection

1. SuperSignal West Femto Maximum Sensitivity Substrate (Pierce, Rockford, IL).
2. X-ray film (Kodak) or Imaging System (e.g., Gel Logic 2200; Kodak).

Others

1. BCA protein assay kit (Pierce), PVDF membrane (Millipore, Billerica, MA or Invitrogen), filter paper (Pierce), cell scrapers (Nunc), 6-well TC-treated plates (Nunc).

2.7. Protection Assay

1. Hydrogen Peroxide (H_2O_2, Sigma).
2. CellTiter-Glo® Luminescent Cell Viability Assay (Promega).
3. 24-well TC-treated plates (Nunc).
4. 96-well solid white flat bottom TC-treated microplates (Corning Incorporated).

3. Methods

In genome-wide screens, each plate should contain a positive and negative control. Constitutively active phosphatidylinositol 3-kinase (PI3K), PI3K* *(9)*, which is known to activate the ARE in IMR-32 cells *(10)*, can be used as a positive control; empty vector is a suitable negative control. However, instead of relying on a single well (or even quadruplicate wells) as the negative control, the average luminescence readout of the plate can be used as background control in genome-wide screening (**Fig. 26.1**). This is acceptable since only relatively few cDNAs will exert an effect

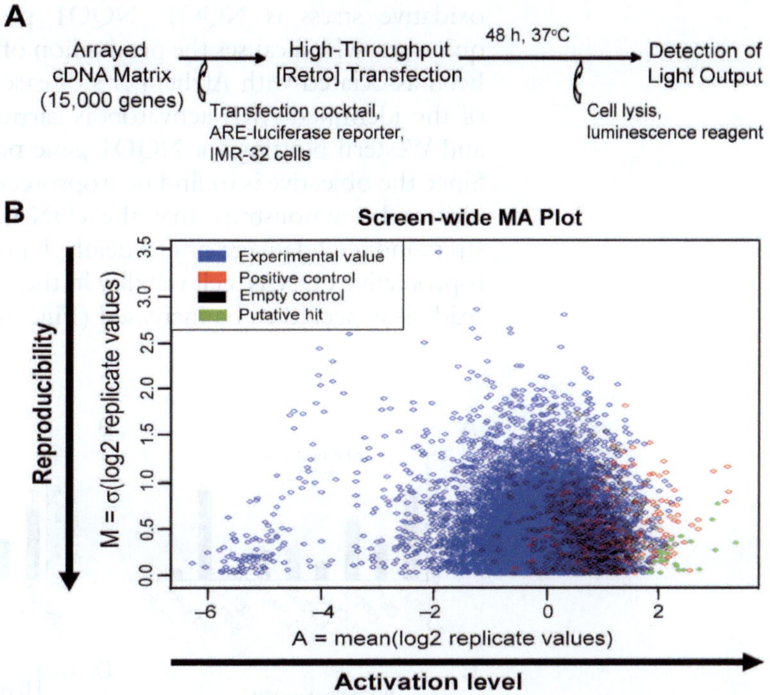

Fig. 26.1. Genome-wide cDNA overexpression screen for ARE activators. (**A**) General high-throughput screening procedure. Approximately 15,000 expression cDNAs, normalized and arrayed in 384-well plates, were transfected into IMR-32 human neuroblastoma cells along with an ARE-luciferase reporter construct. Following 48 h incubation, luciferase activity was assessed by measuring luminescence output per well. (**B**) Screen-wide MA plot. The screen was carried out in duplicate and M (a measure of screen-to-screen variation; σ = standard deviation) plotted as a function of A (a measure of the mean ARE activation from both screens). The cDNAs that strongly activated the ARE in a reproducible manner were investigated further (indicated in green; lower right corner). (Copyright 2007 National Academy of Sciences, U.S.A.; reproduced from reference 7 with permission).

in the ARE reporter gene assay when using a random cDNA collection. However, empty vector (or a cDNA that has no effect) should be used as negative control for confirmation and validation studies.

ARE activators, identified by high-throughput genome-wide cDNA library screening, should be subjected to thorough validation, accounting for well-to-well differences in transfection efficiency by cotransfecting another reporter gene construct that is not responsive to ARE activators (**Fig. 26.2A**). To ensure that the cDNAs do not induce oxidative stress, which would also lead to ARE activation, ARE activity is determined in the presence of excess antioxidant (e.g., *N*-acetyl cysteine). It is then necessary to correlate the reporter assay data with the induction of endogenous target genes. One enzyme that is highly responsive to ARE activation and also plays an important role in the protection from oxidative stress is NQO1. NQO1 prevents the reduction of quinones, which causes the production of radical species; it has also been associated with Alzheimer's disease *(11)*. Further validation of the identified ARE activators is carried out by real-time PCR and Western blotting for NQO1 gene products (**Fig. 26.2B**, **C**). Since the objective is to find neuroprotective cDNAs, one needs to ultimately demonstrate that the cDNAs can attenuate oxidative stress-induced damage or cell death. Upon overexpression of neuroprotective cDNAs, cell viability in the presence of hydrogen peroxide is expected to be increased (**Fig. 26.2D**).

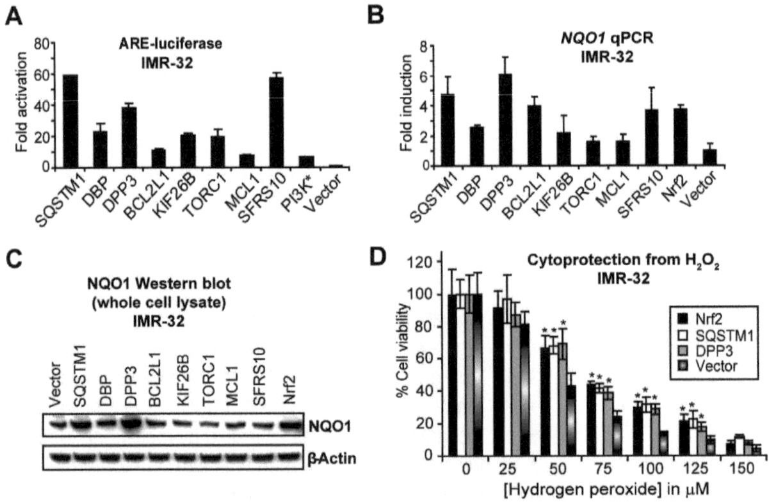

Fig. 26.2. Confirmation and validation of putative screening hits in IMR-32 cells. (**A**) Transcriptional ARE activation in IMR-32 cells. IMR-32 cells were cotransfected with cDNAs, the ARE-luciferase reporter, and actin-*lacZ* for normalization in 24-well plates. Luminescence was detected 48 h later; normalized values are given ($n = 6$). Eight cDNAs showed equal or higher activity than the positive control, PI3K*. (**B**) Effect of cDNA overexpression in IMR-32 cells on *NQO1* transcript levels as analyzed by quantitative real-time PCR (qPCR). The cDNAs were introduced by lipofection and then total RNA was isolated 48 h later, reverse-transcribed to cDNA, and subjected to TaqMan analysis ($n = 3$). Glyceraldehyde-3-phosphate dehydrogenase (*GAPDH*) expression was used as internal control for normalization. Overexpression of several cDNAs increased *NQO1* levels to the same extent as Nrf2 overexpression. (**C**) Induction of NQO1 upon cDNA overexpression in IMR-32 cells as analyzed by Western blot analysis. cDNAs were transfected using lipofection, and proteins were isolated 48 h later, resolved by SDS-PAGE, and subjected to Western blot analysis for NQO1. SQSTM1 and DPP3 induced NQO1 most strongly and to a comparable extent as Nrf2. A representative blot ($n = 4$) is shown. (**D**) Nrf2, SQSTM1, and DPP3 mediate protection from oxidative stress in vitro. In 24-well plate format using siLentFect, IMR-32 cells (180,000) were transfected with corresponding cDNAs and with vector control (0.5 μg). After 48 h, cells were treated with various concentrations of hydrogen peroxide ($n = 4$). Cell viability was assessed 6 h later using the CellTiter-Glo® assay kit (Promega). Overexpression of Nrf2, SQSTM1, or DPP3 resulted in a shift of the IC_{50} for hydrogen peroxide against IMR-32 cells. * Indicates statistical significance compared to vector of $p < 0.05$. (Copyright 2007 National Academy of Sciences, U.S.A.; reproduced with minor modifications from reference 7 with permission).

3.1. Reporter Gene Assay-Based Identification of ARE Activators

3.1.1. High-Throughput Genome-Wide Screening

1. cDNAs are arrayed in 384-well plates (**Fig. 26.1A**); each well contains pre-spotted 62.5 ng of cDNA (*see* **Note 3**). The plates are stored at −80°C.
2. The cDNA-containing plates and Opti-MEM are equilibrated to room temperature (*see* **Note 4**).
3. Transfection mixture (*see* **Note 5**) is prepared based on the ratio of 3 (μL, FuGENE 6) to 1 (μg, cDNA and reporter gene). In a sterile bottle, Opti-MEM is added first, then FuGENE 6 is added, gently mixed by inversion (~five times), and then ARE-luciferase reporter construct is added, again followed by repeated inversion to allow gentle mixing (*see* **Notes 6** and 7). For one 384-well plate, the transfection mixture contains the following: Opti-MEM (7.68 mL; 20 μL/well), ARE-luciferase reporter construct (19.2 μg; 50 ng/well), FuGENE 6 (129.6 μL; 0.3375 μL/well).
4. 20 μL of transfection mixture is dispensed into each well and incubated for 30 min (*see* **Note 8**).
5. During the incubation period, cells are prepared (*see* **Note 9**). The medium is carefully removed from the tissue culture dishes by aspiration and 2 mL of Versene is added to each 10-cm dish to detach IMR-32 cells. The cells are spun down and resuspended in 20% FBS-containing DMEM at a density of 400,000 cells/mL.
6. Around 20 μL of the resuspended cells is dispensed into each well (8000 cells/well).
7. The cells are incubated in a cell culture incubator (37°C, 5% CO_2) for 48 h (**Fig. 26.1A**).
8. Detection reagent (Bright-Glo) is prepared immediately before use by mixing the lysis buffer and luciferase substrate and equilibrated to room temperature. Assay plates are equilibrated to room temperature (~10 min).
9. Around 40 μL of Bright-Glo is dispensed into each well.
10. The plate is mixed on a plate shaker for ~1 min and the luminescence is read within 5 min after adding Bright-Glo (*see* **Note 10**).
11. The fold activation over plate average is calculated for each well (**Fig. 26.1B**).
12. cDNAs that showed high activation in duplicate assays are selected for confirmation studies.

3.1.2. Hit Confirmation

1. This study is performed in a 24-well plate format.
2. Transfection mixture (*see* **Note 5**) is prepared in a sterile microcentrifuge or Falcon tube (*see* **Note 6**). Each tube (for four wells: two for *N*-acetyl cysteine treatment, two for solvent control) contains the following: Opti-MEM (1 mL; 250 μL/well),

cDNA or CMV empty vector or CMV-*PI3K** (6.24 μg; 1.56 μg/well), ARE-luciferase reporter construct (2.5 μg; 625 ng/well), actin-*lacZ* (800 ng; 200 ng/well; for normalization), and FuGENE 6 (28.62 μL; 7.155 μL/well).

3. The transfection mixture is incubated for 30 min (up to 1 h) at room temperature.
4. The same method as described above (*see* **Section 3.1.1**) is used to prepare cells at a density of 800,000 cells/mL in 20% FBS-containing DMEM (*see* **Note 9**).
5. Resuspended cells (1 mL) are added to each tube and tube content is mixed by inverting or tapping.
6. The mixture is transferred to a 24-well plate (each tube's content is split into four wells).
7. The plate is incubated in a cell culture incubator (37°C, 5% CO_2).
8. 12 h after transfection, cells are treated with excess of antioxidant, *N*-acetyl cysteine (1 mM final concentration), or solvent (DMEM).
9. The cells are incubated in the cell culture incubator for another 36 h.
10. Bright-Glo (0.5 mL; the same volume as the medium volume) is added to half of the wells (*N*-acetyl cysteine treated and solvent treated), well content mixed by pipetting or on a plate shaker, and transferred to a 96-well white solid bottom plate to analyze luciferase activity by luminescence.
11. Gal-Screen system (0.5 mL; the same volume as the medium volume) is added to the other half of the wells. The mixture is homogenized by pipetting or on a plate shaker and incubated for approximately 60–90 min, then transferred to a 96-well white solid bottom plate. Luminescence is read as a measure of β-galactosidase activity, which correlates with transfection efficiency.
12. The ARE activity (=luciferase activity) is normalized for transfection efficiency (=β-galactosidase activity) by determining the ratio of both activities.
13. The cDNAs, which activate the ARE to the same extent with or without antioxidant are used for the validation study (**Fig. 26.2A**).

3.2. Validation Study

1. Overexpression of cDNAs in IMR-32 cells
   1.1 Transfection is performed in 6-well plate format. The transfection mixture (*see* **Note 5**) is prepared in a sterile microcentrifuge or Falcon tube. Each tube (for each well) contains the following: Opti-MEM (1.0 mL), cDNA or CMV empty vector (1.6 μg),

3.2.1. Induction of Endogenous ARE-Regulated Genes on Transcription Level (Real-Time PCR)

CMV-*GFP* (0.4 μg, to monitor transfection efficiency), siLentFect (4.0 μL, based on the ratio of 2:1, *see* **Note 11**). The transfection mixture is incubated for 30 min at room temperature. Cells are prepared at a density of 600,000 cells/mL in 20% FBS-containing DMEM (*see* **Note 9**).

1.2 Resuspended cells (1.0 mL) are added to each tube, and tube content is mixed by inverting or tapping.

1.3 The mixture is transferred to a 6-well plate.

1.4 Cells are incubated in a cell culture incubator (37°C, 5% CO_2) for 48 h.

2. Real-time PCR assay

   2.1 Total RNA extraction – This step should be performed according to the handbook of the RNeasy kit (*see* **Note 12**).

   2.2 First strand cDNA synthesis – First strand cDNA is synthesized by using 2 μg of total RNA according to the manual for SuperScript™ II Reverse Transcriptase (Invitrogen).

   2.3 Real-time PCR

      2.3.1 The PCR reaction mixture is prepared for each sample separately as follows (*see* **Note 6**). For 96-well plate assays, 25 μL reactions are prepared for each well. Each reaction contains either 1.25 μL human *NQO1* or human *GAPDH* (as endogenous control for normalization) TaqMan Gene Expression Assay (20X), 11.25 μL of the mixture of first strand cDNA (10–100 ng) and sterile H_2O, and 12.5 μL of TaqMan Universal PCR Master Mix (2X). Assays are carried out in triplicate.

      2.3.2 Quantitative PCR is monitored on the ABI 7300 system according to the manufacturer's instructions for TaqMan analysis.

      2.3.2 Relative *NQO1* transcript levels in cDNA-transfected cells versus vector control-transfected cells are determined (**Fig. 26.2B**).

3.2.2. Induction of Endogenous ARE-Regulated Genes on Protein Level (Immunoblot Analysis)

1. Overexpression of cDNAs in IMR-32 cells
   The same method as described in **Section 3.2.1** (Real-time PCR) is used.
2. Western blotting

   2.1 Cell lysates (*see* **Notes 12** and **13**) are collected according to the lysis buffer manual.

   2.2 Protein concentrations are measured using a BCA protein assay kit.

2.3 Protein samples (20–50 μg) are separated by gel electrophoresis at 200 V for ~40 min.

2.4 Proteins are transferred from the separating gel to a PVDF membrane at 30 V for 1–2 h at room temperature.

2.5 The PVDF membrane is incubated in 50 mL blocking buffer for 1 h at room temperature on a rocking shaker.

2.6 The membrane is incubated with primary antibody of a 1:500 (anti-NQO1) or 1:1000 (anti-β-actin) dilution in blocking buffer at 4°C overnight on a rocking shaker.

2.7 The primary antibody is then removed and the membrane washed three times for 15 min each with 50 mL TBS-T.

2.8 The membrane is then incubated with secondary antibody of a 1:2000 dilution in blocking buffer at room temperature for 1 h.

2.9 The secondary antibody is removed and the membrane washed three to five times for 15 min each with TBS-T.

2.10 The signal is detected with SuperSignal West Femto Maximum Sensitivity Substrate (or other substrates for HRP) with X-ray film or imaging system (**Fig. 26.2C**).

3.3. Protection Assay

3.3.1. Overexpression of cDNAs in IMR-32 Cells

Overexpression is performed in 24-well plate format. The same transfection method as described above (*see* **Section 3.2.1**) is used. The transfection mixture (for each well) consists of the following: Opti-MEM (250 μL), cDNA or CMV empty vector (0.5 μg), CMV-*GFP* (0.1 μg, to monitor transfection efficiency), and siLentFect (1.2 μL).

3.3.2. Oxidative Stimulation

After transfection, cells are placed into the incubator for 48 h to allow cDNA overexpression and consequent induction of ARE-regulated target genes (*see* **Note 14**). Cells are treated with various concentrations of H_2O_2 for 6 h in quadruplicate (*see* **Note 15**).

3.3.3. Cell Viability Assay

1. CellTiter-Glo® reagent is prepared shortly before use and 500 μL added to each well (*see* **Note 16**).
2. The content is mixed for 2 min on a plate shaker to induce cell lysis.
3. The product mixtures are transferred to 96-well plates (200 μL per well) and plates incubated at room temperature for 10 min to stabilize the luminescence signal.
4. Luminescence is recorded and viability calculated for each well (**Fig. 26.2D**).

4. Notes

1. cDNAs encoding other known ARE activators such as Nrf2 (>200-fold activation) can be used alternatively. However, reliable detection of *PI3K** (5- to 10-fold activation) as a hit would guarantee that other weak activators could be detected as well.
2. Alternatively, buffers can be made according to standard recipes or obtained from other commercial suppliers.
3. cDNA amount can be reduced as desired. For example, assay-ready plates from Open Biosystems contain 35 ng of DNA per well.
4. The assay plates are centrifuged for 1–2 min at 1000 g to make sure that the DNA is on the bottom of the wells. Approximately 30 min are needed to equilibrate cDNA-containing assay plates and Opti-MEM to room temperature.
5. All transfection mixtures can be prepared either as described in this protocol or, alternatively, by pre-adding DNA into the tube, and then adding Opti-MEM followed by transfection reagent. This would not affect the transfection efficiency for IMR-32 cells.
6. Extra amount is prepared due to volume loss occurring between each dispensing or pipetting step.
7. For library screening, plates and dispenser should be kept as sterile as possible and the time calculated carefully. For example, two people may be needed when dealing with larger plate numbers (e.g., 30+). One person may then dispense the transfection mixture into the wells while the second person starts with the preparation of cells.
8. The transfection mixture could be incubated from 20 min to >1 h (depending on cell type and transfection reagent). Usually, 30–45 min is advised.
9. The confluency of the IMR-32 cells mentioned in this protocol should be 50–80%.
10. All the luminescence readings are performed within 5 min after adding Bright-Glo. The half-life of Bright-Glo is approximately 30 min in IMR-32 cells. Other luciferase detection reagents with more stable signal and from different suppliers may be used.
11. To transfect IMR-32 cells, the ratio of FuGENE 6 (μL) to DNA (μg) amount is 3:1. For siLentFect, the ratio is 2:1. The ratio should be optimized before starting the experiment by cotransfecting CMV-*GFP* to monitor transfection efficiency. Transfection efficiency varies depending on cell type and transfection reagent.

12. The rinse step is optional and depends on cell types. To remove medium from IMR-32 cells, the plate should be tilted and residual medium removed by pipetting instead of rinsing with PBS. IMR-32 cells attach only loosely to the plate and may otherwise be washed away easily.
13. Other lysis buffers can be used. PhosphoSafe or other phosphatase inhibitor containing lysis buffers retain the phosphorylation status of the proteins, potentially allowing the investigation of phosphorylation changes upon cDNA overexpression.
14. Optimal time points may be cDNA dependent. Generally 48–72 h should be suitable.
15. Optimal treatment time must be determined empirically. Longer (chronic) exposure times and other oxidative insults such as mitochondrial complex I inhibitors (e.g., rotenone) or complex II inhibitors (e.g., 3-nitropropionic acid) may also be considered.
16. This reagent measures ATP content as an indicator of cell viability. Other cell viability reagents such as MTT (Promega), AlamarBlue (Invitrogen), or sulforhodamine B (Sigma) may also be suitable. Dynamic range, which depends on the combination of cell line and detection reagent, should be considered in the reagent selection.

Acknowledgments

We would like to thank the Evelyn F. and William L. McKnight Brain Institute of the University of Florida and the American Heart Association for funding this research (H.L.) and Prof. Jeffrey A. Johnson for providing the ARE-luciferase reporter gene construct.

References

1. van Muiswinkel, F. L., and Kuiperij, H. B. (2005). The Nrf2-ARE signalling pathway: promising drug target to combat oxidative stress in neurodegenerative disorders. *Curr. Drug Targets CNS Neurol. Disord.* **4**, 267–281.
2. Kensler, T. W., Wakabayashi, N., and Biswal, S. (2007). Cell survival responses to environmental stresses via the Keap1-Nrf2-ARE pathway. *Annu. Rev. Pharmacol. Toxicol.* **47**, 89–116.
3. Nguyen, T., Sherratt, P. J., and Pickett, C. B. (2003). Regulatory mechanisms controlling gene expression mediated by the antioxidant response element. *Annu. Rev. Pharmacol. Toxicol.* **43**, 233–260.
4. Li, J., Lee, J. M., and Johnson, J. A. (2002). Microarray analysis reveals an antioxidant responsive element-driven gene set involved in conferring protection from an oxidative stress-induced apoptosis in IMR-32 cells. *J. Biol. Chem.* **277**, 388–394.
5. Lee, J. M., Calkins, M. J., Chan, K., Kan, Y. W., and Johnson, J. A. (2003). Identification of the NF-E2-related factor-2-dependent genes conferring

protection against oxidative stress in primary cortical astrocytes using oligonucleotide microarray analysis. *J. Biol. Chem.* **278**, 12029–12038.

6. Johnson, D. A., Andrews, G. K., Xu, W., and Johnson, J. A. (2002). Activation of the antioxidant response element in primary cortical neuronal cultures derived from transgenic reporter mice. *J. Neurochem.* **81**, 1233–1241.
7. Liu, Y., Kern, J. T., Walker, J. R., Johnson, J. A., Schultz, P. G., and Luesch, H. (2007). A genomic screen for activators of the antioxidant response element. *Proc. Natl. Acad. Sci. USA* **104**, 5205–5210.
8. Moehlenkamp, J. D., and Johnson, J. A. (1999). Activation of antioxidant/electrophile-responsive elements in IMR-32 human neuroblastoma cells. *Arch. Biochem. Biophys.* **363**, 98–106.
9. Hu, Q., Klippel, A., Muslin, A. J., Fantl, W. J., and Williams, L. T. (1995). Ras-dependent induction of cellular responses by constitutively active phosphatidylinositol-3 kinase. *Science* **268**, 100–102.
10. Lee, J. M., Hanson, J. M., Chu, W. A., and Johnson, J. A. (2001). Phosphatidylinositol 3-kinase, not extracellular signal-regulated kinase, regulates activation of the antioxidant-responsive element in IMR-32 human neuroblastoma cells. *J. Biol. Chem.* **276**, 20011–20016.
11. SantaCruz, K. S., Yazlovitskaya, E., Collins, J., Johnson, J., and DeCarli, C. (2004). Regional NAD(P)H:quinone oxidoreductase activity in Alzheimer's disease. *Neurobiol. Aging* **25**, 63–69.

Chapter 27

Method for Conducting Microarray Study of Oxidative Stress Induced Gene Expression

Sita Subbaram, Juan A. Melendez and Sridar V. Chittur

Abstract

Cellular systems produce reactive oxygen species during the process of metabolism. Oxidative stress results in the activation or repression of many genes in important signaling pathways. DNA microarrays allow for a high throughput evaluation of the changes in gene expression levels in any biological system. In this study, we describe a method to employ gene expression microarrays to study the transcriptional changes in redox-engineered cell lines that will overexpress MnSOD and/or catalase in the mitochondria.

Key words: Oxidative stress, gene regulation, microarray expression profiling, microarray, array, biomarker.

1. Introduction

All aerobic organisms produce reactive oxygen species as a byproduct of metabolism. In order to counteract their harmful effects, aerobic cells have developed a variety of antioxidant defenses. One of the major free radicals generated from the mitochondrial respiratory chain is superoxide anion (O_2^-). During the process of oxidative phosphorylation, electrons leak during the transfer reaction between complexes I and II to complex III and react with molecular oxygen to form O_2^-. Superoxide can spontaneously dismutate to hydrogen peroxide (H_2O_2), but in the mitochondria manganese superoxide dismutase (MnSOD) carries out this reaction very efficiently resulting in formation of H_2O_2 and oxygen. H_2O_2 is an electron neutral reactive species whose levels are kept in check by

From: *Methods in Molecular Biology, vol. 477: Advanced Protocols in Oxidative Stress I,*
Edited by: D. Armstrong, DOI: 10.1007/978-1-60327-517-0_27, © Humana Press, New York, NY

the activity of either glutathione peroxidase in the mitochondria or catalase in the peroxisomes. In addition to MnSOD (Sod2), superoxide dismutase occurs as a cytosolic CuZn containing (Sodl) and an extracellular isoform (Sod3). These antioxidant enzymes are responsible for keeping the levels of these highly reactive oxygen species (ROS) at a minimum and restrict cellular damage. However, decreases in the levels of antioxidant enzymes or increase in free radical formation as seen in many disease conditions can shift the redox balance toward a more pro-oxidative state resulting in increased oxidative damage and cell death.

The role of ROS has now evolved from a cytotoxic cellular byproduct to that of a second messenger, signaling molecule, and mitogen *(1)*. Their role in promoting various signaling pathways and genes is in the process of being elucidated. Experiments using externally added H_2O_2 have been commonly used to evaluate the contribution of ROS to various signaling responses. However, the cell rarely sees H_2O_2 at such high micromolar conditions and leads to either senescence or death. We have designed a series of redox-engineered cell lines that will overexpress MnSOD alone or in combination with catalase in the mitochondria. These cell lines have been generated using HT1080 fibrosarcoma cells and have been extensively characterized *(2)*. Using many redox-sensitive fluorophores as well as biochemical approaches, it has been demonstrated that overexpression of MnSOD in the mitochondria results in an increase in H_2O_2 levels two fold over that of control cells transfected with empty vector (54 pM vs 90 pM) *(3)*. Co-transfection with catalase reverses the increase in the steady state production of H_2O_2 to that of control cells. These levels are well within the range of H_2O_2 generated physiologically in response to receptor signaling, ionizing radiation, or other pathological conditions.

2. Materials

The reagents described here have been used with mammalian cells and tissue samples. However, we imagine that these methods could be easily adapted or modified to samples from other organisms. Throughout these methods, we suggest that all cell culture techniques use cell-culture grade media. Reagents for biochemical manipulations of nucleic acids and proteins should be purchased as DNase-free, RNase-free, pyrogen-free, and 0.2 μm filtered.

2.1. Equipment

1. Hemocytometer
2. BD FACS Calibur flow cytometer
3. Bioanalyzer 2100 system, Agilent Technologies, Santa Clara, California
4. Nanodrop ND-1000 UV-Vis spectrophotometer
5. Affymetrix Genechip system
6. BioRad iCycler iQ5 system

2.2. Materials for Cell Culture

1. HT1080 fibrosarcoma cells transfected with empty vector (Cmv or control cell line)
2. HT1080 fibrosarcoma cells transfected with vector containing MnSOD (Sod2)
3. HT1080 fibrosarcoma cells co-transfected with vectors containing MnSOD as well as catalase (Sod2CAT)
4. HT1080 fibrosarcoma cells transfected with vector containing catalase (CmvCAT)
5. Phosphate buffered saline (PBS) – 137 mM Nacl, 2.7 mM KCl, 4.3 mM $Na_2HPO_4 \cdot 7H_2O$, 1.4 mM KH_2PO_4, pH 7.4

2.3. Materials for Characterization of Oxidant Stress in the Above Mentioned Cell Lines

1. 2′,7′-dichlorodihydrofluorescein diacetate (H_2DCFDA, Molecular Probes, catalog# D-399) (*see* **Note 1**)
2. DMSO
3. Hanks buffered salt solution (Cellgro, catalog# 21–023) containing added glucose 4.7 g/l
4. Flow cytometer compatible tubes
5. BD FACS Calibur Flow cytometer
6. PBS-EDTA (Phosphate buffered saline – 137 mM Nacl, 2.7 mM KCl, 4.3 mM $Na_2HPO_4 \cdot 7H_2O$, 1.4 mM KH_2PO_4, 1 mM EDTA).
7. Hemocytometer

2.4. Materials for RNA Isolation

1. All tips, tubes, and reagent bottles must be DNAse and RNase free (*see* **Note 2**)
2. Tri-reagent (Molecular Research Inc, cat#TR118) or TRIzol (Invitrogen cat#15586-026)
3. 1-bromo-3-chloropropane (Molecular Research Inc, cat# BP151) or chloroform
4. Isopropanol
5. We recommend the use of nuclease-free water (Ambion cat# AM9932) to prepare all buffers and solutions
6. RNeasy mini RNA isolation kit (Qiagen cat# 74104)

7. DNase I (Ambion cat#AM2222)
8. RNase Zap (Ambion cat # AM9780)

2.5. Materials for RNA QC and Microarray Experiment

1. Agilent Bioanalyzer 2100 system
2. Agilent RNA 6000 Nanokit (cat#5067-1511)
3. Nanodrop ND-1000 UV-Vis spectrophotometer
4. Affymetrix Genechip® System
5. Affymetrix One Cycle target labeling and Control reagents (cat#900493)
6. Affymetrix GeneChip® human genome U133 Plus 2.0 arrays (cat# 900466)

2.6. Reagents for SYBR Green Quantitative PCR

1. SuperScript® III First-Strand Synthesis System for RT-PCR (Invitrogen cat# 18080-151)
2. Forward and Reverse Primers (100 μM stock)
3. iQ SYBR Green Supermix (Biorad cat# 170-8882) (*see* **Note 3**)
4. BioRad iCycler iQ5 system

3. Methods

The present cell culture system is only presented as an example. This method is highly adaptable to any other biological system one may be using to evaluate oxidative stress. One can also add varying concentrations of oxidants such as H_2O_2 or oxidant-generating compounds such as N,N′-Dimethyl-4,4′-bipyridinium dichloride (paraquat) for superoxide or Diethylenetriamine NONOate (Deta NONOate) for NO to different cellular systems. These compounds have been widely used in the literature to study the role of free radicals in cell biology.

3.1. Measuring Oxidant Levels by Flow Cytometry

1. Cells were harvested after treatment with PBS-EDTA (*see* **Note 4**) and washed once with PBS and counted using a hemocytometer (*see* **Note 5**).
2. The cells were then resuspended in Hanks Buffered Salt solution to make a final count of 5×10^5 cells/ml.
3. H_2DCFDA was added to make a final concentration of 10 μM.
4. Rate of oxidation of the dye was monitored by FACS analysis using wavelengths between 580 and 620 nm for a period of 300 s (*see* **Note 6**).

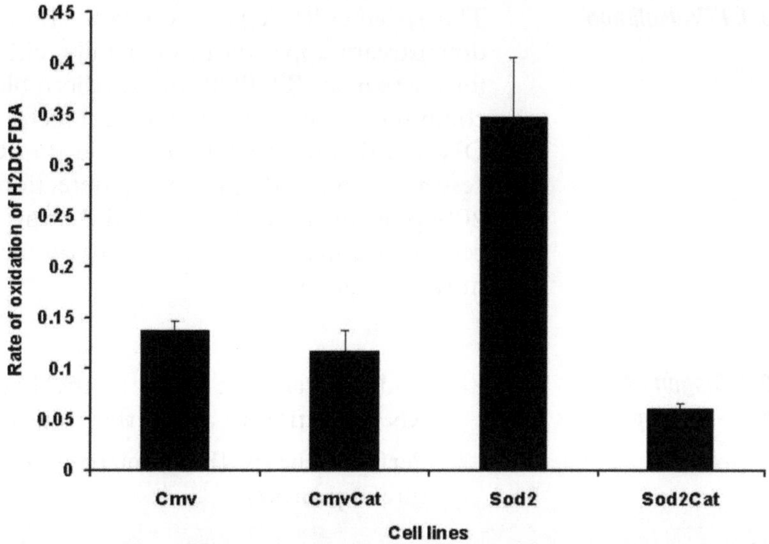

Fig. 27.1. Rate of oxidation of H2DCFDA of various redox-engineered cell lines by FACS analysis.

5. **Figure 27.1** depicts a representative measurement of rate of oxidation of H2DCFDA by FACS analysis. Here we see that Sod2 cells overexpressing MnSOD2 have a higher rate of oxidation of H2DCFDA, which is primarily oxidized by H_2O_2. Upon co-expression of catalase, we see that Sod2Cat cells have a lower level of oxidation of the dye as compared to Sod2 cells.

3.2. Cell Culture and Harvesting of Cells for RNA Isolation

1. All indicated cell lines are maintained in 25-cm^2 or 75-cm^2 flasks in a 37°C incubator under 5% CO_2, in MEM containing 10% fetal calf serum, 1000 U/ml penicillin, 500 μg/ml streptomycin, and 1 mg/ml Neomycin (Gemini) for the Cmv and Sod2cell lines and additional 50 μg/μl zeocin (Invitrogen) for the Sod2CAT cell lines, which require double selection.
2. Wash the cells with PBS to remove any residual media prior to harvesting.
3. Add 3 ml Tri-reagent or TRIzol directly to the cells in the flask. DO NOT trypsinize the cells prior to treatment with tri-reagent or TRIzol. (*see* **Note** 7). Move the TRIzol around the flask and gently tap to slough off all attached cells. Pipette into a clean tube and store at −20°C till further use.

3.3. RNA Isolation

The specific RNA isolation you choose will depend on your downstream application. Generally, either method is acceptable for microarray, RT-PCR, or Northern blotting. The Qiagen spin-column cleanup offers the advantage of performing an optional DNase I digestion while purifying the RNA so that further processing is avoided. However, detection of RNA molecules of 200bp or smaller will be limited if using the Qiagen cleanup procedure and hence not advised if you intend to use the RNA for miRNA analysis.

3.4. Qiagen RNEasy Mini-Cleanup

1. Add nuclease free water to the 150–200 μl RNA from the RNA isolation to adjust the volume to 200 μl (*see* **Note 8**).
2. Perform the RNEasy micro-cleanup as per the manufacturer's protocol.

3.5. Assessment of RNA Quality

1. Using a NanoDrop® spectrophotometer, measure the optical absorbance characteristics of the sample. (*see* **Note 9**). The A260/A280 as well as the A260/A230 ratio will ideally be close to 2.0, signifying the purification of nucleic acids away from protein and other organics, respectively. If either ratio is lower than 1.6, expect problems with downstream applications of the RNA (*see* **Note 10**).
2. Performance of a NanoChip assay using Agilent's BioAnalyzer allows for measurement of the molecular weight profile of the isolated RNA. In this way, you may evaluate the 28S/18S ratio measurements. A total RNA ratio between 1.8 and 2.0 is desirable; however, ratios 1.6–1.8 may be acceptable. An RNA Integrity number (RIN) score should be between 7 and 10 if the samples are to be used in a microarray or QPCR experiment downstream. (*see* **Note 11**).
3. **Figure 27.2** depicts a BioAnalyzer output trace displayed as an electropherogram. An insilico gel like view of corresponding total RNA samples is also shown

3.6. Expression Analysis of mRNA from Cells

1. We have identified many gene targets of oxidative stress using cDNA/genomic arrays and have found that the Affymetrix GeneChip® array platform and the Eppendorf Dual Chip cDNA Expression Array platform are excellent for conducting these analyses. However, other array platforms have worked with varying success.
2. If gene expression analysis is performed using custom spotted glass arrays that utilize Cy3 and Cy5 labeling, we typically increase the amount of starting material to 10 μg.

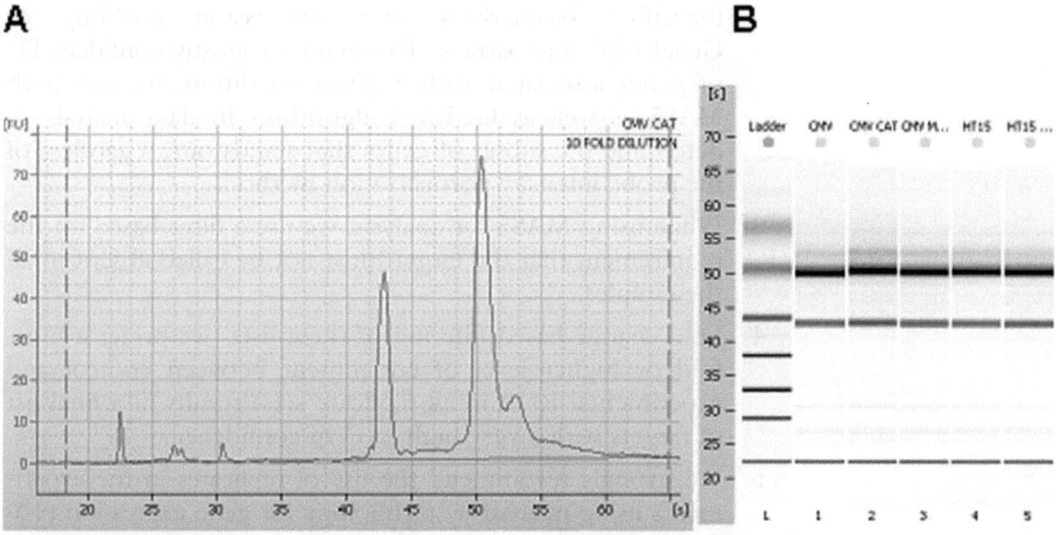

Fig. 27.2. (**A**) Bioanalyzer electropherogram shows the distribution of the mRNA. The ratio of the 28S/18S subunits is used as a measure for QC of total RNA. (**B**) A gel-like image of the samples shows the traditional 2 band (28S, 18S) profile as seen in formaldehyde gels.

3.7. Synthesis of Labeled cRNA and Microarray Hybridization

1. We have successfully used 10 ng–2 μg of total RNA for microarray studies using the Affymetrix GeneChip® plaftform. When starting with less than 1 μg total RNA (e.g. LCM samples or flow sorted samples), we recommend a two-cycle amplification and labeling protocol as described elsewhere *(4)*.
2. We recommend using 1–2 μg of good quality total RNA material if available. The RNA is first converted to T7-oligo (dT) primed double stranded cDNA using the Affymetrix one-cycle cDNA synthesis kit as per the manufacturer's protocol. (*see* **Note 12**).
3. This is then converted to complimentary RNA (cRNA) by in vitro transcription using the Affymetrix IVT labeling kit wherein biotinylated UTPs are incorporated resulting in labeled, amplified cRNA.
4. The cRNA (15 μg) is fragmented using a metal-induced hydrolysis step to obtain 25–200bp fragments that are then hybridized to the GeneChip® U133 plus 2.0 arrays as per manufacturer's protocol. (*see* **Notes 13** and **14**).

3.8. Analysis of Gene Expression Data Obtained Using GeneChip® Microarrays

1. Traditional microarray analysis methods present a steep learning curve for the average user *(5, 6, 7)*. The problem resides primarily in the normalization techniques used to distribute the signal intensities on the array. We have developed a strategy to overcome this problem and have successfully

identified biomarkers from expression profiling of GeneChip® microarrays. To obtain a robustly confident list of genes associated with a given condition, we use both MAS5 as well as GCRMA algorithms in data analysis to determine the subset of genes that stands out, regardless of the probe intensity normalization method.

3. While using MAS5 for analysis, we use a filter based on the requirement that the target must not be called absent in all the samples.
4. Fold change based filtering of these lists (we prefer a four-fold or higher level of enrichment between control and experimental RNA in all replicate sets) results in a final list of targets with a very high level of confidence.
5. We strongly recommend the use of replicates in the experiments using microarray technology for gene expression profiling. While we realize that these experiments can be cost prohibitive, confidence in that data from microarray experiments requires the use of at least 2–3 biological replicates. While generating preliminary data, one could resort to pooling of multiple samples to neutralize the biological variance; however, this could lead to loss of meaningful important data. When using 2-color spotted arrays, an addition requirement for technical replicates that evaluate dye bias is necessary.
6. The output of the Affymetrix protocols results in a high level of signal enrichment over traditional methods like Northern Blots. Consequently, the protocol creates the impression that a large number of targets are being expressed by the "negative control."
7. In our analysis, we use the "negative" targets as an estimate of our background noise, subtracting their signals from the true signals to adjust the confidence of high-background probes. For instance, in the example discussed here, we subtract the signal values for the vector control samples (Cmv and CmvCAT) from those of the experimental (Sod2and Sod2CAT) to remove background noise.

3.9. Quantitative PCR Validation of Microarray Data

1. We recommend that all microarray data be validated by independent techniques such as Northern blots (time consuming), RT-PCR (cheap, but less quantitative), or QPCR (SYBR green or Taqman). Alternately, one can run the same RNA on a different microarray platform for high-throughput validation.
2. In this study, we used SYBR green assays to validate the expression profiles of genes that were identified from the microarray experiment.

Table 27.1
Sequences for primers used for quantitative PCR validation shown in Fig. 27.3

Primer	Sequence
MMP-1 mRNA forward	AGTGACTGGGAAACCAGATGCTGA
MMP-1 mRNA reverse	GCTCTTGGCAAATCTGGCGTGTAA
IL-1 alpha mRNA forward	AATGATCAGTACCTCACGGCTGCT
IL-1 alpha mRNA reverse	TGGTCTTCATCTTGGGCAGTCACA
Actin mRNA forward	ACCAACTGGGACGACATGGAGAAA
Actin mRNA reverse	TAGCACAGCCTGGATAGCAACGTA

3. The RNA (1 μg) is first converted to single strand DNA using the manufacturer's protocol for the Superscript III First strand cDNA synthesis kit.
4. Two microliters of the reaction mix from the cDNA synthesis step was used as template in the qPCR reaction along with 2X SYBR Green Mastermix and primers (400 nM final concentration) (*see* **Table 27.1**) using the following thermocycler parameters:
   Cycle 1, (1x) step 1 94°C for 3 min
   Cycle 2, (40x) – step 1 94°C for 30 s, step 2 62°C for 30 s, step 3 72°C for 30 s
   Cycle 3, (1x) step 1 94°C for 1 min
   Cycle 4, (1x) step 1 62°C for 1 min
   Cycle 5, (71x) step 1 57°C for 10 s, increasing set point temperature by 0.5°C for melt curve analysis
5. **Figure 27.3** shows that MMP1 and IL-1 alpha expression was upregulated in the Sod2 overexpressing cells as compared to the others. Actin expression was used as an internal control.

4. Notes

1. H_2DCFDA is made up in DMSO as a vehicle; the dye is extremely light sensitive and should be stored under inert Nitrogen gas in a dark container. After the solution is made up, it should be generally used on the same day as preparation

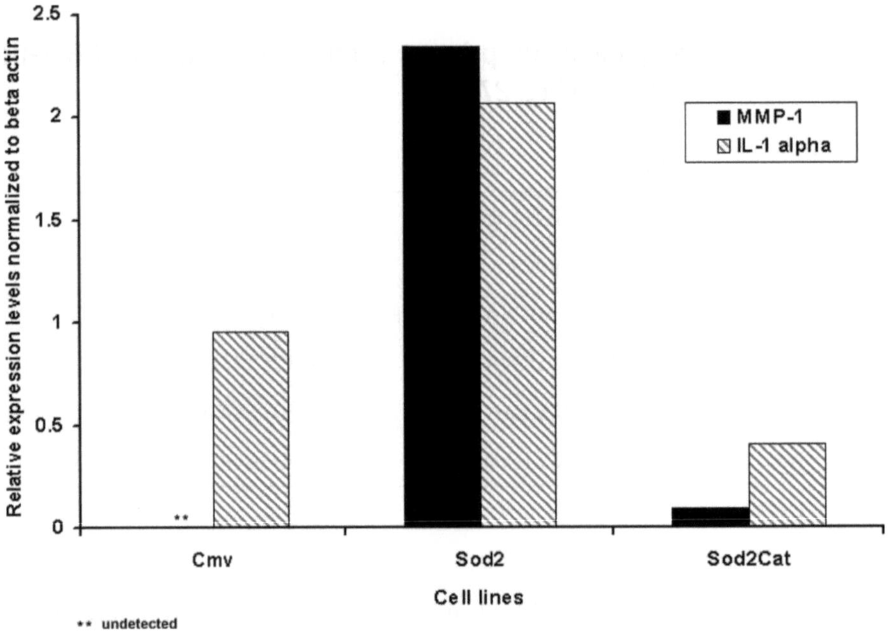

Fig. 27.3. Quantitative PCR validation of microarray data. MMP1 and IL-1α are upregulated in SOD2 cells as compared to other cell types.

and be covered in foil to protect it from light until addition to the cells before analysis.

2. All instruments, glassware, and plastic-ware that touch cells or cell lysates should be certified DNase-free and RNase-free or should be pre-washed with RNase Zap (Ambion, cat. #9780; 9782) or RNase Away (Molecular BioProducts cat. #7001) followed by DEPC water and allowed to air dry.
3. Please check the specifications of the instrument being used for quantitative PCR. Instruments like the iCycler iQ do not use ROX as an internal reference but use fluorescein for dynamic well factor collection. If using ABI 7900 Sequence detection system or other instruments that require ROX as an internal reference dye for fluorescent signal normalization and correction of well-to-well optical variations, use the iTaq SYBR Green supermix with ROX (cat#172-5850)
4. These cells can be lifted by PBS-EDTA because they require calcium to maintain cell-cell contacts. If your cell line cannot be lifted with PBS-EDTA, scraping them gently with a cell scraper can be used to make a cell suspension.
5. Cells should be counted and equal number of cells should be used to account for variability in dye uptake due to difference in cell number.

6. Cell suspensions should be placed in ice prior to addition of dye and measurement by FACS.
7. The number of cells required for each microarray experiment can vary from cell type to cell type. Typically, we utilize a 25 or 75 cm^2 flask of confluent cells per condition. This corresponds to about 2–10 $\times$ 10^6 cells and provides enough material for both the microarray experiment as well as other validation and QC experiments.
8. If using this RNA for any miRNA analysis, AVOID the Qiagen cleanup step since it results in loss of small RNAs.
9. If limited in the amount of available sample, one can analyze the RNA via NanoDrop® and then recover material to use for BioAnalyzer runs.
10. Ambion and Affymetrix protocols and technical literature (and our experience) suggest that samples failing to meet either (or both) of these criteria may (or will) perform poorly in molecular techniques, which are based on reverse transcription followed by amplification. This is likely due to the interference of protein, carbohydrate, or phenolic contaminants on the reverse transcription process.
11. The Agilent BioAnalyzer is a preferable substitute to MOPS-formaldehyde agarose gel analysis due to the reduced sample required, increased sensitivity, and reduced exposure to toxic reagents.
12. The Agilent 2100 Expert software provides a RIN or RNA integrity number *(8)* for the RNA nano and pico assays (series II). It is recommended that this RIN number be between 7 and 10 if the RNA sample is to be used in a microarray experiment. We generally use the RIN number as a secondary QC criteria along with 260/280, 260/230, and 28S/18S ratios.
13. All the reagents for this protocol are supplied in the Affymetrix one-cycle cDNA synthesis kit and IVT kit. However, if needed, some of these reagents (5X First Strand Buffer, 0.1 M DTT, 10 Mm dNTP Mix, Superscript II, 5X Second Strand Buffer, E. coli DNA Ligase, E. coli DNA Polymerase) can be substituted with those from Invitrogen. It is recommended that polyA RNA controls be spiked into the starting RNA samples since this will allow to QC for any degradation occurring during the protocol. The signals from these spikes can also be used for normalization.
14. It is recommended that hybridization controls be prepared from a master mix. The signal from the controls (bioB, bioC, bioD, and Cre) can be used to qualitatively compare chips being hybridized over time.

Acknowledgments

We wish to thank Jaya Dasgupta and David Frank for suggestions and technical support.

References

1. Sundaresan M, Yu Z, Ferrans VJ, Irani K and Finkel T. Requirement for generation of H_2O_2 for platelet-derived growth factor signal transduction. Science 1995;270:296–299.
2. Rodriguez AM, Carrico PM, Mazurkiewicz JE, Melendez JA. Mitochondrial or cytosolic catalase reverses the MnSOD-dependent inhibition of proliferation by enhancing respiratory chain activity, net ATP production, and decreasing the steady state levels of H(2)O(2). Free Radic Biol Med 2000;29:801–813.
3. Dasgupta J, Subbaram S, Connor KM, et al. Manganese superoxide dismutase protects from TNF-alpha-induced apoptosis by increasing the steady-state production of H2O2. Antioxid Redox Signal 2006;8:1295–1305.
4. Affymetrix. GeneChip Expression Analysis Technical Manual, 701025 Rev. 6. http://www.affymetrix.com/support/technical/manual/expression_manual.affx
5. Pepper SD, Saunders EK, Edwards LE, Wilson CL, Miller CJ. The utility of MAS5 expression summary and detection call algorithms. BMC Bioinformatics 2007;8:273.
6. Wu Z, Irizarry RA, Gentleman R, Martinez-Murillo F, Spencer F. A model-based background adjustment for oligonucleotide expression arrays. J Am Stat Assoc 2004;99:909–917.
7. Millenaar FF, Okyere J, May ST, van Zanten M, Voesenek LA, Peeters AJ. How to decide? Different methods of calculating gene expression from short oligonucleotide array data will give different results. BMC Bioinformatics 2006;7:137.
8. Schroeder A, Mueller O, Stocker S, et al. The RIN: An RNA integrity number for assigning integrity values to RNA measurements. BMC Mol Biol 2006;7:3.

Chapter 28

A Systems Approach Demonstrating Sphingolipid-Dependent Transcription in Stress Responses

Alan J. Wilder and L. Ashley Cowart

Abstract

Microarray hybridization allows genome-wide screening of changes in mRNA levels under stress conditions. In *Saccharomyces cerevisiae*, this approach has demonstrated that responses to heat stress, oxidative stress, nutrient deprivation, and other stress signals are highly overlapping and mRNA levels of a core group of genes, termed 'Environmental Stress Response' (ESR) genes, respond similarly to many stressors. In addition to changes in mRNA levels, stress responses induce wide changes in cell metabolic pathways and metabolite levels. Microarrays coupled with chemical inhibition of these pathways and/or using organisms with genetic mutations in enzymes in the pathways of interest allow determination of the roles of specific metabolites in gene expression. In cases where high-throughput '-omics' strategies are available for determining changes in a spectrum of metabolites, these datasets can be integrated with gene expression data to obtain a systems view of regulations and functions of a given pathway. We have used these approaches to determine the regulation and functions of sphingolipid synthesis in *Saccharomyces cerevisiae*. Microarray hybridization and sphingolipidomic analysis experiments were performed on two yeast strains bearing mutations in enzymes of sphingolipid metabolism (and their respective parental strains), under normal conditions and during heat stress. These strategies have revealed diverse roles for sphingolipids in regulating stress response genes, and moreover, could be applied to numerous biological systems and thus provide a method to elucidate activities for a vast array of biomolecules, the metabolic pathways by which they are generated, and their cellular functions.

Key words: Microarray, Genomic, Sphingolipid, Stress responses, Systems biology, lipidomics.

1. Introduction

Sphingolipids comprise a large group of lipid mediators important in eukaryotic cell biology. In mammalian cells, sphingolipids regulate diverse processes including cell division, differentiation,

From: *Methods in Molecular Biology, vol. 477: Advanced Protocols in Oxidative Stress I,*
Edited by: D. Armstrong, DOI: 10.1007/978-1-60327-517-0_28, © Humana Press, New York, NY

apoptosis, migration, and others (reviewed in *[1, 2, 3, 4]*). In *Saccharomyces cerevisiae*, previous studies have demonstrated roles for sphingolipid metabolism in the regulation of the transient cell cycle arrest and nutrient permease degradation associated with the heat stress response, the ability to grow on nonfermentable carbon sources, the regulation of actin cytoskeletal dynamics, and the modulation of endocytosis and key signaling pathways *(4, 5, 6, 7, 8, 9)*. Though much is known regarding the diverse cellular activities of this lipid class, mechanistic insights remain few.

Microarray studies were initiated to investigate potential sphingolipid-mediated gene expression programs. Since previous data demonstrated an increase in sphingolipid synthesis in response to heat stress in yeast *(5, 6, 10)*, it was proposed that the heat stress model could serve as a platform by which to dissect out such programs. To connect gene expression to sphingolipids, the response of yeast strains deleted for specific sphingolipid metabolic pathways to heat stress was evaluated with respect to responses of their nonmutant parental strains.

In this study, we utilized yeast strains with mutations in the LCB1 gene, which encodes a subunit of serine palmitoyl transferase, the initial and rate-limiting step in de novo sphingolipid synthesis, and in the ISC1 gene, which encodes the yeast phosphosphingolipid phospholipase C (**Scheme 28.1**). Deletion of LCB1 is lethal, however, the *lcb1-100* temperature-sensitive strain effectively prevents the *de novo* synthesis

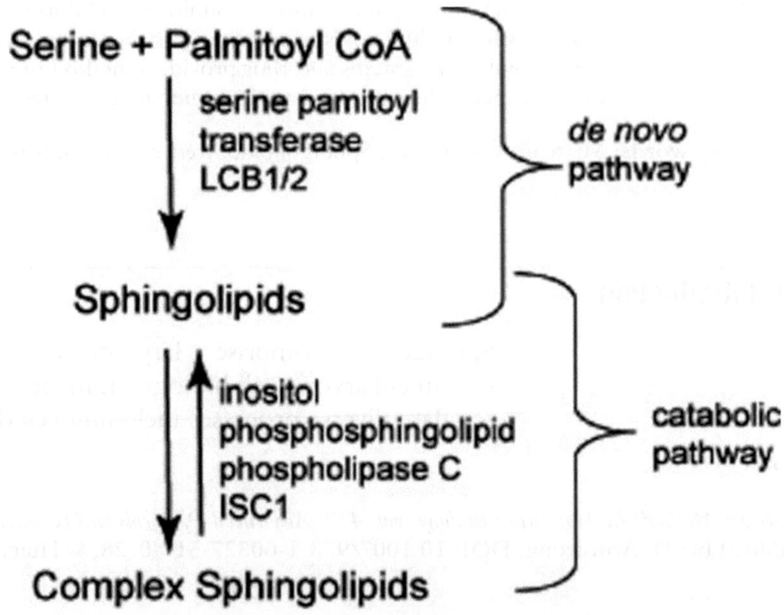

Scheme 28.1. Routes of sphingolipid synthesis in yeast.

of sphingolipids from serine and palmitoyl-CoA when cultures are subjected to heat stress (*i.e.*, non-permissive) temperature, while allowing enzyme function at normal (*i.e.*, permissive) temperature, thus preserving viability *(7)*. The *isc1*Δ prevents catabolic generation of ceramide (and its downstream metabolites) by blocking hydrolysis of complex sphingolipids, which comprise up to 10% of total membrane phospholipids in yeast *(11)*. Thus, the roles of each of these pathways (*de novo* versus catabolic synthesis, **Scheme 28.1**) in alteration of lipid profiles during heat stress were determined by analysis of lipid levels in these mutants. Coupling these findings with microarrays performed under similar experimental conditions allowed the development of novel hypotheses addressing roles of distinct metabolic pathways in regulating gene expression during stress responses. This strategy allows a systems view of stress-induced changes in lipid levels and subsequent alteration of key transcriptomic programs.

2. Materials

2.1. Equipment

1. New Brunswick Inova 3100, water bath shaker, for culturing cells at constant 30°C or at heat stress temperature of 39°C (New Brunswick, Edison, NJ).
2. Beckman Coulter Allegra X-15r centrifuge (Beckman Coulter, Fullerton, CA).
3. Beckman Coulter DU-800 spectrophotometer (Beckman Coulter, Fullerton, CA).
4. Baxter SP Vortex mixer (VWR Scientific Products).
5. Affymetrix Fluidics Station reagents (Affymetrix, La Jolla, CA).
6. Techne Dri-Block DB-3A, heating block for lipid analysis (Techne, Cambridge, UK).
7. The Meyer N-Evap, evaporator for drying lipid extracts under a stream of N_2 gas (Organomation, Berlin, MA).

2.2. Cell Culture and Isolation

1. Difco YPD Broth (BD, Sparks, MD) used for culturing *S. Cerevisiae.*
2. Difco Granulated Agar (BD, Sparks, MD), for agar plates used in preparation of cell cultures.
3. Ethyl Alcohol USP, 190 proof (AAPER, Shelbyville, Kentucky), used with dry ice for ultra-cold ethanol bath to flash freeze cell cultures.
4. 50 mL Centrifuge tubes (USA Scientific, Ocala, FL), for cell culturing.

5. New Brunswick Inova 3100 water bath shaker, for culturing cells at constant 30°C or at heat stress temperature of 39°C (New Brunswick, Edison, NJ)
6. Beckman Coulter Allegra X-15r centrifuge (Beckman Coulter, Fullerton, CA).
7. Beckman Coulter DU-800 spectrophotometer (Beckman Coulter, Fullerton, CA).

2.3. RNA Extraction

1. RNeasy Mini Kit (Qiagen Inc., Valencia, CA), used to extract RNA from isolated cell cultures.
2. Ethyl Alcohol, 140 proof, reagent used in RNA extraction procedure.
3. Baxter SP Vortex mixer (VWR Scientific Products)
4. Acid Phenol:Chloroform, 5:1 solution, pH 5.4 (Ambion, Austin, TX), for RNA extraction.
5. Chloroform 99+%, stabilized with amylenes (Sigma-Aldrich, Boston, MA)
6. RNAse-free water, commonly obtained by treating distilled water with 0.1% diethylpyrocarbonate overnight, followed by autoclaving at 115°C for 15 min.
7. TES solution (10 mM Tris Cl, pH = 7.5, 10 mM EDTA, 0.5%SDS) made with RNAse-free water.
8. 3 M sodium acetate, pH 5.3, made with RNAse-free water.

2.4. Microarray Sample Preparation and Hybridization

1. GeneChip Expression 3′ Amplification Reagents one cycle cDNA synthesis kit (Affymetrix, La Jolla, CA).
2. GeneChip Expression 3′ Amplification Reagants for IVT labeling (Affymetrix, La Jolla, CA).
3. GeneChip sample cleanup module (Affymetrix, La Jolla, CA).
4. Bio-Rad Agarose, for agarose gel electrophoresis confirmation of RNA fragmentation (Bio-Rad, Hercules, CA).
5. YG-S98 yeast genome chip or the updated Yeast 2.0 version (Affymetrix, La Jolla, CA).
6. Affymetrix Fluidics Station and hybridization reagents (Affymetrix, La Jolla, CA).

2.5. Preparation for Lipidomic Analysis

1. Techne Dri-Block DB-3A, heating block for lipid analysis (Techne, Cambridge, UK).
2. 70% isopropanol:ethylacetate:pyridine:25% ammonia (60:20:2:0.5 by volume), for sphingolipid extraction.
3. The Meyer N-Evap, evaporator for drying lipid extracts under a stream of N_2 gas (Organomation, Berlin, MA).

4. Chloroform-Methanol (2:1, v/v), for Bligh-Dyer lipid extraction
5. Methanol A.C.S. certified reagent (Fischer Scientific, Waltham, MA), for Bligh-Dyer lipid extraction
6. Chloroform 99+%, stabilized with amylenes (Sigma-Aldrich, Boston, MA), for Bligh-Dyer lipid extraction.
7. 1 mM NaH_2PO_4, for phosphate standards.
8. Screw-top Borosilicate tube, 18 (x150 mm (Fischer Scientific, Waltham, MA)
9. Ashing buffer: 10 M H2SO4/70%HClO4/H2O, 9:1:40, by volume, for overnight ashing of total phospholipids.
10. 0.9% (w/v) ammonium molybdate, for phosphate determination.
11. 9.0% (w/v) ascorbic acid, for phosphate determination.

3. Methods

A time course appropriate to the specific stress response must be determined (*i.e.*, the experimental design must take into account the time from the initiation of the stress treatment to the time when changes in mRNA levels occur). Early time points are more likely to reveal primary effects of the stressor, while data from later time points may include secondary and tertiary effects; thus, we consider early time points more valuable in initial studies. Moreover, while a greater number of time points provide greater analytical capabilities, it also increases cost and difficulty of data analysis. Therefore, selection of biologically relevant time points in the experimental model must take into account available resources one the one hand, and obtaining sufficient data, on the other hand.

Most transcriptional adaptation to heat during the heat stress response in yeast takes place within 1 h of transfer of log-phase yeast cultures from normal to elevated temperature *(12)*. In this experiment, cultures of *lcb1-100* yeast and their background strain, RH406, were grown to mid-log phase at 30°C, and then aliquots were transferred to 39°C for 15, 30, or 60 min. Cultures of the *isc1Δ* mutant and its background strain JK93d(α were grown to mid-log phase at 30 (°C, and then aliquots arewere transferred to 39 (°C for 30, or 60, 90, and 120 min. This later time course was selected because metabolism of complex sphingolipids occurs at later time points. Both of these time courses provide ample time to stress, centrifuge, decant, and snap-freeze the samples. Previous experiments have used as many five mutant strains and as many as 7 time points to determine altered gene expression over time, which began as early as 5 min after stimulation of stress to as late as 2 h (Wilder, A.J, and Cowart, L. A., unpublished data).

3.1. Cell Culture and Lysis

1. RH406, *lcb1-100*, JK93dα, and *isc1*Δ cultures are grown to saturation in 5 mL liquid cultures of YPD. From these cultures, 100 mL YPD cultures in 500 mL flasks are seeded to be in mid-log phase (*i.e.*, an A_{600} of 0.4–0.7) at least 6 h after seeding (when the experiment will be performed). Cultures are grown in a shaking water bath at a temperature of 30°C and 225 RPM.
2. When the cells are nearing the appropriate O.D., prepare a dry ice-ethanol bath in a cold-resistant vessel for snap-freezing the cell pellets during the time course.
3. The growth phase of the cultures is monitored by checking the A_{600}. When the concentration of all the cultures reaches an A_{600} of 0.4–0.7 (mid-logarithmic growth), pipette 10 mL of culture into pre-labeled 50 mL centrifuge tubes. Prepare two cultures for each time point. Clearly label both the caps and the tubes.
4. Except for the control, insert the samples into a water bath shaker preheated to 39°C. Start the timer. Proceed to steps 5 and 6 with the 0 time point samples.
5. At the appropriate time point, remove the tubes labeled for this time point and pellet the cells at 3500× g at 25°C for 3 min. Discard the supernatant.
6. Insert the bottom of the tube into the dry ice-ethanol bath so that the portion containing the cell pellet is completely submerged. Once the pellet is completely frozen (in ~1–2 min), remove the tube at −80°C. Frozen cell pellets appear white and opaque.
7. Repeat steps 5 and 6 for the remaining time points.

3.2. RNA Extraction

1. Re-suspend cells in 1 mL ice-cold water. Transfer to a clean 1.5 mL microcentrifuge tube. Centrifuge for 10 s at 4°C, and remove supernatant.
2. Re-suspend pellet in 400 μL TES solution. Add 400 μL acid phenol and vortex vigorously for 10 s. Incubate 30 to 60 min at 65°C with occasional, brief vortexing.
3. Place on ice for 5 min. Microcentrifuge for 5 min at top speed, 4°C.
4. Transfer aqueous phase to a clean 1.5 mL microcentrifuge tube and add 400 μL acid phenol. Vortex vigorously. Repeat step 3.
5. Transfer aqueous phase to a clean 1.5 mL microcentrifuge tube and add 400 μL chloroform. Vortex vigorously and microcentrifuge for 1 min at top speed, 4°C.

6. Transfer aqueous phase to a new tube, precipitate RNA by addition of 40 μL 3 M sodium acetate, pH 5.3 and 1 mL ice-cold 100% ethanol. Microcentrifuge for 5 min at top speed, 4°C. Wash RNA pellet by vortexing briefly in ice-cold 70% ethanol. Microcentrifuge as before to pellet RNA.
7. Proceed to the RNA cleanup protocol found in the Qiagen RNeasy kit protocol to purify the total RNA.
8. The sample concentration and purity are determined by measuring the A_{260} and A_{280} using a UV-vis spectrophotometer. Samples were diluted 1:100 in RNase-free water in order to obtain an accurate reading. A reading of A_{260} of 1.0 corresponds to a concentration of 40 μg/mL. A yield of at least 1 μg of total RNA is required for use in the microarray procedures. An A_{260}/A_{280} ratio of 1.7–2.0 indicates a sample purity acceptable for use in microarray preparation procedures.

3.3. Microarray Sample Preparation and Hybridization

1. Total RNA samples are converted to fragmented, biotin-labeled cRNA, as recommended by Affymetrix (La Jolla, CA), using the kits listed in the Materials section and the protocols provided. Sample concentrations are determined as before by measuring the A_{260} and A_{280}. Fragmentation is confirmed by agarose gel electrophoresis using ethidium bromide for visualization, as described in the Affymetrix manual. This protocol yields biotinylated target RNA fragments in amounts directly proportional to the RNA levels in the original sample.
2. Target samples are then hybridized to the YG-S98 yeast genome chip (Affymetrix), stained with streptavidin-phycoerythrin, and washed in the Affymetrix Fluidics Station exactly as indicated in the Affymetrix Expression Analysis Technical Manual. Many research institutions provide hybridizations and scanning on a fee-for-service basis at a core facility. Furthermore, these core facilities often accept samples from outside their home institution. Due to the complexity of the protocol and the many points for the introduction of error, we recommend consistent use of the same facility.

3.4. Preparation for Lipidomic Analysis

1. Samples are taken as described in the Cell Culture section.
2. Internal standards are added to frozen pellets. Sphingolipids are extracted in a one-phase neutral organic solvent 70% isopropanol:ethylacetate:pyridine:25% ammonia (60:20:2:0.5 by volume) as used in *(13)*. Reserve a third of total extract from each sample for phosphate determination (see step 5). Analyze the remaining samples using a Surveyor/TSQ 7000 liquid chromatography–MS system, as described *(14)*.

3. Samples are quantitatively analyzed on the basis of calibration curves generated with synthetic standards. The mass of each species is normalized to total phospholipid levels measured in the sample fraction taken previously, to account for variations in extraction efficiency and cell number.
4. Lipid phosphate determination: Samples are dried and phosphates are extracted by the method of Bligh and Dyer *(15)*:
   a. Add 3 mL chloroform:methanol, vortex.
   b. Add 0.8 mL water. If you get a premature phase break, add 0.5 mL methanol. Transfer to a new glass tube and leave at 4°C overnight.
   c. Pellet debris at 3000 g and transfer the supernatant to a new glass tube.
   d. Add 1 mL of chloroform and 1 mL of water, vortex.
   e. Allow a few minutes for the phases to separate, and spin for 15 min at 3500 g.
   f. Aspirate most of the aqueous phase, and then carefully transfer the lower organic phase using a Pasteur pipette to a new tube.
   g. Dry down the organic phase using a nitrogen evaporator or a speed vac. The nitrogen is considerably faster.
   h. Dissolve the dried lipids in 0.300 mL of chloroform.
   i. Prepare the following standards from 1 mM NaH_2PO_4: 0, 5, 10, 20, 40, 60 μL (0, 5, 10, 20, 40, and 60 nmol, respectively) in labeled glass tubes appropriate for use in the heating block you will be using.
   j. Add 0.6 mL of ashing buffer to the dried lipid samples and the standards. It is not necessary to dry the standards.
   k. Vortex and incubate overnight in a heating block at 150–160°C. It is important to start the incubation while the heating block is cold. This incubation should be performed under a fume hood, due to hazardous chemicals in the ashing buffer.
   l. In the morning, remove tubes and allow to cool. Add 0.9 mL H_2O to the ~0.1 mL of sample that is left. Vortex.
   m. Add 0.5 mL of 0.9% (w/v) ammonium molybdate. Vortex.
   n. Add 0.2 mL of 9.0% (w/v) ascorbic acid. Vortex. This solution should be prepared immediately prior to use.
   o. Incubate for 30 min at 45°C.
   p. Measure the A_{820}. Generate a standard curve using values from the standards of NaH_2PO_4. Normalize lipid levels measured with LC-MS to the phosphate levels determined for each sample.

3.5. Data Analysis

1. Analysis begins with normalization of signal values and calculation of signal *p*-values using the Affymetrix MAS 5.0 software. Probe signals are scaled to have an average signal value of 250 for each array, in order to facilitate comparison between arrays. Probe sets are filtered to include only those which have a *p*-value of 0.05 or less, and thus a high probability of representing a signal above background. We have routinely used Microsoft Excel for these manipulations. Probe sets which have absent signals (*p*-value of 0.05 or less) in all samples are removed.
2. Genes of interest are then selected from these signals by sorting the data based upon the signal ratio of stressed to unstressed cultures (*i.e.* $\text{Signal}_{\text{Stress}}/\text{Signal}_{\text{Basal}} = \Delta\text{Signal}$) and/or the change ratio between parental and mutant cultures (*i.e.* $\Delta\text{Signal}_{\text{Mutant}}/\Delta\text{Signal}_{\text{WT}} = \text{Change Ratio}$). This is done by creating a new column in the spreadsheet and generating an equation that calculates the appropriate values for each probe set across all samples.
3. Genes of interest are identified as having a ratio of mutant/wild type that is over 2 or under 0.5. This results in significant filtering of the large dataset, but more stringent criteria can be set. For example, probe sets with lower signals (<100) can be filtered out, as evidence suggests they may be less reliable. Special attention is paid to genes that are induced or repressed by stress in wild type, but fail to change in mutant cells.
4. The list of probe set IDs is then converted to systematic names using the Affymetrix website (www.affymetrix.com). From the homepage, select the 'Netaffx' tab. Select the option 'retrieve annotations for a probe list.' Many annotation formats are available. For yeast, the most widely used annotation format is the SGD systematic name (*Saccharomyces* Genome Database, www.yeastgenome.org). Export the new list of gene annotations for use in data analysis.
5. These lists of genes of interest are analyzed for cellular functions that are statistically over-represented in the sample in the list, using the GOstat program developed in *(16)*. This data analysis software groups the over- and under-expressed genes into functional groups, simplifying analysis of stress – and lipid – induced gene expression changes. For information on Gene Ontology, see *(17, 18)* and www.geneontology.org.
6. Behaviors of specific gene subsets are analyzed with respect to lipid profiles. For example, gene changes that occur in the parental, but not in the mutant are investigated with respect to lipid changes absent in the mutant. From these relationships, hypotheses can be developed to test the lipid-mediated regulation of transcription of specific gene sets.

4. Notes

- Many yeast deletion mutants are made by generating a DNA fragment bearing a selectable marker (for example, Kanamycin resistance gene) flanked by 50–100 base pair regions of homology to the gene of interest. Transformation of this fragment into yeast by standard methods (for example, the widely used lithium chloride method) followed by plating on selective media routinely results in many recombinant clones. Correct recombination is then verified by PCR of genomic DNA using a forward primer upstream of the sequence in the transformed fragment, and a reverse primer in the selectable marker. Amplification of a fragment of expected size, coupled with lack of amplification from DNA from a parental control, is sufficient to confirm appropriate deletions. Importantly, all viable mutants are available in the European *Saccharomyces Cerevisiae* Archive for Functional Analysis (EuroSCARF), in the BY4741 (a-type) or BY4742 (α-type) background strains.
- The above technique is useful only for nonlethal mutations. For lethal mutations, a conditional mutant strain (e.g., a temperature-sensitive strain) is required. These strains contain mutations in the desired gene that results in functional protein at permissive temperature (*i.e.* 30°C), but nonfunctional at a higher, nonpermissive temperature (*i.e.* 39°C). The generation of these strains is beyond the scope of this manuscript. It should be noted that the yeast community is very generous in the sharing of mutant strains.
- Maintain consistency of the growth phase of the cultures being used. Gene expression changes continuously throughout yeast culture growth in order to thrive in the changing environment (*i.e.*, gradual depletion of nutrients and increasing concentrations of metabolic by-products). In particular, the 'diauxic shift', when the cells switch from glucose to ethanol as a carbon source, which can begin as early as at A_{600} of 0.7, causes massive genome remodeling. In general, care should be taken to have the separate cultures as close to the same growth phase (*i.e.* A_{600}) as possible prior to the execution of the experiment. For this reason, it is important to know the doubling times of the different yeast strains utilized prior to execution of the experiment.
- The cell density and volume of the cultures used in this experimental method are selected to provide enough cells for lipid measurements and RNA measurements. This volume also facilitates the use of a 50 mL centrifuge tube for both stressing and snap-freezing the samples. If more cells are needed for lipid measurements or if more RNA is

needed than just for the microarray hybridization (*i.e.* for RT-PCR), a larger volume of culture can be used, preferably in a larger centrifuge tube. In this case, different methods of RNA extraction should be investigated to optimize yields for larger volumes.

- When designing the time course, it is important to realize that primary and secondary effects will occur as a result of stress. A well-designed time course will capture the earliest observable alterations in lipid and gene expression levels. Focus should be kept on primary effects of stress, as they are more easily interpreted.
- The experimental procedures that take place post-stress should be optimized to minimize time of sample preparation prior to snap-freezing. Centrifuge temperature should be as close to stress temperature as possible. Speed and time of centrifugation should be minimized. We have noted that 3 min at 3500× g suffices for pelleting of yeast from culture volumes denoted herein. Snap-freezing, while not entirely necessary, minimizes any cold stress-induced changes that could occur if samples were placed directly in the freezer. To minimize potential RNA degradation, the RNA extraction procedure should be performed without breaks and as quickly as possible.
- Many protocols for RNA extraction are available. We recommend the hot phenol method, as it maximizes yield and minimizes selective RNA degradation. If the Qiagen RNeasy kit is used, enzymatic lysis, requiring a 1 h digestion, is not recommended, due to concerns about selective RNA degradation during this incubation. Mechanical disruption of the cells is acceptable, but yields are greater and RNA degradation less when the hot phenol method is used.
- When handling RNA, care must be taken to eliminate possible RNase contamination. RNase-free water must be used in any experiment that involves RNA. Gloves must be worn when handling tubes containing samples of RNA. Tubes should be specifically RNase-free.
- When observing the fragmentation of the RNA, a standard protocol for DNA electrophoresis using a tris-acetate-EDTA solution (1% agarose/TAE) can be used. Denaturing conditions are not required. This is only in reference to the verification of RNA fragmentation at the end of the microarray sample preparation protocol.
- Previous protocols report RNA sample purity acceptable at A_{260}/A_{280} of 1.9–2.1. Our experience is that cDNA and cRNA prepared using the Affymetrix kits typically have a sample purity of 1.5–1.8, when dissolved in RNase-free

water. The M.U.S.C. microarray core facility has confirmed that these levels are acceptable for successful microarray hybridization, using the Affymetrix platform.

- Scaling the average chip signal to 250 is one effective method of signal normalization. However, other methods include the addition of standards to the hybridization cocktail or the addition of poly-A controls during the cRNA preparation. Another controversial method is to normalize all signals on each array to a housekeeping gene on the same array. While the method presented herein provides satisfactory results, other methods have been published and can be referred to.
- Other open-source and commercially available software platforms for analyzing microarray datasets are available for use in data analysis. Additionally, other statistical methods can be used to determine statistical significance of changes in gene expression. We have routinely used Robust Multichip Analysis, in addition to the analysis presented above. Different analytical methods have produced different, but overlapping, lists of significant gene sets.
- The solvent for lipid extraction used in this protocol has been optimized for use especially with yeast sphingolipids. For sphingolipid analysis in mammalian cells, see *(14)*.
- While the method for the identification of lipid species by mass spectrometry is very broad, it is not comprehensive. New methods to quantify other lipid species are continuously being developed.
- Many methods can be used for lipid normalizations. For example, in addition to normalization to total phospholipid levels, normalization to total cell protein or cell number can be used. Normalization is further improved with the addition of internal standards to the sample prior to analysis, specifically, a nonnatural lipid (for example C17 sphingosine).

References

1. Futerman, A.H. and Y.A. Hannun, *The complex life of simple sphingolipids.* EMBO Rep, 2004. **5**(8): p. 777–782.
2. Hannun, Y.A. and L.M. Obeid, *The Ceramide-centric universe of lipid-mediated cell regulation: stress encounters of the lipid kind.* J Biol Chem, 2002. **277**(29): p. 25847–25850.
3. Dickson, R.C., C. Sumanasekera, and R.L. Lester, *Functions and metabolism of sphingolipids in Saccharomyces cerevisiae.* Prog Lipid Res, 2006. **45**(6): p. 447–465.
4. Cowart, L.A. and L.M. Obeid, *Yeast sphingolipids: recent developments in understanding biosynthesis, regulation, and function.* Biochem Biophys Acta, 2007. **1771**(3): p. 421–431.
5. Jenkins, G.M. and Y.A. Hannun, *Role for de novo sphingoid base biosynthesis in the heat-induced transient cell cycle arrest of Saccharomyces cerevisiae.* J Biol Chem, 2001. **276**(11): p. 8574–8581.
6. Jenkins, G.M., et al., *Involvement of yeast sphingolipids in the heat stress response of*

Saccharomyces cerevisiae. J Biol Chem, 1997. **272**(51): p. 32566–32572.

7. Munn, A.L. and H. Riezman, *Endocytosis is required for the growth of vacuolar H(+)-ATPase-defective yeast: identification of six new END genes*. J Cell Biol, 1994. **127**(2): p. 373–386.
8. Chung, N., et al., *Sphingolipids signal heat stress-induced ubiquitin-dependent proteolysis*. J Biol Chem, 2000. **275**(23): p. 17229–17232.
9. Vaena de Avalos, S., et al., *The phosphatidylglycerol/cardiolipin biosynthetic pathway is required for the activation of inositol phosphosphingolipid phospholipase C, Isc1p, during growth of Saccharomyces cerevisiae*. J Biol Chem, 2005. **280**(8): p. 7170–7177.
10. Dickson, R.C., et al., *Sphingolipids are potential heat stress signals in Saccharomyces*. J Biol Chem, 1997. **272**(48): p. 30196–30200.
11. Cowart, L.A., et al., *Distinct roles for de novo versus hydrolytic pathways of sphingolipid biosynthesis in Saccharomyces cerevisiae*. Biochem J, 2006. **393**(Pt 3): p. 733–740.
12. Gasch, A.P., et al., *Genomic expression programs in the response of yeast cells to environmental changes*. Mol Biol Cell, 2000. **11**(12): p. 4241–4257.
13. Kitagaki, H., et al., *Isc1 regulates sphingolipid metabolism in yeast mitochondria*. Biochem Biophys Acta, 2007. **1768**(11): p. 2849–2861.
14. Bielawski, J., et al., *Simultaneous quantitative analysis of bioactive sphingolipids by high-performance liquid chromatography-tandem mass spectrometry*. Methods, 2006. **39**(2): p. 82–91.
15. Bligh, E.G. and W.J. Dyer, *A rapid method of total lipid extraction and purification*. Can J Biochem Physiol, 1959. **37**(8): p. 911–917.
16. Beissbarth, T. and T.P. Speed, *GOstat: find statistically overrepresented Gene Ontologies within a group of genes*. Bioinformatics, 2004. **20**(9): p. 1464–1465.
17. Ashburner, M., et al., *Gene ontology: tool for the unification of biology. The Gene Ontology Consortium*. Nat Genet, 2000. **25**(1): p. 25–29.
18. Beissbarth, T., *Interpreting experimental results using gene ontologies*. Methods Enzymol, 2006. **411**: p. 340–352.

Chapter 29

Bioluminescence: Imaging Modality for *In Vitro* and *In Vivo* Gene Expression

Ruxana T. Sadikot, and Timothy S. Blackwell

Abstract

Molecular imaging offers many unique opportunities to study biological processes in intact organisms. Bioluminescence is the emission of light from biochemical reactions that occur within a living organism. Luciferase has been used as a reporter gene in transgenic mice but, until bioluminescence imaging was described, the detection of luciferase activity required either sectioning of the animal or excision of tissue and homogenization to measure enzyme activities in a conventional luminometer. Bioluminescence imaging (BLI) is based on the idea that biological light sources can be incorporated into cells and animal models artificially that does not naturally express the luminescent genes. This imaging modality has proven to be a very powerful methodology to detect luciferase reporter activity in intact animal models. This form of optical imaging is low cost and noninvasive and facilitates real-time analysis of disease processes at the molecular level in living organisms. Bioluminescence provides a noninvasive method to monitor gene expression *in vivo* and has enormous potential to elucidate the pathobiology of lung diseases in intact mouse models, including models of inflammation/injury, infection, and cancer.

Key words: Bioluminescence, *In vivo* imaging, Inflammation, Infection, Gene expression.

1. Introduction

In the recent years, noninvasive molecular imaging has emerged as a powerful tool to monitor cellular and molecular events in vivo. Molecular imaging techniques represent an important advance in our ability to study structural and functional relationships in biology by combining the disciplines of molecular and cellular biology and imaging technology. Bioluminescence is the main form of optical imaging that has been used for gene expression imaging studies, particularly in transgenic mice

From: *Methods in Molecular Biology, vol. 477: Advanced Protocols in Oxidative Stress I,*
Edited by: D. Armstrong, DOI: 10.1007/978-1-60327-517-0_29, © Humana Press, New York, NY

and in studies of in vivo gene transfer. Pathogens or mice are engineered to express genetically encoded luciferase enzymes from bacteria or insects and bioluminescent imaging is used to study a variety of biological processes. Over the last decade, the number of studies that have used bioluminescence imaging has grown exponentially.

Several forms of molecular imaging have been developed over the last decade, including nuclear imaging techniques *(1, 2, 3)* and magnetic resonance imaging techniques. Compared to these technologies, optical imaging offers distinct advantages, including relative ease of use, lower cost, reproducibility, sensitivity, short imaging time, and the ability to study animals at multiple time points. Optical imaging with bioluminescence at the emission wavelength of firefly luciferase (560 nm peak) can be imaged as deep as several centimeters within tissue, which allows at least organ-level resolution. Other forms of optical imaging, including microscopy of green fluorescent protein *(4, 5)* or near-infrared fluorescence *(6)* are excellent for cell-based systems but have limited depth potential in vivo.

Bioluminescence imaging is based on the sensitive detection of visible light produced during enzyme (luciferase)-mediated oxidation of a molecular substrate when the enzyme is expressed in vivo as a molecular reporter. Firefly luciferase catalyzes the oxidation of its molecular substrate luciferin in the presence of ATP and oxygen to emit light. Bioluminescence at the emission wavelength of firefly luciferase (560 nm) can be imaged as deep as several centimeters within tissue. BLI is simple to execute and enables monitoring throughout the course of disease, allowing localization and serial quantification of biological processes without killing the experimental animal. The ability to image luciferase activity directly in the living transgenic animal has enabled the monitoring of expression patterns of specific genes. This technique can reduce the number of animals required for experimentation because multiple measurements can be made in the same animal over time, allowing the use of the same animal as its control and minimizing the effects of biological variation. Bioluminescence has been applied in studies to monitor transgene expression, progression of infection, tumor growth and metastasis, transplantation, toxicology, viral infections, and gene therapy.

1.1. In Vivo Imagers

In recent years, several systems that are commercially available have been developed for the in vivo detection of luciferase and fluorescence activity. These systems are capable of detecting the light emitted from cells or small animals that contain the luminescent cells. Intensified-charged couple device cameras (ICCD) were initially used for the in vivo imaging

(C2400-32; Hamamatsu, Bridgewater, NJ). These devices use bialkali photocathodes that are blue sensitive and suffer from low quantum efficiency. Thus, these imagers have limited ability to detect light that emanates from animals that express the luciferase reporter gene. Furthermore, intensified CCD detectors generally have a limited dynamic range such that methods for controlling signal intensity must be used. More recently, cooled integrating CCD cameras have been developed. These detectors have higher quantum efficiencies and are more sensitive to light in the red and near infrared spectral regions. Cooling of the CCD chip can significantly reduce background dark current signals, although the read noise on the detector can still present a problem for measurements of very weak signals. Thus, it is recommended that background measurements be taken and subtracted from experimental data when using these systems.

These imaging systems include: (1) a light-tight chamber in which the cells or animals are placed, (2) a sensitive charged coupled device camera to detect the light that is emitted, and (3) a computer controller that acquires the image and allows for quantification and analysis of the image. One of the commercially available systems that is specifically designed to measure light emission from small animals is the IVIS imaging system (Xenogen Corporation, Almeda, CA), which uses an ultracooled CCD camera. In this system, a back-illuminated cooled integrating CCD camera is mounted on top of a light impermeable imaging chamber. Within the chamber, there is a moveable heated stage that allows different fields of view to be obtained. This permits the imaging of several animals simultaneously. The heated stage also aids in keeping the animals warm. The stage also has a manifold with nose cones through which anaesthetic gas such as isoflurane can be administered to animals during imaging. This allows animals to remain comfortably sedated through the imaging period.

1.2. Luciferase Detection by Luminometer

Light from the cells that express the luciferase reporter gene is emitted when luciferase is exposed to the appropriate luciferin substrate in the presence of ATP. Photon emission can be detected by a light-sensitive apparatus such as a luminometer. Several luminometers are commercially available that have a highly sensitive on-board computer and software platform for data analysis and reporting. The luminometers are either of the plate reading or single tube injector variety.

A single injector luminometer apparatus has a light-tight box into which a tube containing a sample and a reagent are combined to produce a luminescent reaction. The operation of this equipment is simple. The sample is inserted in a test tube along with the substrate (Luciferin) just prior to the measurement of luciferase activity and the sliding door is closed. The built-in software provides complete sample documentation and transferring of data to a PC for further analysis. The luciferase activity is normalized to the protein content of the sample. The plate reading luminometers use a 96 well plate for the measurement of the luciferase activity instead of the single tube allowing the reading of multiple samples at the same time.

1.3. Transgenic Cells and Animals

To introduce a reporter gene into an organism, the reporter gene and the regulatory sequence of interest are put in the same DNA construct, which is inserted into the cell or organism. Lines of transgenic mice containing luciferase reporter constructs can be generated using standard techniques for in vivo reporter assays. For example, we have produced lines of transgenic mice expressing firefly luciferase complementary DNA under the control of two different NF-κB-dependent promoters in order to examine NF-κB activation in disease models *(7, 8)*. These reporter mice have allowed us to develop a convenient, quantitative method for evaluating NF-κB activation over time to examine the consequences of NF-κB activation in multiple organs in vivo.

In addition to the firefly luciferase system, the bacterial lux operon system has been used to develop genetically engineered bioluminescent bacteria. The *lux* operon from bacteria such as *Photorhabdus luminescens* or *Xenorhabdus luminescens* encodes genes to synthesize a luciferase enzyme as well as its substrate, luciferin; therefore, bacteria expressing the *lux* operon constitutively produce light without the need for an exogenous substrate. The *lux* operon has been transferred stably to different species of bacteria and parasites for imaging studies.

2. Materials

2.1. Equipment

1. In Vivo Imaging System (C2400-32), Hamamatsu, Bridgewater, NJ
2. IVIS System, Xenogen Corp., Almeda, California
3. Luminometer; plate reader, or single injector

2.2. Reagents and Buffers

1. D-Luciferin, D-Luciferin methyl ester, D-Luciferin sodium salt, D-Luciferin potassium salt (Synchem)
2. Tissue Luciferase Activity, final pH should be 7.3
   a. PBS, Mg^{+2}, and Ca^{+2} free
   b. 137 mM NaCl
   c. 2.7 mM KCl
   d. 4.3 mM Na_2HPO_4
   e. 1.4 mM KH_2PO_4

2.3. Lysozyme

1. Add one volume of 1 M K2HPO4, pH 7.8–20 mM EDTA in nine volume water.
2. Adjust to final concentration of 5 mg/ml, vortex until clear; prepare fresh daily.

2.4. Luciferase Cell Culture Lysis Reagent

1. 25 mM Tris phosphate buffer, pH 7.8
2. 2 mM dithiothreitol
3. 1,2-diaminocyclohexane-N,N,N,N-tetraacetic acid
4. 10% Glycerol
5. 1% Triton X-100

2.5. Lysis Mix (Alternative to Lysis Reagent)

1. Cell Culture lysis reagent (CCLR)
2. 1.25 mg/ml lysozyme
3. 2.5 mg/ml BSA

2.6. Tissue Lysis Buffer

1. 25 mM glycylglucine
2. 15 mM $MgSO_4$
3. 4 mM EGTA
4. 1% Triton X-100
5. 1 mM dithiothreitol

2.7. Luciferase Assay Buffer

1. 25 mM glycylglycine
2. 15 mM $MgSO_4$
3. 4 mM EGTA
4. 15 mM KPO_4
5. 3 mM dithiothreitol
6. 3 mM ATP

3. Methods

3.1. Preparation of Animals

Depending on the coat color of the mice and the area where bioluminescence is to be measured, hair may be removed from the skin over the area of interest. Hair can be removed by shaving or using a depilatory. If shaving is used, mice are briefly anaesthetized with isoflurane. White mice usually do not require removal of hair as the absorption of light is minimal; however, a substantial reduction in signal can occur in black mice if hair is not removed.

3.2. Anesthesia and Luciferin Administration

If a system such as IVIS is used, which has a moveable heated stage with a manifold for gas anesthesia, mice may not require injectable anesthetic. Alternatively, if another intensified-CCD camera system is used, sedation with an anesthetic such as ketamine and xylazine or phenobarbitol during the period of imaging is required. Mice are given D-Luciferin 1 mg per mouse (or approximately 30 mg/kg) dissolved in PBS (100 μl) via the intraperitoneal or intravenous route prior to imaging. Luciferin rapidly distributes throughout most mouse tissues after intraperitoneal or intravenous injection. Therefore, substrate availability is not typically a limitation for in vivo imaging with firefly luciferase. For imaging with the IVIS system, mice are placed into a clear plexiglass anesthesia box (2.5–3.5% isoflurane) that allows visual monitoring of the animals. The tube that supplies anesthesia to the box is split so that the same concentration of anesthesia is plumbed to the anesthesia manifold in the imaging box. After mice are fully anesthetized, they are placed on the stage in the light tight box for the detection of bioluminescence.

3.3. Imaging and Quantification of Photon Counts

A grayscale reference image of the animal is acquired under weak illumination, and then in complete darkness the bioluminescent signal is captured. This can take from a few seconds to several minutes depending on the brightness of the bioluminescent signal and the sensitivity of the detector. The signal intensity is then represented as a pseudocolor image and superimposed on the grayscale reference image. The magnitude of the signal is then measured from specified regions using a 'region of interest' (ROI) function that determines photon counts over the specified region. The timing of peak bioluminescence after luciferin injection depends on the route of administration of the luciferin and the tissue of interest, but typically occurs between 5 and 30 min after luciferin injection. After the experiment, mice may be euthanized and the tissues may be homogenized for the measurement of luciferase.

3.4. Luciferase Assays

Tissues are removed, placed in 1 ml of lysis buffer (25 mM glycylglycine, 15 mM $MgSO_4$, 4 mM EGTA, 1% Triton X-100, and 1 mM dithiothreitol), and homogenized with a tissue homogenizer. Around 300 μl of the homogenate is then removed and centrifuged for 5 min at 15,000× *g*, and 200 μl of the supernatant is assayed for luciferase activity.

Luciferase assays are performed in a luminometer. Samples are injected with 100 μl of luciferin (0.75 mM) and 100 μl of assay buffer (25 mM glycylglycine, 15 mM $MgSO_4$, 4 mM EGTA, 15 mM KPO_4, 3 mM dithiothreitol, and 3 mM ATP), and luminescence is measured for 20 s and reporter as relative light units. The relative light units can be normalized for sample size by correcting for total protein content.

3.5. Applications

Transgenic reporter mice can be used to monitor the promoter activity of individual genes or determine the activity of specific transcription factors. The latter application involves creating

3.5.1. Monitoring Transcriptional Activity in Disease Models

reporter mice with simple promoter-luciferase constructs where the promoter is driven by a particular transcription factor. In this chapter, we will describe applications of this technology for the study of two transcriptional pathways – NF-κB and TGF-β. Of these, NF-κB reporter mice have been most extensively used by us and several other investigators.

3.5.2. Detection of NF-κB Activation

Nuclear factor-κB (NF-κB) is a ubiquitous transcription factor that is a master regulator of inflammation and plays an important role in immune response to various microbes and microbial products. Transcription of a variety of proinflammatory mediators, including cytokines, chemokines, and adhesion molecules, is regulated by NF-κB *(9)*. To develop a convenient method for examining NF-κB activation in vivo, we have engineered a line of transgenic mice (referred to as HLL [HIV-LTR/Luciferase] mice) that carries the proximal 5′ HIV-1 long terminal repeat (LTR), driving the expression of *Photinus* luciferase complementary DNA. We have shown that luciferase activity in cells and tissues from these transgenic mice reflects NF-κB activation over time. Using these reporter mice, we have determined the timing and distribution of NF-κB activity in response to *E. Coli* LPS *(8)*, hepatic cryoablation *(10)*, *P. aeruginosa* pneumonia *(11)*, and pancreatitis *(12)*. These models have proven to be valuable for measuring activation of NF-κB in real time and have helped overcome limitations of other methods of detecting NF-κB activation, such as electrophoretic mobility shift assay and western blot analysis. In several different studies, we have shown an excellent correlation between tissue luciferase activity and bioluminescent detection of luciferase activity in HLL mice.

In addition to detecting the activation of NF-κB in response to a stimulus, we have also shown that bioluminescent in vivo imaging can be used to assess whether specific interventions affect NF-κB activity in the lung. We have used adenoviral vectors to activate or inhibit NF-κB in the lung epithelium and shown that the host response to *P. aeruginosa* can be altered in vivo *(13)*. In these experiments, activation of NF-κB was detected by bioluminescent imaging in reporter mice (**Fig. 29.1**). Representative bioluminescent images from HLL NF-κB reporter mice after treatment with adenoviral vectors that activate or inhibit NF-κB (Ad-RelA or Ad-IκBdn, respectively) and controls are shown (in the upper panel). Mice were treated with adenoviral vectors (10^9 PFU) for 48 h and luciferase activity was measured by bioluminescence following i.p. luciferin injection. Mice were then infected with *P. aeruginosa* (PA103) and reimaged 24 h later following luciferin injection (lower panel). Thus, bioluminescent imaging indicated that treatment with adenoviral vectors resulted in modulation of NF-κB activity in the lungs in this pneumonia model *(14, 15, 16)*.

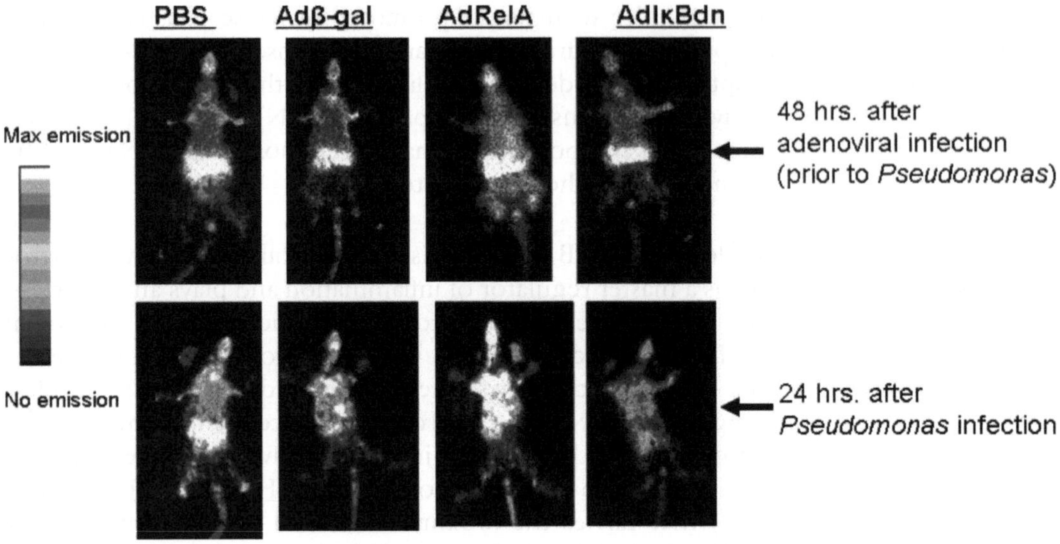

Fig. 29.1. Treatment with adenoviral vectors alters NF-κB activity in lungs following *P. aeruginosa* infection. *a*, Representative bioluminescent images from HLL NF-κB reporter mice after treatment with PBS, Adβ-gal, RelA, or AdIκBdn are shown. Mice were treated with PBS or adenoviral vectors (10^9 PFU) for 48 h and luciferase activity was measured by bioluminescence following i.p. luciferin injection. Mice were then infected with *P. aeruginosa* (PA103) and reimaged 24 h later following luciferin injection.

Other investigators have also used similar NF-κB transgenic reporter mice to study lung and systemic NF-κB–dependent inflammatory responses *(17, 18, 19)*. In a recent study, Dohlen et al. investigated the transcriptional regulation of key proteins related to oxidative stress and inflammation in pregnancy using an NF-κB reporter mouse model *(20)*. NF-κB reporter mice have also been used for the evaluation of host–biomaterial interaction by the implantation of genipin-cross-linked gelatin conduit (GGC) and bacterial endotoxin-immersed GGC in the dorsal region of mice. The correlation between the bioluminescence imaging and histological changes indicated that noninvasive imaging technique could be used to monitor inflammation real-time in this model *(21)*.

3.5.3. Detection of Smad Activation

Transforming growth factor-β (TGF-β) has been shown to have important roles in a variety of processes, including wound healing, fibrosis, and inflammation. In response to receptor binding by TGF-β, Smad2/3 translocates to the nucleus and regulates transcription of TGF-β responsive genes. Transgenic mice containing a firefly luciferase reporter gene fused to a Smad2/3-binding element have been created and used to study TGF-β responses in inflammation and wound healing models *(22, 23)*. After systemic LPS, Smad activation was identified in the intestines and brain *(22)*. In a localized wound healing model, Chong et al. used these reporter mice to show that wound healing occurs with delayed TGF-β signaling kinetics. By bioluminescence imaging, Smad activity peaked at 17 days after injury *(23)*. These studies indicate that important information regarding the biology of the TFG-β pathway can be obtained by bioluminescence imaging techniques using this reporter system.

3.5.4. Monitoring Inflammatory Cell Migration

Inflammatory cell recruitment and trafficking are important components of inflammation that can be studied using in vivo bioluminescence techniques. Costa et al. used retroviral vectors to introduce a luciferase reporter gene into CD4+ T cells and follow the trafficking of these cells in experimental autoimmune encephalomyelitis *(24)*. More recently, Azadhniv and colleagues used the human CD2 promoter to express firefly luciferase in T cells in a transgenic reporter mouse and investigated the immune response to an ova peptide encoded by a herpes simplex virus (HSV)-based amplicon vector. The investigators were able to track the sites of T cell migration and showed that activated T cells were first seen in the local lymph node, and then at the site of HSV vector injection. Subsequently, systemic dispersal of the T cells was observed with a high proportion of the cells migrating to the abdomen *(25)*.

3.5.5. Monitoring Infectious Organisms

The feasibility of detecting microbially generated luminescence within a living mouse was shown by Contag and colleagues using *Salmonella typhimurium* transformed with the *lux* operon from *Photorhabdus luminescens.* Following this application, several studies have been published using bioluminescent Gram-negative and Gram-positive bacteria *(26, 27)*. We have used bioluminescent *P. aeruginosa* to determine the bacterial load in immunocompromised mice with NADPH oxiadase (p47phox) deficiency *(11)*. Using the luminescent *P. aeruginosa*, we demonstrated that p47phox deficient mice had reduced clearance of *P. aeruginosa* compared to wild type mice (**Fig. 29.2**).

Genetically engineered parasites with a luciferase reporter have also been used to localize infection and determine the response to treatment *(28, 29, 30)*. Viruses and fungi can also be engineered to encode luciferase *(31, 32, 33)*. Luciferase-producing nontypeble *H. Influenza* (NTHI) have been used to identify the factors that lead to formation of biofilm in vivo *(34)*. Together, published studies indicate that luminescent microbes can be used to monitor infectious processes in living animals and also determine the effects of specific interventions used to treat infections.

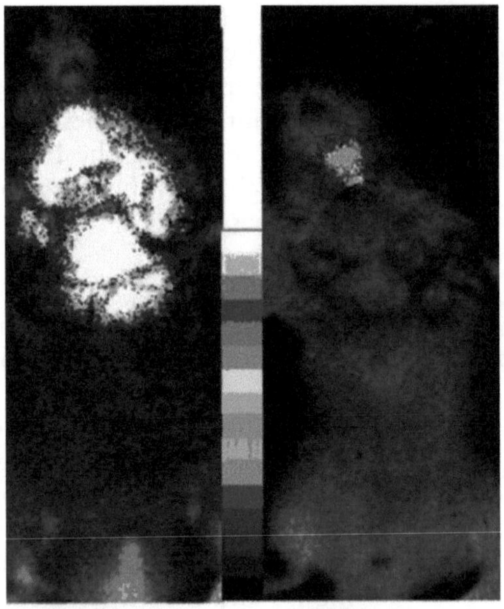

Fig. 29.2. Bioluminescence measurements of bacterial load. Representative images of labeled *P. aeruginosa* (XEN-5) dose of 10^6 in HLL and p47^{phox}−/−HLL mice reveal a higher signal in p47^{phox}−/−HLL compared with HLL mice at 24 h.

4. Notes

1. Cooled integrating CCD cameras are preferred for bioluminescence.
2. Imaging of animals with dark fur may require removal of hair to increase the sensitivity of detection of bioluminescence.
3. The region of interest for calculation of photonic counts has to be selected such that the area is uniform with each measurement.
4. Sodium salts of luciferin are preferred as a substrate as potassium salts can be toxic.
5. Luciferin, when stored should be protected from light.
6. Luciferase assay reagent can be purchased ready to use or prepared by reconstituting luciferase assay substrate with luciferase assay buffer.
7. Avoid exposure of the luciferase assay reagent to multiple freeze thaw cycles by dispensing the reconstituted reagent into working aliquots.

References

1. Tjuvajev JG, Stockhammer G, Desai R, Uehara H, Watanabe K, Gansbacher B, Blasberg RG (1995). Imaging the expression of transfected genes in vivo. *Cancer Res.* **55**:6126–6132.
2. Tjuvajev JG, Finn R, Watanabe K, Joshi R, Oku T, Kennedy J, Beattie B, Koutcher J, Larson S, Blasberg RG (1996). Noninvasive imaging of herpes virus thymidine kinase gene transfer and expression: a potential method for monitoring clinical gene therapy. *Cancer Res.* **56**:4087–4092.
3. Gambhir SS, Barrio JR, Phelps ME, Iyer M, Namavari M, Satyamurthy N, Wu L, Green LA, Bauer E, MacLaren DC (1999). Imaging adenoviral-directed reporter gene expression in living animals with positron emission tomography. *Proc. Natl. Acad. Sci. USA* **96**:2333–2339.
4. Yang M, Baranov E, Moossa AR, Penman S, Hoffman RM (2000). Visualizing gene expression by whole-body fluorescence imaging. *Proc. Natl. Acad. Sci. USA* **97**:12278.
5. Yang M, Baranov E, Jiang P, Sun FX, Li XM., Li L, Hasegawa S, Bouvet M, Al-Tuwaijri M, Chishima T (2000). Whole-body optical imaging of green fluorescent protein-expressing tumors and metastases. *Proc. Natl. Acad. Sci.* USA **97**:1206–1213.
6. Weissleder R, Moore A, Mahmood U, Bhorade R, Benveniste H, Chiocca EA, Basilion JP (2000). In vivo magnetic resonance imaging of transgene expression. *Nat. Med.* **6**:351–359.
7. Blackwell TS, Yull FE, Chen CL, Venkatakrishnan A, Blackwell TR, Hicks DJ, Lancaster LH, Christman JW, Kerr LD (2000). Multiorgan nuclear factor kappa B activation in a transgenic mouse model of systemic inflammation. *Am. J. Respir. Crit. Care Med.* **162**:1095–1101.
8. Everhart MB, Han W, Sherrill TP, Arutiunov M, Polosukhin VV, Burke JR, Sadikot RT, Christman JW, Yull FE, Blackwell TS (2006) Duration and intensity of NF-kappaB activity determine the severity of endotoxin-induced acute lung injury. *J. Immunol.* **176**:4995–5005.
9. Christman JW, Sadikot RT, Blackwell TS (2000) The role of nuclear factor-kappa B in pulmonary diseases. *Chest* **117**:1482–1487.
10. Sadikot RT, Wudel LJ, Jansen DE, Debelak JP, Yull FE, Christman JW, Blackwell TS, Chapman WC (2002). Hepatic cryoablation-induced multisystem injury: bioluminescent detection of NF-kappaB activation in a transgenic mouse model. *J. Gastrointest. Surg.* **6**:264–270.
11. Sadikot RT, Zeng H, Yull FE, Li B, Cheng DS, Kernodle DS, Jansen ED, Contag CH, Segal BH, Holland SM, Blackwell TS, Christman JW (2004). p47phox deficiency impairs NF-kappa B activation and host defense in Pseudomonas pneumonia. *J. Immunol.* **172**:1801–1808.
12. Gray KD, Simovic MO, Chapman WC, Blackwell TS, Christman JW, Washington MK, Yull FE, Jaffal N, Jansen ED, Gautman S, Stain SC (2003). Systemic nf-kappaB activation in a transgenic mouse model of acute pancreatitis. *J. Surg. Res.* **110**:310–314.
13. Sadikot RT, Blackwell TS (2005). Bioluminescence imaging. *Proc. Am. Thorac. Soc.* **2**:537–540.
14. Sadikot RT, Jansen ED, Blackwell TR, Zoia O, Yull F, Christman JW, Blackwell TS (2001). High-dose dexamethasone accentuates nuclear factor-kappa b activation in endotoxin-treated mice. *Am. J. Respir. Crit. Care Med.* **164**:873–878.
15. Sadikot RT, Han W, Everhart MB, Zoia O, Peebles RS, Jansen ED, Yull FE, Christman JW, Blackwell TS (2003). Selective I kappa B kinase expression in airway epithelium generates neutrophilic lung inflammation. *J. Immunol.* **170**:1091–1098.
16. Sadikot RT, Zeng H, Joo M, Yull F, Li B, Christman, JW, Blackwell TS (2006). Targeted immunomodulation of the NF-κB pathway in airway epithelium impacts host defense against *P. aeruginosa*. *J. Immunol.* **176**:4923–4930.
17. Contag PR, Olomu IN, Stevenson DK, Contag CH (1998). Bioluminescent indicators in living mammals. *Nat. Med.* **4**:245–247.
18. Carlsen H, Moskaug JO, Fromm SH, Blomhoff R (2002). In vivo imaging of NF-kappa B activity. *J. Immunol.* **168**: 1441–1446.
19. Hubbard AK, Timblin CR, Shukla A, Rincon M, Mossman BT (2002). Activation of NF-kappaB-dependent gene expression by silica in lungs of luciferase reporter mice. *Am. J. Physiol. Lung Cell. Mol. Physiol.* **282**:L968–L975.
20. Dohlen G, Odland HH, Carlsen H, Blomhoff R, Thaulow E, Saugstad OD (2002). Antioxidant activity in the newborn brain: a luciferase mouse model. *Neonatology* **93**:125–131.

21. Ho TY, Chen YS, Hsiang CY (2007). Noninvasive nuclear factor-kappaB bioluminescence imaging for the assessment of host-biomaterial interaction in transgenic mice. *Biomaterials* **28**:4370–4377.
22. Lin AH, Luo J, Mondshein LH, ten Dijke P, Vivien D, Contag CH, Wyss-Coray T (2005). Global analysis of Smad2/3-dependent TGF-beta signaling in living mice reveals prominent tissue-specific responses to injury. *J. Immunol.* **175**:547–554.
23. Chong AK, Satterwhite T, Pham HM, Costa MA, Luo J, Longaker MT, Wyss-Coray T, Chang J (2007). Live imaging of Smad2/3 signaling in mouse skin wound healing. *Wound Repair Regen.* **15**:762–766.
24. Costa GL, Sandora MR, Nakajima A, Nguyen EV, Taylor-Edwards C, Slavin AJ, Contag CH, Fathman CG, Benson JM (2001). Adoptive immunotherapy of experimental autoimmune encephalomyelitis via T cell delivery of the IL-12 p40 subunit. *J. Immunol.* **167**:2379–2387.
25. Azadniv M, Dugger K, Bowers WJ, Weaver C, Crispe IN (2007). Imaging CD8+ T cell dynamics in vivo using a transgenic luciferase reporter. *Int. Immunol.* **10**:1165–1173.
26. Francis K, Joh D, Bellinger-Kawahara C, Hawkinson M, Purchio T, Contag P (2000). Monitoring bioluminescent *Staphylococcus aureus* infections in living mice using a novel luxABCDE construct. *Infect. Immun.* **68**:3594–3600.
27. Francis KP, Yu J, Bellinger-Kawahara C, Joh D, Hawkinson MJ, Xiao G, Purchio TF, Caparon MG, Lipsitch M, Contag PR (2001). Visualizing pneumococcal infections in the lungs of live mice using bioluminescent *Streptococcus pneumoniae* transformed with a novel Gram-positive lux transposon. *Infect. Immun.* **69(5)**:3350–3358.
28. Dellacasa-Lindberg I, Hitziger N, Barragan A (2007). Localized recrudescence of Toxoplasma infections in the central nervous system of immunocompromised mice assessed by in vivo bioluminescence imaging. *Microbes Infect.* **9**:1291–1298.
29. Franke-Fayard B, Waters AP, Janse CJ (2006). Real-time in vivo imaging of transgenic bioluminescent blood stages of rodent malaria parasites in mice. *Nat. Protoc.* **1**:476–485.
30. Franke-Fayard B, Janse CJ, Cunha-Rodrigues M, Ramesar J, Büscher P, Que I (2005). Murine malaria parasite sequestration: CD36 is the major receptor, but cerebral pathology is unlinked to sequestration. *Proc. Natl. Acad. Sci. USA* **102**:11468–11473.
31. Cook S, Griffin D (2003). Luciferase imaging of a neurotropic viral infection in intact animals. *J. Virol.* 77:5333–5338.
32. Burgos JS, Guzman-Sanchez F, Sastre I, Fillat C, Valdivieso F (2006). Non-invasive bioluminescence imaging for monitoring herpes simplex virus type 1 hematogenous infection. *Microbes Infect.* **8**:1330–1338.
33. Doyle TC, Nawotka KA, Kawahara CB, Francis KP, Contag PR (2006). Visualizing fungal infections in living mice using bioluminescent pathogenic *Candida albicans* strains transformed with the firefly luciferase gene. *Microb. Pathog.* **40**:82–90.
34. Jurcisek JA, Bookwalter JE, Baker BD, Fernandez S, Novotny LA, Munson RS Jr, Bakaletz LO (2007). The PilA protein of non-typeable Haemophilus influenzae plays a role in biofilm formation, adherence to epithelial cells and colonization of the mammalian upper respiratory tract. *Mol. Microbiol.* **65**:1288–1299.

Part IV
Stem Cells

Chapter 30

Reactive Oxygen Species and Upregulation of NADPH Oxidases in Mechanotransduction of Embryonic Stem Cells

Heinrich Sauer, Carola Ruhe, Jörg P. Müller, Maike Schmelter, Rochelle D'Souza, and Maria Wartenberg

Abstract

Deciphering the differentiation pathway of embryonic stem (ES) cells is a challenging task not only for basic research, but also for clinicians who intend to use ES cells for cell transplantation approaches. We have shown that reactive oxygen species (ROS) play a primordial role in the differentiation of mouse ES cells toward the cardiovascular cell lineage. During differentiation, ES cells robustly generate ROS, which interfere with signaling pathways that direct cardiac and vascular commitment. Differentiating ES cells expression of Nox-1, Nox-2, and Nox-4 has been demonstrated. We have shown that mechanical strain application to embyoid bodies grown from ES cells initiates the cardiovascular differentiation program. Under these conditions, a burst of ROS generation occurs which is followed by induction of Nox-1 and Nox-4 and a feed-forward upregulation of ROS production.

Key words: Reactive oxygen species, NADPH oxidase, Embryonic stem cells, Mechanical strain.

1. Introduction

Embryonic stem (ES) cells differentiated in vitro are commonly accepted as a promising source of cells for transplantation and tissue engineering *(1, 2, 3)*. In animal studies, ES cell-derived cardiomyocytes have been shown to form stable intracardiac grafts *(4)*, and have been successfully used to improve heart function after cardiac infarction *(5)*. Cardiomyocytes derived from pluripotent ES cells differentiate in successive maturation

From: *Methods in Molecular Biology, vol. 477: Advanced Protocols in Oxidative Stress I*,
Edited by: D. Armstrong, DOI: 10.1007/978-1-60327-517-0_30, © Humana Press, New York, NY

steps *(6, 7, 8, 9, 10, 11)*, and have been discussed to exert considerable proliferative activity for several days after isolation from multicellular embryoid bodies, thus making them a useful tool to investigate the molecular mechanisms of cardiac cell proliferation control as well as gain of cardiomyocyte function *(12)* (**Fig. 30.1**). Besides the potential to differentiate cardiac cells, ES cells have been shown to form vascular networks that may be used in approaches of therapeutic angiogenesis to dissipate oxygen gradients in ischemic tissues (**Fig. 30.1**). Currently, not much is known about the signaling mechanisms that initiate differentiation of ES cells toward the cardiovascular cell lineage.

Mechanical forces that are either exerted toward cells and tissues or are generated through themselves are essential signals that mediate cell structure, survival, proliferation, and differentiation. Mechanical forces arise in growing tissues, e.g., during embryonic development or during cell function, e.g., in beating heart and contracting muscle or through cell volume regulation in anisosmotic media. Currently localized and decentralized models of mechanotransduction have been discussed. In the localized model, the cellular signal is generated in the spatial vicinity of the cell membrane. In the decentralized model, mechanical forces, which are generated at the cell membrane are transduced to more distant locations via the cytoskeleton *(13)*. In addition to affecting mechanosensitive ion channels and the integrin cytoskeleton, ROS have been proposed to play a role in mechanotransduction *(13)*. ROS are generated mainly via a cell membrane-associated NADPH oxidase or within the respiratory

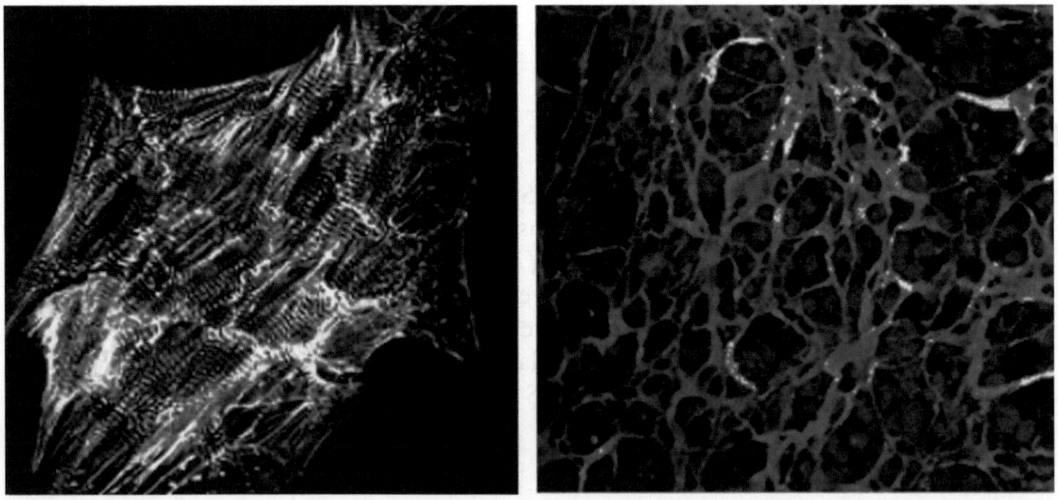

Fig. 30.1. Differentiation of ES cells toward cardiac cells (left) and capillary-like structures (right). Cardiac cells were stained with an antibody directed against α-actinin. Staining of capillary structures was performed by use of an antibody directed against PECAM-1 (CD31).

chain of mitochondria. NADPH oxidases have been found to be involved in the regulation of vascular tone, smooth muscle growth, inflammatory responses, and matrix metalloproteinase activity. They have been implicated in hypertension, atherosclerosis, heart failure, diabetic vascular disease, and restenosis *(14, 15)*. Recently, homology searches in human genome databases resulted in the discovery of six novel NADPH oxidase enzymes: Nox-1, Nox-3, Nox-4, Nox-5, Duox-1, and Duox-2. They all have at least partially similar structure and generate ROS in response to various stimuli *(16, 17)*; their physiological function is currently under intensive investigation. Recently, it has been shown by us that mechanical strain induces cardiovascular differentiation of ES cells, which may represent one of the basic principles of stem cell differentiation *(18)*. Notably, it is known for long times that differentiation of ES cells requires growth within the three-dimensional tissue of embryoid bodies in which the growing cells are subjected to mechanical forces in the expanding tissue. It was observed that mechanical strain resulted in the generation of ROS within minutes and a long-term upregulation of Nox enzymes. Comparably exogenous addition of either hydrogen peroxide *(19, 20, 21)* or generation of ROS by electrical fields *(22)* stimulated cardiac and vascular differentiation of ES cells. Interestingly, exogenous addition of prooxidants resulted in upregulation of Nox enzymes indicating that a short termed burst in ROS generation initiates a feed forward cycle of Nox activity resulting in prolonged elevation of ROS generation. Undifferentiated ES cells display only low levels of ROS generation, whereas expression of Nox enzymes as well as Nox activity is upregulated during the differentiation of ES cells within the three-dimensional tissue of embryoid bodies, thus suggesting that ROS may interfere with signaling pathways involved in differentiation processes. Recent research has demonstrated that differentiating ES cells express Nox-1, Nox-2, and Nox-4 in distinct time windows. Maximum expression of Nox-1 was found between day 6 and 8 of differentiation, i.e., just during the time where cardiomyogenesis occurs. In contrast, Nox-2 expression peaked at day 12 of differentiation. Nox-4 mRNA was continuously expressed with a maximum around day 14 of cell culture. The impact of Nox enzymes for cardiac differentiation of ES cells was recently convincingly shown in experiments where Nox-4 was downregulated in ES cell by a siRNA approach. Inhibition of Nox-4 expression totally abolished cardiomyogenesis concomitant with a decreased generation of ROS *(23)*, thus corroborating previous experiments of our group, which showed that elimination of ROS by free radical scavengers interferes with cardiomyogenesis of ES cells *(19)*. In this chapter, we offer methods for mass culture of embryoid bodies in spinner culture and communicate protocols for assessing Nox enzyme expression and function in differentiating ES cells. We describe how cardiovascular differentiation and ROS generation can be stimulated by application of mechanical strain and offer methods for downregulating Nox enzymes by an shRNA approach.

2. Materials

2.1. Equipment

1. CO_2 incubator (HeraCell, Themoscientific, Langenselbold, Germany, cat. no. 369436)
2. Centrifuge (Allegra 25R, Beckmann Coulter, Krefeld, Germany, cat. no. 369436)
3. Liquid nitrogen freezer (MIDSCI, St. Louis, MO, USA, cat. no. 50LDB)
4. Flexercell strain unit, (Model FX-4000T, Flexercell International Corporation, McKeesport, PA, USA)
5. Bioflex 6 well culture plates (Dunn Labortechnik, Asbach, Germany, cat. no. BF-3001C)
6. Confocal microscope (Leica DM IRE2, Leica Microsystems, Heidelberg, Germany)
7. Spectrophotometer (Nanodrop-1000, Peqlab Biotech, Erlangen, Germany, cat. no. 91-ND-1000)
8. Cell Spin (Integra Biosciences, Fernwald, Germany)
9. Spinner flasks (Integra Biosciences, cat. no. 182026)
10. IQ5 Real Time Cycler (Biorad, Munich, Germany, cat. no. 170–9750)

2.2. Reagents and Buffers

1. Basal Iscoves medium with glutamine (Biochrom AG, Berlin, Germany, cat. no. F0465)
2. Glasgow minimal essential medium (Sigma Aldrich, Deisenhofen, Germany, cat. no. G5154)
3. Fetal calf serum (Sigma Aldrich, Deisenhofen, Germany, cat. no. F7524)
4. Glutamine (Biochrom AG, Berlin, Germany, cat. no. K0283)
5. Nonessential amino acids (Biochrom AG, Berlin, Germany, cat. no. K0293)
6. Sodium pyruvate (Biochrom AG,Berlin, Germany, cat. no. L0473)
7. 2-mercaptoethanol (Sigma Aldrich, Deisenhofen, Germany, cat. no. M7522)
8. Penicillin/streptomycin (PAA Laboratories, Cölbe, Germany, cat. no. P11–010)

9. Leukemia inhibitory factor (ESGRO (LIF); Millipore, Billerica, MA, USA, cat. no. ESG 1106)
10. Trypsin-EDTA 1x (Invitrogen, Karlsruhe, Germany, cat. no. 25300–062)
11. Dimethyl sulfoxide (DMSO) (Calbiochem, San Diego, CA, USA, cat. no. 317275)
12. Dulbeccos PBS 1x (PAA Laboratories, Cölbe, Germany, cat. no. H15–002)
13. Gelatin (Stem Cell Technologies, Grenoble, France, cat. no. 07903)
14. Sigmacote (Sigma Aldrich, Deisenhofen, Germany, cat. no. SL2)
15. PEI, Polyethylenimine, stock solution 10 μg/μl in H_2O (Sigma-Aldrich, Deisenhofen, Germany, cat. no. 40872-7)

2.2.1. Cell Line

The most suitable cell line is the CGR8 mouse embryonic stem cell line, which can be cultivated in the absence of feeder layers (European Collection of Cell Cultures (ECACC), Wiltshire, UK, cat. no. 95011018).

2.2.2. Cell Culture Media for the Culture of ES Cells

1. Complete growth medium for monolayer culture (1x stored at 4°C).

To 10 ml Glasgow minimal essential medium, add 1 ml inactive fetal calf serum (*see* **Note 1**), 0.005 ml 2-mercaptoethanol (working solution 0.1 M), 0.1 ml glutamine (working solution 200 mM), 0.1 ml LIF (working solution 10^5 Units/ml).

2. Complete growth medium for spinner flask culture (1x stored at 4°C).

To 500 ml basal Iscoves medium, add 100 ml inactive fetal calf serum, 0.5 ml 2-mercaptoethanol (working solution 0.1 M), 6.25 ml nonessential amino acids (working solution 100x).

2.3. Stains and Supplies

1. SYTOX Green nucleic acid stain (Molecular Probes, Eugene, OR, USA, cat. no. S-7020)
2. Dichlorodihydrofluorescein diacetate (H_2DCFDA) (Molecular Probes, Eugene, OR, USA, cat. no. D-399)
3. Dihydrorhodamin 123 (Molecular Probes, Eugene, OR, USA, cat. no. D-632)

2.4. RNA Isolation

1. RNA later ICE (Ambion Inc, Austin, TX, USA, cat. no. 7030)
2. PeqGOLDTrifast (Peqlab Biotech, Erlangen, Germany, cat. no. 30–2020)
3. Isopropanol (J.T. Baker, Deventer, The Netherlands, cat. no. 8067)

4. Molecular biology grade water (Applichem, Darmstadt, Germany, cat. no. A 2864)
5. DNAse I (Invitrogen, Karlsruhe, Germany, cat. no. 18068–015)
6. Chloroform (Merck, Darmstadt, Germany, cat. no. 2445)

2.5. cDNA Synthesis

1. oligo dT (Roche Diagnostics, Mannheim, Germany, cat. no. 10814270001)
2. DTT (Invitrogen, Karlsruhe, Germany, cat. no. Y000147)
3. RNAsin (Promega, Madison, WI, USA, cat. no. N2611)
4. MMLV-RT (Invitrogen, Karlsruhe, Germany, cat. no. 10297-0117)
5. dNTPs (Invitrogen, Karlsruhe, Germany, cat. no. 18427-013)

2.6. Polymerase Chain Reaction (PCR)

1. iQ SYBR green Super mix (Biorad, Munich, Germany, cat. no. 170–8882)
2. Primers (Invitrogen, Karlsruhe, Germany)
   2.1 Target genes:
   Nox-1
   Fwd: 5′-AAT GCC CAG GAT CGA GGT-3′
   Rev: 5′-GAT GGA AGC AAA GGG AGT GA-3′
   Nox-2
   Fwd: 5′-ACC TTA CTG GCT GGG ATG AA-3′
   Rev: 5′-TGC AAT GGT CTT GAA CTC GT-3′
   Nox-4
   Fwd: 5′-GAT CAC AGA AGG TCC CTA GCA G-3′
   Rev: 5′-GTT GAG GGC ATT CAC CAA GT-3′
   2.2 Housekeeping genes:
   BACT
   Fwd: 5′-GAT GAC CCA GAT CAT GTT TGA G-3′
   Rev: 5′-CCA TCA CAA TGC CTG TGG TA-3′
   GAPDH
   Fwd: 5′-TCG TCC GGT AGA CAA AAT GG-3′
   Rev: 5′-GAG GTC AAT GAA GGG GTC GT-3′

2.7. Immunohistochemistry

1. anti Mox-1 (Nox-1) (Santa Cruz Biotechnology, Santa Cruz, CA, USA, cat. no. sc-10183)
2. anti Nox-4 (Santa Cruz Biotechnology, Santa Cruz, CA, USA, cat. no. sc-21860)
3. anti p22-phox (Santa Cruz Biotechnology, Santa Cruz, CA, USA, cat. no. sc-11712)
4. anti p47-phox (Santa Cruz Biotechnology, Santa Cruz, CA, USA, cat. no. sc-7660)

5. anti p67-phox (Santa Cruz Biotechnology, Santa Cruz, CA, USA, cat. no. sc-15342)
6. anti gp91-phox (Santa Cruz Biotechnology, Santa Cruz, CA, USA, cat. no. sc-5827)
7. Cy-5 rabbit anti-goat IgG (Fab2-fragment) (Dianova, Hamburg, Germany, cat. no. 305-175-047)
8. Cy-5 goat anti-rabbit IgG (Fab2-fragment) (Dianova, Hamburg, Germany, cat. no. 111-175-006)
9. PBST (1%, 0.1%, 0.01%) Dulbeccos PBS 1x (PAA Laboratories, Cölbe, Germany, cat. no. H15-002) supplemented with 1%, 01%, 0.01% Triton X-100 (Sigma-Aldrich, Deisenhofen, Germany, cat. no. T-8787), respectively.
10. Fluoromount-G (Southern Biotech, Birmingham, USA, cat. no. 0100-01)

2.8. shRNA Experiments

1. HEK293FT-based amphotropic Phoenix packaging cells (Phoenix-Ampho, Invitrogen, Karlsruhe, Germany, cat. no. R700-07), capable of producing gag-pol and envelope protein for amphotropic viruses,
2. 6-well tissue culture plates (Nunc, Wiesbaden, Germany, cat. no. 140675)
3. 12-well tissue culture plates (Nunc, Wiesbaden, Germany, cat. no. 150628)
4. Dulbecco modified Eagle medium (DMEM) (Invitrogen, Karlsruhe, Germany, cat. no. 1965-039)
5. Penicillin/Steptomycin – L-Glutamine (Cambrex IA, East Rutherford, NJ, USA cat. no. 17-718R)
6. Defined fetal calf serum (Sigma-Aldrich, Deisenhofen, Germany, cat. no. F-7524)
7. 0.2 μm low-protein binding syringe filter holders (Sartorius, Göttingen, Germany, cat.no. 16534 K)
8. PEI, Polyethylenimine, stock solution 10 μg/μl in H_2O (Sigma-Aldrich, Deisenhofen, Germany, cat. no. 40872-7)
9. Polybrene (1,5-dimethyl-1,5-diazaundecamethylene polymethobromide, Sigma-Aldrich, Deisenhofen, Germany, cat. no. AL-118, cat. no # 10,768–9)
10. Puromycin (10 mg/ml stock solution) (Sigma-Aldrich, Deisenhofen, Germany, cat. no. P-8833)
11. EndoFree Plasmid Maxi Kit (Qiagen, Hilden, Germany, cat. no. 12362) for plasmid isolation
12. pLKO.1-puro derivative plasmids carrying shRNA targeting NADPH oxidase 1 (NM_172203) (MISSION™ TRC shRNA Target Set, clones NM_172203.1-882s1c1 targeting sequence

CGTGATTACCAAGGTTGTCAT), NM_172203.1-444s1c1 (target sequence GAAAGAAGATTCTTGGCTAAA), and NM_172203.1-1236s1c1 (target sequence CTTGAAATC-TATCTGGTACAA) (*see* **Note 2**)

13. MISSION™ Lentiviral Packaging Mix (Sigma-Aldrich, Deisenhofen, Germany, cat. no. SHP001)
14. MISSION™ nonTarget shRNA control vector (Sigma-Aldrich, Deisenhofen, Germany, cat. no. SHC002)
15. MISSION™ eGFP Control Vector (Sigma-Aldrich, Deisenhofen, Germany, cat. no. SHC005)

3. Methods

3.1. Mass Culture Techniques for Embryoid Bodies – The Spinner Flask Technique

Routine screening of embryotoxic and/or anti-angiogenic agents as well as transplantation of organ-specific cells requires large numbers of embryoid bodies differentiated from ES cells. In most laboratories, the hanging drop method is applied, which was previously developed by Wobus and coworkers *(24)* and has been proven useful for the characterization of the developmental aspects of cardiomyogenesis *(11)*, skeletal- *(25)*, and smooth muscle *(26)* development as well as neurogenesis *(27)*. However, this technique is limited by the low yield of embryoid bodies, since a single hanging drop has to be prepared manually for each embryoid body. To achieve mass cultures of embryoid bodies with a high efficiency of differentiation, a spinner flask culture technique (CellSpin, Integra Biosciences, Fernwald, Germany) was previously developed in our laboratory *(28)*. It was observed that by use of this technique large amounts of embryoid bodies (about 1,000/spinner flask) could be cultivated. Embryoid bodies cultivated in spinner flask culture grew to a significantly larger size as compared to cultures of embryoid bodies in methylcellulose-supplemented cell culture medium or cultivation using the liquid overlay technique, thereby indicating an improved supply with oxygen and nutrients. The spinner flask technique proved efficient for the differentiation of cells of the endothelial cell and the cardiac cell lineage.

3.1.1. Resuscitation of Frozen CGR8 ES Cells and ES Cell Culture

All culture incubations should be performed in a humidified 37°C, 5% CO_2 incubator unless otherwise specified. The media may require different levels of CO_2 to maintain pH 7.4.

1. Remove vial from liquid nitrogen freezer and immediately place it into a 37°C water bath. Agitate vial continuously until medium is thawed (*see* **Note 3**).
2. Transfer thawed cell suspension into a sterile centrifuge tube containing 4 ml warm complete medium (Glasgow minimal

essential medium). Centrifuge for 4 min at 800× *g*, room temperature. Discard supernatant (*see* **Note 4**).

3. Gently resuspend cell pellet in small amount (1 ml) of complete medium and transfer to gelatin-coated culture plate containing the appropriate amount of medium.
4. Check cultures after 24 h to ensure that cells have attached to the plate.
5. Change medium every day.

3.1.2. Procedure for Spinner Flask Preparation Prior to the Addition of ES Cells

1. Wash clean spinner flasks with excessive Milli-Q-plus water and dry for 1 h at 60°C.
2. Siliconize spinner flasks by moistening the interior as well as the mallets with Sigmacote. Excessive Sigmacote is removed from the flasks using a 10 ml glass pipette.
3. Dry the silicon coat in an oven for 1 h at 120°C.
4. Rinse spinner flasks three times with 250 ml Milli-Q-plus water and autoclave subsequently.
5. Moisten the interior of the flasks with 20 ml of complete Iscoves medium prior to the addition of ES cells; exchange the medium for 125 ml of complete Iscoves medium.

3.1.3. Procedure for the Cleaning of the Spinner Flask at the End of the Experiment

After the end of the experiment, the spinner flasks have to be cleaned prior to the inoculation with fresh ES cells.

1. Remove the old medium with residual embryoid bodies and cells.
2. Wash the flasks with 70% ethanol and subsequently with 1 l water.
3. Remove the silicon coat by adding 250 ml 5 N NaOH to the spinner flasks for a maximum of 12 h.
4. Remove NaOH and wash the flasks with at least 5 l water. The interior of the flasks is thoroughly cleaned with a brush. Subsequently, rinse the flasks with 1 l of Milli-Q-plus water.

3.1.4. Inoculation of Spinner Flasks with ES Cells

1. Wash ES cells grown in 6 cm cell culture Petri dishes once with 0.2% trypsin and 0.05% EDTA in PBS.
2. Remove the trypsin solution and incubate ES cells for 5 min with 2 ml 0.2% trypsin and 0.05% EDTA.
3. Triturate the cells with a 2 ml glass pipette until the ES cell clusters are dissociated and a single cell suspension is achieved (control under a microscope).
4. Prepare spinner flasks (as described under **Section 3.1.2**) and seed ES cells at a density of $1 \cdot 10^7$ cells/ml in 125 ml complete Iscoves cell culture medium. Stir at a speed of 20 rotations/min. The stirring direction is reversed every 1,440 degrees.

5. Add 125 ml Iscoves complete cell culture medium after 24 h to yield a final volume of 250 ml.
6. Exchange 125 ml of the cell culture medium every day.
7. For the differentiation of cardiomyocytes, embryoid bodies are carefully removed from the spinner flask by the use of 10 ml plastic pipettes to avoid any cell injury.
8. They are subsequently plated to 10 cm cell culture dishes filled with 20 ml complete Iscoves cell culture medium. After 24–48 h, spontaneous beating activity of the embryoid bodies indicates the differentiation of cardiac cells.

3.2. Application of Mechanical Strain to ES Cell-Derived Embryoid Bodies

Cells are subjected to mechanical strain using the Flexercell Strain Unit (model FX-4000T, Flexercell International Corporation, McKeesport, PA). In this system, mechanical strain is obtained using 6-well, flexible-bottomed, collagen-coated culture plates (Bioflex culture 6-well plates, Dunn Labortechnik, Germany, cat. no. BF-3000IC) with a hydrophilic surface, capable of up to 20% stretch (**Fig. 30.2**). Frequency, strain rate, and degree of elongation of the culture substratum are controlled with a computer by regulating the rate of evacuation (vacuum level) and rate of air influx to the cell culture plate bottoms.

1. Plate of ES cell-derived embryoid bodies on Bioflex membranes. For the experiments 10 to 15 3-day-old embryoid bodies are plated in complete cell culture medium into 6-well, flexible-bottomed culture plates, which are filled with 3 ml complete cell culture medium (**Fig. 30.2**).
2. Incubate for 24 h to allow the embryoid bodies to attach and grow out on the surface of the Bioflex membrane (**Fig. 30.3**).
3. After 24 h, outgrown embryoid bodies are subjected to mechanical strain by elongation of the flexible membranes by either 5%, 10%, or 20%. Either static or cyclic strain protocols may be used to apply mechanical strain to the cells.
4. Change cell culture medium in mechanical strain-treated and control embryoid body cultures and place cultures in a conventional CO_2 incubator.
5. Assess cell vitality after mechanical strain application. Mechanical strain application should not impair cell vitality. Therefore, cell viability after mechanical strain application is controlled by the lethal cell dye SYTOX green.
6. At different times after mechanical strain application, 0.1 μM of SYTOX green dissolved in complete cell culture medium is added to the strained cell cultures. SYTOX green is a high-affinity nuclear acid stain that easily penetrates cells that have

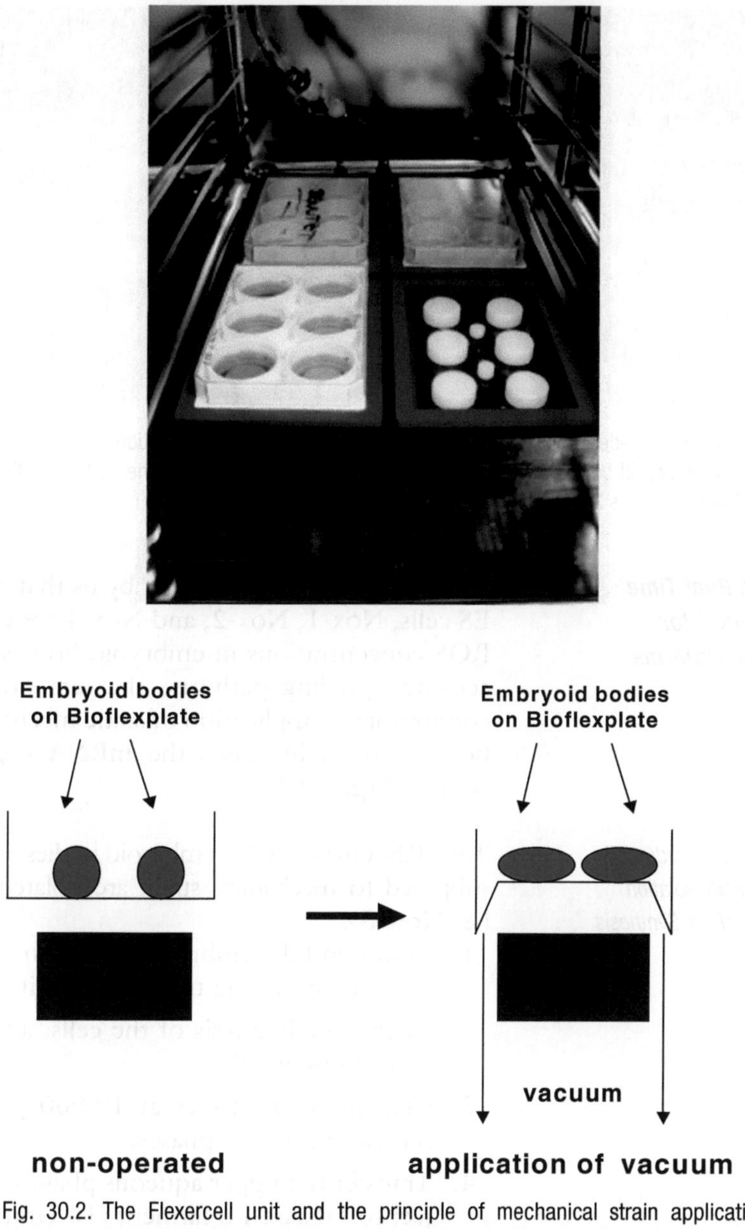

Fig. 30.2. The Flexercell unit and the principle of mechanical strain application. Embryoid bodies grown from ES cells are plated in 6-well-plates, the bottom of which is formed by a flexible silicone membrane. Application of vacuum to the system results in mechanical strain of the membrane across pistils, which are located underneath the flexible membranes.

compromised plasma membranes and yet will not cross the membranes of living cells.

7. Incubate embryoid body cell cultures for 20 min with 0.1 μM SYTOX green and count SYTOX green-positive cell nuclei by confocal laser scanning microscopy using an excitation wavelength of 488 nm and an emission wavelength of >515 nm.

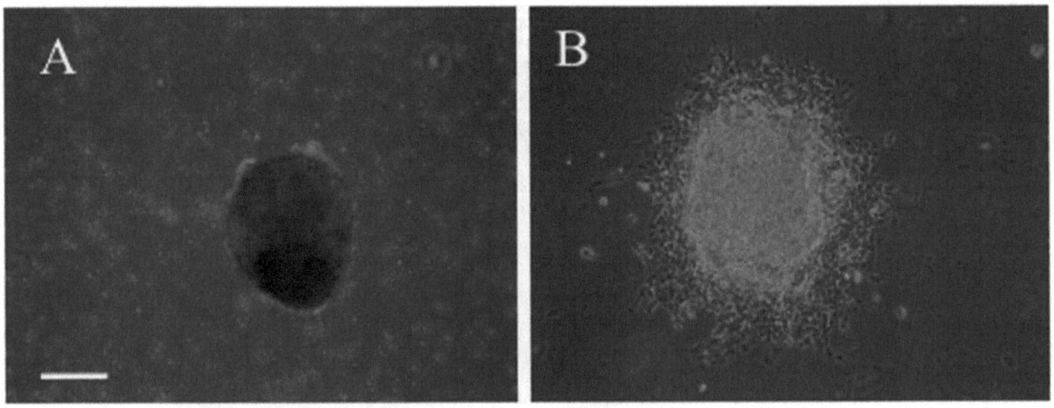

Fig. 30.3. ES cell-derived 3-day-old embryoid bodies shortly after addition to the Bioflex plates (**A**) and 24 h thereafter (**B**). The embryoid bodies with a size of 300–400 μm easily grow on the surface of the membranes and attach firmly. The bar represents 200 μm.

3.3. Real Time RT-PCR for Nox-Isoforms

It has been previously shown by us that during differentiation of ES cells, Nox-1, Nox-2, and Nox-4 are expressed, which control ROS concentrations in embryoid bodies (**Fig. 30.4**). ROS may activate signaling pathways that are involved in cardiovascular commitment. Application of static mechanical strain to embryoid body cultures increases the mRNA expression of Nox-1 and Nox-4 (**Fig. 30.4**).

3.3.1. Procedure for RNA Isolation and cDNA Synthesis

Total RNA from CGR8 embryoid bodies, which had previously been subjected to mechanical strain are isolated using PeqGOLDTrifast (*see* **Note 5**).

1. Resuspend the embryoid bodies in 1 ml of PeqGOLDTrifast in microcentrifuge tubes and stir if needed (*see* **Note 6**).
2. After complete lysis of the cells, add 200 μl of chloroform and agitate gently.
3. Centrifuge the tubes at 12,000 g for 20 min at 4°C for separation of two phases.
4. Transfer the upper aqueous phase into a fresh tube and subject to DNAse I treatment (10 Units for every μl of sample) for 1 h at 37°C.
5. Subject samples to chloroform treatment mentioned under point 2 until no precipitate is seen at the junction of the two layers.
6. Transfer the aqueous phase into a fresh tube and add a few drops of 3 M sodium acetate and two volumes of ethanol.
7. Incubate the mixture at −20°C for 1 h and then centrifuge at 12,000 g for 20 min at 4°C (*see* **Note 7**).
8. Wash the pellet with 70% ethanol at 12,000 g for 20 min and dry at room temperature.

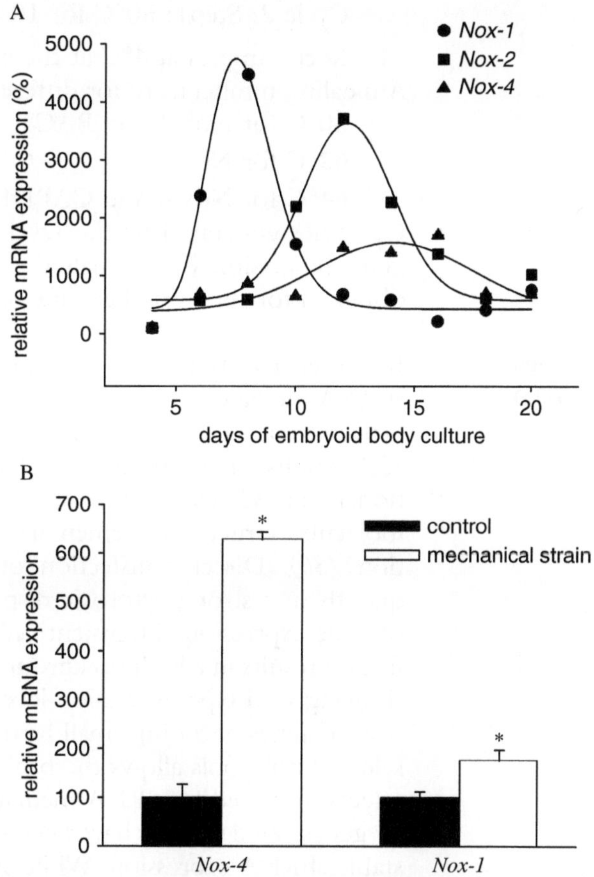

Fig. 30.4. Expression of Nox-1, Nox-2, and Nox-4 during the time course of embryoid body differentiation (**A**), and effects of mechanical strain application (10% elongation of the membrane) on mRNA expression of Nox-1 and Nox-4 (**B**). It is evident that upon mechanical strain, Nox enzymes are upregulated. Reproduced with permission from *(20)*.

9. Resuspend the pellet in 10 μl of molecular biology grade water. Determine the concentration of RNA spectrophotometically with Nanodrop ND-1000.
10. Perform cDNA synthesis with 2 μg of RNA in 20 μl reaction with MMLV-RT.

3.3.2. Semiquantitative PCR

Primer concentration in a 20 μl reaction is 10 pM and the reaction is carried out in triplicates. The following protocol is used for PCR:

1. Cycle1: step 1 – denaturation at 95°C for 15 min
2. Cycle 2:
   Step 1: denaturation at 95°C for 30 s
   Step 2: specific annealing temperature for 30 s
   Step 3: extension at 72°C for 30 s

Cycle 2 is repeated 45 times.

3. Cycle 3: Step1: 50°C for 10 min.
4. Keep samples at 4°C at the end of the experiment.

Annealing temperature for different genes are as follows:

60°C for Nox-1 and BACT

62°C for Nox-2

64°C for Nox-4 and GAPDH

C_T values generated by the Icycler are used to calculate relative gene expression by normalizing the values with the C_T values obtained for the housekeeping gene using the $\Delta\Delta C_T$ method.

3.4. Stable Down Regulation of NOX Gene Expression

In order to study the function of particular proteins, stable shRNA-mediated down-regulation of gene expression has been proven to be a powerful tool. The process called RNA interference *(29)* results in the reduction of proteins through posttranscriptional gene silencing, when short small interfering RNA (siRNA) molecules target complementary mRNA molecules for degradation *(30)*. Direct transfection of siRNA molecules results frequently in a strong cytotoxic response *(31)* and down-regulation of gene expression is transient only. In addition, inefficient transfection results in a high background caused by nontransfected cells showing wild type expression level of target genes. Stable expression of genes encoding small hairpin RNA (shRNA) molecules of selected cell pools allows the background-free study of long-term effects mediated by siRNA-mediated down-regulation of particular genes. Viral systems have been shown to be successfully used for stable shRNA expression. While standard retroviral systems allow the efficient introduction of gene expression cassettes into proliferating cells, lentiviral systems can be used for typically difficult cell lines, such as primary cells or nondividing cells. The lentiviral expression system comprises the shRNA cassette usually expressed by a RNA polymerase III H1 or U6 promoter, the LTR regions necessary for their stable integration into the chromosome, and a selective marker for the selection of infected cell pool (**Fig. 30.5**). For selection of infected cells, expression of fluorescence markers (i.e., green fluorescence protein and its derivatives) or antibiotic resistance cassettes are commonly used. In addition, a fluorescence labeled expression cassette can serve to optimize transfection and

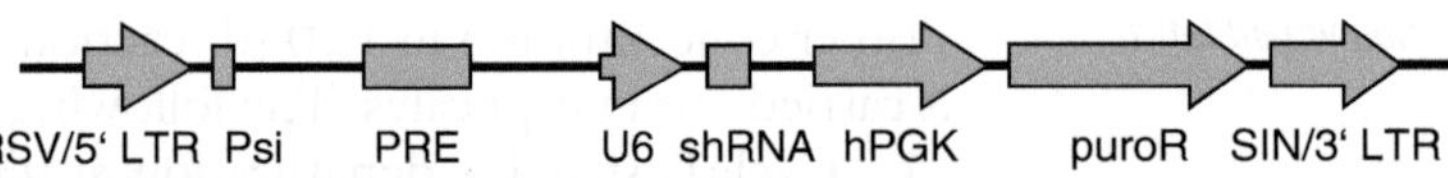

Fig. 30.5. Lentiviral shRNA-adapted expression vector. Lentiviral expression vector comprises a 5′ long terminal repeat (RSV/5′LTR), a RNA packaging sequence (Psi), a Rev response element (RRE), a U6 promoter (U6), the shRNA gene, the human phosphoglycerate kinase promoter (hPGK), a puromycin resistance gene (puroR), and a 3′ self inactivating long terminal repeat (SIN/3′ LTR). The domains are placed on a bacterial vector allowing the amplification in *E. coli*.

infection efficiency when using a cell line for the first time. Currently, numerous companies offer expression systems of stable shRNA production. We successfully used and describe here the MISSION shRNA system from Sigma Aldrich, which is based on a lentiviral packaging system. Similar systems can be applied with the same methodological repertoire. The procedure involves the production of pseudoviral particles by transfection of plasmids into a packaging cell line, the isolation of viral particles, and the subsequent infection and selection of a target cell line.

3.4.1. Production of Lentiviral Particles

To obtain high viral titers, it is important to frequently passage Phoenix packaging cells. A 100% confluence should be prevented. The cells should remain in a fast growth phase while producing viral particles. In order to obtain high viral titers make sure that the efficiency of transfection was above 80%.

1. Seed 3×10^5 to 5×10^5 Phoenix packaging cells per well of a 6-well plate resulting in 40% to 70% confluency. Make sure that cells get evenly distributed over the well.
2. Incubate the plate at 37°C overnight (*see* **Note 8**).
3. Prewarm DMEM medium without fetal calf serum to room temperature. Dilute 2 μg lentiviral expression plasmid (pLKO.1-derivative plasmids encoding the shRNA genes) and the ready prepared lentiviral packaging mix in 0.25 ml DMEM. In parallel, prepare pLKO.1 derivative plasmid encoding a nontargeting shRNA (*see* **Note 9**).
4. Dilute PEI stock solution in 0.25 ml DMEM medium, take 2.5 μg PEI per 1 μg DNA.
5. Add diluted PEI solution to the DNA solution, vortex immediately, and incubate mixture at room temperature for 20–30 min.
6. Replace fetal calf serum-containing DMEM medium from the packaging cell line and with 2 ml prewarmed DMEM medium. Prevent detachment of cells.
7. Add DNA-PEI solution drop-wise to the cells. Carefully spread solution over the entire surface of the well. Incubate 4–8 h.
8. Replace PEI-containing DMEM medium by DMEM supplemented with 10% FCS. Incubate cells for 24–48 h.
9. Viral particles can be harvested after 24–48 h posttransfection. Best titers were obtained by harvesting virus-containing media 24 h and a second time 48 h posttransfection. After removing media from the cells, carefully add prewarmed fresh fetal calf serum-containing DMEM medium. Again handle cells with care to prevent detachment.
10. Collect virus-containing medium by transferring media into a syringe and filter it through the sterile low-protein binding

0.2 μm filter unit into collection tubes (i.e. safe seal Eppendorf tubes or something comparable). Alternatively, cell-free virus containing supernatant can be obtained by centrifugation of the medium for 5 min at 500× *g* and subsequent collection of the supernatant.

11. The viral particles can be stored at 4°C for short time storage (up to several days). Since the particles are rather unstable, for long-term storage, freeze at −80°C. Aliquot samples to prevent refreezing.

3.4.2. Infection of Target Cells, Selection of Stable Clones

1. Seed the targeting cell line into a 12 well plate; cell concentration should be seeded sparsely of about 2×10^4 to 5×10^4 per well. Use medium for optimal growth conditions of the recipient cells (*see* **Note 10**).
2. Add polybrene (final concentration 8 μg/ml) to each well. Gently swirl the plate to homogenize (*see* **Note 11**). Prevent cells from detaching.
3. Add lentiviral particles to the wells. Gently distribute in the medium. Amount of virus-containing supernatant depends on viral titer (*see* **Note 12**).
4. Titrate amount of virus particles between 5–500 μl if you transduce a cell line for the first time.
5. Subsequently, centrifuge plates at 500× *g* for 60 min and further incubate plates at 37°C in a humidified incubator in an atmosphere of 5% CO_2.
6. Around 6–8 h postinfection, remove polybrene and virus-containing medium and replace by fresh growth medium optimized for the cells used. Make sure that the cells do not detach from the well surface. In order to raise the efficiency of transduction, a second and a third infection round can be carried out by adding new virus particles to the cells and the plates spun again. Incubate cells at 37°C in a humidified incubator in an atmosphere of 5% CO_2 (*see* **Note 13**).
7. After 24–48 h, replace cell growth medium and add puromycin for selection. Resistance level against puromycin varies for cell type to cell type. Usually, 2–10 μg/ml is sufficient for selection. If a cell line is used for the first time, determine resistance level by titrating puromycin concentration between 0.5 and 20 μg/ml. To monitor efficiency of selection, leave one well uninfected and select with the same amount of puromycin.
8. Selection against puromycin should be maintained from 10 to 14 days. If no viable cells are visible in the uninfected control wells, selection can be stopped.

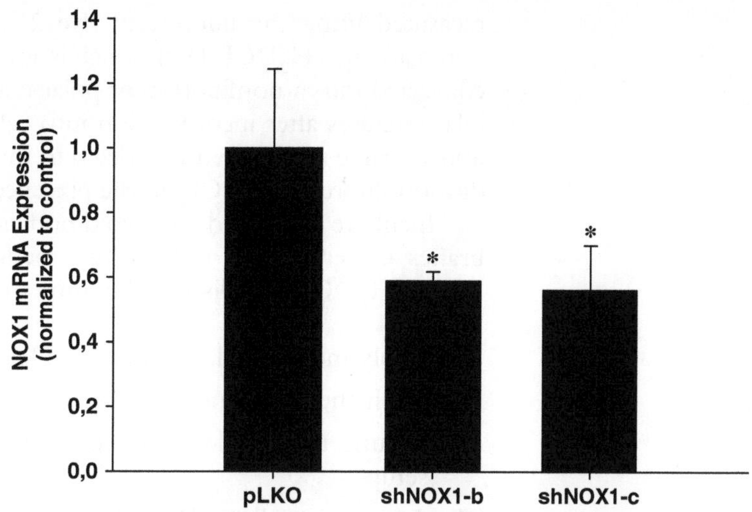

Fig. 30.6. Nox-1 gene silencing by sh-RNA technology. Gene silencing was assessed by quantitative, real-time RT-PCR analysis using NOX-1 primers. *Expression of Nox-1 was normalized to the mean expression of housekeeping gene.*

3.4.3. Validation of Gene Suppression

For validation of efficiency of shRNA-mediated suppression of gene expression, characterization of the mRNA level by using RT-PCR (**Fig. 30.6**) or if an antibody is available – immunological analysis of the protein level should be carried out.

3.5. Measurement of ROS Generation

During differentiation, ES cell-derived embryoid bodies endogenously generate ROS through NADPH oxidase activity *(19, 22)*. Application of mechanical strain raises intracellular ROS concentrations within minutes (**Fig. 30.7**). Intracellular ROS levels are

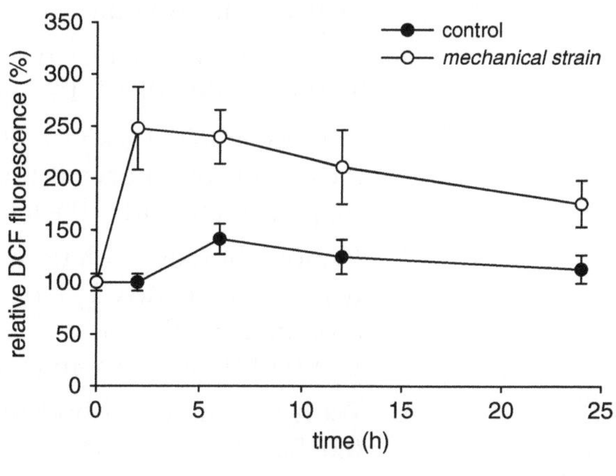

Fig. 30.7. ROS generation following mechanical strain application by use of the ROS-sensitive fluorescence indicator H_2DCF-DA. Mechanical strain (10% elongation of the membrane) was applied at time 0 and fluorescence of oxidized DCF was monitored by confocal laser scanning microscopy at times indicated. *Reproduced with permission from (18).*

measured using the fluorescent dye 2′7′-dichlorodihydrofluorescein diacetate (H_2DCF-DA), which is a nonpolar compound that is converted into a nonfluorescent polar derivative (H_2DCF) by cellular esterases after incorporation into cells. H_2DCF is membrane impermeable and is rapidly oxidized to the highly fluorescent 2′,7′-dichlorofluorescein (DCF) in the presence of intracellular ROS.

Incubate 4-day-old embryoid bodies plated on Bioflex membranes in serum-free cell culture medium supplemented with 20 μM H_2DCF-DA dissolved in dimethyl sulfoxide (DMSO) for 20 min.

1. Apply mechanical strain in the Flexercell System.
2. Wash the plate once with serum-free cell culture medium.
3. Mount the Bioflex plate onto the stage of the confocal setup.
4. Assess intracellular DCF fluorescence directly after mechanical strain application for online recording of ROS generation. DCF fluorescence (corrected for background fluorescence) is evaluated in 3600 μm^2 regions of interest using an overlay mask. For fluorescence excitation, the 488 nm band of the argon ion laser of the confocal setup is used. Emission is recorded using at wavelengths >LP515 nm (*see* **Note 14**).

3.6. Immunohistochemistry of NADPH Oxidase Subunits

Immunohistochemistry is performed with whole mount embryoid bodies plated onto Bioflex membranes. We have shown that mechanical strain results in upregulation of NADPH oxidase subunits (**Fig. 30.8**).

1. Apply mechanical strain to 4-day-old embryoid bodies plated on day 3 of cell culture on Bioflex membranes.
2. After 24 h of postincubation, fix cells in ice-cold (−20°C) methanol for 60 min.
3. Wash embryoid bodies twice in PBS supplemented with 1% Triton X-100 (PBST 1%) and keep in PBST for 15 min.
4. Remove PBST 1% and block against unspecific binding by incubation of embryoid bodies for 60 min in PBST 0.01% supplemented with 10% fetal calf serum.
5. Remove blocking solution and incubate with primary antibodies (either Mox-1, Nox-2, Nox-4, p22-phox, p47-phox, p-67-phox, gp91-phox) at a dilution of 1:20 in blocking solution for 90 min at room temperature (*see* **Notes 11** and **12**).
6. Wash the embryoid body cultures for three times in PBST 0.01%.
7. Incubate embryoid bodies with secondary antibodies dissolved at a concentration of 1.5 μg/ml in blocking solution for 60 min.
8. Wash embryoid bodies for five times in PBST 0.01%.

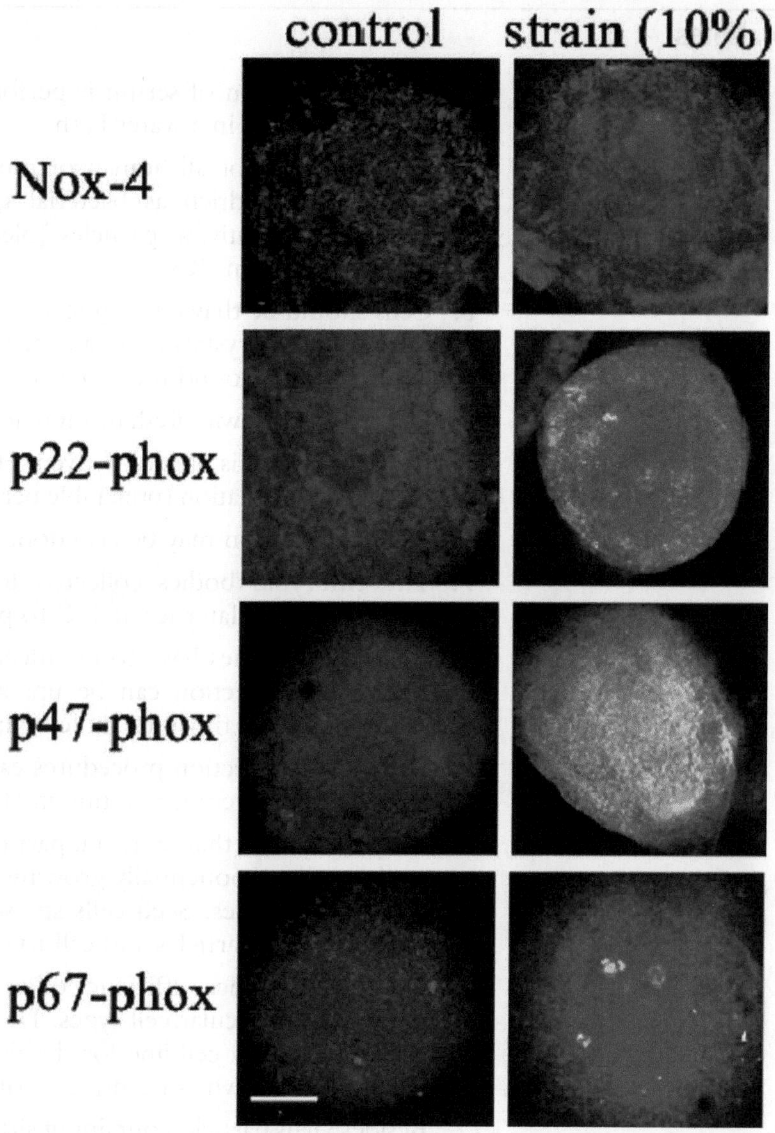

Fig. 30.8. Immunohistochemistry of NADPH oxidase subunits 24 h after application of mechanical strain (10% elongation of the membrane for 2 h). Whole mount embryoid bodies outgrown on Bioflex membranes were immunostained against Nox-4, p22-phox, p47-phox, and p67-phox. The bar represents 300 μm. *Reproduced with permission from (18).*

9. Remove PBST 0.01% and aspirate residual solution with a paper towel.
10. Transfer the Bioflex membranes to object slides and add 20 μl of Fluoromount-G solution.
11. Deck the cells with 20 × 20 mm coverslips and incubate for 24 h in the dark at room temperature to harden the Fluoromount-G solution.

4. Notes

1. Heat inactivation of serum is performed by incubation for 30 min at 56°C in a water bath.
2. Targeting sets for all human or murine genes are available from Sigma Aldrich as bacterial glycerol stocks, purified plasmids, or lentiviral particles (please check: http://www.sigmaaldrich.com/RNAi).
3. Cells should be thawed as quickly as possible to prevent formation of ice crystals that can cause cell lysis. Try to avoid getting water around the cap of the vial.
4. Cells are washed with fresh medium to remove residual DMSO.
5. PeqGold Trifast is phenol-derived. Hence care must be taken during RNA isolation (preferable performed in a fume hood).
6. The precipitation may be also done overnight.
7. The embryoid bodies collected from the plates may be stored in RNA, later ice at 4°C to preserve RNA integrity.
8. If several cell lines have to be infected with the virus particles, the transfection can be upscaled to 100 mm dishes resulting in two times 10 ml lentiviral supernatant.
9. Different transfection procedures can be used. We use PEI-mediated transfection resulting in efficient transfection rates.
10. Despite the fact that lentiviral particles can infect nonproliferating cells, exponentially growing cultures result in higher transduction rates. Seed cells sparsely to have a high ratio between virus particles and cell to be infected.
11. Polybrene enhances efficiency of virus infection, but is detrimental for particular cell types. Determine cell sensitivity if you work with a cell line for the first time. If cells lose viability, titrate down concentration of polybrene used.
12. Besides virus particle-containing shRNA, targeting and non-targeting shRNA cassettes of virus particle-containing GFP should be transduced in parallel allowing easy monitoring of the efficiency of infection.
13. If nonadherent cells are used, transfer cells in a polypropylene tube and spin them for 5 min at 500 g. Subsequently resuspend cell pellet in fresh growth medium and transfer cells back into wells.
14. As an alternative to the redox-sensitive dye H_2DCF-DA, the redox-sensitive fluorescence dye dihydrorhodamine-123 (Molecular Probes) has been successfully used. Embryoid bodies are stained for 10 min with 10 μM of the substance. Fluorescence excitation is performed with the 543 nm band of a He/Ne laser. Emission is recorded at 570–630 nm.

15. To spare antibody solution, selected areas of the Bioflex membranes with embryoid bodies attached to them can be cut out using a scalpel. These membrane patches can be incubated with antibody solutions in smaller incubation chambers.
16. Alternatively, staining with primary antibodies can be performed overnight at 4°C.

Acknowledgments

This work was supported by the DFG graduate college 534 and the DFG excellence cluster Cardiopulmonary System.

References

1. Dowell J. D., Rubart M., Pasumarthi K. B., Soonpaa M. H., and Field L. J. (2003) Myocyte and myogenic stem cell transplantation in the heart. *Cardiovasc. Res.* **58**, 336–350.
2. Hassink R. J., Dowell J. D., Brutel D. L. R., Doevendans P. A., and Field L. J. (2003) Stem cell therapy for ischemic heart disease. *Trends Mol. Med.* **9**, 436–441.
3. Nir S. G., David R., Zaruba M., Franz W. M., and Itskovitz-Eldor J. (2003) Human embryonic stem cells for cardiovascular repair. *Cardiovasc. Res.* **58**, 313–323.
4. Klug M. G., Soonpaa M. H., Koh G. Y., and Field L. J. (1996) Genetically selected cardiomyocytes from differentiating embryonic stem cells form stable intracardiac grafts. *J. Clin. Invest.* **98**, 216–224.
5. Behfar A., Zingman L. V., Hodgson D. M., Rauzier J. M., Kane G. C., Terzic A. et al. (2002) Stem cell differentiation requires a paracrine pathway in the heart. *FASEB J.* **16**, 1558–1566.
6. Sachinidis A., Fleischmann B. K., Kolossov E., Wartenberg M., Sauer H., and Hescheler J. (2003) Cardiac specific differentiation of mouse embryonic stem cells. *Cardiovasc. Res.* **58**, 278–291.
7. Hescheler J., Fleischmann B. K., Wartenberg M., Bloch W., Kolossov E., Ji G. et al. (1999) Establishment of ionic channels and signalling cascades in the embryonic stem cell-derived primitive endoderm and cardiovascular system. *Cells Tissues Organs* **165**, 153–164.
8. Hescheler J., Wartenberg M., Fleischmann B. K., Banach K., Acker H., and Sauer H. (2002) Embryonic stem cells as a model for the physiological analysis of the cardiovascular system. *Methods Mol. Biol.* **185**, 169–187.
9. Boheler K. R., Czyz J., Tweedie D., Yang H. T., Anisimov S. V., and Wobus A. M. (2002) Differentiation of pluripotent embryonic stem cells into cardiomyocytes. *Circ. Res.* **91**, 189–201.
10. Wobus A. M., Guan K., Yang H. T., and Boheler K. R. (2002) Embryonic stem cells as a model to study cardiac, skeletal muscle, and vascular smooth muscle cell differentiation. *Methods Mol. Biol.* **185**, 127–156.
11. Wobus A. M., Rohwedel J., Maltsev V., and Hescheler J. (1995) Development of cardiomyocytes expressing cardiac-specific genes, action potentials, and ionic channels during embryonic stem cell-derived cardiogenesis. *Ann. N. Y. Acad. Sci.* **752**, 460–469.
12. Kehat I., and Gepstein L. (2003) Human embryonic stem cells for myocardial regeneration. *Heart Fail. Rev.* **8**, 229–236.
13. Ali M. H., and Schumacker P. T. (2002) Endothelial responses to mechanical stress: where is the mechanosensor? *Crit. Care Med.* **30**, S198–S206.
14. Griendling K. K. (2004) Novel NAD(P)H oxidases in the cardiovascular system. *Heart* **90**, 491–493.
15. Sorescu D., and Griendling K. K. (2002) Reactive oxygen species, mitochondria, and NAD(P)H oxidases in the development and progression of heart failure. *Congest. Heart Fail.* **8**, 132–140.

16. Orient A., Donko A., Szabo A., Leto T. L., and Geiszt M. (2007) Novel sources of reactive oxygen species in the human body. *Nephrol. Dial. Transplant.* **22**, 1281–1288.
17. Geiszt M. (2006) NADPH oxidases: new kids on the block. *Cardiovasc. Res.* **71**, 289–299.
18. Schmelter M., Ateghang B., Helmig S., Wartenberg M., and Sauer H. (2006) Embryonic stem cells utilize reactive oxygen species as transducers of mechanical strain-induced cardiovascular differentiation. *FASEB J.* **20**, 1182–1184.
19. Sauer H., Rahimi G., Hescheler J., and Wartenberg M. (2000) Role of reactive oxygen species and phosphatidylinositol 3-kinase in cardiomyocyte differentiation of embryonic stem cells. *FEBS Lett.* **476**, 218–223.
20. Buggisch M., Ateghang B., Ruhe C., Strobel C., Lange S., Wartenberg M. et al. (2007) Stimulation of ES-cell-derived cardiomyogenesis and neonatal cardiac cell proliferation by reactive oxygen species and NADPH oxidase. *J. Cell Sci.* **120**, 885–894.
21. Sauer H., Bekhite M. M., Hescheler J., and Wartenberg M. (2005) Redox control of angiogenic factors and CD31-positive vessel-like structures in mouse embryonic stem cells after direct current electrical field stimulation. *Exp. Cell Res.* **304**, 380–390.
22. Sauer H., Rahimi G., Hescheler J., and Wartenberg M. (1999) Effects of electrical fields on cardiomyocyte differentiation of embryonic stem cells. *J. Cell Biochem.* **75**, 710–723.
23. Li J., Stouffs M., Serrander L., Banfi B., Bettiol E., Charnay Y. et al. (2006) The NADPH oxidase NOX4 drives cardiac differentiation: Role in regulating cardiac transcription factors and MAP kinase activation. *Mol. Biol. Cell* **17**, 3978–3988.
24. Wobus A. M., Holzhausen H., Jakel P., and Schoneich J. (1984) Characterization of a pluripotent stem cell line derived from a mouse embryo. *Exp. Cell Res.* **152**, 212–219.
25. Rohwedel J., Maltsev V., Bober E., Arnold H. H., Hescheler J., and Wobus A. M. (1994) Muscle cell differentiation of embryonic stem cells reflects myogenesis in vivo: developmentally regulated expression of myogenic determination genes and functional expression of ionic currents. *Dev. Biol.* **164**, 87–101.
26. Drab M., Haller H., Bychkov R., Erdmann B., Lindschau C., Haase H. et al. (1997) From totipotent embryonic stem cells to spontaneously contracting smooth muscle cells: a retinoic acid and db-cAMP in vitro differentiation model. *FASEB J.* **11**, 905–915.
27. Strubing C., Ahnert-Hilger G., Shan J., Wiedenmann B., Hescheler J., and Wobus A. M. (1995) Differentiation of pluripotent embryonic stem cells into the neuronal lineage in vitro gives rise to mature inhibitory and excitatory neurons. *Mech. Dev.* **53**, 275–287.
28. Wartenberg M., Gunther J., Hescheler J., and Sauer H. (1998) The embryoid body as a novel in vitro assay system for antiangiogenic agents. *Lab. Invest.* **78**, 1301–1314.
29. Mello C. C., and Conte D., Jr. (2004) Revealing the world of RNA interference. *Nature* **431**, 338–342.
30. Dykxhoorn D. M., Novina C. D., and Sharp P. A. (2003) Killing the messenger: short RNAs that silence gene expression. *Nat. Rev. Mol. Cell Biol.* **4**, 457–467.
31. Hunter T., Hunt T., Jackson R. J., and Robertson H. D. (1975) The characteristics of inhibition of protein synthesis by double-stranded ribonucleic acid in reticulocyte lysates. *J. Biol. Chem.* **250**, 409–417.

Part V
Biostatistics

Chapter 31

Pooling Data When Analyzing Biomarkers Subject to a Limit of Detection

Leslie Rosenthal and Enrique Schisterman

Abstract

The use of biomarkers to assess exposure and investigate biomedical questions is common in epidemiology. The usefulness of biomarker research, however, is contingent upon the ability to achieve a complete understanding of the role they play within a population. In estimating distributional parameters for a particular biomarker, such as oxidative stress or antioxidant markers, scientists face two main challenges: overcoming the cost of performing a large number of assays and dealing with data subject to a limit of detection. While approaches have been suggested to deal with each of these issues individually, pooling is a strategy that can address both problems.

Key words: Biomarkers, Detection limit, Pooling.

1. Introduction

The use of biomarkers to assess exposure and investigate biomedical questions is common in epidemiology. Implications of exploring the relationship between biomarker levels and outcome can have profound effects on the biomedical community, leading to new research as well as increased diagnostic capabilities. The usefulness of biomarker research, however, is contingent upon the ability to achieve a complete understanding of the role they play within a population. We have previously published protocols for using biomarkers of oxidative stress for discriminating between individuals with a disease and a normal population *(1)* as well as for test performance *(2)*.

From: *Methods in Molecular Biology, vol. 477: Advanced Protocols in Oxidative Stress I,*
Edited by: D. Armstrong, DOI: 10.1007/978-1-60327-517-0_31, © Humana Press, New York, NY

In estimating distributional parameters for a particular biomarker, such as oxidative stress or antioxidant markers, scientists face two main challenges: overcoming the cost of performing a large number of assays and dealing with data subject to a limit of detection. The power gained by a large sample must be weighed against the cost of performing more assays. After reproducibility and variability are established for the biomarker, financial constraints often limit further evaluation to small sets of samples. Instrument sensitivity may also be problematic when studying levels of oxidative stress in biological samples. Some members of the population may have serum levels below a detection threshold, d *(3)*. Under these circumstances, values at or above the detection threshold which is designated (d) are measured and reported, but values below the detection threshold are unobservable, limiting the information one can utilize in his or her analysis.

While approaches have been suggested to deal with each of these issues individually, pooling is a strategy that can address both problems. In this method, two or more specimens are physically combined into a single "pooled" unit for analysis. Thus, a greater portion of the population is assayed for the same price; and, information per assay increases meaning fewer assays are needed to achieve equivalent information *(4, 5, 6, 7)*. Additionally, pooling the specimens reduces the effective variance of the biomarker. This can ultimately decrease the proportion of observations below the detection threshold and increase the amount of information that can be derived from the data. Such results are useful when studying biomarkers that may naturally exist in small quantities, such as oxidative stress markers.

Pooling can be seen as a primary tool for case-control and cohort studies exploring discrete outcomes. Since it minimizes cost as well as the amount of information lost due to the detection threshold, the use of pooled data is preferable (in a context of a parametric estimation) to using all available individual measurements for certain values of d. This chapter is designed to explain how this method can be applied in such studies while discussing the benefits of the pooling strategy and the circumstances under which it is most useful.

2. Materials

Statistical software to perform analysis with data subject to limits of detection is available upon request at schistee@mail.nih.gov.

3. Methods

3.1. Formulas and Terminology

1. Suppose we have biologic specimens from a patient population, A, consisting of N individuals and wish to analyze any oxidative stress biomarker. The population, $A = \{A_1, A_2, \ldots, A_N\}$, has test results $X = \{X_1, X_2, \ldots, X_N\}$. In the pooling strategy, samples from patient population A are randomly combined into n pooled specimens of size p, where $n = N/p$. The n pooled assays are considered the average of the contributing individual results, i.e.:

$$X^{(p)} = \{X_1^{(p)}, X_2^{(p)}, \ldots, X_n^{(p)}\}$$
$$= \Big(X_{k_{11}} + \ldots + X_{k_{1p}}\Big), \frac{1}{p}\Big(X_{k_{21}} + \ldots + X_{k_{2p}}\Big), \ldots,$$
$$\frac{1}{p}\Big(X_{k_{n1}} + \ldots + X_{k_{np}}\Big)\Big\},$$

where $\{k_{1i}, i = 1, \ldots, p\}, \ldots, \{k_{ni}, i = 1, \ldots, p\}$ are some disjoint subsequences of set $\{1,2,3, \ldots, N\}$.

2. Random sampling, another method of cost-effective sampling, selects a random sample of the patient population $A^{(r)} = \{A_{k1}, A_{k2}, \ldots, A_{kn}\} \in A$, where $n(\leq N)$ is determined by a power calculation and $\{k_i, i = 1, \ldots, n\}$ is a subsequence of set $\{1,2,3, \ldots, N\}$ where assays are performed on the subset of specimens with observed results $\{X_{k1}, X_{k2}, \ldots, X_{kn}\}$.

3. In practice, serum levels of a biomarker of interest may fall below a detection threshold resulting in unavailable test results. When looking at population A's results, instead of X, we observe $Z = \{Z_1, Z_2, \ldots\}$, such that:

$$Z_i = \begin{cases} X_i, & \text{if } X_i \geqslant d; \\ \text{Not Available (N/A)}, & \text{if } X_i < d, \end{cases}$$

where d is the value of the detection threshold.

Likewise, in the pooling design, we observe $Z^{(p)} = \{Z_1^{(p)}, \ldots, Z_n^{(p)}\}$, where:

$$Z_i^{(p)} = \begin{cases} X_i^{(p)}, & \text{if } X_i^{(p)} \geqslant d; \\ \text{N/A}, & \text{if } X_i^{(p)} > d. \end{cases}$$

3.2. Statistical Background of Pooling with Data Subject to a Detection Threshold

1. When dealing with data that is subject to a detection threshold, the pooling strategy is more or less beneficial depending on the location of d in relation to the mean, μ. In determining the effectiveness of this method, it is important to consider the following three cases:

1.1 When the detection threshold is below the mean, $d < \mu$.

1.2 When the detection threshold is above the mean, $\mu < d$.

1.3 When the detection threshold is far above the mean, $\mu << d$.

2. First, consider the case of a population, X, normally distributed around a mean of 0 that is subject to a detection threshold somewhere below the mean $X \sim N(\mu = 0, \sigma_X^2 = 1)$, $\mu > d$. In this case, pooling takes advantage of the statistical properties of averages through physical implementation, i.e., the value of pooled specimens is the mean of the individual biomarker values. More numeric observations are available because the pooled distribution $X^{(p)}$ with $\text{var}(X^{(p)}) = \sigma_X^2/p$ is more concentrated around the expectation $\mu = 0$.
3. Next, consider the case where the detection threshold is above the mean $X \sim N(\mu = 0, \sigma_X^2 = 1)$, $\mu < d$. In this case, pooling can be detrimental. With over half of the data below the detection threshold, more pooled samples have values below d than unpooled samples.

   Never the less, in this situation, the pooling strategy might still be more efficient than random sampling. Intuitively, the pooled observations might be more informative than the unpooled observations because each pooled observation is based on more than one test result.
4. Lastly, consider the case where the detection threshold is far above the mean $d >> \mu = EX$. When the detection threshold is much greater than the mean biomarker value, the pooling strategy is completely inefficient because the pooled data are based upon substantially less numeric information than a random sample of unpooled data.
5. For clarity, the above cases assumed X has a normal distribution; however, the conclusions from this section are true for most commonly used distributions, including gamma.

3.3. Applications

1. **Figure 31.1A** plots the density function of the normally distributed biomarker X with a detection threshold at $d = -1$. The shaded area corresponds to values of X below d where missing values would be reported. The un-shaded area corresponds to reportable numeric values of X. In this case, since $\Pr\{X_1 < -1\} \approx 0.16$, the expected proportion of observations below d is approximately 16%. Pooling the specimens reduces the effective variance of biomarker X. The variance of the pooled samples is $\text{var}(X^{(p)}) = \sigma_X^2/p$ *(4)*. Assuming $p = 2$, $\Pr\{X_1^{(p)} < -1\} \approx 0.08$ leaving only approximately 8% of the pooled observations that are below d as shown in **Fig. 31.1C**. Thus, pooling the samples cut the amount of unobservable values by half.
2. **Figure 31.1B** and d depict when the location of the detection threshold is above the mean of X. As shown in

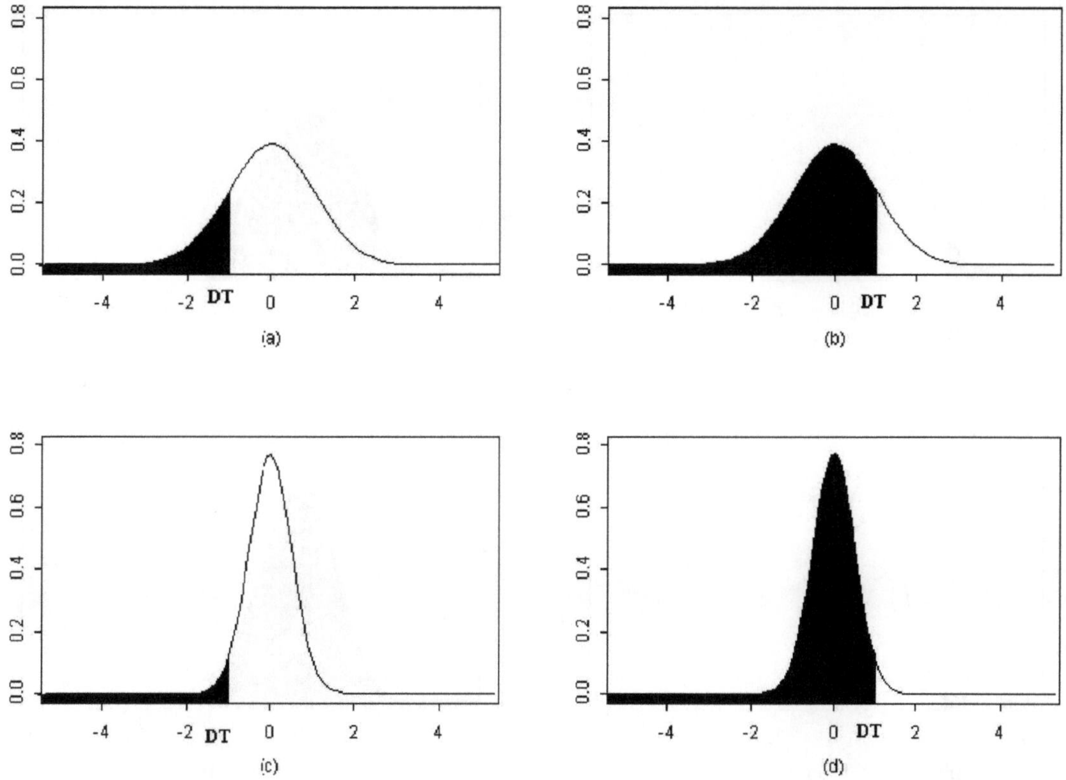

Fig. 31.1. Normally distributed data constrained by a detection threshold (shaded area represents unobserved data).

Fig. 31.1B, the amount of unobserved data (shaded area) is smaller in the unpooled data than in the pooled data.

3. **Figure 31.2** shows an example of pooling with gamma data ($\chi^2_{(20)}$ for unpooled data and $\chi^2_{(40)}$ for pooled data). Clearly, pooling with gamma data leads to similar conclusions as found in working with Normal distributions.

4. Notes

1. The efficiency of the pooling design is dictated by the location of the detection threshold, but is independent of the distributional assumptions (e.g. Gamma, t-distribution, Lognormal, etc).
2. One is able to stratify the pooled samples by confounders in order to retain confounding and covariate information in the pooled samples.
3. If $d = -\infty$ the maximum likelihood estimators of μ based on full data Z and pooled data $Z^{(p)}$ have equal efficiency *(4)*.

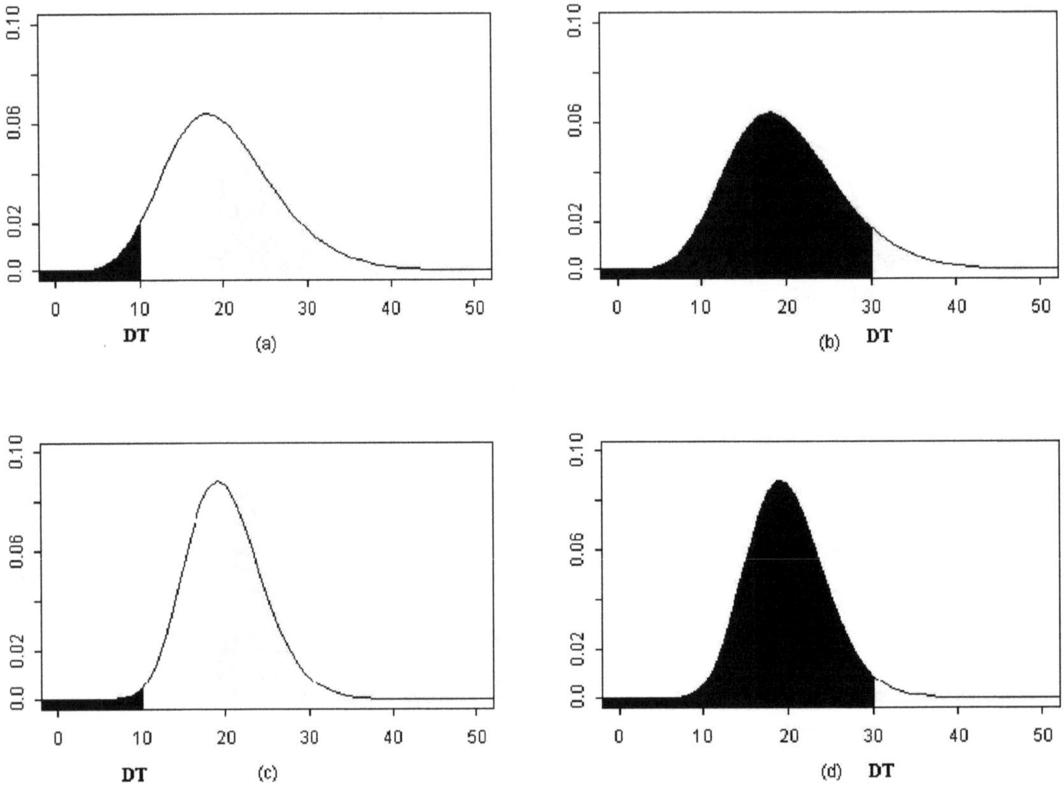

Fig. 31.2. Chi-square distributed data constrained by a detection threshold (shaded area represents unobserved data).

References

1. Schisterman E. (2002) "Statistical Correction of the Area Under the ROC Curve in the Presence of Random Measurement Error and Applications to Biomarkers of Oxidative Stress." Methods in Molecular Biology. Volume 186. New Jersey: Humana Press.
2. Schisterman E. (2002) "Statistical Analysis: Receiver Operating Characteristic (ROC) Curve and Lipid Peroxidation." Methods in Molecular Biology. Volume 196. New Jersey: Humana Press.
3. Helsel D. (2005) Nondetects and Data Analysis: Statistics for Censored Environmental Data. Hoboken, New Jersey: John Wiley & Sons, Inc.
4. Faraggi D, Reiser B, Schisterman EF. (2003) ROC curve analysis for biomarkers based on pooled assessments. Statistics in Medicine; 22:2515–2527.
5. Liu A, Schisterman EF. (2004) Sample size and power calculation in comparing diagnostic accuracy of biomarkers with pooled assessments. Journal of Applied Statistics; 31:41–51.
6. Liu A, Schisterman E. (2003) Comparison of diagnostic accuracy of biomarkers with pooled assessments. Biometrical Journal; 45:631–644.
7. Schisterman EF, Perkins NJ, Liu A, Bondell H. (2005) Optimal cut-point and its corresponding Youden Index to discriminate individuals using pooled blood samples. Epidemiology; 16:73–81.

INDEX

G

H

I

L

M

N

MIX
Papier aus verantwortungsvollen Quellen
Paper from responsible sources
FSC® C105338

If you have any concerns about our products,
you can contact us on
ProductSafety@springernature.com

In case Publisher is established outside the EU,
the EU authorized representative is:
Springer Nature Customer Service Center GmbH
Europaplatz 3, 69115 Heidelberg, Germany

Printed by Libri Plureos GmbH
in Hamburg, Germany